LAYERED
IGNEOUS ROCKS

LAYERED IGNEOUS ROCKS

L. R. WAGER

M.A., Sc.D., F.R.S., F.G.S.
Professor of Geology, University of Oxford

and

G. M. BROWN

M.A., B.Sc., D.Phil., F.G.S.
Lecturer in Petrology, University of Oxford

W. II. FREEMAN AND COMPANY

SAN FRANCISCO

Library of Congress Catalog Card Number : 67-30382

Printed and published in Great Britain by
Oliver and Boyd Ltd. Edinburgh

PREFACE

Layering may seem to be rather a superficial feature on which to base a major grouping of igneous rocks. Its significance, however, lies in the fact that layered igneous rocks were not formed simply by the congelation of magma to give rocks of the same bulk composition as the liquid, but by the accumulation of crystals as they formed and settled through the cooling body of liquid. This crucial petrogenetic process, known as crystal fractionation, can be studied to advantage in layered intrusions, in which the successive batches of crystals are preserved where deposited, as conformable sheets or layers.

The term ' layering ' might be used to describe features of igneous, sedimentary, or metamorphic rocks although so far it has been employed mainly for the former. When the Skaergaard intrusion of East Greenland was originally mapped in 1935, the stratigraphical term ' bedding ' was used to describe the sheet features of the rocks but it was subsequently discarded in favour of layering. Geological nomenclature is made unnecessarily complicated by the existence of synonyms, and it is suggested that the term *layering* should be kept for the high-temperature sedimentation features of igneous rocks, leaving bedding and stratification for use in connection with the usual sedimentary rocks.

The varied phenomena of layered intrusions form the subject of this book, the first part of which deals with the Skaergaard intrusion and the second part with other layered intrusions. One of us (L.R.W.) was chiefly responsible for writing Part I and the other author (G.M.B.) for Part II, the exceptions being most of Chapter III (G.M.B.) and of Chapter XVI (L.R.W.). The work stemmed and developed from our mutual interest in all layered igneous rocks, and few aspects were described without the benefit of each other's advice, generally entailing extensive discussion and re-writing. It will be clear to the reader that the greater part of our concerted study of fundamental principles went into the synthesis of work done on the Skaergaard intrusion, in collaboration with some of our colleagues and research students and on materials provided for specialist studies in other institutions. Since about 1953 (when we last visited the intrusion) this work has resulted in so much revision of the data and ideas presented in the 1939 Memoir that the full account given in Part I was felt desirable. In Part II, the small Rhum intrusion is described in detail, partly because work on it in 1950–53 necessitated alternative hypotheses relevant to the formation of layered igneous rocks, from which stage we began a re-consideration of the Skaergaard mechanisms. Chiefly, these hypotheses involved the concepts of replenished magma chambers and of adcumulus growth, both requiring a re-consideration of variability in rhythmic layering, cryptic layering, crystallization processes, and nomenclature.

The large Stillwater and Bushveld intrusions are of particular importance amongst layered intrusions, and are described at some length in Part II. For the Stillwater (Ch. XIII) we have summarized chiefly the significant contributions made by H. H. Hess and E. D. Jackson, although one of us (G.M.B.) also visited the intrusion in 1955 and benefited from discussions with Professor Hess during a year at Princeton. For the Bushveld (Ch. XIV) we both had the advantage, in 1958, of seeing and collecting the rocks in the field and this, aided by discussions with B. V. Lombaard and J. Willemse, helped in a critical assessment of the vast amount of published information. There remained a large number of basic layered intrusions differing widely in their characteristics and in the quality of the published information, and we decided to give examples in Ch. XV of those of which we either had first-hand knowledge or of which we felt that the information available to us was such that a fairly concise account could now be given. The granitic and syenitic layered rocks then clearly formed the basis for a separate treatment in Ch. XVI. In Ch. XVII a list is given of other major intrusions, which is still not complete and will also be extended as further layered intrusions are discovered. Generalizations should not be drawn hastily from a glance at this list, because there are bound to be some intrusions that we have overlooked for reasons other than inadequate information or strong doubts as to their layered origins. The list includes examples described only briefly in the literature, those where a layered origin is in slight doubt, and some (e.g. Muskox) that are being studied extensively at present and for which a full description at this stage would be premature. The inclusion of a brief description of sills (Ch. XVIII) was a difficult decision to make, as discussed in that chapter, but in the end we felt that because many aspects of their crystallization history had a direct bearing on the problems under consideration, they should be discussed here. The final conclusions (Ch. XIX) are rather brief and deal only with certain selected topics. The reason for this is that no layered intrusion shows such a complete picture, or has yet been studied in the same detail as the Skaergaard, so the particular features on which to base comparisons and generalizations are necessarily limited. The reader will, therefore, find most of the discussion on the principles of layered igneous rocks under the various headings of Part I, where we have tended to make some comparisons and generalizations as we went along. Similarly, in Part II the reader will find analogies drawn between intrusions wherever a particular aspect under discussion warrants this, it being understandable that in this section of the book the extensive data available on the Skaergaard are often used as a kind of standard or yardstick. Nevertheless, in presenting the brief conclusions in Chapter XIX we have aimed at providing a discussion on many of the obvious or likely similarities and differences shown by this great variety of igneous rocks.

Many geologists have given valuable help in our investigations of layered intrusions, and to these we wish to express our grateful thanks. First among these is Professor W. A. Deer (University of Cambridge) who collaborated with one of us (L.R.W.) in the longest continuous spell of fieldwork on the Skaergaard intrusion (1935–6) and in laboratory work that resulted in the first detailed description in 1939. Later, spectrographic trace-element investigation with Dr R. L. Mitchell (Macaulay Institute for Soil Research, Aberdeen) and other chemical work on Skaergaard rocks and minerals at Oxford with Professor E. A. Vincent formed the basis of our more recent knowledge on the chemistry of the Skaergaard, without which much of the new petrological interpretation would have been impossible. In addition, many research students and others have assisted in our investigations of layered intrusions and related problems, and particularly we wish to thank F. B. Atkins, P. E. Brown, J. M. Carr, J. A. V. Douglas, A. C. Dunham, C. H. Emeleus, E. C. Fountaine, E. I. Hamilton, R. R. Harding C. J. Hughes, D. R. C. Kempe, J. F. Lewis, G. D. Nicholls, D. N. Ojha, R. Phillips, A. T. V. Rothstein, B. G. J. Upton, W. J. Wadsworth, H. G. Wager, D. S. Weedon, and P. Zinovieff.

For valuable discussions on problems of layered intrusions and for help in many other ways we are deeply grateful to H. H. Hess and E. D. Jackson (Stillwater); B. G. Worst (Great Dyke); B. V. Lombaard, F. C. Truter, and J. Willemse (Bushveld); S. W. Morse (Kiglapait); S. Goldich (Duluth); and also to A. F. Buddington, H. D. Holland, I. D. Muir, H. Neumann, A. Noe-Nygaard, H. Sørenson, F. H. Stewart, C. E. Tilley, and H. S. Yoder, Jr.

For help in preparation of the manuscript and diagrams we are grateful to Miss A. B. Onions, Mrs J. Cooke and Mrs D. C. Rex, and to those other members of the technical staff, Department of Geology and Mineralogy, Oxford, who helped on numerous occasions.

The diagrams and photographs are our own, or were available in our Department, except where we have given credits to individual authors. The courtesy of the following publishers and organizations is also acknowledged: Ministry of Defence (Air), *American Journal of Science*, Clarendon Press, Oxford (for the loan of blocks), Geological Society of America, Geological Survey of Canada, Grønlands Geologiske Undersøgelse, *Meddelelser om Grønland* (for the loan of blocks), Mineralogical Society, Mineralogical Society of America, Royal Society of London, United States Geological Survey.

Oxford, October 1965   L. R. WAGER
                                                                                        G. M. BROWN

On 20th November 1965, Lawrence R. Wager died suddenly at the age of 61. At that time, with the exception of Chapter XIX and the minor details, the book had been written and most of it was with the publishers. The study of layered igneous rocks had been Wager's primary geological interest for about 30 years, and no intrusion was better suited to his particular talents for grappling both with exacting field conditions and with demanding laboratory studies than the Skaergaard intrusion of East Greenland.

He was always the first to acknowledge the contributions made by his several collaborators, but surely the science of petrology can count as one of its greater fortunes the fact that the intrusion was discovered by this man, who was then prepared to devote most of his life's work to a continuous study which grew from each new discovery, and who was able to present an array of facts and ideas of fundamental importance to igneous petrology, mineralogy, and geochemistry. The writing of this book was at his suggestion, and undoubtedly it was his vigour and enthusiasm that carried it to completion.

Oxford, March 1966   G. M. BROWN

CONTENTS

PART I

THE SKAERGAARD INTRUSION, EAST GREENLAND

PART II

EXAMPLES OF VARIOUS TYPES OF LAYERED INTRUSIONS

CHAPTER I

CRYSTAL SETTLING AND THE DIVERSITY OF IGNEOUS ROCKS

' The sinking of crystals through a viscid substance like molten rock, as is unequivocally shown to have been the case in the experiments of M. Drée, is worthy of further consideration, as throwing light on the separation of the trachytic and basaltic series of lavas.'

CHARLES DARWIN (1844, p. 118).

Unless there be clear evidence to the contrary, it has usually been assumed that when magma crystallizes about randomly-spaced nuclei, the crystals neither floated nor sank in the liquid. However, crystals will have the same density as the magma only by rare coincidence, and normally they will have a tendency to float or sink. The effectiveness of such movement will increase with the fluidity of the magma and the time available before consolidation.

The possibility that some of the diversity among igneous rocks is the result of sinking of crystals has been realized for over a hundred years. Darwin had it in mind during the voyage of the *Beagle* (1828-33), and in his book *Geological Observations on the Volcanic Islands* (1844, pp. 117-24) he reviewed the ideas current at that time, especially those of Von Buch, which were based on observations in the Canary Islands. Darwin himself suggested that the porphyritic felspars in lava filling crater on James Island in the Galapagos Group had sunk, enriching the lower layers. It is not known whether the sinking of felspars is the right explanation for the particular case described by Darwin, but he clearly envisaged that the accumulation of early crystals at the bottom of a liquid is one possible factor in the development of diversity among igneous ricks. It is interesting to note that he also considered that certain dikelets which he found cutting a granite might have formed by the more fluid parts of the partially-cooled granite oozing into fissures in the already solidified part. The mechanism proposed by Darwin is essentially the one now commonly described as filter-press differentiation. During his five years as naturalist on the *Beagle*, Darwin had much time for observation and reflection, and he formulated what are still considered to be the two chief mechanisms by which diversity is developed among igneous rocks.

For a hundred years after Darwin's voyage the concept that crystals might sink or

A

float in the magma was current among geologists, but there was little unambiguous field evidence to indicate that it had actually occurred. N. L. Bowen's experimental examination of the question in 1915 was, therefore, of particular importance. In a paper which is nowadays too little known, Bowen (1915a) described experiments in which melts were maintained for varying lengths of time, at temperatures such that a small proportion of crystals were present in the liquid. In one series of experiments he used a melt of 56% diopside and 44% forsterite at 1430° C, at which temperature 4% of forsterite crystals were present in the liquid; different charges were held at this temperature in a platinum crucible for times varying from 15 to 80 minutes. After each run the contents of the crucible were quenched and then examined to see how far the olivine crystals had sunk in the time available. Despite the difficulty of sectioning the strained, glassy material, Bowen was able to make some small thin-sections and one of

his illustrations is reproduced as Fig. 1 (Bowen, 1915a, fig. 1, p. 178). This shows that in this particular melt, after 15 minutes, most of the olivine crystals had sunk a centimetre or two and formed a precipitate at the bottom of the crucible. It shows, also, an interesting size distribution of the crystals; the first to reach the bottom were small, apparently because they did not have time to grow as large as the more slowly settling crystals. In melts of other bulk compositions, sinking of pyroxenes and floating of tridymites were seen to occur. Bowen also considered the influence of viscosity on the rate of sinking, and the order of magnitude of the temperature gradient which might be expected to produce convection currents in the liquids. His firm

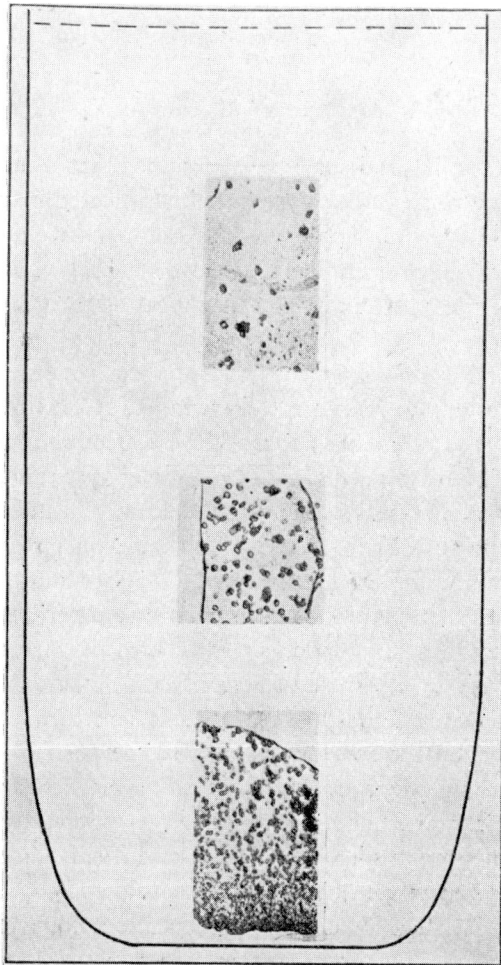

FIG. 1

Illustration of N. L. Bowen's experiment on the sinking of olivine crystals in a melt. The relative positions in the crucible of three thin sections cut from the quenched melt are shown. The abundance of the olivine increases downwards but the grain-size decreases. (From Bowen, 1915a, fig. 1.)

conclusion was that ' we cannot avoid assigning a general importance to sinking of crystals in the differentiation of igneous rocks '.

Favourable conditions for observing the effects of the sinking of crystals in natural magma would not be expected in lavas, where flow of the congealing liquid would produce disturbing effects; on the other hand, in thick sills, which were emplaced rapidly and cooled slowly and tranquilly, the effects should be more obvious. The Palisades Sill of New Jersey, U.S.A., provides what is usually accepted as clear evidence for the sinking of the first-formed olivines. In early accounts by Lewis (1907, p. 125 & pp. 129–34) and by Bowen (1915a, pp. 186–90), an olivine diabase layer is described as occurring about 40–50 ft above the base of the sill (Fig. 2), the rest being free from olivine, except for very small amounts in the rapidly-chilled upper and lower selvedges. The generally accepted interpretation is that after some chilling of the magma, giving the fine-grained top and bottom rocks, slower cooling allowed time for the small amount of olivine crystals, formed in the liquid, to sink, giving an olivine-rich layer near the base of the sill. During the subsequent cooling, olivine ceased to form and the rest of the sill is olivine-free. A re-study of the sill by Walker (1940) confirmed the hypothesis of the sinking of olivines during the early stages, and variations in the modal amounts of the pyroxenes and felspars are also ascribed, with less certainty, to the effects of gravity. Sinking of olivine crystals has been demonstrated in other sills (see Ch. XVIII), and it is probably a common phenomenon. On the other hand there are many sills of considerable thickness, particularly of quartz dolerites, where crystal sinking has taken place only to a limited extent or not at all, while intermediate and acid sills, carrying porphyritic crystals, do not show the effects of sinking, presumably because of the high viscosity of the liquids.

When we turn from minor intrusions to the large plutonic masses, evidence for the sinking or floating of crystals is usually by no means obvious. Direct observation of the sinking of crystals in plutonic bodies is, of course, not possible; all that can be done is to show, from the rocks produced by the solidification of the magma, that certain features can best be explained by this hypothesis. Only slowly is it becoming apparent that the process of crystal sinking must be a common event.

In many basic and ultrabasic intrusions a structural feature, which used to be called banding but now, more usually, layering, is widespread and conspicuous. For instance, it is finely displayed in the central part of the Tertiary Cuillin gabbro on the Isle of Skye (Fig. 3), and Geikie (1897, vol. 2, p. 327) recorded that when he first saw these rocks he thought he was looking at banded Precambrian gneisses of the Lewisian (see also Geikie, 1894, p. 317). Later, he and Teall (1894) showed that at any one level, the different bands of the gabbro were composed of the same minerals, present in different proportions. This highly significant observation has since proved to be generally true of banding or layering in basic igneous rocks. Subsequent mapping by Harker (1904, pp. 90–2) showed that the banding is regularly disposed, the sheets dipping towards a single focus. Following Geikie and Teall, Harker suggested that the

Fig. 2. The lower part of the Palisades sill, New Jersey, showing the weathered olivine-rich layer. The effect of the weathering is to suggest sharper boundaries to the layer than is true. The chilled base of the sill is not shown but is only a few feet below the normal dolerite.

banding resulted from the flow of heterogeneous magma just prior to consolidation, a concept not nowadays easy to accept for this case.

Many basic plutonic intrusions have been found to show a great regularity in their banding so that individual sheets may be traced for great distances, and the whole intrusion viewed as built up of a succession of broadly conformable layers. The most impressive example of layering so far known is shown by the Bushveld intrusion of South Africa, where layers can be traced for 20 or 30 miles and where one highly characteristic, thin horizon, the Merensky Reef, is probably continuous over a distance of 110 miles. The layering of the Bushveld intrusion, together with associated texture due to the parallelism of tabular minerals, was described originally 'pseudo-stratification' or 'rifting', the latter term having reference to the easy fracture parallel to the layers (Hall, 1932, pp. 264–6). The layering is actually a form of stratification and should not be described as 'pseudo'. The easy fracturing or 'rifting' is a textural feature since called igneous lamination (Wager & Deer, 1939); it is a common feature of layered rocks but not universally present. Soon after the appearance of Hall's memoir, sheet structures in the Bay of Islands intrusion in Newfoundland (Ingerson, 1935; Cooper, 1936) and the Stillwater Complex in Montana (Peoples, 1936; Buddington & Hess, 1937), were described as *layering* and this term was adopted for

similar features of the Skaergaard intrusion in East Greenland, which were being described at about the same time.

The term stratification or bedding could have been used equally well for the sheet-like structures of igneous rocks. However, these terms have priority for features of sedimentary rocks and, as already pointed out in the Preface, it is desirable to restrict their use to such rocks, leaving the term layering for similar features of igneous rocks. The term banding, although strictly describing the appearance of a two-dimensional feature on a surface exposure, should perhaps be kept for any streaky or roughly planar heterogeneity in igneous rocks, whatever its origin.

FIG. 3

Layering in the Cuillin gabbro, Druim an Eidhne, Isle of Skye. (From Geikie & Teall, 1894.)

Layering, which is the common characteristic of the igneous intrusions just mentioned, and of many others which have been discovered subsequently, is believed to be the result of crystals settling out of the magma during slow cooling. No one feature of the layered intrusions is sufficient, by itself, to prove that they originated by crystal settling, but when all the available facts are considered, this is clearly the picture which emerges. In his classic book *The Evolution of the Igneous Rocks* (1928), Bowen devotes much space to a consideration of rocks produced by what he called crystal sorting, that is, the sorting of crystals during settling under the influence of gravity. In particular, Bowen argued that peridotites and dunites were generally to be ascribed to the accumulation of discrete olivine crystals, separating from a basic magma. Although the problem of the origin of all ultrabasic rocks is by no means finally disposed of by the hypotheses of crystal settling, it is, no doubt, the explanation of many examples. Bowen considered several types of monomineralic rocks to be the result of crystal sorting but he did not realize the extent to which many ordinary looking eucrites and gabbros are to be ascribed to the same process. Rocks believed to result from crystal settling were described by Bowen as ' accumulative rocks '; this term has since been shortened to ' cumulates ' (Wager, Brown & Wadsworth, 1960), so that it may more conveniently be used as a group name and be prefixed by mineral names to give a new nomenclature for layered rocks.

If a rock is the result of discrete crystals settling out of the magma, the crystal precipitate must be, to begin with, a loose pile of crystals surrounded by liquid, and some layered rocks show evidence from their textures of the former existence of such a stage. In other cases, during solidification, the liquid surrounding the discrete crystals behaves in a more complicated way, and the evidence for an early stage when the crystals were surrounded by liquid becomes obscure. When there is good reason to believe that crystal accumulation has been involved in the formation of a particular rock, then the crystals which formed the sediment at the bottom of the liquid may be called the *cumulus crystals*, and the liquid which once surrounded them, the *intercumulus liquid*. To make a precise distinction, in any particular rock, between the cumulus crystals and the crystalline material which, on cooling, formed from the intercumulus liquid is often difficult, but there are various indirect methods which can help in deciding this point.

Although the bottom accumulation of crystals (or the top accumulation, if they are lighter than the liquid) is the cause of the fundamental similarity of the intrusions described as layered, the actual process of crystal settling cannot be observed directly, and should not be used in a definition of layered intrusions. Layered intrusions can, perhaps, be defined as plutonic complexes which, on the basis of their structural or mineralogical character, are divisible into a succession of extensive sheets lying one above the other. The successive sheets should not show cross-cutting relationships or chilled margins. In addition, most layered intrusions show systematic changes in the nature and composition of the mineral phases present (cryptic variation), which are related to structural height in the succession of sheets. This phenomenon, first clearly identified in the Skaergaard intrusion, is often of great regularity and, in such cases, virtually eliminates the possibility of the layering being due to successive, sill-like injections.

Layered intrusions are not yet widely recognized among plutonic rocks; they are, however, being found in increasing abundance and some degree of crystal accumulation will probably come to be recognized as commonplace, at least among the basic and alkaline intrusions. The varying composition of lavas extruded from volcanoes, and of the inter-related rocks of many plutonic complexes, are often considered to be the result of fractional crystallization and may have been brought about by the formation of a deep layered intrusion, not yet exposed (see Ch. XII). The layered intrusions thus have an importance greater than their limited abundance would suggest, because they provide the clearest natural examples of the effects of fractional crystallization of magmas.

A layered intrusion may be likened to the crucible, greatly magnified, of the experimental petrologist. However, because of its vastly greater scale, new phenomena not seen in the crucible become important. Among these are convection currents in the liquid, effects due to different hydrostatic pressures at different levels in the liquid, the variable sites of nucleation and growth of the crystals in the magma, and special processes connected with the crystallization of the interstitial liquid between the cumulus crystals. Fundamentally, however, the varied phenomena of layered intrusions are to be traced back to the results of gravity acting on a crystallizing magma as heat is lost slowly to the surroundings.

PART I

THE SKAERGAARD INTRUSION
EAST GREENLAND

CHAPTER II

GENERAL FEATURES OF THE SKAERGAARD INTRUSION

The Skaergaard intrusion was found by the senior author during an expedition to East Greenland in 1930–1 led by H. G. Watkins. It is situated on the east coast of Greenland opposite Iceland, in a region which is difficult of access and had been visited only once before, in 1900, by a Danish expedition on which there was no geologist. When first seen from a distance, the layered rocks of the intrusion looked like a series of reddish sandstones and, in view of geological conditions elsewhere in Greenland, it seemed not unreasonable that sedimentary rocks should occur overlying the basement metamorphic complex and underlying the Tertiary plateau basalts. However, on landing it was not a little surprising to find that the supposed sediments were, in fact, regularly 'banded' igneous rocks. Investigations begun on this expedition were continued in 1932, in 1935–6, during which time the main mapping was done with the help of W. A. Deer, and in 1953, when a party including the two authors made further detailed collections, especially along measured traverses. The early work was carried out only with sketch maps but in 1935 a Danish map (1:250 000), made from oblique air photographs, was available. Fjords and glaciers cover some of the intrusion but glaciated or steep rock surfaces, free from moraine and vegetation provide exposures of a kind not normally available outside the Arctic (Pl. I, II, & III). The original description of the intrusion by Wager and Deer was published in 1939 and was re-issued without change in 1962. Since the original description, further work has appeared in various papers which are listed in the re-issue and are, for the most part, also given here in the list of references.

The Skaergaard intrusion is one of a series of Tertiary plutonic complexes following the East Greenland coast for 1000 km and having a spacing similar to that of the contemporaneous British Tertiary centres (Fig. 4). At the present level of erosion, some of the plutonic complexes are only in contact with the basement metamorphic complex but since they contain extensive basaltic inclusions it is clear that they extended upwards into the former, more extensive spread of Tertiary basalts. The Skaergaard intrusion, on the other hand, is seen cutting the metamorphic complex, the overlying thin Cretaceous sediments, and the Tertiary basalt and tuff series which is here 6 km

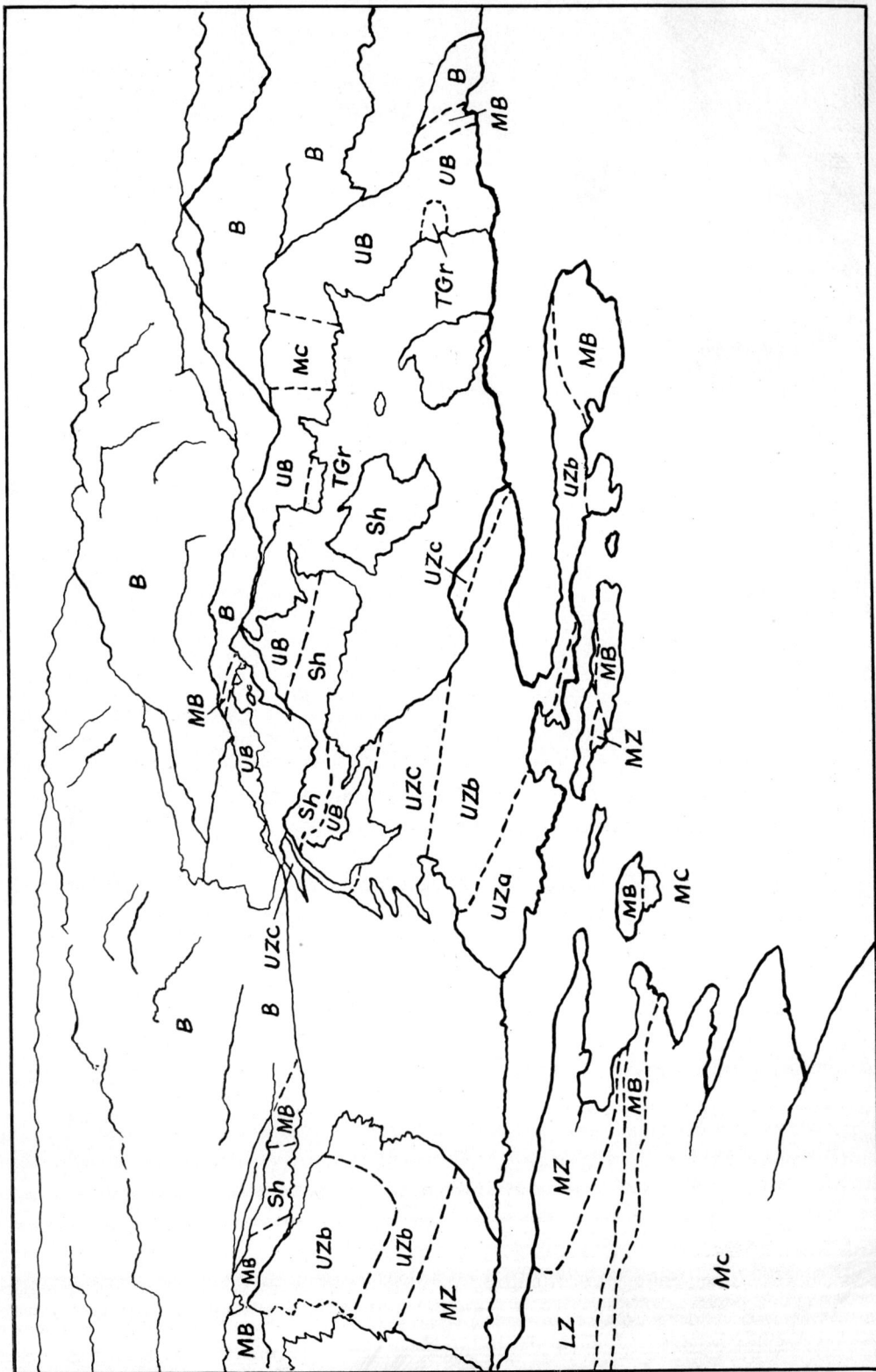

MC, metamorphic complex; *B*, basalts and tuffs; *MB* marginal border group; *LZ, MZ, UZa, UZb* and *UZc*, zones of the Layered Series; *UB*, upper border group; *TGr*, Tinden granophyre sill; *Sh*, Basistoppen Sheet. (After Wager & Deer, 1939.)

PLATE II

Northern part of the Skaergaard Intrusion from the west.
Air photograph by H. G. Watkins, August 1930.

thick. The Kangerdlugssuaq syenite intrusion, in the neighbourhood of the Skaergaard intrusion and belonging to the same group, has been dated isotopically by E. I. Hamilton (personal communication, 1965) as 52 ± 2M, which is Lower Eocene on the time-order scale. The mean age of the Tertiary epigranites of Skye (Moorbath & Bell, 1965, p. 62) is 54 ± 3M and of the Lundy Island granite (Dodson & Long, 1962) is 52 ± 2M. Rather exact contemporaneity of some of the Tertiary igneous phenomena around the North Atlantic is, therefore, indicated by the present isotopic age determinations.

Parallel to the coast and the general line of plutonic complexes there is a major tectonic structure, the so-called coastal flexure, which controls the position of the coast

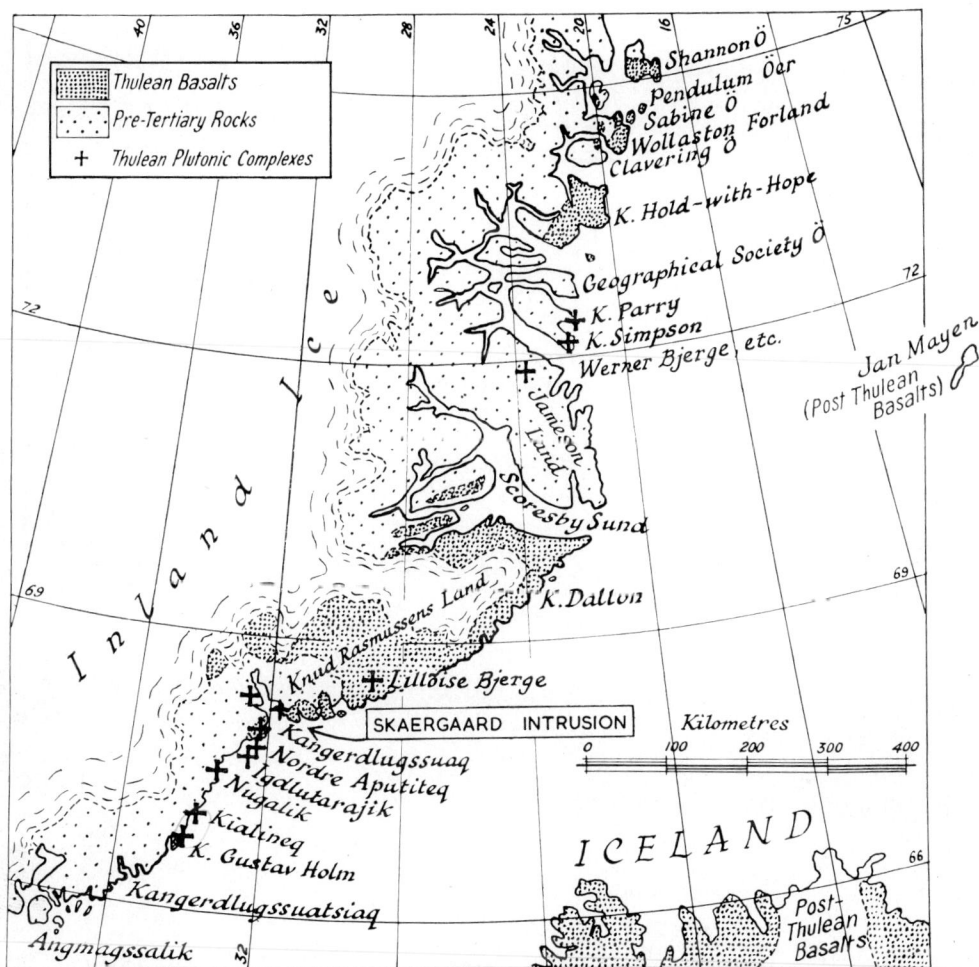

FIG. 4. The distribution of Thulean basalts and plutonic complexes in East Greenland. North of Scoresby Sound, the basalt series is relatively thin and the areas indicated include some sheets intrusive into Mesozoic sediments. (After Wager, 1947.)

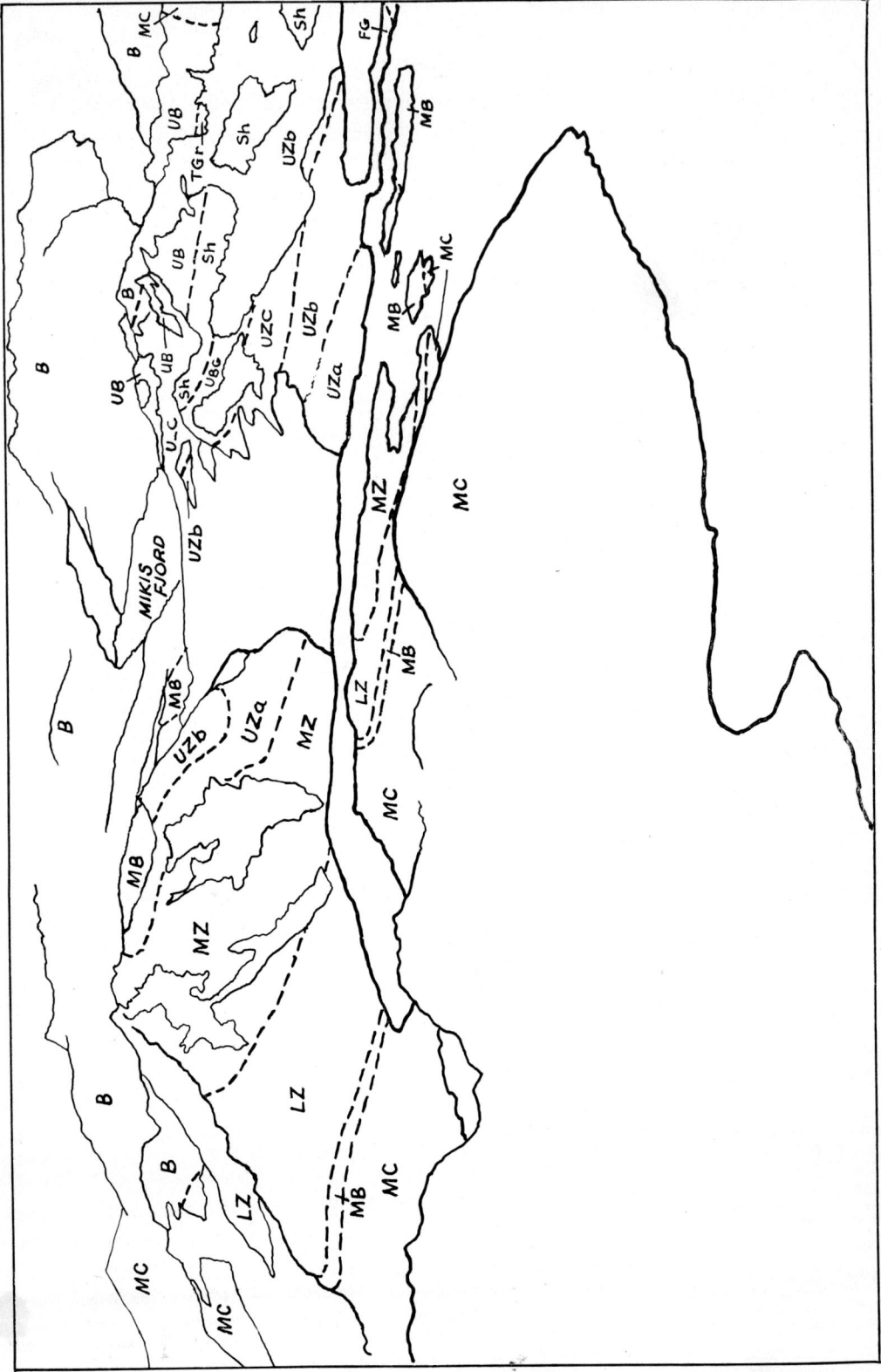

MC, metamorphic complex; *B*, basalts and tuffs; *MB*, marginal border group; *UB*, upper border group; *LZ*, *MZ*, *UZa*, *UZb* and *UZc*, zones of the Layered Series; *TGr*, Tinden granophyre sill; *Sh*, Basistoppen Sheet. (After Wager & Deer, 1939.)

PLATE III

Southern part of the Skaergaard Intrusion from the west.
Air photograph by H. G. Watkins, August 1930.

and the lie of the basalts (Fig. 5). Inland, where preserved, the basalts have dips of only 0°–3° but in the neighbourhood of the Skaergaard intrusion the dips are 10°, increasing to 50° on the capes of the outer coast. An impressive dyke swarm has been injected into the crest of the flexure, the dykes being approximately at right angles to the tilted lavas. Where densest, the dyke material is equal to, or greater in amount than the basalt or metamorphic complex into which it is injected (Fig. 6). The Skaergaard intrusion was emplaced before most of the flexuring had occurred, and the massive intrusion has been tilted and injected by dykes, apparently with the same ease as the plateau basalts. Although the flexure complicates the structural picture, it has the advantage of allowing 2500 m of layered rocks and another 1000 m of upper border group rocks to be examined; without the flexure, the thickness of accessible rock could not have exceeded the present relief of 1300 m.

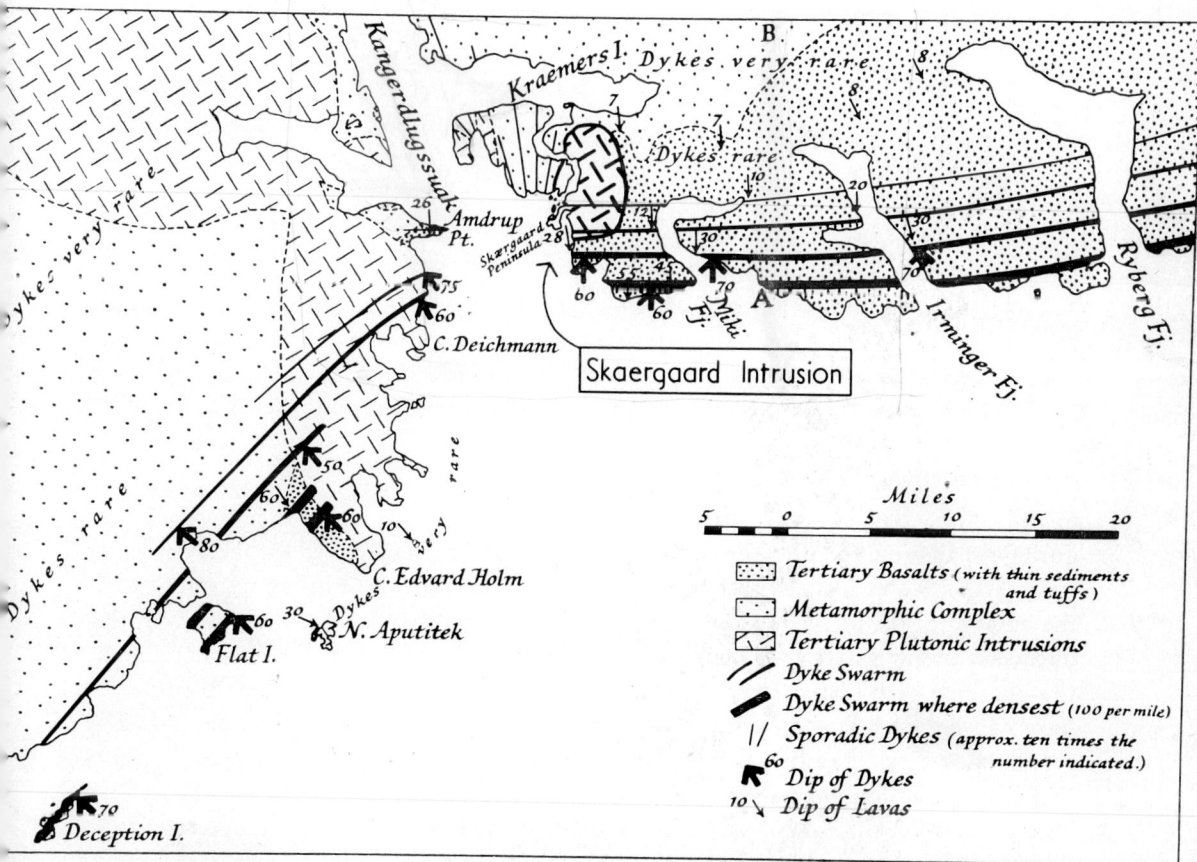

FIG. 5. The coastal dyke swarm in relation to the Skaergaard intrusion (which is shown by thickened boundary and pattern). (After Wager & Deer, 1939.)

DIVISION INTO THE LAYERED SERIES AND THE BORDER GROUPS

At an early stage in the mapping of the intrusion, the marginal and upper border groups were distinguished from the rocks of the layered series, lying within them (see map, Pl. XI, and cross-sections, Fig. 9). The marginal border group (MBG) is found on all sides of the intrusion. The outer part consists of 4 m of a fine-grained gabbro, the result of chilling of the magma against the country rock; inwards from the chilled material, various kinds of banded rocks are found, dipping at 60° or 70° inwards except along the northern border where the dip is about 45° (cf. Fig. 9). The layered series

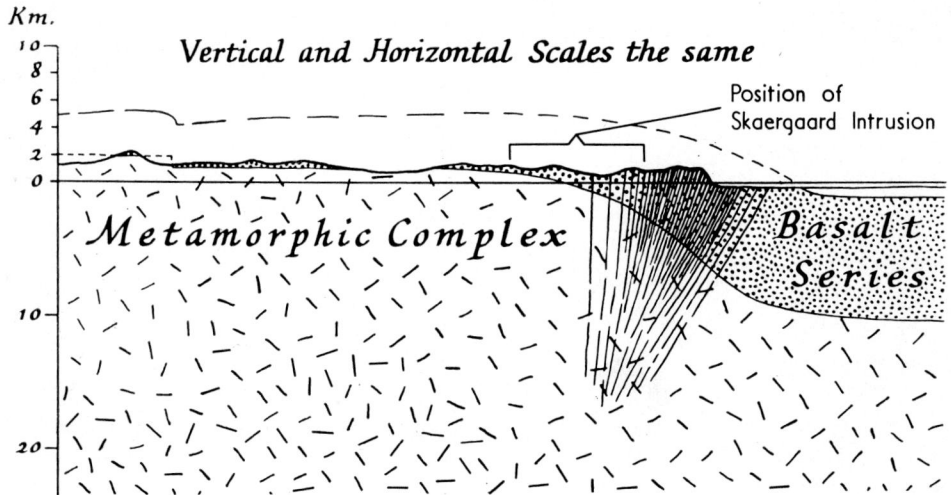

Fig. 6. Vertical section along A–B of Fig. 5, showing the observed thickness of the lavas and the relation of the dyke swarm to the crustal flexure. (After Wager & Deer, 1939.)

(LS), which lies within the envelope of border rocks, usually has dips of about 15 to 30° to the south. On extensive surfaces, produced by fracture along the weaker layers during weathering, the dips of the layering can often be measured to within a degree or less. When it was realized that the intrusion had been involved in the coastal flexure, the amount of tilt in zones across the intrusion was estimated from the observed dips of the surrounding lavas, which at one time must have been approximately horizontal: the observed dips of the layering were then converted to the values that they would have had before flexuring, using a stereographic projection method (Fig. 7a, b). The recalculated dips show that before the flexuring, the layered series was essentially flat lying, except around the edges where the dips increased to 20° or 30°; thus the various sheets of the layered series originally resembled saucers stacked one above the other. The thicknesses of the layers are apparently fairly constant, and approximate structural heights above the lowest accessible layers have been determined from the mapping, just as with a series of sedimentary rocks.

The inner contact between the marginal border group and the layered series can be defined within a metre along the western side but elsewhere, even where the exposures are good, the contact is less definite. There is no chilling of either group against the other. The character of the marginal border group is maintained up to the inner contact, which is either inward-dipping or vertical. The layered series shows some changes as the inner contact is approached; thus the rocks become more melanocratic, have steeper dips, show a false-bedded type of layering, and are distinguished from the main part by being called the cross-bedded belt. In several exposures the layered series appears to be banked up against the border group, and this is believed to be the result of the crystal cumulate having been laid down against the contemporary, steeply dipping surface between the marginal border rocks and the liquid.

The upper border group (UBG) is preserved mainly on the upper parts of the mountains in the southern third of the intrusion. The rocks are gabbros which are generally coarser than the layered series rocks. There is some layering to be seen (Pl. IX) and sometimes igneous lamination; both structures dip at 25–30° in a southerly direction. The dip of the layering in the upper border group is such that it would have been horizontal before the flexuring. The thickness of the upper border group rocks is about 1000 m and part has certainly been eroded. The rocks are plagioclase-rich gabbros and ferrodiorites and, as will be shown later, the anorthite percentage in the plagioclases increases in an upward direction, which is contrary to the change occurring in the layered series. The mapping of the UBG is complicated by a heterogeneity resulting from abundant inclusions of the metamorphic complex, and also by indefinite, unchilled granophyre sheets which are regarded as forming an integral part of the intrusion. A further complication is caused by a differentiated basic intrusion, the Basistoppen Sheet (formerly called a Raft), which was intruded near the junction of the LS and the UBG during the period of cooling of the main intrusion.

In the early paper, the upper border group rocks were described as banded. Using the definition of layering now adopted, it is apparent that the upper border group can also be described as layered, although the structures are not nearly so obvious as in the layered series proper. In describing the Skaergaard intrusion, however, the term layered series will be kept for the well layered rocks due to the bottom accumulation of crystals. The overlying, poorly layered rocks, different in appearance and origin, as will be shown later, will be distinguished as the upper border group.

THE INVERTED-CONE FORM OF THE INTRUSION

The dips of the outer margin of the intrusion can be obtained in various places and they can be modified, as with the dips of the layering, to obtain the values before flexuring (Fig. 7b). When treated in this way, they are found to define a space having the form of an inverted cone with its central axis inclined southwards. The marginal and upper border groups form a lining to the space so defined, and have a lesser volume

B

than the layered series within. Dips in the marginal border group are parallel to the nearby margins of the intrusion and dips in the upper border group are parallel to the presumed roof of the intrusion which, like the original layered series, was horizontal

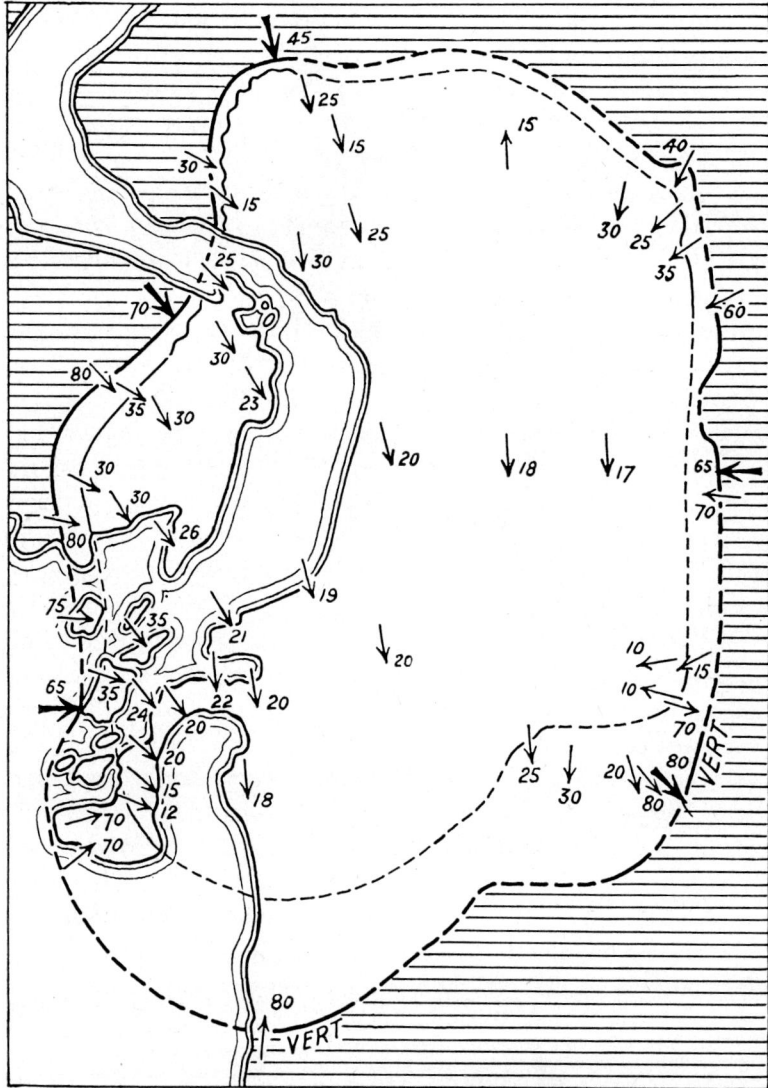

FIG. 7a. Dips of the margin (thick arrows), of the fluxion structures in the border group, and of the plane of layering (thin arrows). (From Wagner & Deer, 1939.)

before the flexuring. The north–south section of the intrusion (Fig. 9) shows the effect of the flexuring, while the east–west section (Fig. 9) is unaffected, in appearance, by the flexuring and therefore represents the disposition of the rocks as they were before the flexure. A hypothetical north–south section of the intrusion and the country rocks, as they are believed to have been before flexuring, is given in Fig. 8. In this section, the present-day sea-level is a line curving upward in passing from north to south, the curve being the antithesis of the post-intrusion flexure.

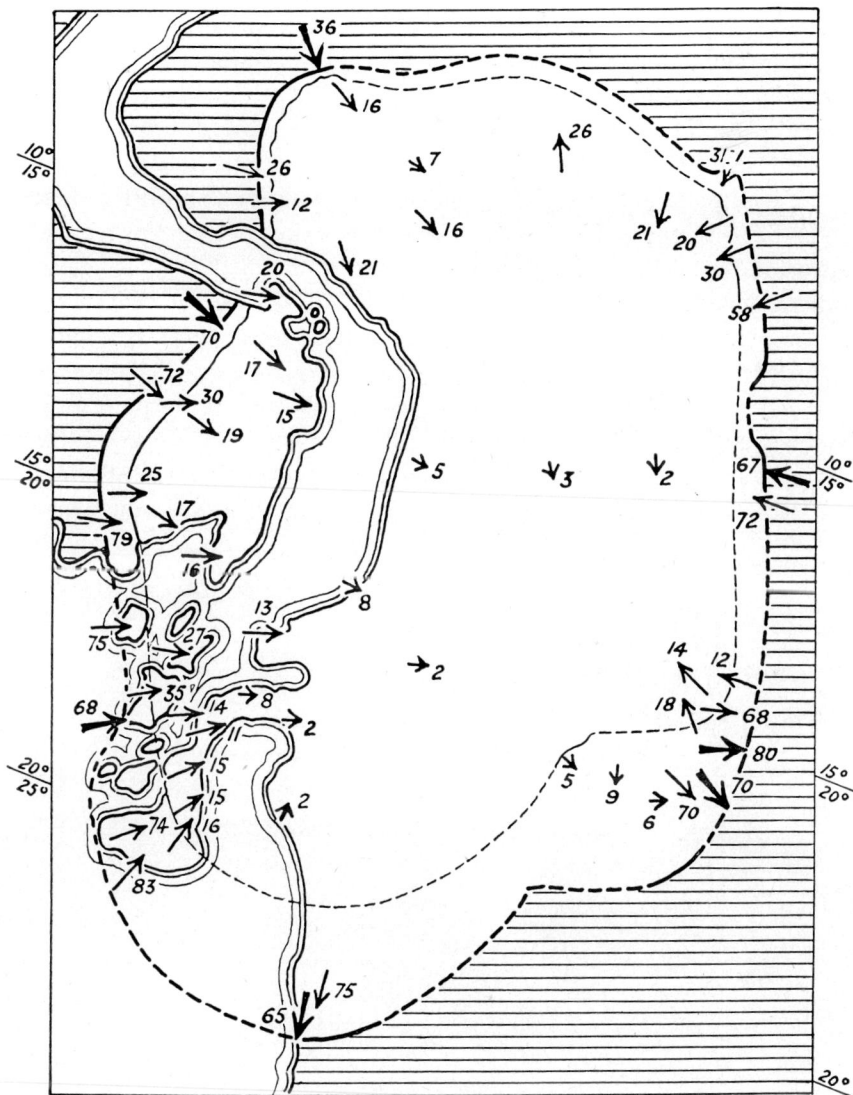

FIG. 7b. Dips shown in Figure 7a modified to the values which probably existed before flexuring. The ranges of tilt due to flexuring are shown at the margins of the diagram. Dips of less than 10° are indicated by short arrows since the direction of such low dips may not be significant. (From Wager & Deer, 1939.)

Because of the inwardly inclined margin of the intrusion, the metamorphic complex and the overlying basalts and sediments, formerly occupying the place of the intrusion, must have been pushed out of the way in an upward direction. The fine-grained olivine gabbro, of constant texture and composition around the circumference of the intrusion, is considered to be the original magma of the intrusion, chilled against the country rocks. The uniformity of the chilled magma and its continuity round the intrusion together give grounds for believing that the intrusion was produced by a single, grand episode in which basic magma, estimated as 500 km³ in volume, was emplaced rapidly in the upper crust. The rocks formerly occupying the place of the intrusion are pictured as having been shattered and blown out by the initial act of intrusion, and to have formed a capping of basalt and gneiss agglomerate over the intrusion and the surrounding area, as shown in the section (Fig. 8). No direct evidence of the material blown out at the time of the intrusion is available, except for a large mass of the metamorphic complex, about one kilometre long and half a kilometre thick, found on the face of Tinden near the southern margin. This large mass now occurs at least 2 km above its original position and is regarded as one of the blocks which was not completely expelled during the explosive emplacement of the intrusion; it serves as an indication of the presumed direction in which the rest of the material was expelled. At Rieskessel in Germany, huge blocks of basement rocks are jumbled in and around a circular depression among flat-lying Mesozoic strata (cf. Daly, 1933, pp. 382–4) and it is inferred from gravity determinations that there

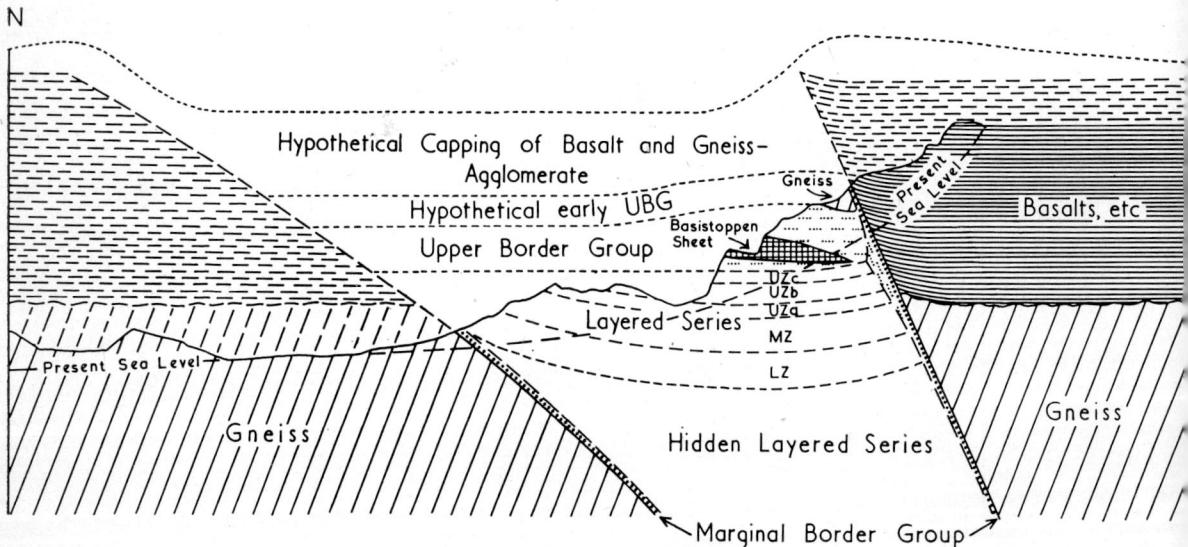

FIG. 8. Hypothetical, north–south section of the original Skaergaard intrusion showing its appearance before the monoclinal flexuring and later erosion (cf. Figs. 9 & 125). The present disposition can be inferred by relating to horizontal sea-level.

is a basic intrusion at depth. Perhaps the Rieskessel is the surface manifestation of a funnel intrusion resembling the Skaergaard in mode of emplacement, although recent work indicates that it may be the result of a meteoric impact—an alternative explanation which is, however, not accepted by Bucher (1963).

Rhythmic Layering and Associated Features of the Layered Series

There is a conspicuous type of layering throughout the whole layered series which is due to variations in the proportions of the chief minerals. Some layers are rich in plagioclase felspar and others rich in the melanocratic minerals; this gives the maximum contrast between the layers, while other layers are fairly average mixtures of the dark and light constituents. The layering is readily appreciated from the extensive rock surfaces such as those shown in Pl. IV and VII but it is not impressive in hand specimens, unless it is on a small scale (Fig. 13). In the original description of the Skaergaard intrusion (Wager & Deer, 1939, pp. 36–7) this kind of layering, the commonest type in basic and ultrabasic plutonic intrusions was given the name *rhythmic layering*. In the Skaergaard and many other layered intrusions, rhythmic layering may result in adjacent thin layers being of such contrasted rock types that in current petrographic

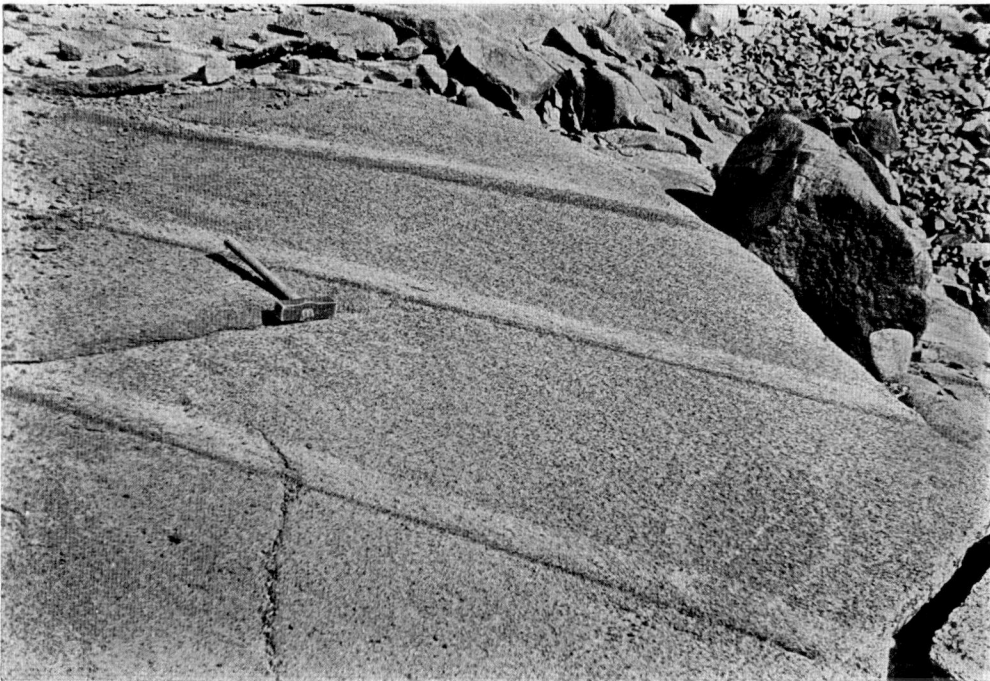

Fig. 10. Gravity-stratified layers separated by layers of average rock, 400 yds WNW of Main House (see map for geographical localities). (From Wager & Deer, 1939.)

nomenclature they would be given completely different rock names, and would often be classified in widely separated petrographic groups.

The variation in the proportions of the minerals giving the rhythmic layering in the Skaergaard layered series is frequently of an especial kind which is related to the direction of gravity. Thus some units have a well-defined base and consist of a lower part rich in heavy, and usually dark minerals which passes gradually upwards into a rock rich in less dense and light-coloured felspar (Fig. 10–13). The arrangement of the heavy and light constituents immediately suggests to the observer that at some stage, crystals of different densities have settled under the influence of gravity. The explanation of this type of layering, to which Buddington (1936) gave the name *gravity stratification*, is apparently that during settling, the heavier crystals sank faster through the magma to the temporary floor and are, therefore, most abundant in the lower part of the unit, while the lighter crystals, which sank more slowly, are more abundant in the upper part of the unit. When gravity-stratified layers occur in a plutonic intrusion there is visual evidence for bottom accumulation of the material, but it is not safe to assume that crystal settling has not occurred where such evidence is lacking. In the Skaergaard case, uniform layers are found between the units showing gravity stratification but they are believed to be equally the result of crystal settling (Fig. 10). It seems that some extra factor besides straightforward crystal settling is needed to produce the special phenomenon of gravity stratification.

FIG. 11. Layering showing gravity stratification, in this case shown by the distribution of cumulus iron ore.

PLATE IV

Rhythmic layering on the southwest face of Gabbrofjaeld (Peak 1227 m). The *triple group* is conspicuous, and below is a finer-scale layering. Height of the face about 300 metres.

[face p. 22]

Besides showing rhythmic layering, the Skaergaard layered rocks may have some degree of fissility due to an arrangement of the platy minerals parallel to the plane of layering. As was originally suggested by Grout (1918d, p. 453), this texture is believed to be due to deposition of crystals of tabular habit from a moving magma, rather as mica flakes in a micaceous sandstone may be laid parallel to the bedding by water currents. To avoid possible confusion with metamorphic and sedimentary textures the term *igneous lamination* was used for this feature in describing the Skaergaard rocks (Wager & Deer, 1939, p. 37). The phenomenon is usually the result of parallelism of tabular plagioclase crystals but occasionally, elongated olivines and pryoxenes with a wide spread of orientations within the plane of layering may contribute. The incidence of igneous lamination in the layered series of the Skaergaard intrusion is sporadic. The plagio- clases of the lower observable layered rocks have a strongly platy habit but one layer may have strong igneous lamination while others, with equally abundant and platy felspars, may be massive. The differences may be a function of the strength of the currents at the time of deposition of the crystals, as discussed later. In the high layered rocks, igneous lamination is rare and eventually absent. In the original account, the topmost rocks, having no igneous lamination, were distinguished from the rest as the unlaminated layered series, a subdivision which is now discarded. Platy parallelism of the plagioclase crystals is also a feature of certain layers of the upper border group; it is believed that here, also, the texture is the result of currents, in this case acting on tabular crystals collecting at the top of the magma and laid parallel to the lower surface of the overlying, congealed material.

In this book the term layering will be limited to features of plutonic rocks which are

FIG. 12. Gravity stratification, both correct way up (left) and with photograph inverted (right).

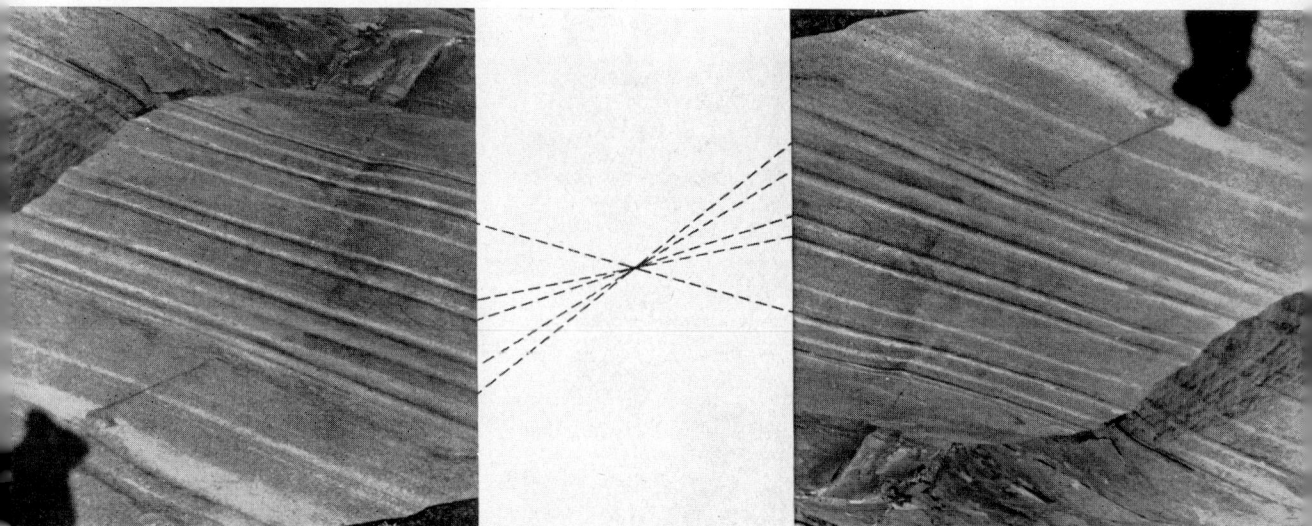

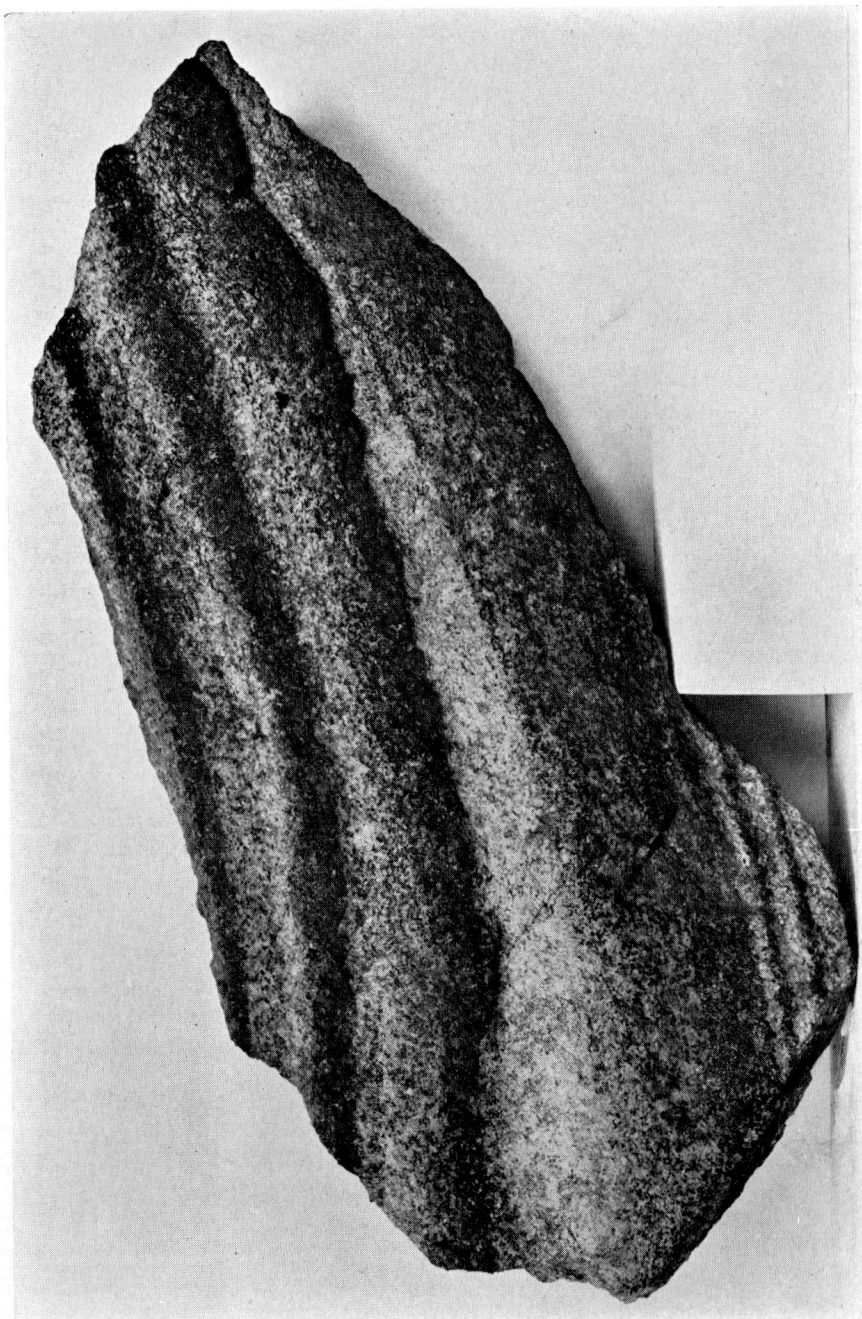

FIG. 13. Specimen showing extreme crystal sorting and gravity stratification (arranged as in original field orientation). Height of specimen 56 cm. Collected near Main House, from trough-band G (see Fig. 58). (From Wager & Deer, 1939.)

to be regarded as the result of the deposition of layers of material, one upon the other. Most layering is apparently the result of deposition of crystals at the bottom of a pool of magma and may be called *crystal-settling layering*, but some types of igneous layering seem to be the result of the congelation of successive sheets around the top and sides of a magma pool. The layering in the marginal and upper border groups of the Skaergaard intrusion, described on pp. 122–37, is apparently of the latter kind.

The term rhythmic layering began as a descriptive term for the conspicuously different layers of the Skaergaard intrusion that are due to differences in mineralogical composition, or texture, or both. However, in some cases, successive layers of the same composition and texture may be laid down on the floor, or deposited on the walls of an intrusion. The term layering should include the laying down of successive layers of similar material equally with the deposition of successive layers of different compositions. If it is necessary to make a distinction, the layering due to conspicuous differences in the mineral assemblages will be called *rhythmic layering*, while if there are no conspicuous differences it will be described as *uniform layering*. In the case of uniform layering, igneous lamination is often present, suggesting an origin by crystal settling.

Cumulus Crystals and Cryptic Layering

At any particular level in the layered series of the Skaergaard intrusion, there are certain minerals which can vary conspicuously in amount and give rise to rhythmic layering. These are the cumulus minerals, which are regarded as having collected together in the form of a precipitate at the bottom of the magma. The plagioclase crystals of the layered series, in thin section, usually show an inner part (varying from half to nine-tenths of the whole) which is essentially unzoned, while outside this is a normally zoned part of increasingly sodic plagioclase. The inner, uniform parts of each of the plagioclase crystals must have formed at a constant temperature, and from a large amount of magma, so that their separation from the magma did not appreciably alter its composition. These early, virtually unzoned crystals, termed *primocrysts*, presumably formed in the main body of magma. The primocrysts become the cumulus crystals when they have settled out, while the zoned, outer parts of the crystals, as will be shown later, formed from the magma trapped in the interstices of the crystal cumulate.

Since the cumulus crystals are unzoned (except in rare cases; see p. 37) their composition may be obtained by optical methods with considerable precision, and it is found that there is a regular variation with height in the layered series. The upward change in the plagioclases is from calcic to more sodic compositions, and in the olivines and pyroxenes from magnesium-rich to iron-rich compositions (cf. Fig. 14). This kind of variation, which was not readily appreciated in the field, was called *cryptic variation* in the original paper (Wager & Deer, 1939, pp. 36 & 37), and the layering to which it gave rise was described as *cryptic layering*.

From experimental data it is now well established that the more sodic plagioclases have lower temperatures of crystallization than the more calcic, and the more iron-rich ferromagnesian minerals, lower temperatures of crystallization than the more magnesium-rich. Thus the observed cryptic variation of the minerals belonging to solid solution series implies a change from high temperatures to low temperatures of precipitation in proceeding upwards in the layered series.

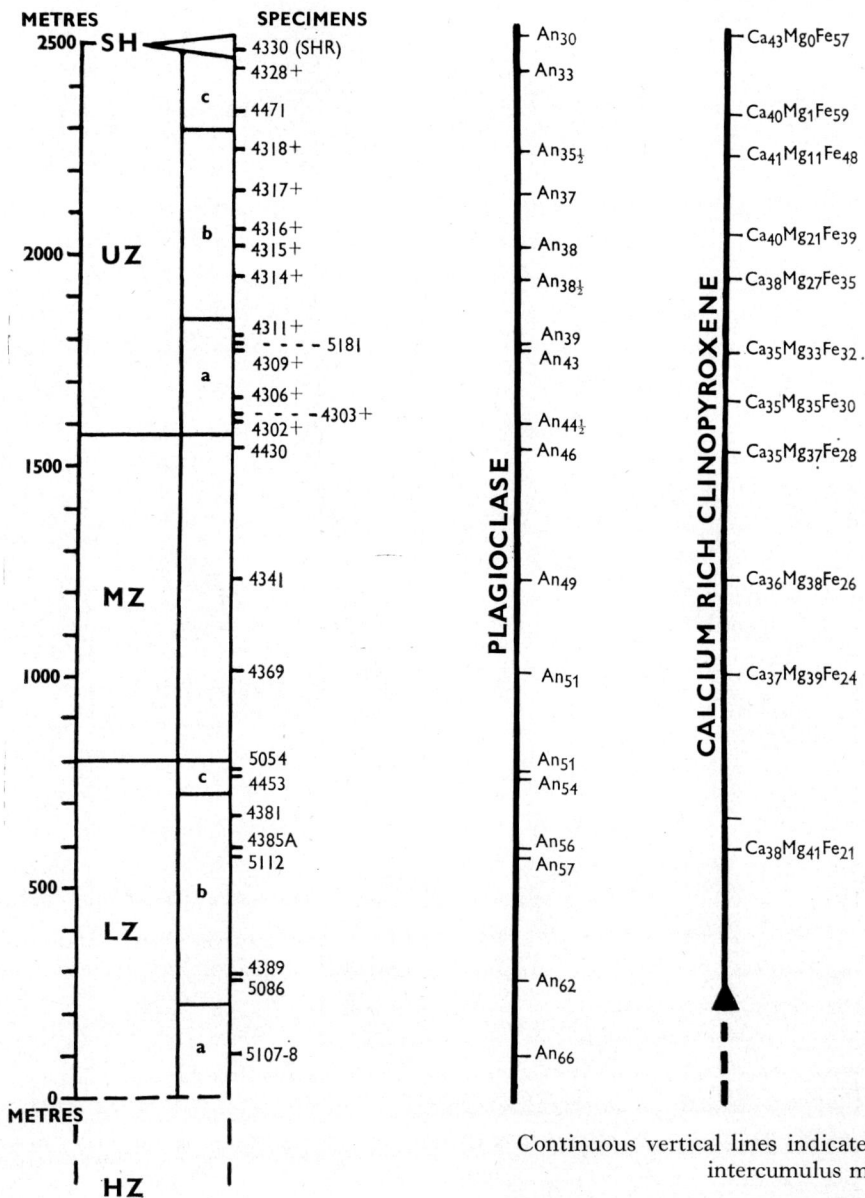

Continuous vertical lines indicate cumulu
intercumulus minerals o

FIG. 14. Minerals present in the various rock

Related to the steady upward change in composition of the solid solution minerals there are certain abrupt entrances and exits of minerals, and these are equally regarded as the result of the steady lowering of the temperature of precipitation which took place as the layered series was slowly built up. This may be illustrated by reference to

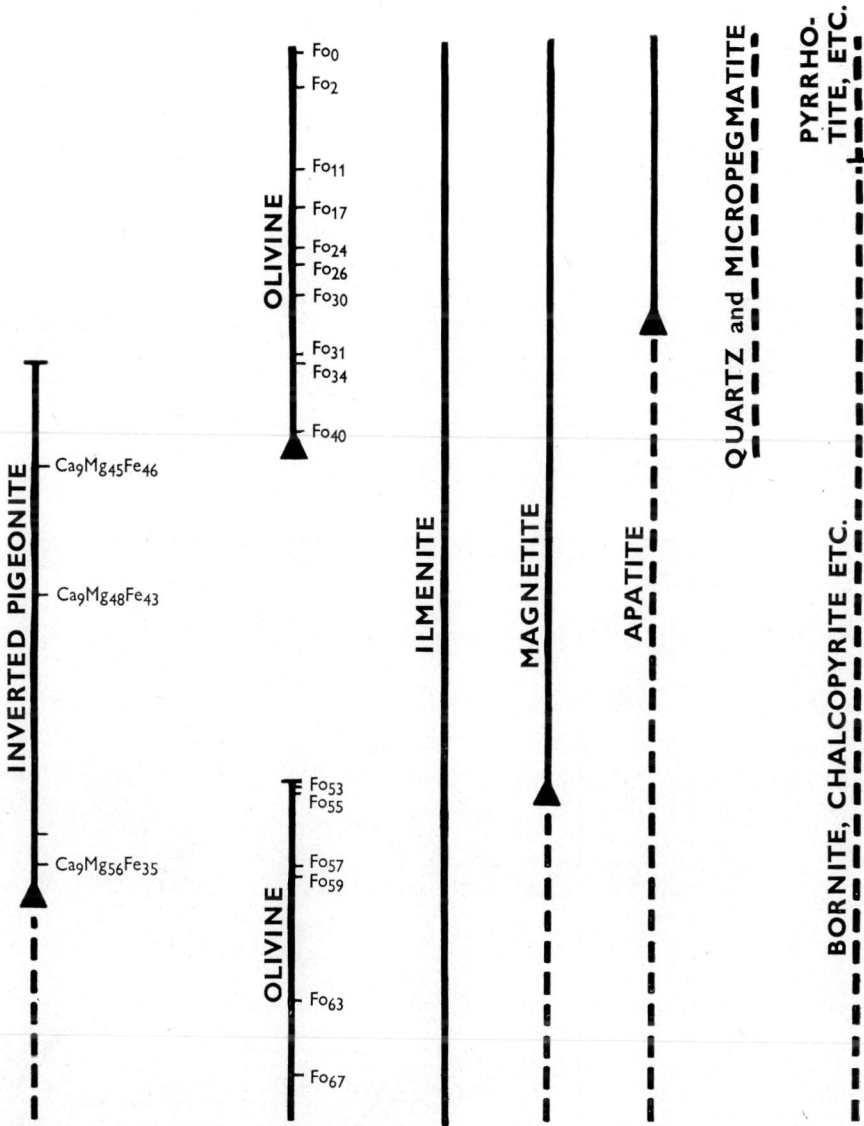

inerals; broken vertical lines relate to ose of indeterminate status.

the Layered Series, and their compositions.

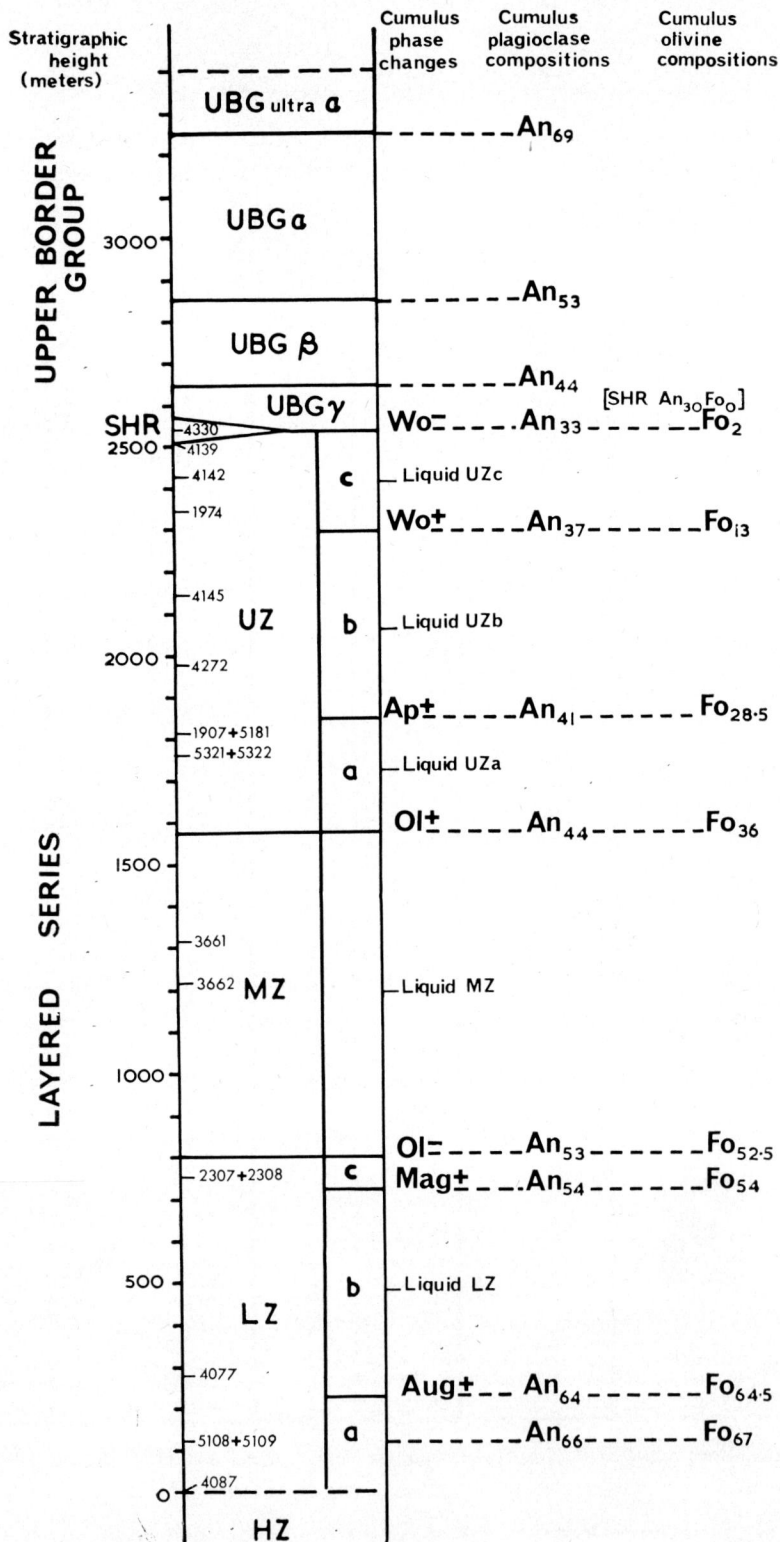

FIG. 15. Revised classification and thicknesses of the Layered Series and Upper Border Group, with heights of the analyzed rocks. The cumulus mineral compositions at each zone-boundary were read from the graph (Fig. 16), and are therefore generalized values (cf. Fig. 14).

magnetite, which is present in only small quantities in the lower observable layered rocks but which, at a height of 700 m on the arbitrary scale, suddenly becomes an abundant constituent, capable of marked variation in amount in adjacent layers. It is considered that by this stage the magma, now at a lower temperature, had changed in composition by fractionation such that direct precipitation of magnetite, as a cumulus mineral, occurred. The small amount of magnetite present in the rocks below this horizon is attributed to crystallization from the intercumulus liquid, which took place after the deposition of the cumulus minerals. The abrupt incoming or outgoing of particular mineral phases is interpreted as just as much the result of crystal fractionation as the continuous changes in the solid solution phases, and it is therefore also described as cryptic variation or cryptic layering. Thus there are two slightly different kinds of cryptic layering:

1. The steady upward change in composition of those minerals which form solid solution series.
2. The abrupt entrance or exit of cumulus minerals at particular horizons, in passing upwards in the layered series, called phase layering by Hess (1960, pp. 51 & 132–3).

These two kinds of variation are related, respectively, to what Bowen called Continuous and Discontinuous Reaction Series (1928, pp. 54–62).

If crystals, slowly forming in a cooling magma, are continuously collected together as an accumulate at the bottom of the liquid, then the upward succession of the layered rocks so formed would be expected to show both types of cryptic layering, and as a result of the removal of the early-formed crystals from the liquid there will be a continuous change in its composition. For this kind of process Becker (1997b, p. 258) introduced into petrology, from chemistry, the term fractional crystallization.

The cryptic variation of plagioclase and olivine throughout the layered series is summarized in Fig. 16, in which the minerals present, and their compositions, are plotted against structural height. From the graphs, the composition of the cumulus plagioclase or olivine, at any level, may be read (Fig. 15). In Fig. 14 the nature of all the cumulus minerals present throughout the layered series is shown. In these diagrams, minerals which are not found as cumulus crystals are not given, although as extra, inter-cumulus phases they are interesting constituents of the rocks; nor are the sulphide minerals, usually present in only very small amounts, given, since they have a petro-genetic status different from that of both the cumulus and the extra, intercumulus minerals (see Ch. III).

If the magma from which the cumulus crystals were settling were changed in com-position, at some time, by the addition of a fresh influx of the original liquid or an influx of some other magma, there would be a repetition of, or some irregularity in, the cryptic layering. Examples of this are found in the Rhum, Bushveld, and other layered intrusions described in Pt. II. In the Skaergaard case, the cryptic variation is apparently not complicated by such processes.

CLASSIFICATION OF THE LAYERED SERIES AND THE MARGINAL AND UPPER BORDER GROUPS

The entrance and exit of mineral phases, formed directly from the main bulk of the magma as primocrysts and collected as cumulus crystals, provides the best method of subdivision of the layered series. Strict use of such cryptic layering has resulted in an improvement in the classification from that used in the original account, and the new version is given in Fig. 15. For the thickness of layered rocks developed while a specific cumulus mineral or assemblage of cumulus minerals was forming, the term *zone* is used. This name seems to be appropriate, for the characteristic cumulus minerals have some analogy with the fossils characteristic of particular zones of the Phanerozoic sedimentary rocks. The major grouping in the new classification is into a lower zone (LZ) in which cumulus olivine is present, a middle zone (MZ) from which cumulus magnesian olivine is absent, and an upper zone (UZ) in which cumulus olivine is again present, although more iron-rich in composition than in the lower zone. Using the same principle, the zones have been further divided into sub-zones. It should be clearly understood that the subdivisions are based on the presence or absence of certain minerals as cumulus crystals, and we are not here concerned with the minerals formed from the intercumulus liquid. In deciding as to which zone any particular layered rock belongs, the distinction between cumulus and intercumulus material has to be made and sometimes this may be a difficulty. In particular cases an ambiguity may also arise because the crucial cumulus mineral may be absent, not because the mineral was not capable of forming from the magma at that time, but because some kind of sorting action, or temporary non-nucleation of the mineral, resulted in it being absent from the particular thin layer. Recently it has been established that there were some deviations from uniformity in the temperature or composition of the magma in different parts of the chamber, that gave minerals of slightly different compositions from the average, and produced other irregularities in the cryptic variation. For example, cases have been found in which the incoming mineral at the junction of two zones may be sporadically present as much as 100 m below the proper horizon at which it is a usual or constant feature. The boundary between zones is, therefore, defined either as the horizon where a particular critical cumulus phase is normally present, though sporadic occurrences of the mineral may be found some way above or below, or as the horizon where a critical cumulus mineral is normally absent, though here again it may occur sporadically in nearby layers.

The marginal and upper border groups were formed from the same fractionating magma as the layered series, and a rather similar method of subdivision is possible, except for the outer part of the marginal border group (*ca.* 70 m). In the latter, which does not show cryptic variation and is separated off as the *tranquil division*, there is a chilled olivine gabbro at, or close to the margin, and a coarser gabbro further in, both of which were formed from the undifferentiated original magma. In the tranquil division there are also sheets of picrite* containing abundant early magnesian olivines, and of perpendicular-felspar rock with striking, long blades of inward-pointing plagioclases. The remainder of the marginal border group exhibits fluxion banding parallel to the outer contact, and is called the *banded division*. This division shows changes inwards which are analogous to cryptic layering, and it is subdivided into lower-, middle- and upper-zone types, based on the presence, absence, and presence again of cumulus olivine crystals, as in the layered series.

The upper border group is contaminated by inclusions of acid gneiss and for this reason, and probably others, olivines are sporadic in their occurrence; the pyroxenes also are somewhat different from those of the layered series. The main subdivision of the UBG cannot, therefore, be satisfactorily based on the presence or absence of olivine. However, the unzoned, or little-zoned centres of the large plagioclase crystals of the upper border group, which are believed to be large cumulus crystals and which become increasingly rich in albite downwards, provide a basis of classification that can be made to correspond to that of the layered series. In the layered series, the discontinuous cryptic variations used in its subdivision can be correlated with the continuous cryptic variation in composition of the felspars. For example, at the top of the lower zone of the layered series, defined as the horizon where olivine ceases temporarily to be a cumulus phase, the plagioclase is An_{53}, and at the top of the middle zone, where olivine comes in again, it is An_{44}. High upper border group rocks collected from the more southerly Skaergaard mountains have plagioclases ranging, in descending, from An_{69} to An_{53}, and this range corresponds to that in the lower zone of the layered series; these rocks are accordingly grouped as UBGα. The rocks below them have plagioclases ranging from An_{53} to An_{44}, which correspond to the felspars of the middle zone of the layered series, and these are grouped as UBGβ, while the lowest of the upper border group rocks, with plagioclase more sodic than An_{44}, are grouped as UBGγ. There is a possibility that the uppermost border group rocks on the summits of some of the southern mountains, not yet climbed, may contain plagioclases more anorthite-rich than An_{69} and would thus correspond to part of the hidden layered series.

Late in the process of solidification of the Skaergaard intrusion there must have been a stage when the upward-growing layered series either came up against the downward-growing upper border group, or was separated by a thin layer of liquid from which rocks developed that are not to be regarded either as typical layered series or as typical upper border group rocks. This junction is called the *sandwich horizon*, and where a thin layer of rock of extreme composition, unlike layered series or upper border group rocks is found, it has been described as a sandwich horizon rock (SHR). Although the sandwich horizon rocks belong strictly neither to the LS nor to the UBG, they are considered for convenience along with the layered series.

The subdivisions of the layered series, as finally decided upon, are shown on the

---

* Formerly called gabbro picrite but not strictly gabbroic since the felspar is An_{77}.

map (Pl. XI, facing p. 244). The boundaries of the zones and subzones were sometimes mapped for short distances by careful scrutiny of the rocks in the field, in order to decide whether or not the critical, index minerals were present, but usually these minerals could only be identified satisfactorily in thin section. Thus the boundaries of the zones have been put on the map largely by extrapolation from rocks classified in the laboratory, taking into consideration the observed dip and strike of the layering at the places where the rocks were collected. The subdivisions of the marginal and upper border groups also are shown on the map but they must not be assumed to be as precise as in the case of the layered series. In particular, the boundary between the layered series and the upper border group, i.e. the sandwich horizon, is not well established since the highest layered series rocks and the lowest upper border group rocks converge in composition. The sandwich horizon rocks, where present, although extreme in chemical composition, also partake of the character of both the overlying and the underlying rocks. Around the sandwich horizon there are, also, complications due to the indefinite, intrusive sheets of intermediate and acid rocks.

There can be no doubt that the layered series persists below the lowest rocks at present exposed, and it is convenient to refer to these rocks as the *hidden layered series* (HLS) while the accessible rocks, when necessary, are distinguished as the exposed layered series (ELS). Some indirect information about the bulk and overall composition of the HLS rocks can be obtained from a consideration of the likely shape and bulk composition of the whole intrusion, and also from the mineralogy of the tranquil division of the marginal border group (see Ch. III, V & VII). The total volume of the hidden layered series is probably greater than that of the exposed layered series, and is estimated to be of the order of $350 \, km^3$ (p. 204). Drilling, to investigate the hidden layered series rocks, will be of considerable scientific interest, and is planned for an expedition to this part of East Greenland in 1966.

CHAPTER III

THE MINERALS OF THE SKAERGAARD INTRUSION

The cumulus minerals of the layered series (plagioclase, olivine, augite, orthopyroxene, pigeonite, titaniferous magnetite, ilmenite and apatite) are virtually unzoned and show a highly regular variation in composition with structural height. Certain minerals of the marginal border group also are largely unzoned, and are regarded as representing the compositions of crystal phases in equilibrium with the magma at an earlier stage than is represented by the exposed layered series. Thus in the marginal border group, the picrite contains magnesium-rich olivines and pyroxenes which are believed to be early minerals in equilibrium with the original Skaergaard magma; the earliest cumulus minerals of the hidden layered series were probably similar in composition. On the other hand, the plagioclases of the perpendicular-felspar rock also have large unzoned cores but are believed to be plagioclase which separated from the original magma under conditions of considerable undercooling.

The sulphides, such extra intercumulus minerals as the alkali felspars, quartz and zircon, certain hydrothermal minerals occurring with the latter, and certain late-stage minerals of the intermediate and acid rocks, are considered only briefly.

PLAGIOCLASE

Cumulus plagioclase persists throughout the exposed layered series and it varies in composition from approximately An_{69} in the lowest, to An_{34} in the uppermost horizon. The inner part or core of the discrete cumulus crystals is essentially unzoned, and normal zoning, often to fairly sodic plagioclase, is often to be seen outside the core. The proportion of essentially unzoned core to outer part varies from about 50% to 95%, or even more. The central 50 or 60% of the crystals is regarded as the original cumulus crystal, while any other unzoned part, having the same composition and therefore the same temperature of formation, is material added to the cumulus crystals under the special conditions discussed in Ch. IV (p. 64). Confining ourselves to the unzoned inner parts of the crystals, we find a fairly steady change in composition with height, the essential facts being presented in the graph (Fig. 16). The compositions of the felspars were determined on the Universal Stage (using the method of maximum symmetrical

C

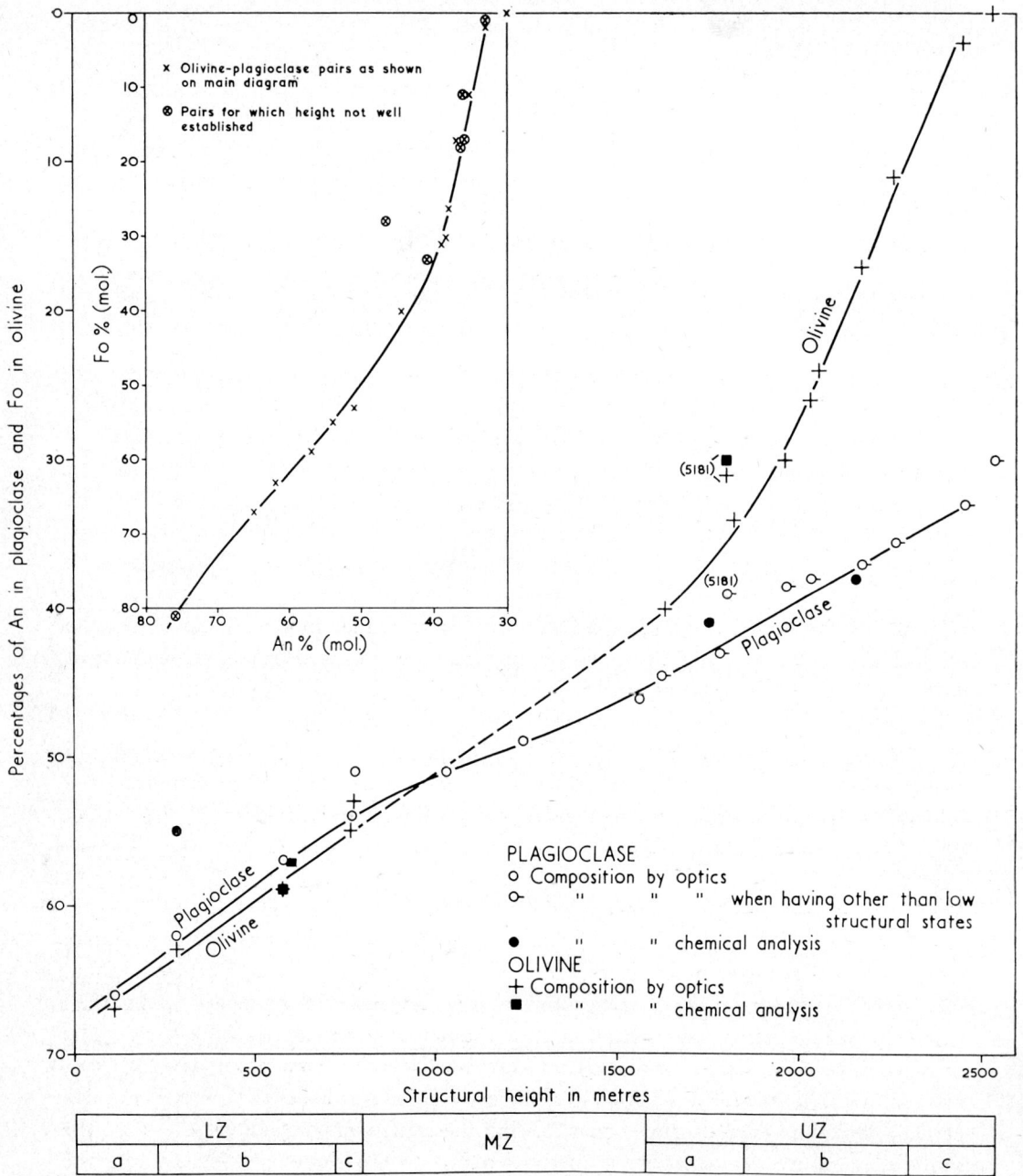

FIG. 16. Graph showing the compositions of the cumulus plagioclases and olivines plotted against structural height in the Layered Series. The inset graph gives the same compositions of cumulus olivines and plagioclases plotted against each other.

extinction and Winchell's curves, 1951) and, in the case of three specimens, by analyses of the separated minerals. The data derived by the different methods are given different symbols on the graph, and it can be seen that the plotted points for the chemically analysed specimens tend to lie above the general trend line of compositions obtained from extinction angle measurements on the unzoned parts of the crystals. This result is believed to be due to the material which was chemically analysed having included some of the outer, more sodic zones. However, some of the compositions estimated from optical data seem to show minor, but real irregularities from the average for a particular horizon. The existence of erratic entrances and exits of phases in relation to height has already been mentioned (p. 30) so that some irregularity in the cumulus felspar composition is not surprising.

The most calcic plagioclase of the intrusion, An_{77}, occurs in the tranquil division of the marginal border group. This plagioclase is considered to be that which was in equilibrium with the initial magma. The composition of the cumulus plagioclase of the hidden layered series is, therefore, believed to vary from An_{77} to An_{69} while that of the exposed layered series is seen to vary from An_{69} to An_{34}.

The cumulus plagioclase crystals in the LZ are square tablets flattened parallel to (010); they are usually about $1 \times 1 \times 0.2$ cm in size. As the layered series is ascended

FIG. 17a. Zoning in plagioclases of a typical plagioclase orthocumulate of LZa (the light part of Fig. 32). Cross-nicols. $\times 22$.

Fig. 17b. Almost unzoned plagioclases in a plagioclase–augite–olivine–magnetite meso-cumulate from UZa. Cross-nicols. $\times 22$.

the crystals, on the average, become smaller. They remain tablets, flattened parallel to (010), but become relatively thicker and also elongated along the *a* crystallographic axis; a typical size and shape in the UZb is 0·6×0·3×0·2 cm. The remarkable, bladed plagioclases of the perpendicular-felspar rock from the tranquil division of the marginal border group show an extremely elongated habit, typical dimensions being 5×0·3×0·1 cm. The size of the plagioclase crystals at any one horizon of the layered series, determined roughly with a pair of dividers, is about doubled in proceeding from the margin to the centre of the intrusion.

The zoning of the cumulus plagioclase has been studied in some detail but the meaning of it is not yet fully understood. There is normally an inner part or core of broadly constant composition, but minor, unexplained irregularities in extinction are often seen. The core is often, but by no means always, tabular in shape, and this part is taken to be the original cumulus crystal. The original cumulus crystal may be extended by adcumulus material of the same composition which cannot be distinguished with certainty from the true core. Zones of decreasing anorthite content, formed from the trapped magma, are, however, usually present discontinuously around the relatively unzoned cores. A typical plagioclase from low in the LZ is shown in Fig. 17a and one from the UZa, which is almost completely unzoned, is shown in Fig. 17b. Occasionally, crystals showing several oscillatory zones (Fig. 18) are found (Carr, 1954) while interesting reversed zoning attributed to the effects of supercooling (see below, p. 112) is found in the tranquil division of the marginal border group.

Three plagioclases have been separated from the layered rocks and chemically analysed. Some of the outer, more sodic zones seem to have been included in the samples analysed, because the compositions of the unzoned cores from plagioclases in nearby rocks, estimated from optics, are higher in anorthite content (see Fig. 16). Potassium is low in all three plagioclases and is not significantly different over the range from labradorite to andesine. The FeO and Fe_2O_3 contents of the plagioclases are believed to be due to iron in the crystal structure rather than to accidental impurities (Wager & Mitchell, 1951, pp. 145–7). In thin sections, acicular, orientated inclusions of an opaque mineral, probably exsolved iron ore, are sometimes seen. So far as present observations go, the incidence of such orientated, opaque inclusions is erratic.

Recent optical and X-ray work by Gay and Muir (1962) has shown that there are variations in the structural state of the plagioclases. Throughout the lower and middle zones of the layered series the plagioclase is in a low structural state, while from about the beginning of the UZ there is an irregular change, with increasing height in the intrusion, into a transitional and then into a high structural state. In the UZ layered rocks, however, there are cases where the plagioclase has a low, or near low, structural state. Gay and Muir suggest that the retention of the high and transitional, metastable structural states in the higher rocks is a function of composition, the more albitic plagioclases achieving the lower temperature state less readily than the more anorthitic. In the upper rocks, where inversion to the low structural state has occurred, there are differences in the type of twinning and other textural features which suggest the influence of some special factor not yet determined.

In the chilled marginal gabbro the plagioclases are in either low or transitional structural states, whereas in the coarser gabbros of the tranquil division of the marginal border group they are usually in a low structural state. As these border group plagioclases have compositions comparable with the plagioclases of the LZ rocks which all have low structural states, Gay and Muir suggest that cooling at the margins was sometimes sufficiently rapid to allow retention of the transitional, metastable structural state.

The composition of the plagioclase is the most significant single factor in any natural classification of igneous rocks, the reason being that its composition is broadly related to the temperature of crystallization. For the same reason, the plagioclase composition proves to be the best index of the fractionation stage at which any particular rock of the Skaergaard intrusion was formed. The zoning of the plagioclases is also of great interest because it records details of the successive stages in their formation; part of the unzoned core may be the original primocryst, and part may be due to adcumulus growth subsequent to deposition, while outer zones, where present, are the result of growth from the trapped interstitial liquid. Some slight variation in the composition of the cores is found, which may be the result of the individual crystal moving through the liquid to places of variable pressure or temperature, with resulting differences in composition of the plagioclase solid solutions.

OLIVINE

The compositions of the cumulus olivines of the layered series, determined by chemical analysis or estimated from optic axial angle or X-ray measurements,* are

* In the case of two Skaergaard olivines, X-ray measurements by Yoder and Sahama (1957) showed that they did not fall on the general curve relating X-ray data and composition. However, new analyses of the olivines have been made (Vincent, Douglas & Bown, 1964) and using these, the Skaergaard olivines now fall satisfactorily on the X-ray curves.

FIG. 18

A cumulus plagioclase crystal in hortonolite ferrodiorite of UZa, showing zoning of oscillatory-normal type. The darker zones are the more calcic. Three reversals, each followed by normal zoning, are to be seen. The reversals are considered to have happened while the crystal was a primocryst suspended in the magma, and not after deposition. Cross-nicols. × 22.

plotted against structural height in Fig. 16. Some slight zoning has been detected in the olivines, and in the optical measurements the central parts of the crystals were used. Chemical analysis of the olivine at the top of the layered series shows that it contains only 1% of MgO, but it has over 2% of CaO and 1% of MnO; estimates of the composition of the iron-rich olivines from optical data cannot be precise because the effects of the calcium, manganese, etc. are not yet known. The most magnesium-rich olivine, Fo_{81}, is from the picrite of the marginal border group and is considered to be the composition of the olivine in equilibrium with the initial magma. The Skaergaard intrusion contains a wide range of olivine compositions, and it was from work on this series that a rationalization of olivine nomenclature, following the pattern for the plagioclases, was first suggested (Deer & Wager, 1939). In the Skaergaard intrusion, strong crystal fractionation has resulted in a temporary period of non-crystallization of olivines, giving a gap in the olivine compositions from about Fo_{52} to Fo_{36}. It was suggested by Bowen and Schairer (1935), from the evidence provided by experimental work on the system $MgO-FeO-SiO_2$, that fractionation of certain liquids in this system would result in the temporary cessation of olivine crystallization.

It is noteworthy that as a result of fractionation, the olivines in the highest layered rocks are about as iron-rich as possible, while the plagioclase is by no means as soda-rich, having only reached An_{34}.

In the inset of Fig. 16 the composition of cumulus olivine is plotted against that of cumulus plagioclase from the same rock. With one or two exceptions, this gives a smooth curve and indicates that there is a close correlation in the changing composition of these two important minerals as the result of crystal fractionation. The minor discrepancies are believed to be significant but have not yet been fully investigated.

THE PYROXENES

In the Skaergaard intrusion, the varied stable associations of members of the pyroxene group, the compositional changes within each group, and the exsolution and sub-solidus inversion effects have all been extensively investigated.

a) *Augite, ferroaugite, and ferrohedenbergite*

The early work on a few samples of the calcium-rich series of pyroxenes (augite varieties), by W. A. Deer, showed that there is continuous variation from magnesium-rich to iron-rich members throughout the layered series (Wager & Deer, 1939, pp. 240–261). Further analyses by Muir (1951) defined part of the trend more clearly, but it was not until analyses of particularly pure fractions from both the layered and border group rocks were made that the compositional trend, in terms of Ca-Mg-Fe contents, was finally established (Brown, 1957; Brown & Vincent, 1963).

The full range, as shown in Table I and Fig. 19, is from $Ca_{42}Mg_{48}Fe_{10}$ in the picrite to $Ca_{42.5}Mg_{0.4}Fe_{57.1}$ in the sandwich horizon ferrogabbro, and is primarily a progressive replacement of Mg by Fe^{++}. The Ca content throughout the series remains fairly constant, between Ca_{42} and Ca_{35}, the minimum calcium content occurring in the middle fractionation stage as defined by Brown and Vincent (1963). Viewed in relation to the percentage of magma crystallized (Fig. 113), there is a marked acceleration in iron-enrichment as fractionation nears the end of its course; the change from Nos. 1 to 3 is inferred, from the marginal border group evidence, to have taken place in the hidden layered series, that is while 70% of the liquid crystallized.

The Skaergaard suite of augites, covering a wide range of compositions, has provided useful information on the effect of $Ca:Mg:Fe^{++}$ substitution on their physical pro-

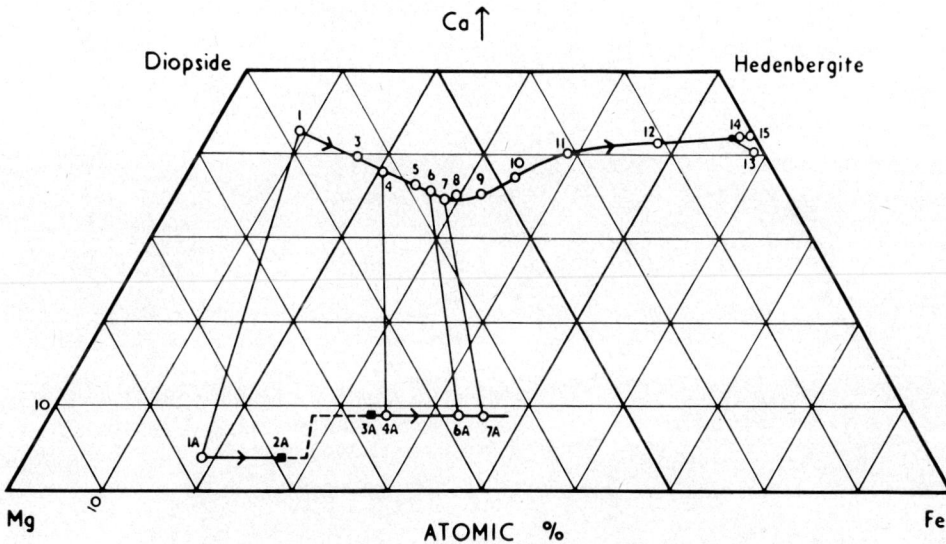

FIG. 19. Crystallization trend for the Layered Series and Marginal Border Group pyroxenes, based on eighteen chemical analyses (open circles) and two optical determinations (squares). Likely projection of No. 13, free from exsolution on inversion, also shown (filled circle).

1. Augite from picrite of tranquil division, MBG (4526A).
1A. Bronzite from picrite of tranquil division, MBG (4526A).
2A. Bronzite from perpendicular-felspar rock of tranquil division, MBG (1851).
3. Augite from LZa gabbro (4392).
3A. Inverted pigeonite from LZa gabbro (4392).
4. Augite from LZb gabbro (4385A).
4A. Inverted pigeonite from LZb gabbro (4385A).
5. Augite from MZ gabbro (4369).
6. Augite from MZ gabbro (4341).
6A. Inverted pigeonite from MZ gabbro (4341).

7. Augite from UZa ferrodiorite (4430).
7A. Inverted pigeonite from UZa ferrodiorite (4430).
8. Augite from UZa ferrodiorite (4306).
9. Augite from UZa ferrodiorite (4309).
10. Ferroaugite from UZb ferrodiorite (4314).
11. Ferroaugite from UZb ferrodiorite (4316).
12. Ferroaugite from UZb ferrodiorite (4318).
13. Inverted ferriferous β-wollastonite from UZc ferrodiorite (4471).
14. Green ferrohedenbergite from UZc ferrodiorite (1881).
15. Brown ferrohedenbergite from SHR ferrodiorite (4330).

perties. Recent re-investigation of the optical properties has been made by Muir (1951), Brown (1957), and Brown and Vincent (1963); their unit cell dimensions have been studied by Brown (1960), and their magnetic properties by Chevallier and Matthieu (1958) and Chevallier and Martin (1959).

The exsolution textures have been studied microscopically, along with those in the pigeonites and orthopyroxenes (Brown, 1957); the character of the thin, exsolved pyroxene lamellae, as inferred from their orientation within the host, has since been confirmed, in most cases, by single-crystal X-ray studies (Bown and Gay, 1960). The more magnesian augites from the marginal border group rocks mostly contain exsolution lamellae of orthopyroxene parallel to (100), while the LZ and MZ augites contain exsolution lamellae parallel to (001) (Fig. 21b) which, according to Bown and Gay (1960), consist of pigeonite in the un-inverted state. The UZa ferroaugites are characterized by very fine-scale exsolution lamellae parallel to (001) which impart to

FIG. 20a. Inverted ferriferous β-wollastonite (cloudy with inclusions), in association with fayalite (almost opaque with inclusions). UZc ferrodiorite (1974; similar to No. 13, Fig. 19). ×18.

FIG. 20b. The same as Fig. 20a, under cross-nicols. Each crystal of inverted ferriferous β-wollastonite can be seen now to consist of a mosaic of small, variously orientated crystals of ferrohedenbergite. ×18.

the crystal pale schillerization colours, to be seen under the microscope in ordinary light. In the pyroxenes from higher horizons, no exsolution has been detected by optical or X-ray studies, with the exception of a ferrohedenbergite from UZc, inverted from ferriferous β-wollastonite, which probably contains an exsolved ferriferous pigeonite according to Bown and Gay (1960).

The regular changes in compositions, exsolution textures and physical properties show that the augites form part of a single isomorphous series. However, at a height of 2300 m in the layered series a marked change takes place in the character of the ferriferous augite, which is used to define the base of UZc; pale greenish grains, cloudy with inclusions, are seen under crossed nicols to consist of a mosaic of small pyroxene crystals (Fig. 20). Such a mosaic texture is taken to indicate that some other crystalline phase was precipitated from the magma and that, on further cooling, it inverted to the observed mosaic of ferrohedenbergite crystals. In rocks a little above this level a similar greenish pyroxene occurs but it has the form of extensive poikilitic crystals rather than a mosaic of small crystals. In the case of the mosaic, ferrohedenbergite inversion seems to have taken place about a large number of centres, while in the case of the poikilitic variety it took place about a few, widely spaced centres. A probable reason for this has been discussed in some detail by Brown and Vincent (1963, p. 189). By analogy with results obtained from the experimental study of the system CaO-FeO-SiO$_2$ (Bowen, Schairer & Posnjak, 1933), it was suggested in the original memoir (Wager & Deer, 1939, pp. 111-2) that the phase directly precipitated from the extremely iron-rich, residual magma was an iron-rich, β-wollastonite. Although no trace of this mineral in the uninverted state has been found in the Skaergaard rocks, Yoder, Tilley and Schairer (1963) have found that by heating a specimen of the mosaic-textured ferrohedenbergite (Analysis 13, Table 1) an iron-wollastonite was produced, the inversion temperature being 970±5° C. The natural inversion to ferrohedenbergite probably occurred at a temperature only slightly lower than that of the β-wollastonite crystallization, because in some cases a brown pyroxene, free from inversion features, forms a rim to the green inverted variety, and this apparently crystallized directly from the trapped liquid. The latest rocks of the layered series (the sandwich horizon rocks) contain this same brown pyroxene as the phase originally precipitated from the liquid, and it is usually rimmed by a late-stage green variety, believed to be a lower temperature outer zone and not an inversion product from β-wollastonite. The pyroxene from inversion of β-wollastonite, although not found in the very latest residual rocks, is the most iron-rich clinopyroxene present in the intrusion (Fig. 19). The clinopyroxene trend line is difficult to define precisely near the Ca–Fe side of the triangular diagram, but it has been tentatively drawn so as to end with the brown ferrohedenbergites (Brown & Vincent, 1963, fig. 3; see also Fig. 19). This means that the inverted, ferriferous β-wollastonite lies slightly off the trend, containing slightly less calcium and more iron than the brown variety. It may be that if it were possible to analyse these inverted β-wollastonites, free from the abundant inclusions believed to have been produced on inversion, they would fall on the trend line of the directly-precipitated ferrohedenbergites.

b) *Orthopyroxene, pigeonite, and inverted pigeonite*

Calcium-poor pyroxenes are present in the marginal border group rocks and, as further cumulus minerals, from about 500 m up to 1700 m in the layered series (Brown, 1957). The calcium-poor pyroxenes exist in equilibrium with calcium-rich varieties; representative co-existing pairs have been analyzed (Table 1) and are joined by tie-lines in Fig. 19. A bronzite (1A) co-exists with augite in the picrite of the marginal border

TABLE 1

Analyses and optical properties of selected, representative, Skaergaard pyroxenes

(Numbers refer to those plotted in Fig. 19)

	1	6	11	13	15	1A	6A
SiO_2	51·17	51·26	49·44	48·28	46·71	53·6	50·9
Al_2O_3	3·22	1·98	1·31	0·79	0·93	2·3	1·8
Fe_2O_3	1·53	1·25	0·88	1·25	0·59	1·3	0·7
FeO	4·54	14·49	21·64	30·78	31·48	10·8	25·1
MnO	0·13	0·35	0·42	0·93	0·26	0·3	0·3
MgO	16·68	12·85	6·92	0·36	0·14	28·7	16·4
CaO	20·54	16·91	18·23	17·67	18·75	2·0	4·2
Na_2O	0·65	0·26	0·29	0·12	0·26	0·2	0·1
K_2O	0·05	0·02	0·03	0·08	0·03	...	...
TiO_2	0·97	0·84	0·83	0·19	0·95	0·5	0·5
Cr_2O_3	0·42	0·01	...	...	...	0·3	...
Total	99·90	100·21	99·99	100·45	100·10	100·00	100·00
Atomic %							
Ca	42·4	35·8	40·0	40·3	42·5	3·9	8·9
Mg	47·9	37·8	21·0	1·2	0·4	77·6	48·2
Fe	9·7	26·4	39·0	58·5	57·1	18·5	42·9
β	1·691	1·702	1·718	1·743	1·742	...	...
γ	...	...	...	...	...	1·689	1·723
2V	47°(+)	41°–44°(+)	48°(+)	55°(+)	56°(+)	75°(−)	47°(−)
%Al in Z	5·0	3·9	2·6	1·8	2·1	3·7	2·9
%Ti in Z	...	...	...	...	0·3	...	...

See caption to Fig. 19 for description of specimens. Analyst: E. A. Vincent.

Note: (1) The 2V for No. 6A is for the orthopyroxene inversion product.

(2) Analyses 1A and 6A are obtained by subtracting small amounts (5 %) of augite (compositions 1 and 6) present as impurity in the samples, and recalculating to 100 %.

group, and a slightly more ferriferous orthopyroxene (2A) occurs in the perpendicular-felspar rock. These orthopyroxenes contain a relatively low calcium content, now present chiefly in the fine-scale exsolution lamellae of augite parallel to (100). In contrast to the early calcium-poor pyroxenes of the marginal border group, those of the lowest exposed part of the layered series are hypersthenes with abundant blebs, or relatively thick orientated lamellae, of exsolved augite. Because of the higher overall calcium content, and the fact that the exsolution lamellae, parallel to (001), were formed from an original monoclinic crystal (Fig. 21c), these hypersthenes from the layered series are regarded as inverted pigeonites. The inverted pigeonites co-exist with augite in the LZ, MZ and UZa rocks (3A–7A, Fig. 19) and become progressively enriched in iron with fractionation. In the MZ, where inverted pigeonites are most abundant, the discrete augite and pigeonite crystals form a composite aggregate of grains (Fig. 22), the significance of which is not yet understood. Towards the top of the MZ some of the pigeonite crystals are either uninverted or only partially inverted. Some 200 m above the base of the UZ, pigeonite ceases to be precipitated and the rocks contain only one pyroxene, a ferroaugite.

The relative distribution of the two kinds of calcium-poor pyroxene, together with the character of their sub-solidus exsolution textures, have been interpreted according to the hypothesis advanced by Hess (1941), and later by Poldervaart and Hess (1951), for pyroxene assemblages in gabbroic rocks. Thus, the presence of a bronzite in the earlier differentiates and a hypersthene inverted from pigeonite in the later differentiates could be attributed to crystallization temperatures lying within the orthopyroxene stability field for the more magnesian varieties, and within a monoclinic pigeonite field for the more ferriferous varieties. In addition, the augite, included in both phases, is considered to be material exsolved at sub-solidus temperatures. Examined in detail, however, the exsolution and inversion characteristics of the Skaergaard pigeonites show some features which were attributed by these earlier workers to relatively quick cooling, such as occurred in volcanic or hypabyssal rocks. These features can be discussed with the help of Fig. 23, which depicts a hypothetical crystallization curve in relation to a clinopyroxene–orthopyroxene inversion curve* based on that presented by Bowen and Schairer (1935), (Brown, 1957, fig. 5). The marginal gabbro, chilled from the initial magma represented by C, contained a pigeonite, now inverted, which is represented on a calcium-free basis by D. On the other hand, the pyroxene which crystallized in equilibrium with the same magma was the bronzite B, of the marginal border group picrite. The original pigeonite of the chilled marginal gabbro crystallized from a liquid supercooled perhaps 20° C below the temperature of the liquid from which pyroxene B crystallized. If the supercooling had been greater, composition D would have been a hypersthene rather than a pigeonite. After the pigeonite crystallization, the chilled rock remained at a high temperature long enough to permit exsolution of augite,

* These relationships are hypothetical in relation to recent studies on pyroxene polymorphism, being used here only as an illustration of the observed natural relationships.

and inversion of pigeonite to hypersthene. In the main body of the magma, which necessarily cooled slowly, orthopyroxenes, progressively more iron-rich than B, continued to crystallize until pigeonite became the stable phase, at which stage the temperature of inversion to orthopyroxene fell below the pyroxene crystallization temperature. This stage probably occurred during formation of the hidden layered series, and according to the compositions of the pyroxenes from the marginal border group and lowest exposed layered series (2A–3A, Fig. 19) it would seem to have occurred roughly at $Mg_{70}Fe_{30}$. Poldervaart and Hess (1951) postulated that the changeover probably took place at this composition in most mafic intrusions, although exceptions can be found (e.g. Kaerven, Ch. XV). The earlier intercumulus pigeonites, such as those near the base of LZ, are inverted to hypersthene with blebby inclusions of augite (Fig. 21a); the lack of orientation has been attributed (Brown, 1957) to the narrow gap between pigeonite crystallization and inversion temperatures at this stage, most of the augite being exsolved during, rather than prior to, inversion. However, the relatively large size of the blebs, although possibly due to the greater ease of ionic migration at the higher temperatures of magnesian pigeonite inversion, has not been satisfactorily explained. In the hypersthene formed by inversion, exsolution then took place parallel to (100) as in the bronzite precipitated directly from the liquid. The later, more ferriferous pigeonites exsolved augite as broad lamellae parallel to (001) *before* inversion to hypersthene, as shown by the herringbone exsolution pattern formed by the lamellae in sections cut parallel to (100). This pattern indicates that at the time of exsolution parallel to (001), the host minerals were twinned monoclinic crystals (Fig. 21c). The progressive increase in the temperature gap between crystallization and inversion (see Fig. 23) would permit

FIG. 21 (*opposite*)

FIG. 21a. Inverted pigeonite from LZa gabbro (4392). The large blebs are of augite, probably exsolved from pigeonite almost at the same time as inversion, while the fine-scale lamellae ‖ (100) of the orthopyroxene host are of augite exsolved subsequent to the pigeonite inversion. Cross-nicols. × 80.

FIG. 21b. Broad exsolution lamellae of augite in inverted pigeonite (just to the left of, and below, centre) and finer exsolution lamellae of pigeonite in twinned augite (giving herringbone pattern). MZ gabbro (4369; No. 5, Fig. 19). Cross-nicols. × 45.

FIG. 21c. Inverted pigeonite from LZb gabbro (4386). The broad lamellae forming the herringbone pattern are of augite, exsolved ‖ (001) of a twinned pigeonite prior to its inversion from a monoclinic structure. Very thin lamellae can just be detected (centre of photo) ‖ (100) of the orthopyroxene, exsolved after inversion. The (100) plane of the orthopyroxene is parallel to the relict (100) twin-plane of the original pigeonite, suggesting relative ease of inversion. Cross-nicols. × 100.

FIG. 21d. Inverted pigeonite from MZ gabbro (4378B). One twinned crystal of pigeonite, after exsolution of augite ‖ (001), has inverted to two crystals of orthopyroxene, one at extinction. This is taken to indicate difficulty of inversion (see Fig. 23) in the more ferriferous pigeonites; at higher levels in UZa, pigeonite has failed to invert. Cross-nicols. × 150.

more time for exsolution in the more ferriferous pigeonites, and in fact the maximum thickness of exsolved augite lamellae is found in the MZ rocks (Fig. 21b).

Towards the top of the MZ, and in the lower part of the UZ, the pigeonites show evidence of sluggish inversion, single crystals inverting to two or more hypersthenes (Fig. 21d) which fail to have the same crystallographic axes as the original pigeonite.

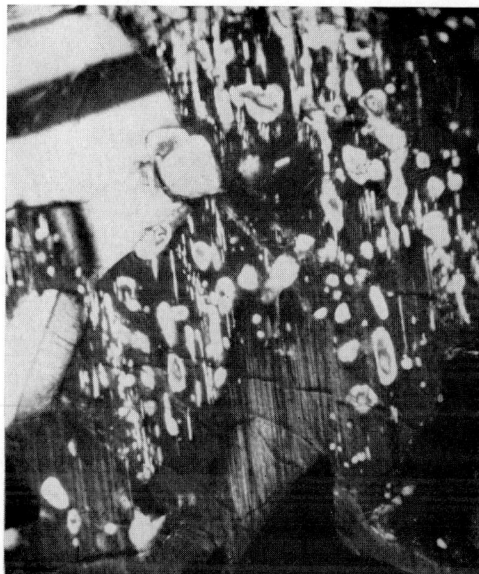
21a

21b

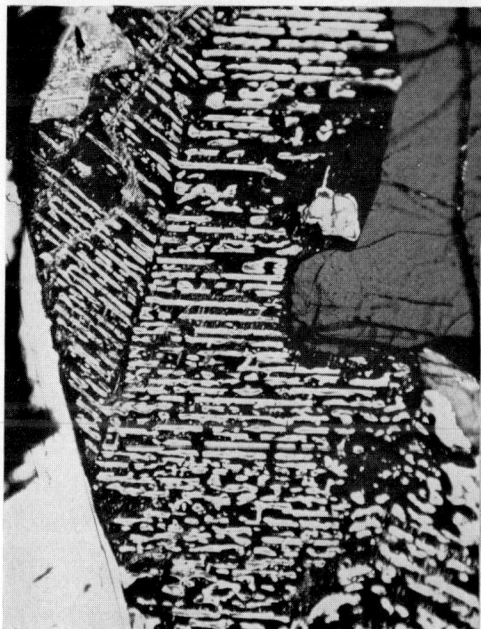
21c

21d

This is attributed to the difficulty of inversion at the relatively low temperatures at which the more ferriferous pigeonites are stable. At slightly higher levels, crystals of uninverted pigeonite occur and these continue to be found sporadically up to about 200 m above the base of UZa, at which level pigeonite ceased to crystallize.

The most ferriferous pigeonite to crystallize as a separate phase is close in composition to $Ca_9Mg_{46}Fe_{46}$ (No. 7A, Fig. 19). The tie-line between this and the co-existing augite has been termed the limit of the two-pyroxene field by Brown (1957), to avoid use of the ambiguous term ' two-pyroxene boundary ', and varies slightly in position according to the composition and temperature of the magmas involved and, possibly, according to the depths of crystallization. Thereafter, the only separate pyroxene phase is a member of the ferroaugite series, but this may contain exsolution lamellae that are probably of ferriferous pigeonite. As discussed by Brown (1957), the Skaergaard pyroxene trends provide evidence as to the compositional relationships at the precise limit of the two-pyroxene field, and in view of this evidence, the hypothesis forwarded by Barth (1962, p. 101) is not believed tenable. The ferroaugites at each side of the two-pyroxene field limit have identical calcium contents (7 & 8, Fig. 19), and it is considered that the pyroxene cotectic has migrated towards the Ca-rich limit of the solvus with fractionation, as initially postulated for the Beaver Bay diabase pyroxenes by Muir (1954).

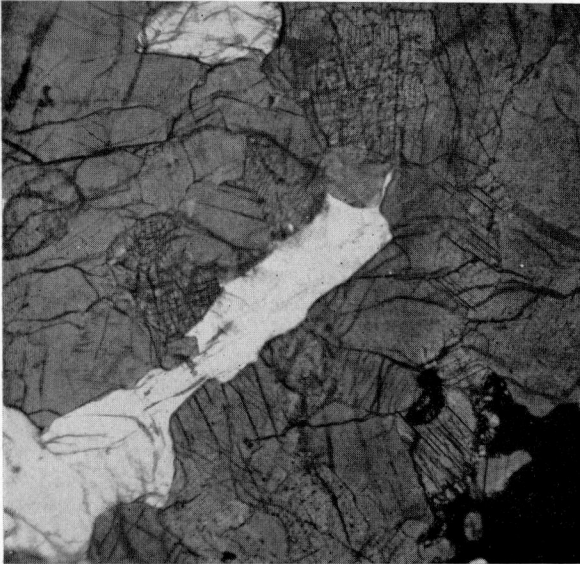

FIG. 22a. MZ gabbro (4341) showing a closely packed aggregate of discrete crystals of augite and inverted pigeonite (Nos. 6 and 6A, Fig. 19). The two pyroxenes differ in cleavage and relief, as can be confirmed by reference to Fig. 22b. ×35.

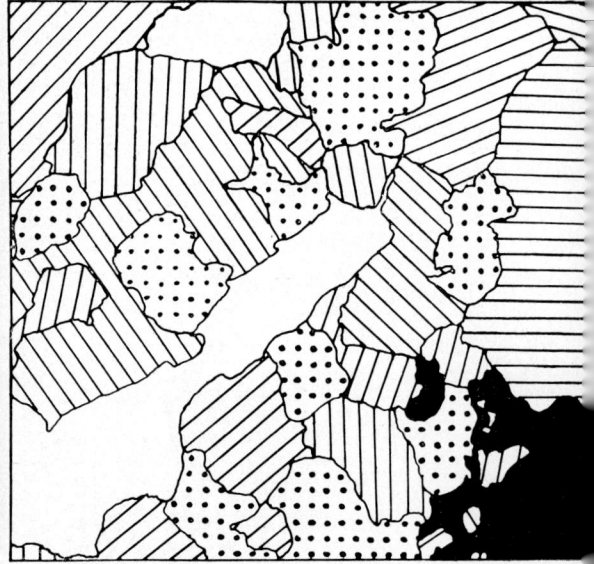

FIG. 22b. Tracing of Fig. 22a to indicate the distribution of the cumulus augites (lined) and cumulus inverted pigeonites (stippled).

The previous discussion is based on the evidence provided by the layered series and marginal border group pyroxenes. In addition, however, pyroxenes are present in the upper border group rocks and are broadly comparable in composition with those of the layered series in that there is iron enrichment with fractionation, from UBGα to UBGγ. Four analysed green ferrohedenbergites from melanogranophyres (believed to be UBGγ by Brown & Vincent, 1963) appear to show iron-enrichment with fractionation, as defined by felspar and olivine compositions. As shown on Fig. 24, the pyroxene trends for these rocks and the layered series converge at a composition close to Nos. 15 and 19. Analysis of an augite from UBGα (Wager & Deer, 1939, pp. 152–3) and optical measurements on augites from various UBGβ rocks by Douglas (1961) also show similarities

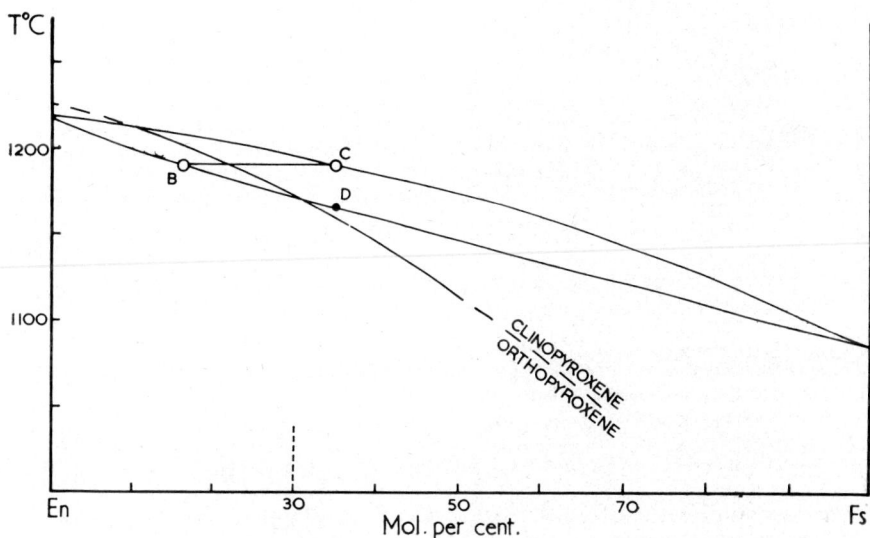

FIG. 23. Possible relationship, according to the natural evidence, between the crystallization and inversion curves for calcium-poor pyroxenes. (For derivation, see Brown, 1957.) D = pigeonite (since inverted) of the chilled marginal gabbro (4507); B = bronzite from the picrite of the Marginal Border Group (4526A). It is suggested that the initial magma, of a composition represented here by C, chilled to give the pyroxene of composition D. From the same magma, slower cooling in the tranquil division of the MBG gave pyroxene B as the earliest phase in equilibrium with the magma. Continued slow cooling and fractionation then gave rise to a series of pyroxenes of increasingly ferriferous composition, the earliest (in the tranquil division of the MBG and probably in the hidden layered series) being bronzites and the later (in the exposed Layered Series) being monoclinic pigeonites. The likely progressive increase in the temperature interval between the crystallization and inversion curves is inferred from the observed increase in the amount of exsolution and the estimated decrease in the ease of inversion for the pigeonites of the Layered Series. The temperature scale is given as a broad estimate of relative values, both crystallization and inversion temperatures, as well as inversion relationships, being given only as a hypothetical interpretation of observed relationships. (After Brown, 1957.)

with those from the LZ and MZ, respectively. The pyroxenes are, on the whole, at a more advanced fractionation stage than the plagioclases in these UBG rocks and this is no doubt because they are intercumulus in character while the plagioclase is essentially cumulus.

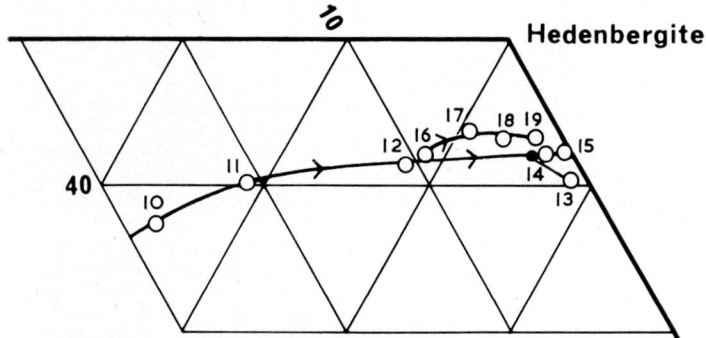

FIG. 24. Trend of crystallization for the late-stage pyroxenes of the layered series, compared with the partly anomalous, but chiefly downward-crystallized trend for the pyroxenes from the Upper Border Group granophyres (believed to be UBGγ rocks, from the pyroxene evidence, by Brown & Vincent, 1963).

 10-15. Pyroxenes from UZb, UZc, and SHR Layered Series rocks. (Key
   as in Fig. 19.)
 16. Ferrohedenbergite from melanogranophyre (5264), Brödretoppen.
 17. Ferrohedenbergite from melanogranophyre (3021), Basistoppen.
 18. Ferrohedenbergite from melanogranophyre (4332), Basistoppen.
 19. Ferrohedenbergite from transitional granophyre (4489), Sydtoppen.

The Iron ores: Ilmenite, Chrome-spinel, Magnetite, and Ulvöspinel

Ilmenite and a complex titaniferous magnetite, which often contains exsolved ulvöspinel, are present abundantly in the Skaergaard rocks; they have been studied particularly by E. A. Vincent (Vincent & Phillips, 1954; Chevallier, Mathieu & Vincent, 1954; Vincent, Wright, Chevallier & Mathieu, 1957; Vincent, 1960; Wright, 1961).

Ilmenite, without chromite or magnetite, occurs in the chilled marginal gabbro. In the picrite of the tranquil division of the northern border group, a small amount of early chrome-spinel occurs but it has not been found in later rocks. Apparently the original Skaergaard liquid had a composition which, under some conditions and for a very short time, precipitated chrome-spinel. Though chrome-spinel occurs associated with the earliest olivine, it did not of course crystallize from the supercooled Skaergaard magma from which the chilled marginal gabbro formed.

In the lowest observable layered rocks of LZa, the iron ore is mainly ilmenite with a somewhat poikilitic habit. A little higher up in LZa and LZb the poikilitic patches of ore are composite, consisting of ilmenite and magnetite, while some idiomorphic ilmenite is found sporadically. It appears that ilmenite was precipitated as a cumulus phase from the Skaergaard liquid earlier than magnetite. The size of the crystals of ilmenite, precipitated at this early stage, is small, perhaps because there was not much Ti in the magma compared with Fe and Mg which largely contribute to the formation of the bulky ferromagnesian minerals. No obvious ilmenite-rich layers have been found that might have confirmed the cumulus status of the ilmenite. However, the absence of such layers is not conclusive, since apatite, which certainly became a cumulus mineral at a late stage in the fractionation, is also never abundant enough to produce layers with more than a few per cent of this mineral.

At a height of 820 m, layers rich in iron ore are found that consist of about equal amounts of ilmenite and complex, titaniferous magnetite grains (Fig. 25A). From this level, which marks the beginning of LZc, magnetite, as well as ilmenite, is a cumulus mineral. It is still not clear why, when titaniferous magnetite became a cumulus phase, ilmenite is also found abundantly in similar, large-sized crystals. One possibility, first suggested by Phillips (Vincent & Phillips, 1954) and further investigated by Wright (1961), is that the ilmenite and titaniferous magnetite, found as clusters of grains, crystallized from the magma as a single crystal of a homogeneous solid solution which, on cooling, unmixed into separate grains of the two phases. On textural grounds, certain isolated ilmenite crystals, surrounded only by silicates and lying outside the ore aggregates, are considered by Wright (1961) to have been separately precipitated as such from the magma. On the other hand, Buddington, Fahey and Vlisidis (1955) doubt the unmixing hypothesis. Another possibility is that ilmenite and magnetite began, by coincidence, to be precipitated simultaneously as cumulus minerals at this particular stage of fractionation, but this hypothesis is probably eliminated by the fact that the chilled marginal gabbro carries ilmenite but no magnetite, suggesting that ilmenite crystallized early from the magma. This problem would be solved if the hidden layered series could be investigated.

By optical and X-ray examination, chemical analysis, and magnetic measurement, the early cumulus titaniferous magnetite has been shown by Vincent (1960) to contain 40–50% of ulvöspinel as a very fine-grained exsolution intergrowth having a box-like form (Fig. 26). The occurrence of ulvöspinel implies a strongly reducing environment. The ilmenite in the LZc and MZ contains sparse, very thin exsolution lamellae of magnetite that also contain very minute units of exsolved ulvöspinel. Cumulus titaniferous magnetite and ilmenite continue to occur in the UZ ferrodiorites. The ilmenite is often found as well-shaped prismatic crystals and the magnetite contains abundant, fairly broad exsolution lamellae of ilmenite which are believed the result of an early, relatively high-temperature exsolution and oxidation. There is also a fine scale exsolution of ulvöspinel, presumably developed later than the exsolution lamellae of ilmenite. This pattern of behaviour is found in magnetites up to a height of 2350 m in the UZc. Above this, exsolved ilmenite lamellae only are found, and these increase in abundance upwards. On the whole, cumulus ilmenite becomes more abundant relative to magnetite in the higher layered rocks; this tendency continues in the late intermediate and acid rocks but, by this stage, none of the minerals can be considered as cumulus crystals.

D

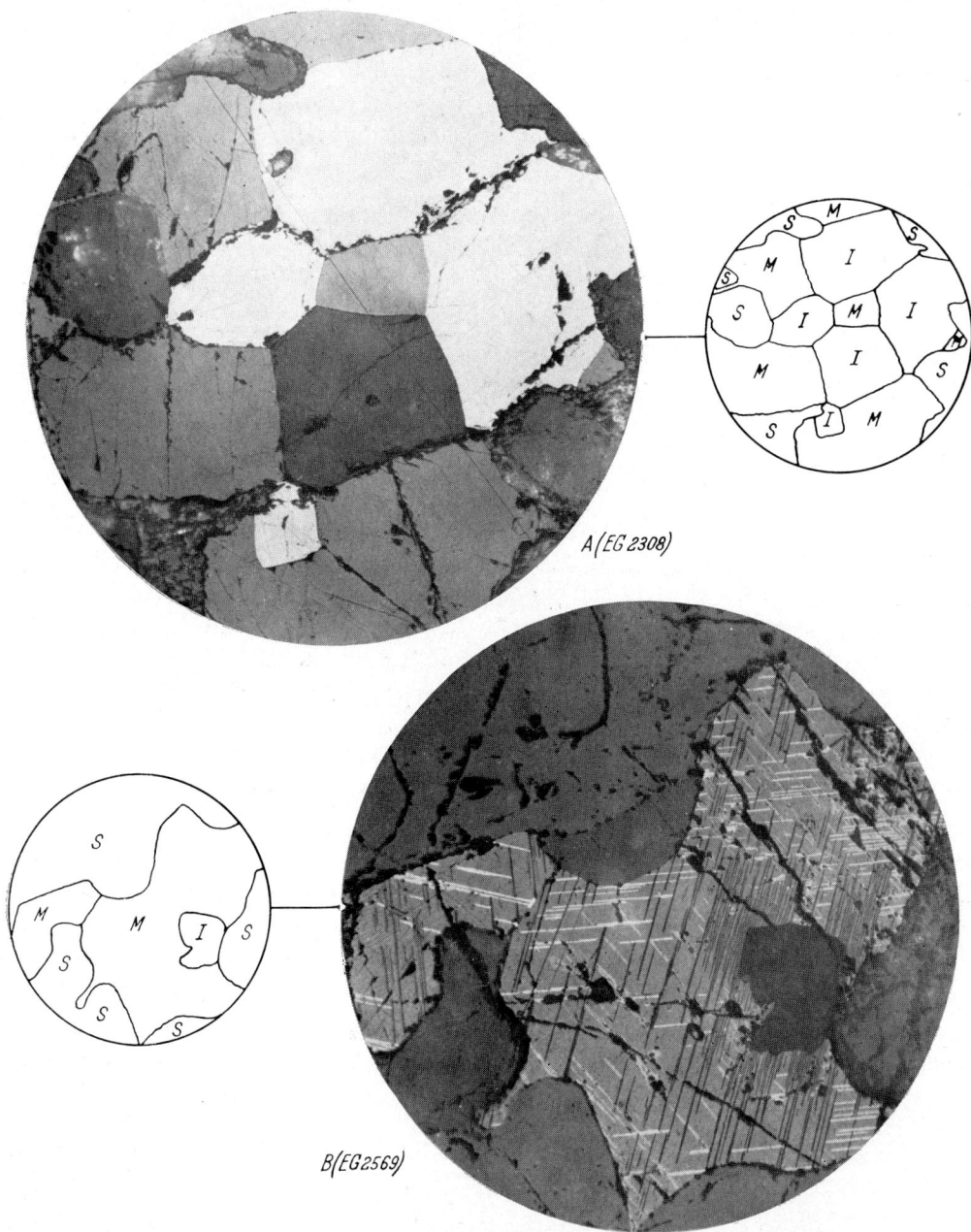

FIG. 25. Polished specimens showing cumulus magnetite and ilmenite in Layered Series rocks.

A. Ore-rich layer from LZc, on Kraemers Island. Key shows : M, magnetite; I, ilmenite; S, silicates. ×45.

B. Ferrodiorite from UZa, House Area. Key shows: M, magnetite with exsolution lamellae of ilmenite; I, ilmenite as independent crystal; S, silicates. Polars 5° off cross-position. ×60.

Chemical analysis shows that the titaniferous magnetites in the melanocratic layers are less oxidized than those in adjacent average and leucocratic layers (Vincent & Phillips, 1954). It was totally unexpected to find cumulus minerals differing in composition at the same horizon. All titaniferous magnetite precipitated as cumulus crystals, at any

FIG. 26. Magnetite–ulvöspinel intergrowth in MZ gabbro (4450). Ulvöspinel (grey) has exsolved with a box-like form. Black dots, usually in the magnetite, are apparently a translucent spinel, also exsolved from the original magnetite. The inset shows magnetite, embedded in an augite crystal, with different proportions of ulvöspinel (grey). ×2200.

one time, should have the same composition, whether present sparsely, in the average and leucocratic layers, or abundantly in the melanocratic layers. The observed anomaly, however, may be explained by post-depositional changes in the state of oxidation of the iron. The amount of trapped liquid, which would be present in approximately equal amounts in the melanocratic, average, and leucocratic rocks, must have had the same initial partial pressure of oxygen, and with lowering temperature this would be able to convert a specific amount of ferrous to ferric iron. In the melanocratic layers, the amount of oxidation would be distributed thinly through a considerable bulk of titaniferous magnetite which would, therefore, be relatively little oxidized. On the other hand, in the average and leucocratic layers the same amount of oxidation would be spread through a small amount of ore which would, consequently, be relatively strongly oxidized. The material being oxidized is apparently the exsolved ulvöspinel and Vincent (1960) has shown that a gradual change from ulvöspinel to ilmenite lamellae takes place in the titaniferous magnetite crystals of many of the leucocratic rocks where oxidation was at a maximum. Buddington et al. (1955) have suggested that the problems of exsolution in the iron ores are as complex as in the felspars. Vincent's work suggests that in certain respects they are even more complicated, because of the additional possibility of oxidation.

APATITE

The distribution and habit of apatite in the intrusion is of considerable petrological importance and is discussed in Chapters IV and VII (pp. 66 & 164). No detailed mineralogical or chemical investigation of the apatite has, however, yet been made. It is presumably a fluor-apatite since this is the type usually found in igneous rocks. Some trace elements have been determined in two examples, each with cumulus status: one separated from low in the UZb zone and the other from the UZc (Wager & Mitchell, 1951, p. 159). Y, La and Pb were found to be considerably enriched in the later apatite.

MINERALS OCCURRING ONLY AS EXTRA INTERCUMULUS PHASES OR IN THE LATE-STAGE, INTERMEDIATE AND ACID ROCKS

All the minerals considered so far (with the exception of ulvöspinel) have formed cumulus crystals, at one stage or another, although they also commonly occur as inter-cumulus minerals. Other minerals, mainly quartz, alkali felspars and zircon, occur only as extra intercumulus minerals while others such as hornblende, biotite, chlorite, serpentine, epidote and zoisite, ilvaite, micaceous decomposition products, zeolites, etc., are lower temperature minerals apparently formed as the trapped intercumulus liquid passed gradually from the magmatic to the hydrothermal stage.

　　Quartz is found in very small amounts as an intercumulus mineral in felspar cumulates of even the lowest exposed layers, but significantly it is not present in the rocks of these horizons when they contain cumulus olivine. It becomes a frequent intercumulus

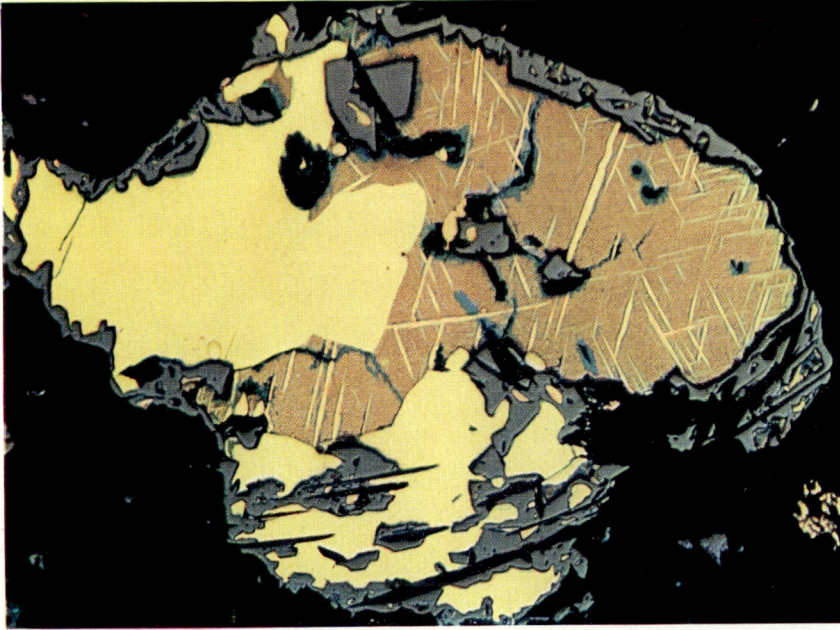

Complex, globular sulphide patch in ferrodiorite (5221) from the Trough Banding horizon of UZa at 1800 m, ¼ mile NNE of the Main House. (As Fig. 28b). The patch consists of chalcopyrite in compact masses and as exsolved lamellae (yellow); bornite (brown) with the exsolved chalcopyrite lamellae and secondary covellite (blue); magnetite (grey); and acicular silicate (black), probably hornblende. The magnetite forms a rim to the sulphide or occurs as idiomorphic crystals within it. ×225.

Sulphide patch within a magnetite crystal. The sulphide consists of bornite (brown) altered extensively to two shades of blue (digenite or covellite). The iron oxide is magnetite with exsolved ilmenite (broad lamellae), the faint pattern being due to exsolution of ulvöspinel. ×250. (Photos by E. A. Vincent.)

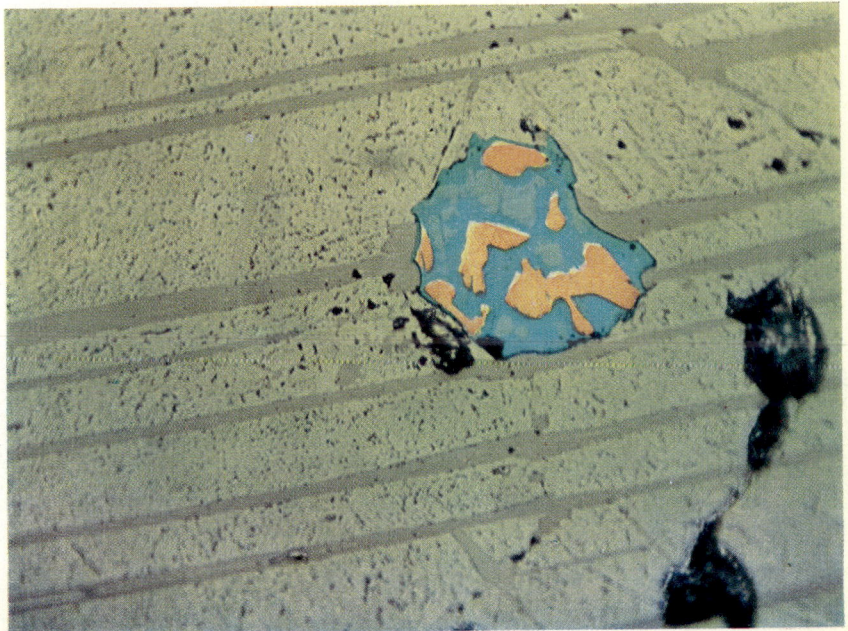

PLATE Vb

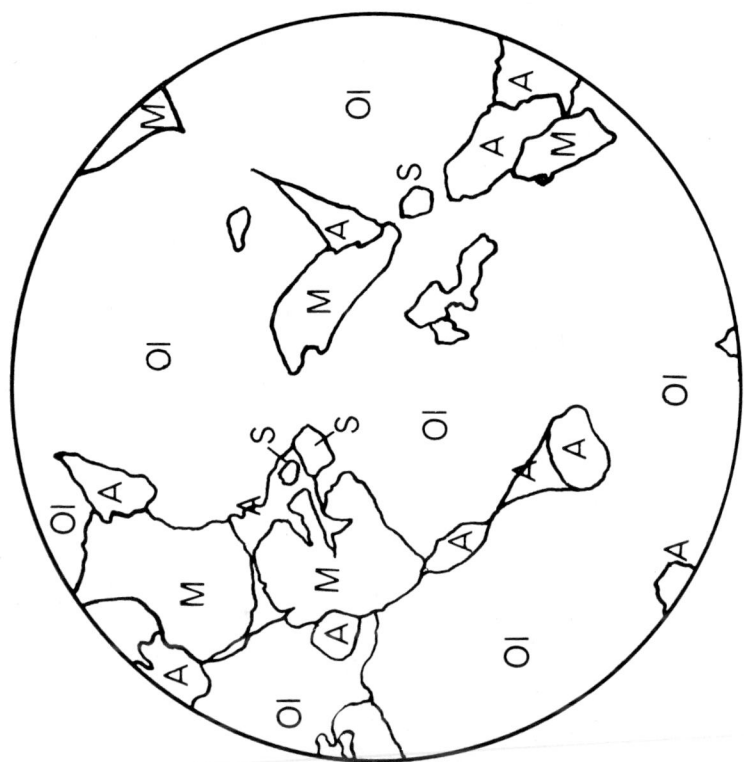

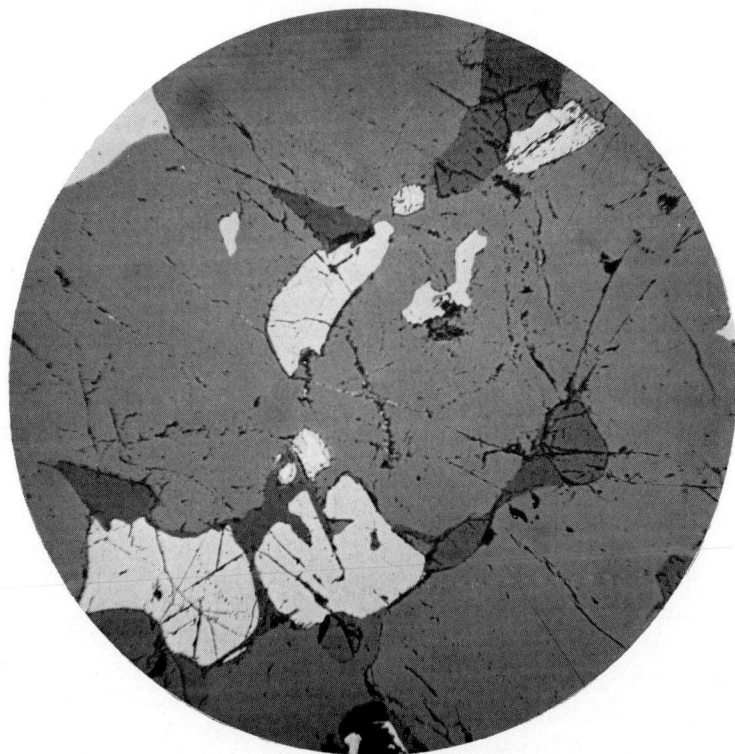

FIG. 27. Photograph and key of a polished specimen showing the small size and rarity of the sulphide globules, even in a relatively sulphide-rich rock. S, bornite and digenite. The rest is largely olivine crystals (Ol) with small amounts of lower-relief ferroaugite (A) and relatively large amounts of higher-relief apatite (A). Melanocratic layer in UZb (5196). ×32.

mineral, often intergrown with alkali felspar, in UZa, where the olivine is iron-rich and therefore in equilibrium with quartz. In certain upper border group rocks underlying the Basistoppen sheet, the form of the quartz indicates that it has formed by inversion of tridymite. The former existence of tridymite may be the result of partial remelting of these rocks at the time of intrusion of the Basistoppen sheet.

Alkali felspar is liable to occur sporadically, and in very small amounts, as an inter-cumulus mineral throughout the layered series but its nature and composition have not been studied. Alkali felspar becomes abundant, in association with quartz, in the late-stage melanogranophyres and acid granophyres.

Biotite, as rare flakes often attached to iron ore, occurs in the LZ rocks. It is less common in the ferrodiorites where alkali felspar is the chief potassium-bearing mineral. Stilpnomelane occurs in the late acid granophyres and is also abundant in certain fused acid gneiss inclusions of the marginal border group (p. 136).

Zircon, if searched for, can usually be found at all horizons along with the late crystallizing minerals from the trapped intercumulus liquid. It is rather abundant, as small ophitic crystals, in the sandwich horizon rocks believed to represent the com-position of the liquid at the latest stage of the layered series. It is also found in the late-stage, intermediate and acid rocks.

Small amounts of hydrothermal-stage minerals often seem to have been directly precipitated from the final dregs of intercumulus liquid, while in other cases they have formed by hydrothermal decomposition of some of the earlier minerals. They appear much the same as the comparable minerals found, in similar situations, in other gabbros and diorites.

FIG. 28 (*opposite*)

Polished specimens of sulphide minerals fram Layered Series rocks.

a. Sulphide patch from the MZ (4429) on Pukugaqryggen (melanocratic part of the upper unit of the Triple Group). Bornite (light grey), with exsolution lamellae of chalcopyrite (white) bordered by digenite (dark grey). A zone across the middle has more extensive exsolution of digenite, which is partly altering to covellite. × 350.

b. Globular sulphide patch in ferrodiorite of UZa (as in Pl. Va). The chal-copyrite is white to light grey, and the bornite is darker grey with light exsolution lamellae of chalcopyrite; the magnetite is a similar grey with dark borders, and the silicates are black. × 150.

c. Sulphide in droplet-form patch in ferrodiorite (4328) of UZc. The patch consists mainly of marcasite with a concentric spongy appearance. Also a little chalcopyrite (light grey as in Fig. 28b) containing two acicular crystals of a translucent mineral (probably hornblende), shown at the lower left-hand corner of patch; and magnetite around part of the margin (right-hand centre). × 330.

d. Part of a sulphide patch, also in a ferrodiorite (5134) of UZc. The patch consists of pyrrhotite (white to light grey) partly replaced by marcasite (con-centric habit) and material with a parallel structure related to some special direction in the pyrrhotite crystal. × 400.

The Sulphides

Sulphides are usually inconspicuous but they have been shown by polished specimen microscopy (Wager, Vincent & Smales, 1957) to be present in all the rocks of the

28a

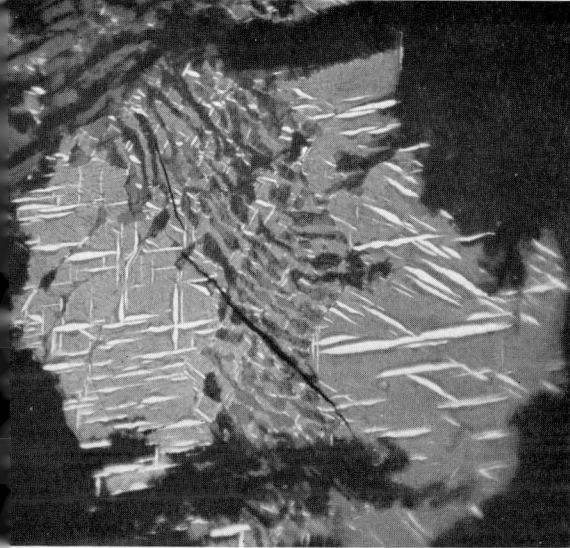

28b

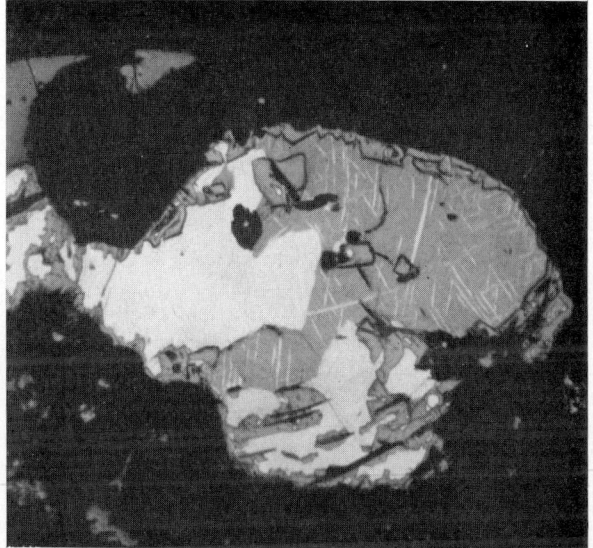

28c

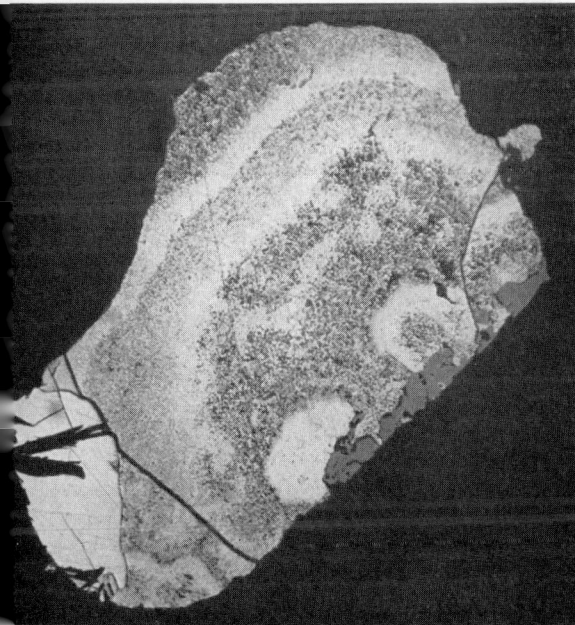

28d

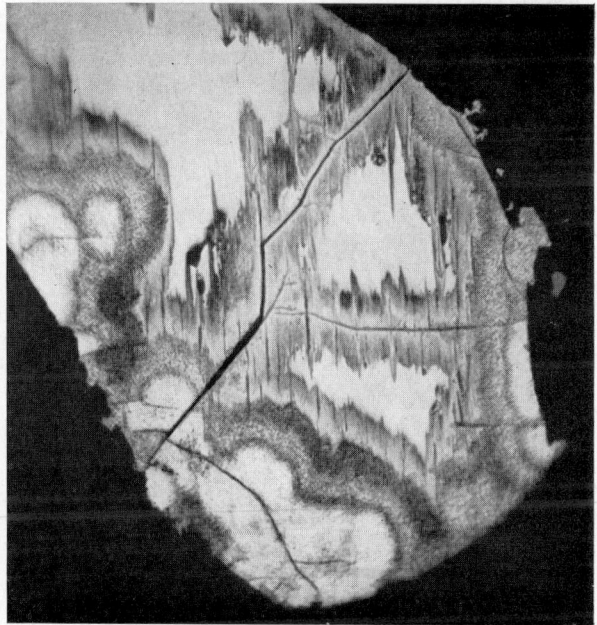

Skaergaard intrusion, although generally in amounts less than 0·05% by volume (Fig. 27). Various copper sulphides occur in the chilled marginal gabbro and in the layered series up to the UZc stage (Fig. 28a, b); these are bornite, digenite or blue chalcocite, and chalcopyrite, together with some covellite which is a secondary product after digenite (Pl. V). The sulphides occur in small, composite units, which often include some magnetite and some acicular silicate mineral (well seen in Pl. Va). From textural considerations, and for other reasons (see p. 188), the composite units are interpreted as

FIG. 29. Thin section in transmitted light, showing pyrrhotite (black) in a granophyric intergrowth of quartz and alkali felspar. Tinden Sill granophyre (5275). Well-shaped quartz–felspar intergrowths apparently grew into the immiscible sulphide liquid before pyrrhotite crystallized. ×28.

the result of crystallization of droplets of an immiscible copper sulphide liquid in which some iron oxide and silicate material was dissolved.

At the UZc horizon the rocks contain more abundant sulphide patches, often amounting to 2% by volume of the rock. The sulphides are mainly pyrrhotite, now partly or wholly replaced by marcasite (Fig. 28c, d), together with a little chalcopyrite. Small amounts of well-shaped magnetite and of crystals of some silicate mineral project into the iron-sulphide patches as into the copper-sulphide patches of the earlier rocks. The composite, iron-sulphide units, like the copper-sulphide, are interpreted as the result of crystallization of droplets of an immiscible sulphide liquid.

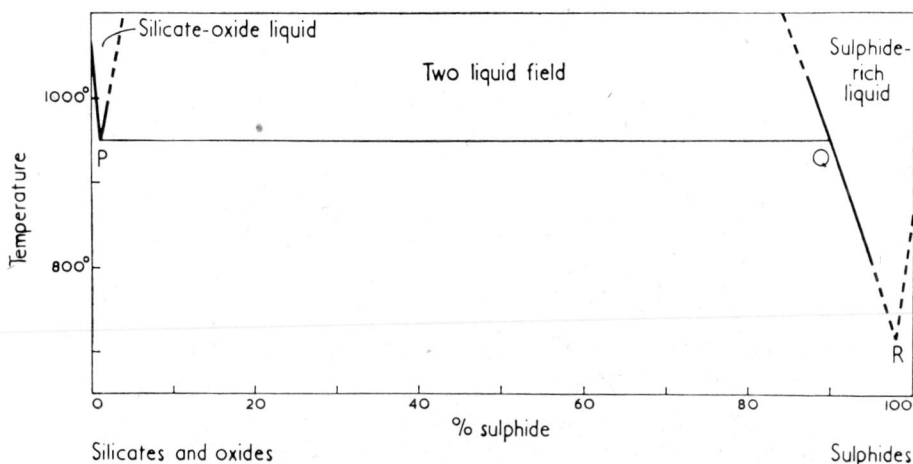

FIG. 30. Generalized diagram for a silicate–oxide and sulphide system, treated as if binary. On the evidence from the Skaergaard intrusion, point P lies between 0·025 and 0·1% sulphide, but is shown here as 1%, for clarity. Point Q is in the region of 90% sulphide and the other 10% is largely of iron oxide. Points P, Q, and R in the natural system would not represent a constant temperature, but would extend over a range.

In the acid granophyre forming the transgressive Tinden Sill, irregular sulphide patches are found, up to a centimetre across; in some rocks these are of pyrrhotite and in others, of pyrite. In such cases as that shown in Fig. 29, the late crystallization of the pyrrhotite is indicated by the silicate crystals projecting into the pyrrhotite masses. The pyrite, on the other hand, crystallized earlier since it is usually idiomorphic towards the quartz–felspar intergrowth. The pyrrhotite from the Tinden Sill has been analyzed and is found to have a surprisingly low amount of nickel (Wager, Vincent & Smales, 1957).

The textures and compositions of the sulphide patches are considered more fully in Chapter VII, where evidence is given for their formation from an immiscible sulphide liquid of steadily changing composition (Fig. 30).

CHAPTER IV

THE SKAERGAARD LAYERED SERIES

THE TEXTURES OF SOME OF THE LAYERED ROCKS IN TERMS OF CUMULUS AND INTERCUMULUS MATERIAL

It has been suggested that the rocks of the layered series can be interpreted as composed of one or more cumulus crystalline phases together with the products of crystallization of the magma which surrounded them at the time of their accumulation. The way in which the textures developed in the rocks support this interpretation may be considered, using the lowest subzone, LZa, as an example. In these rocks, only two cumulus phases are certainly present, i.e. plagioclase, approximately An_{66}, and olivine, approximately Fo_{65}. Layers consisting dominantly of plagioclase or olivine, or more average layers involving both these minerals, occur adjacent to each other (Fig. 31). No layers showing marked differences in the abundance of pyroxene or magnetite, or consisting dominantly of these minerals, are found in LZa, which suggests that these minerals were not forming as cumulus crystals at this stage.

The leucocratic layers (Fig. 31) contain about 80–90% of plagioclase by volume, and they might legitimately be described as anorthosite. The layers richest in plagioclase tend to show igneous lamination. The inner two-thirds of each of the plagioclase crystals are An_{66} and almost unzoned; these parts, usually rectangular in shape in thin section, are taken to be the original primocrysts which sank to give the sediment at the bottom of the liquid.

The bulk of the material between the plagioclase crystals of the leucocratic rocks consists of poikilitic crystals of augite, inverted pigeonite, and iron ore, together with occasional poikilitic olivines (Fig. 32). In addition, there are sporadic, small patches of quartz and apatite, and near these, minerals produced by local hydrothermal alteration are often found (Fig. 35). All these minerals, together with the outer zones of the plagioclase crystals, are interpreted as having formed from the contemporary liquid which surrounded the crystal precipitate at the time of its accumulation.

During the original investigation of the Skaergaard intrusion (Wager & Deer, 1939, p. 122) the poikilitic habit was shown to be characteristic of those crystal phases which only crystallized from the intercumulus magma (then called the interprecipitate magma). The poikilitic texture was explained as due to the very slow and tranquil cooling of the intercumulus liquid, resulting in only relatively few and wide-spaced centres of

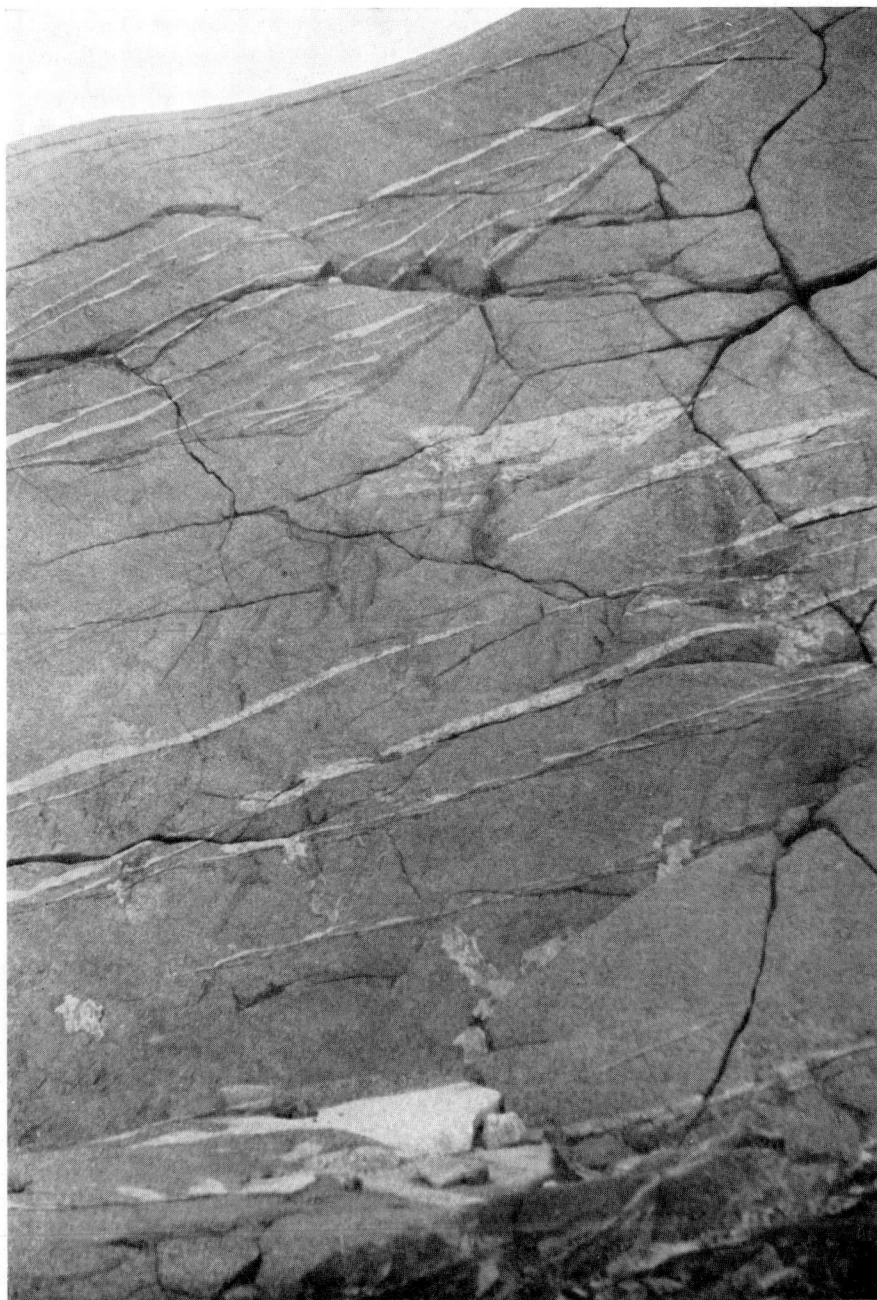

FIG. 31. Leucocratic and less conspicious melanocratic layers in uniform gabbro of LZa. Approximately a strike section, near northern margin on Uttentals Plateau. Some of the features of the layering were produced by slumping.

crystallization being established, which grew into extensive poikilitic crystals because cooling and crystallization was slow, and allowed plenty of time for the diffusion of the appropriate substances towards the sparse centres of crystallization. The crystals of relatively abundant intercumulus minerals, such as pyroxene, reach a centimetre or so in size and have a poikilitic habit because they grew large enough to enclose several of the cumulus crystals. Intercumulus crystals present in relatively small amounts, such as quartz, alkali felspar, and apatite, are not large enough to be properly poikilitic and the term sub-poikilitic is used for the texture they exhibit. Good examples of quartz and apatite showing sub-poikilitic textures are seen in Fig. 50a, b. In some of the highest layered rocks, even zircon is found to have a sub-poikilitic habit.

The melanocratic layers, which are usually much thinner than the leucocratic (10 cm or less), consist of olivine crystals surrounded by poikilitic plagioclase, augite, inverted pigeonite, and iron ore (Fig. 34). Thus, in the melanocratic layers the textural relationship between olivine and plagioclase is reversed. Unlike the leucocratic rock described in the previous paragraph, quartz is not present as an intercumulus mineral. In fact, quartz is present only in the leucocratic layers which contain no cumulus olivines, and its presence in the lower zone rocks had not been observed at the time of writing the original memoir. The local absence of quartz is believed to be due to the small amount of excess silica in the intercumulus liquid, which might ultimately have produced quartz, being used up by reaction with the magnesian olivine to give small quantities of pyroxene or secondary minerals (see Ch. VII, p. 172). Poikilitic iron ore is particularly conspicuous in both the leucocratic and melanocratic rocks (Fig. 32, 34) but it differs in appearance because in the one it is surrounding tabular plagioclase and in the other, rather rounded olivines.

The uniform rock, forming the greater part of the layered series at the LZa horizon (see Fig. 31), looks like an ordinary olivine gabbro in thin section (Fig. 33). Plagioclase and olivine, the latter usually the less abundant, are both present as cumulus crystals, and poikilitic pyroxenes, iron ore, and a little apatite as intercumulus crystals. Quartz is again absent, presumably because of reaction with the magnesian olivine. In the photomicrograph (Fig. 33), extensions of the olivines, formed from the intercumulus liquid, are seen filling, rather conspicuously, the interstices between the plagioclase crystals.

In the LZ rocks at present being considered, there was apparently about 50 or 60% of cumulus minerals and 50 or 40% of intercumulus liquid, the latter having the composition of the magma existing at the time. By the LZ stage, the composition of the magma, as a result of fractionation, was such that the cumulus plagioclase crystallizing from it was An_{66}, and the cumulus olivine Fo_{65}. With the falling temperature, more plagioclase and olivine would crystallize from the trapped intercumulus liquid and would be deposited on the cumulus plagioclase and olivine crystals. Then, as the appropriate temperatures and compositions were reached, pyroxenes, iron ore, quartz, etc., would successively crystallize as extra intercumulus minerals.

For minerals such as plagioclase, which are members of a solid solution series, there are, theoretically, two alternative kinds of behaviour during cooling, as Bowen (1922) first pointed out : either the crystals are made over continuously to compositions in equilibrium with the successive, lower temperature liquids, or material of changing composition, appropriate to the lower temperatures, is deposited as a succession of zones around the first-formed crystals. It might have been anticipated that the slow cooling necessarily prevailing in the relatively large Skaergaard intrusion would have provided the conditions for the first alternative, that is maintenance of equilibrium, but in fact, zones of lower temperature plagioclases were formed showing that, at any rate in the lower layered rocks now being considered, the early plagioclases were not extensively made over to equilibrium solid solutions. Although some limited making over of the plagioclase to lower temperature solid solutions may account for some of the observed irregularities in the arrangement of zones, what chiefly happened was successive deposition of lower temperature zones. For the lower part of the exposed layered series it may confidently be stated that the bulk of the trapped liquid crystallized by depositing zones of successively lower temperature material around existing cumulus crystals, and not by the making over by solid diffusion of the early solid solutions to steadily lower temperature compositions.

The sequence of the different mineral phases forming from the intercumulus liquid should be the same as the sequence of cumulus phases as seen in ascending the layered series, that is, augite, pigeonite, iron ore, and apatite. The textural relations of the minerals, as seen in thin section, confirms this order so far as these criteria can. Late quartz, alkali felspar, and occasional minute zircons—phases which never form cumulus crystals—were also deposited from the trapped liquid (cf. Fig. 50). Finally, some dregs of water-rich liquid, occurring patchily within the rock, produced hydrothermal alteration in small amounts to give micaceous and talcose minerals, epidote, zeolites, and probably additional quartz (Fig. 35). The amount of these is probably less than 2% of the total rock.

The original hypothesis to explain the textures of Skaergaard accumulative rocks is the one just given for the rocks of LZa, namely, the crystallization of the intercumulus liquid having the composition of the magma existing, at the time, around the cumulus crystals,* to give the intercumulus material.† As a first approximation, this simple concept provides an explanation of the composition and textures of the lower zone layered rocks, and for this type of cumulate the name *orthocumulate* has been proposed (Wager, Brown & Wadsworth, 1960, pp. 74–9). In the higher rocks of the Skaergaard layered series, the textures show that this simple explanation is not the whole story and that another process was involved, which is considered in the next section.

---

* Called, originally, the primary precipitate crystals (Wager & Deer, 1939, p. 122).
† Called, originally, the interprecipitate material (Wager & Deer, 1939, p. 122).

Fig. 32. A leucocratic layer in LZa (5109), 30 ft below, and similar to, those shown in Fig. 31. The rock is a plagioclase orthocumulate with intercumulus pyroxene (both augite and inverted pigeonite), olivine (one crystal only), and iron ore. × 10.

Fig. 33. Average rock (5107) of LZa, associated with the leucocratic and melanocratic rocks of Figs. 32 and 34. The rock is a plagioclase–olivine orthocumulate, with the pyroxenes and iron ore as intercumulus crystals. × 10.

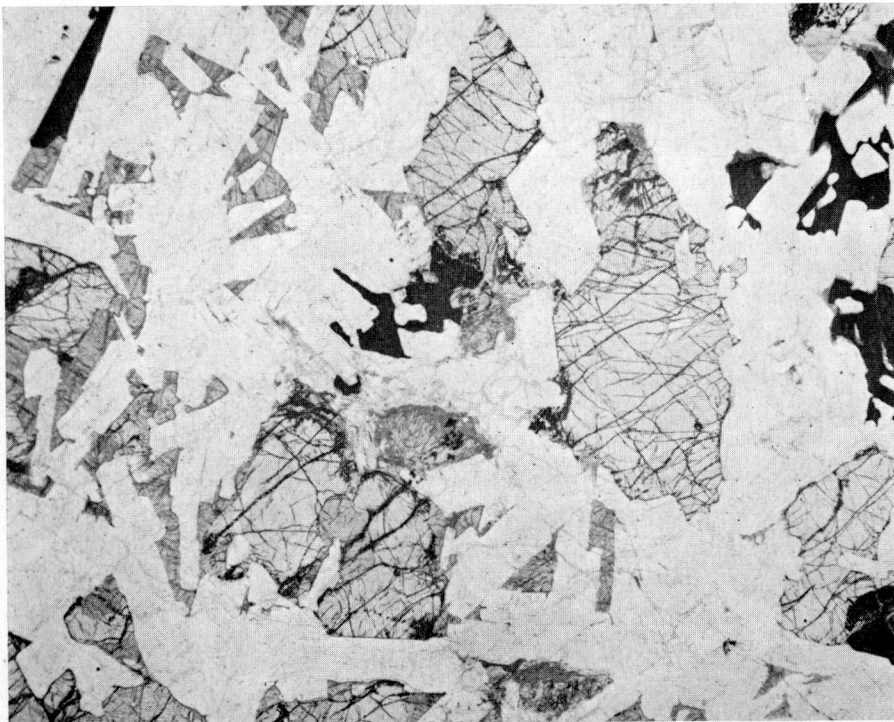

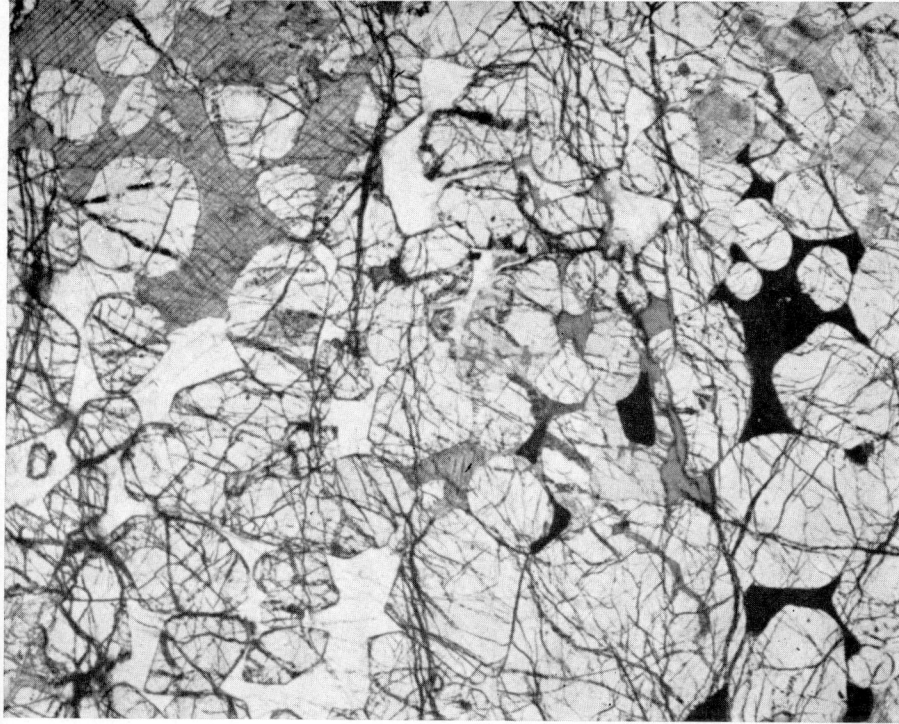

FIG. 34. A melanocratic layer (5108) of LZa, adjacent to the leucocratic layer of Fig. 32. The rock is an olivine orthocumulate, with plagioclase, pyroxenes, and iron ore as intercumulus crystals. ×10.

FIG. 35. Detail of another part of rock shown in Fig. 32, showing intercumulus quartz and alkali felspar, and localized hydrothermal alteration. ×20.

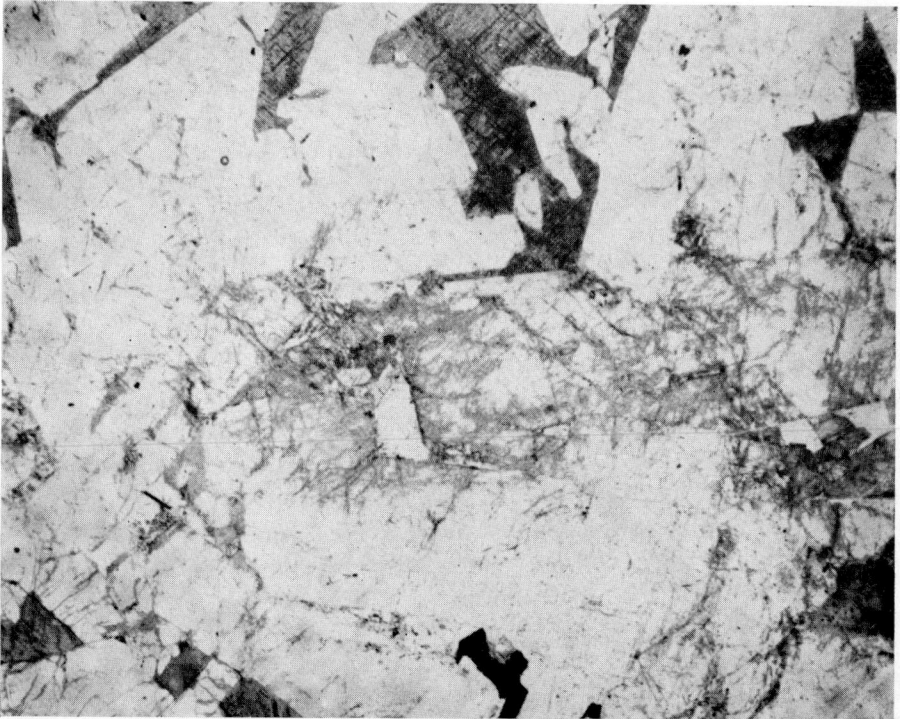

Adcumulus Growth in the Rocks of the Layered Series

Since the original description of the Skaergaard layered rocks, studies of the cumulates of the Stillwater and Rhum intrusions have provided an important new conception of the way in which crystallization may take place after the initial deposition of the cumulus crystals. In explaining some of the Stillwater rocks, Hess, as early as 1939, indicated in a couple of sentences the need to postulate some degree of enlargement of the original cumulus crystals at the same temperature as that at which they originally formed (Hess, 1939, p. 431). He suggested that diffusion of substances took place from the overlying reservoir of magma, through the intercumulus liquid, to growing crystals in the upper layers of the precipitate, and that simultaneously there was diffusion of the unwanted material in the opposite direction. This diffusion mechanism was then used by Brown (1956, pp. 14, 19) to explain various features of the cumulates of Rhum, and especially certain thin layers consisting wholly of unzoned calcic plagioclase which occur at or near the top of some of the rhythmic units.

The exact way in which the cumulus crystals become enlarged by material of the same composition as themselves is still not certain, but it clearly takes place at the same temperature as the formation of the cumulus crystals themselves, because the enlarged parts have the same composition. By the growth of the cumulus crystals, the intercumulus liquid tends to be pushed out, and in extreme cases may be totally eliminated. The pushing out of the intercumulus liquid is a mechanical effect, brought about by the further growth of the cumulus crystals as a result of the diffusion to them of the required substances. The enlargement of the top crystals of the pile would be expected to take place before those below, and to form a solid crust, preventing further diffusion. This consideration seems to show that the enlargement of the cumulus crystals by material of the same composition must have taken place while they formed the uppermost layer of the pile. Any thick layer of such rocks in which the intercumulus liquid was largely eliminated must have formed slowly because the cumulus crystals of each successive layer had to be enlarged while they lay, temporarily, at the top of the crystal pile.

For the process of enlargement of the cumulus crystals by material of the same composition the term *adcumulus growth* has been proposed (Wager, Brown & Wadsworth, 1960). As a result of adcumulus growth all the intercumulus liquid may be eliminated; on the other hand, some may be trapped by the deposition of a further layer of crystals, and this part is conveniently distinguished as the *trapped liquid*. When the trapped liquid crystallizes, the material so formed will be described as the *pore material*. The trapped liquid will necessarily have had the overall composition of the contemporary magma, and thus the composition of the pore material must be that of the contemporary magma. Rocks in which there has been much adcumulus growth have been called *adcumulates*, and those intermediate between orthocumulates and adcumulates are conveniently called *mesocumulates* (Wager, Brown & Wadsworth, 1960). Fig. 36A–C illustrate, diagrammatically, plagioclase ortho-, meso-, and adcumulates.

In the Skaergaard intrusion, evidence for adcumulus growth becomes increasingly clear as the layered series is ascended. In the plagioclase orthocumulate from LZa illustrated in Fig. 32, half of the plagioclases show, in thin section, considerable normal zoning (Fig. 17a) which is the result of crystallization of the trapped liquid over a range

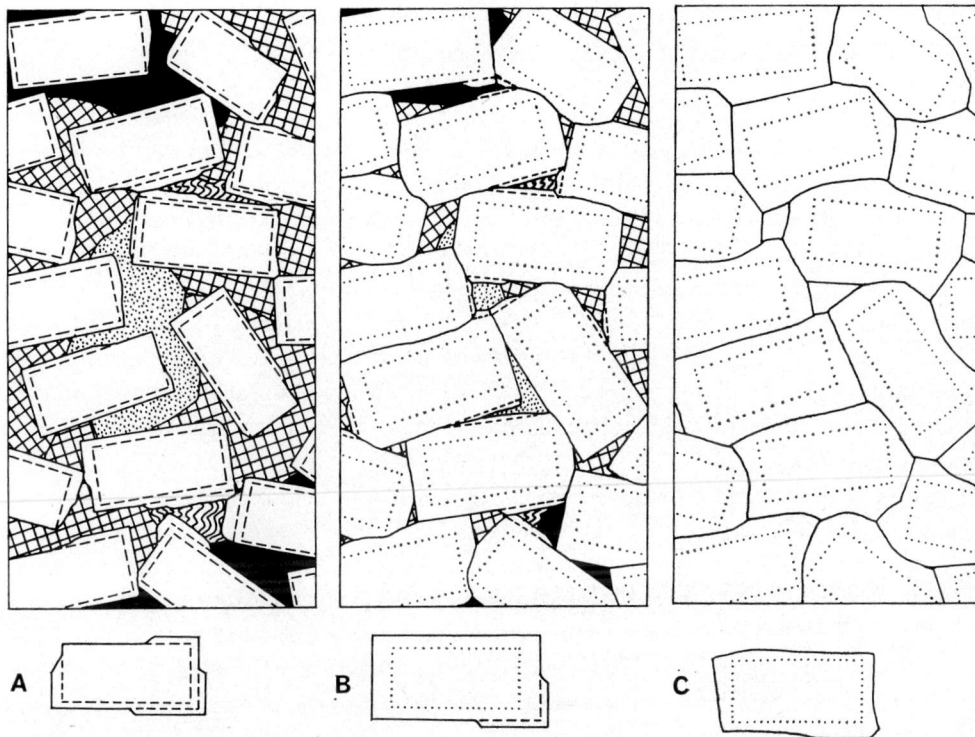

A

B

C

PLAGIOCLASE: Boundary of the cumulus crystals (labradorite) diagrammatically shown by the innermost rectangle. The limits of medium and low temperature zones, where developed, shown outside the cumulus crystal boundaries.

PLAGIOCLASE: Boundary of the cumulus crystals (labradorite) shown by the dotted line. Outside is adcumulate growth of plagioclase of similar composition. In places beyond the broken lines, lower temperature zones are shown.

PLAGIOCLASE: The cumulus part of the crystal is shown within the dotted line. This has been enlarged by growth of more plagioclase of the same composition, which fills the crystal interstices.

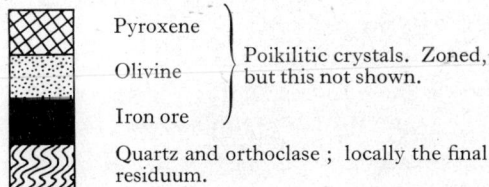

Pyroxene

Olivine   } Poikilitic crystals. Zoned, but this not shown.

Iron ore

Quartz and orthoclase ; locally the final residuum.

FIG. 36

Diagrammatic representation of plagioclase cumulates formed from a gabbroic magma. A, extreme plagioclase orthocumulate; B, plagioclase mesocumulate; C, extreme plagioclase adcumulate.

E

of falling temperatures.   In typical rocks in the UZ, such as the plagioclase–augite–olivine–magnetite cumulate of UZa, illustrated in Fig. 17b, the plagioclases are scarcely zoned, and there must have been much adcumulus growth.

Using tablets of sheet aluminium, comparable in shape with the plagioclase crystals of the lower-layered series, packing experiments were made to decide on the likely proportion between cumulus crystals and intercumulus liquid.   On being allowed to sink in water in a measuring cylinder, a loose packing resulted, in which there was about 50% of tablets and 50% of water.   With shaking, the tablets packed a little closer and the amount of intercumulus liquid was reduced to about 40%; at the same time the tablets arranged themselves so as to resemble the felspars in a rock having igneous lamination.   It is unlikely that closer packing than that giving 60% of cumulus crystals would occur naturally, and hence if a plagioclase-rich cumulate has 90% of plagioclase, as is found in some leucocratic layers low in the LZ, it must be presumed that there has been adcumulus growth, because crystallization from the intercumulus liquid could not account for the extra 30% of plagioclase.   A rock containing 80% of plagioclase is likely to have been produced by a loose packing of 50% of plagioclase tablets, followed by adcumulus growth giving about 20% more plagioclase of the same composition and, finally, by deposition from the trapped liquid of about 10% of lower temperature, more sodic plagioclase.

Except in favourable circumstances, textural observations give only a rough idea of whether a particular layered rock is to be considered an orthocumulate, mesocumulate or adcumulate.   For of some of the Skaergaard layered rocks, a more quantitative estimate of the amount of trapped liquid, and hence of the probable extent of adcumulus growth, has been obtained from the determination of the amount of phosphorus in the rocks.   Phosphorus, so far as is known, is not a constituent of plagioclase, olivine, pyroxene or iron ore, even in trace amounts, and therefore the phosphorus in cumulates made up of any combination of these minerals must be present solely in the apatite of the pore material which has come from the trapped liquid.   If two cumulates at about the same horizon have different phosphorus contents, it is clear that there must have been a difference in the amount of trapped liquid proportional to differences in their phosphorus contents.   Furthermore, if an estimate can be made of the percentage of phosphorus in the contemporary magma, then comparison between the percentage of phosphorus in the rock and in the contemporary magma should give an estimate of the proportion of cumulus crystals (plus adcumulus growth, where present) to the trapped liquid.

Ordinary methods of analysis for phosphorus in the rocks of the lower and middle zones of the Skaergaard intrusion, where there is generally less than 0·2% P_2O_5 and often only a tenth of this amount, were not good enough to enable this hypothesis to be tested.   However, a radioactivation method of analysing for phosphorus, devised by Dr E. A. Vincent, has provided data of greater precision which are plotted in Fig. 37. The graph shows that the amount of P_2O_5 varies from 0·05 to 0·2% until the point at

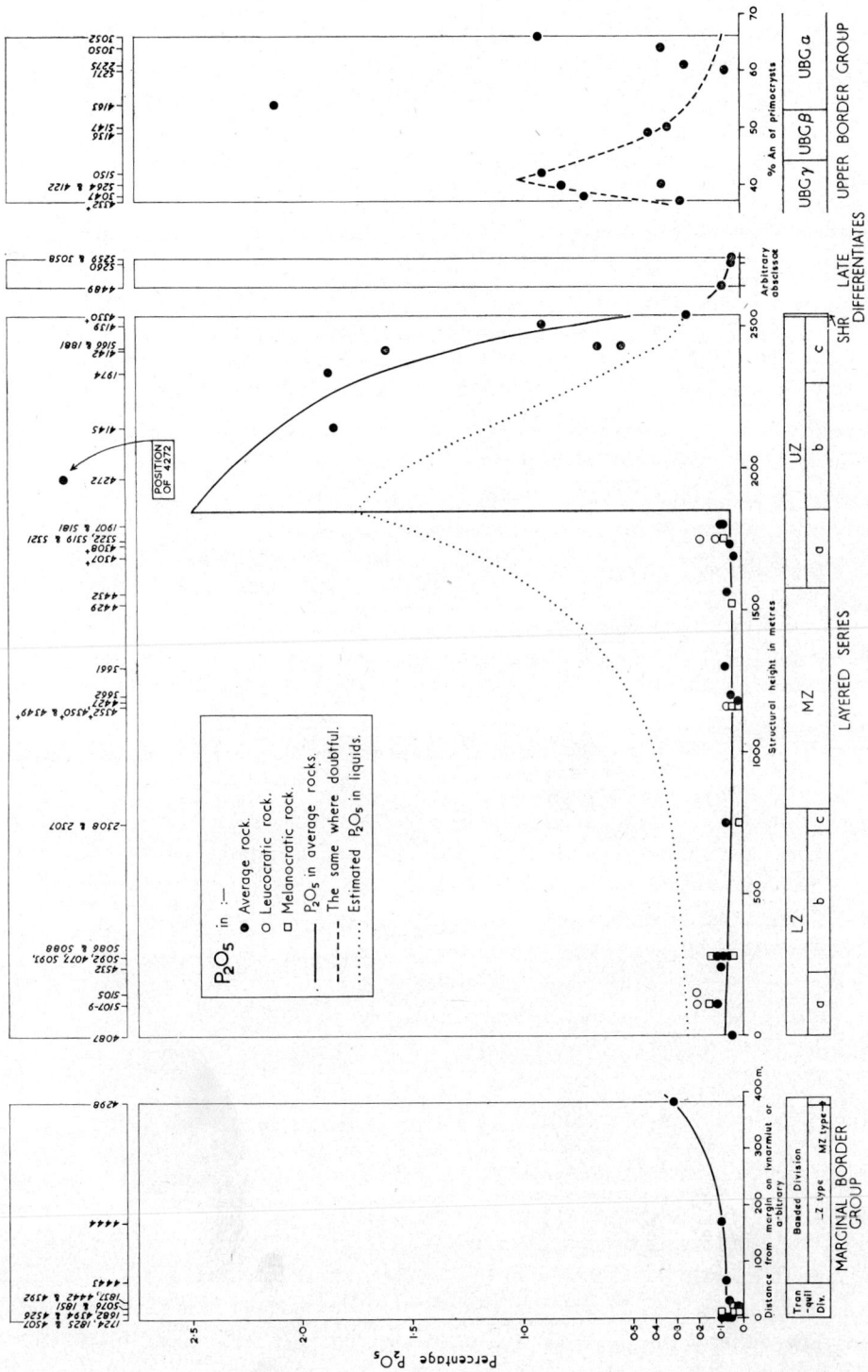

FIG. 37. P_2O_5 in rocks of the Skaergaard intrusion, with estimated curve for P_2O_5 in the successive liquids.

which apatite becomes a cumulus mineral, when it rises abruptly to almost 2.5%. During fractionation, the amount of P_2O_5 in the liquid must have risen until it reached saturation level for apatite at the beginning of UZb. If the amount of trapped liquid remained about constant in the layered rocks below the level of the appearance of cumulus apatite, the amount of P_2O_5 in the rocks should increase because of the increasing P_2O_5 content of the liquid. The plot, however, shows that this is not the case; from the base of the exposed layered series to 1850 m, at which level apatite became a cumulus mineral, the rocks show an overall decrease in P_2O_5. This result indicates that in proceeding upwards in the layered series there was a decreasing amount of trapped liquid, as a consequence of an increasing amount of adcumulus growth. The P_2O_5 content of the leucocratic, melanocratic, and average rocks are indicated separately on the graph. The more leucocratic rocks have relatively high amounts of P_2O_5, the average rocks a relatively low amount, and the melanocratic rocks a more variable amount, but tending to be small. This means that the leucocratic rocks, on the whole, contained most trapped liquid. The felspars of the leucocratic cumulates would probably be of nearly the same density as the magma and would, no doubt, tend to accumulate as a loose, flocculent pile; the felspar cumulates would thus start off with much intercumulus liquid, unless some mechanical movement caused them to develop igneous lamination. In the average rocks, the smaller amount of trapped liquid may be due to the presence of a greater number of different kinds of cumulus crystals on which adcumulus growth could occur, so causing a greater elimination of the original intercumulus liquid.

The general inferences to be drawn from the phosphorus content of the layered rocks is that, on the average, the amount of trapped liquid is greater in the LZ rocks, and the rocks are orthocumulates or mesocumulates; upwards, the amount of phosphorus in the rocks of layered series becomes progressively less and the rocks are predominantly mesocumulates or adcumulates. The detailed causes of variation in the phosphorus content of rocks from different layers at the same level have not yet been satisfactorily determined and work is still in progress. Above 1850 m, where apatite becomes a cumulus mineral, the P_2O_5 content cannot, of course, be used to indicate the amount of trapped liquid but from the more general evidence of the relatively slight zoning of the felspars it is clear that the rocks remain meso- or adcumulates up to the top of the layered series.

The Sequence of the Layered Series Rocks

a) *The lower zone (LZa, LZb, and LZc)*

The lowest exposed layered rocks are found in the northern part of the intrusion, as a result of the tilting giving the coastal flexure (p. 16). The rocks show much rhythmic layering, and frequently this takes the form of gravity stratification (Fig. 38). In other places the rocks have a uniform appearance due to constant proportions of the minerals. They may be either fissile, due to well-developed igneous lamination, or massive.

Rather indefinite, raft-like masses, usually of leucocratic rocks and up to 10 or 20 ft thick (Fig. 31, 39) are common, and some small rafts of melanocratic rocks may be associated with them. The plagioclase and other minerals of the rafts have the same, or closely similar compositions to the minerals in the enclosing layered material. They are either slabs of just consolidated material which have slid from their place of origin near the margin, or masses fallen from the roof.

The lowest rocks of the lower zone are exposed only near the inner margin and are in or near the cross-bedded belt, which everywhere occurs adjacent to the margin (see below, p. 96), and this may be the cause of certain unusual characters. It seems that there are 200 m of lower zone rocks (termed LZa) which have only plagioclase and olivine as cumulus crystals. These are the rocks with poikilitic pyroxenes which were grouped as the transitional layered series in the original account (Wager & Deer, 1939, fig. 18, pp. 87–8). The general appearance in thin section of the average, the leucocratic, and

FIG. 38. A single, 20 cm-thick, gravity-stratified layer in LZb, with uniform rock above and below. Uttentals Plateau.

the melanocratic rocks have already been illustrated (Fig. 32, 33, 34); most of the rocks are ortho- or mesocumulates and the best examples of orthocumulates in the Skaergaard intrusion are in LZa, described previously (pp. 58–61).

At about the 200-m level, augite begins to occur as cumulus crystals and for this reason the rocks are assigned to another subzone, LZb. Some melanocratic layers consist of cumulus olivine and augite with poikilitic, inverted pigeonite, indicating that the latter was not yet forming as a cumulus mineral. The precise level at which cumulus

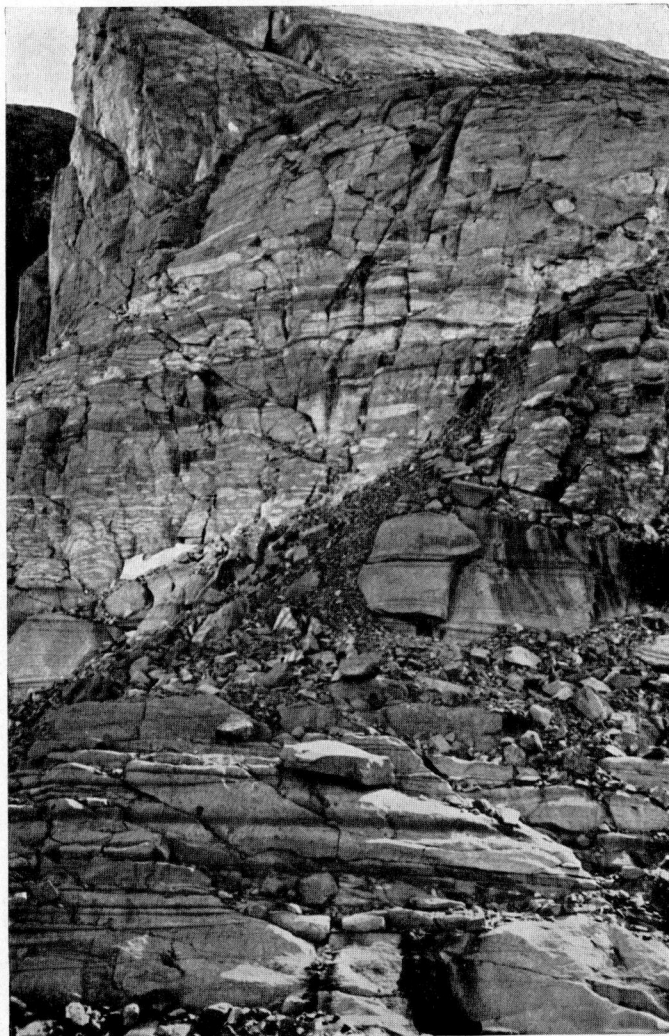

FIG. 39. View of WNW ridge of Gabbrofjaeld, taken from Uttentals Plateau, showing continuous layering in the foreground and slumped rafts associated with the layering in the middle distance. (From Wager & Deer, 1939.)

crystals of pigeonite first formed has not been established, but they certainly occur erratically at about the 500-m level. The lowest cumulus augites found are large crystals enclosing plagioclase; these probably formed as composite primocryst clusters in the main body of the magma. There may have been considerable supersaturation of the magma in the augite components before nucleation happened, followed by rapid growth producing large crystals in which several plagioclase crystals were entangled. Such apparently poikilitic crystals of augite, if they are due to the formation of composite clusters in the magma, are to be viewed as cumulus crystals, yet they cannot always be distinguished with certainty from poikilitic crystals developed from the intercumulus liquid (cf. Ch. XI, Fig. 141).

Magnetite and ilmenite are by no means lacking in the LZa and LZb zones but they have formed only from the intercumulus liquid and are characteristically poikilitic. At about the 700-m level, wispy layers rich in iron ore begin to be seen on the glaciated rock surfaces, due to the sporadic appearance of the ore minerals as cumulus crystals; between ore-rich layers there are still rocks with only intercumulus magnetite but after perhaps a further 50 m, cumulus magnetite is commonly present in the rocks (Fig. 11, 40), although occasional layers may still be devoid of it in the cumulus form. The incoming of magnetite as a cumulus mineral is an important discontinuous phase change (p. 29), and the level of its appearance is used to mark the base of LZc.

b) *The middle zone (MZ)*

The middle zone, from about 800 to 1600 m, is characterized by the absence of olivine as cumulus crystals (Fig. 15). In the LZ the olivine shows a steady change in composition, with increasing height, from Fo_{69} to $Fo_{52.5}$, and then it ceases to occur as a cumulus mineral. Where it returns at the beginning of the upper zone, the composition is Fo_{36}. Apparently, fractionation of the Skaergaard magma behaved in the same way as certain liquids in the system $MgO-FeO-SiO_2$ described by Bowen and Schairer (1935, pp. 209–13), despite the more complex composition of the natural magma.

While the MZ is characterized by the general absence of cumulus olivine there are, in fact, certain sporadic layers containing cumulus olivines for 100 m or so above the level at which it is generally absent. Also, near the top of the MZ, at 1420 m, a thin layer containing cumulus olivine, heralding its reappearance as a cumulus mineral, is found; this is 180 m below the level chosen for the top of the MZ, on the basis of the abundant re-entry of olivine. Thus the bottom and top of the middle zone can be fixed only approximately (see p. 30).

Although olivine is absent as a cumulus mineral from the usual MZ rocks, it is often present as a reaction rim between iron ore and pyroxene (Fig. 41) or sometimes as a mantle to the ore where against plagioclase (Fig. 42). Such rims are not limited to the MZ but occur also in the LZc. They are particularly well developed in the rocks just below the beginning of the UZ. At this horizon, the olivine rims might be thought of

FIG. 40

FIG 41

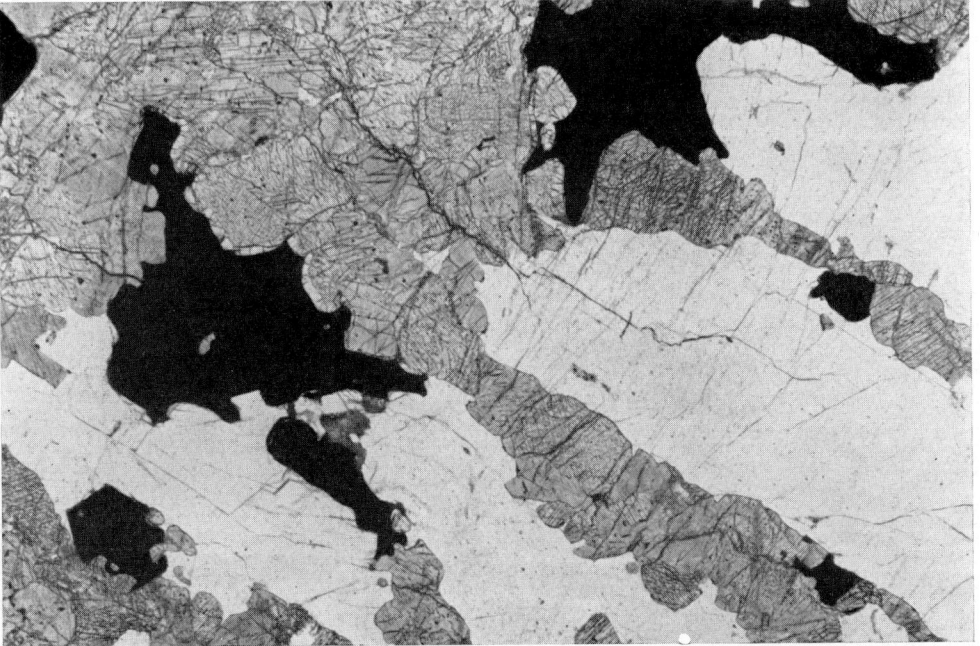

as intercumulus olivine which, however, developed preferentially in the neighbourhood of magnetite.

Another type of reaction rim, between olivine and plagioclase, is often found as a feathery growth of orthopyroxene and spinel, particularly in rocks of LZc and the lower part of MZ (Fig. 42). Their development may have happened at a rather low temperature, involving only very late dregs of the trapped liquid.

The absence of olivine and the presence of cumulus magnetite gives a distinctive appearance to the textures of MZ rocks (Fig. 41). Also, under crossed nicols (e.g. Fig. 21b) the pyroxene areas in these rocks are seen to be composed of granular aggregates of rather small grains of augite and inverted pigeonite (Fig. 22). As mentioned earlier (Ch. III), the reason for these clusters of small grains is not yet understood.

FIG. 40. A melanocratic rock (5069) of LZc, Kraemers Island. The rock is a plagioclase–olivine–augite–magnetite cumulate (variable proportions in the area of a thin-section). ×15.

FIG. 41. Olivine-free gabbro (3661) of MZ, Pukugaqryggen. Plagioclase, augite, inverted pigeonite, and iron ore are cumulus minerals (cf. Fig. 22a). Minor amounts of olivine are present as a reaction rim between the iron ore and pyroxenes. A feathery reaction rim is present occasionally between the iron ore and felspar (see Fig. 42). ×15.

FIG. 42. The feathery reaction rims as in Fig. 41 but much enlarged. An olivine rim around iron ore is present sporadically. ×60.

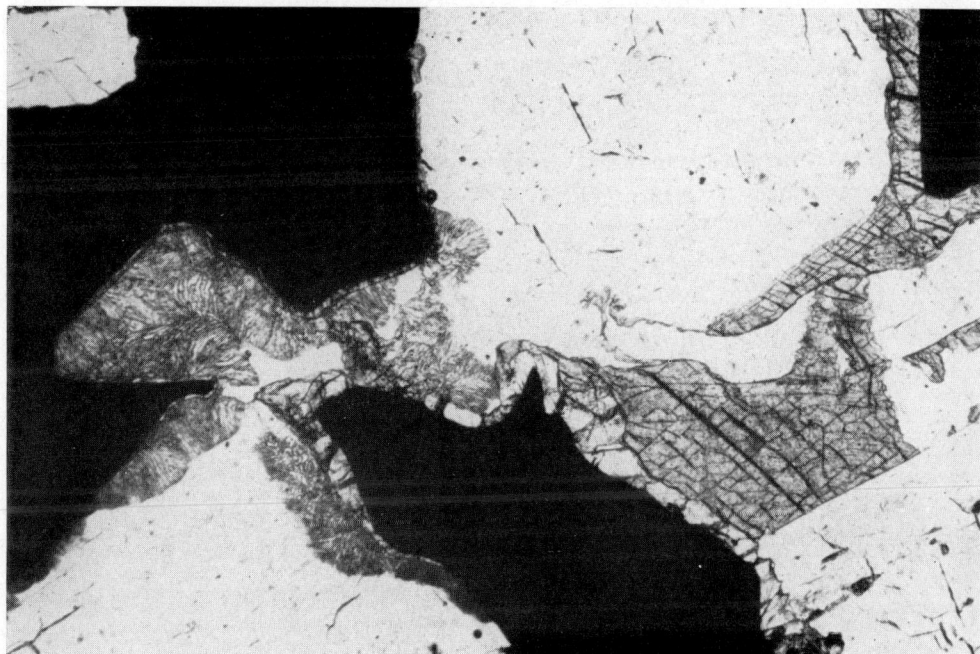

Interstitial areas of quartz and micropegmatite occur sporadically as pore material minerals. There has been some adcumulus growth in most of the rocks which are, therefore, classed as mesocumulates.

In the MZ, the rhythmic layering may be due chiefly to the concentration of cumulus magnetite in thin layers (e.g. Fig. 44). Another type of rhythmic layering, which was called *zebra banding* in the field (because of the distinctive striping on rock surfaces), is due to changes in the relative amount of cumulus pyroxene and cumulus felspar (Fig. 43); variations in the amount of cumulus iron ore are also sometimes involved. Near the top of the MZ, three thick and persistent layers form the so-called *triple group*, the lowest about 20 m thick, the next about the same, and the highest 60 m thick. These layers can be identified over an area of at least 4 km² (Pl. IV, XI; Fig. 45). Each unit is gravity stratified on a very different scale from that usual in the rhythmic layering of the Skaergaard layered series. The triple group layering has some similarities in scale and character with the layered units of eastern Rhum (Ch. X) but it probably has a

FIG. 43. Zebra-banding in MZ, mainly due to layers fairly rich in pyroxene alternating with those rich in plagioclase. The layers have no sharp base or top. SE part of Kraemers Island.

FIG. 44. Thin, ore-rich, gravity-stratified layers in shallow troughs.
The same MZ horizon as in Fig. 43 and only 50 yds distant.

different origin.   There are also some similarities with thick layered units in the Basal Series of the Bushveld (Ch. XV) but again it is doubtful whether they have a similar origin.

At about the middle of the MZ, abundant blocks, up to 100 m thick, interrupt the regular layering (see Fig. 46).   The larger blocks, indicated diagrammatically on the map (Pl. XI), are probably of upper border group material which has fallen from above. These blocks have not only embedded themselves in the already formed layers but they may sometimes have rucked up the layering due to their having had a horizontal component of movement (Fig. 46).   Deposition of cumulus material covered the blocks, giving local high dips.   Several cases were noticed of blocks which had split open vertically, and cumulus material, showing poor layering, had been laid down in the gaping cracks (Fig. 47).   This suggests that the blocks, at the time of the deposition, were formed of only weakly cohering crystals.

Contemporaneous veins, a few inches or a foot wide, have been found occasionally cutting steeply across the layering.   The material of the veins is related to the surrounding

FIG. 45.   The face of Pukugaqryggen and Gabbrofjaeld seen from Connecting Glacier.   The triple group of MZ can be traced by eye for 2½ km.   (From Wager & Deer, 1939.)

FIG. 46. Block of UBG material which has fallen and slid into position, rucking up the layering in MZ rocks. Kraemers Island.

layered rocks and is considered to be essentially a residual liquid from the layered rocks that has been squashed into fractures which are often, also, faults with a throw of a few feet. Three of the five cases noted are in the neighbourhood of large inclusions and are probably the result of movement due to local instability of the accumulating mush of crystals on the irregular floor formed by the inclusions. Occasional slump structures are also found in the layered rocks in the neighbourhood of these large inclusions. Slump structures are also found in the steeper-dipping layered rocks of the cross-bedded belt, and sometimes on the flanks of the trough bands (see pp. 87 and 100).

c) *The upper zone (UZa, Trough banding, UZb, and UZc)*

The upper zone rocks, well seen on the west flank of Basistoppen (Fig. 48), show a tendency to weather to a dark brown colour and in the topmost 200 m they take on a deep purple-brown appearance; in the field, the latter were mapped separately and described as the Purple Band horizon in the original memoir. Rhythmic layering reaches a maximum intensity in the UZa and then decreases, becoming unnoticeable in the UZc. On the other hand, cryptic layering is strong throughout the whole zone, the plagioclases varying steadily upwards from An_{44} to An_{34}, while the ferromagnesian minerals increase in iron content almost to the extreme limits possible, the olivines, for instance, reaching Fo_2.

The rocks of the upper zone contain much iron-rich olivine and pyroxene, and also considerable amounts of iron ore (Fig. 49). As a group, they were called ferrogabbros in the original memoir (Wager & Deer, 1939, pp. 66, 98–9). This has proved a useful word and has been used by other petrologists. However, plagioclase is undoubtedly the most significant mineral for any natural classification of igneous rocks, and the general name diorite is being used increasingly and, we think, rightly, for plutonic rocks in which the felspar is an andesine. When the name ferrogabbro was originally adopted, the possibility of using ferrodiorite instead was considered, but the overall gabbroic appearance of rocks having 50% or so of olivine and augite, the relatively low silica percentage, and the gradational nature of the layered series from gabbros into these rocks, made the authors of the memoir decide to use the term ferrogabbro. This was probably a wrong decision and we consider it desirable to adopt the name ferrodiorite instead of ferrogabbro, a proposal already made in a paper on a ferrodiorite from Skye (Wager & Vincent, 1962, pp. 30–1). It is suggested that the name ferrodiorite be used for those rocks in which the actual (not normative) plagioclase is less calcic than about An_{50}, and the ferromagnesian minerals iron-rich. If olivine is present, we suggest it should be more iron-rich than about Fo_{40}, and other ferromagnesian minerals should be correspondingly iron-rich. Thus the UZ rocks of the Skaergaard intrusion are now referred to as ferrodiorites when a general name is needed; for specific descriptions a name based on the cumulus minerals present is used, as for the gabbroic cumulates of the lower and middle zones.

UZa. The somewhat gradual return of olivine as a cumulus crystal, after the MZ,

FIG. 47. Fallen block (coarse UBG material) which has cracked open and been filled with layered material. (Layering visible at top centre.) MZ on Kraemers Island.

has been commented upon already. The subzone called UZa is taken as beginning at the 1580-m level, where olivine has again become a common cumulus mineral. Within the UZa, at about 1700 m, pigeonite ceases to occur, the only pyroxene in later rocks being ferroaugite. Apatite, quartz, and micropegmatite are common in small quantities as interstitial minerals formed from the trapped liquid (Fig. 50a, b). According to the extent of the pore material, the rocks are either meso- or adcumulates.

FIG. 48. West face of Basistoppen from Ivnarmiut, showing the upper part of the layered series (UZa, b, and c) overlain by the Basistoppen sheet. The prominent shoulder at about 500 m is due to the resistant Purple Band of UZc. Above lie the less resistant UBGγ rocks and then comes the sheet up to the summit. (From Wager & Deer, 1939.)

PLATE VI

Uniform rock in the foreground, passing into a shallow trough-band, which itself passes into continuous layering. 830 m ENE of the northerly Base House.

[face p. 81]

The rhythmic layering in UZa includes strongly gravity-stratified layers alternating with layers of uniform rock (Fig. 10). There are some layers of abnormal coarseness (Fig. 51), the crystals having apparently grown large while forming the top surface of the cumulus pile. The layering is sometimes of the normal planar type but in other cases it has a trough-like form. This so-called trough banding is best developed at the top of UZa and the bottom of UZb. Being of considerable significance in understanding the mechanism of deposition of the layered series, it is described in detail in the next section.

Trough banding. On the glaciated and, largely, vegetation-free surfaces around the three houses on the Skaergaarden, the trough banding structures are seen to perfection, and a three-dimensional concept of their form has been obtained by detailed mapping. Steep rock surfaces, approximately parallel to the nearest margin of the intrusion, show gravity stratification in the form of shallow crescents alternating with uniform layers (Fig. 52, 53, 54). Sections at right angles to these faces show that the crescent structures are persistent in that direction and so have the form of shallow, radially directed troughs. Series of troughs are found to persist through vertical thicknesses of 100–200 m. They are usually about 20 m wide but vary from this by a factor of two. Sometimes a single trough, when traced upwards, breaks down into two or three smaller troughs. The margins of the troughs are seen in Fig. 53 and Pl. VI, and between the troughs,

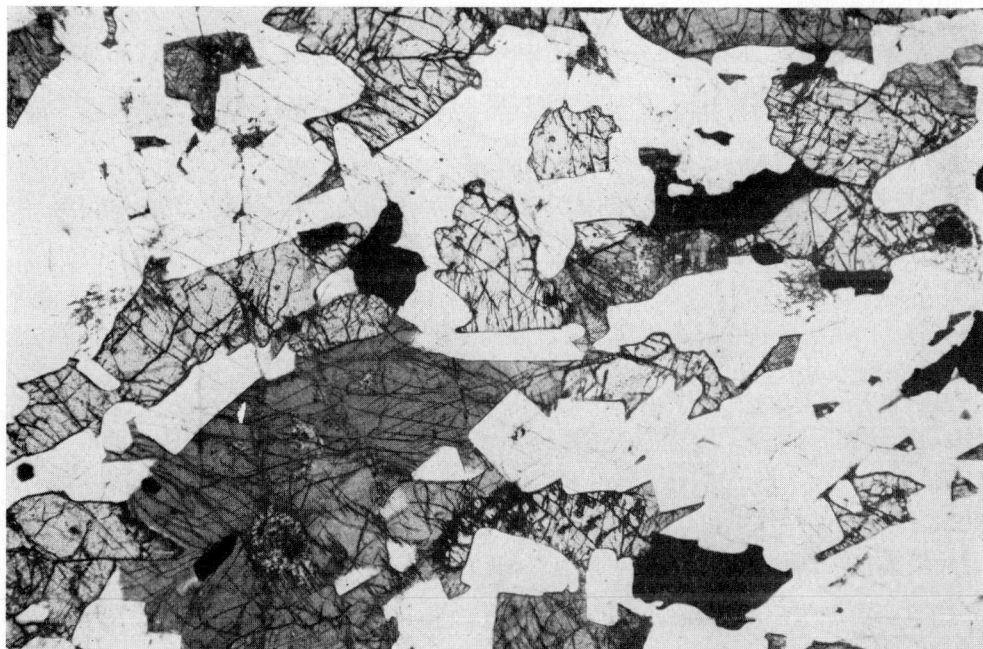

FIG. 49. Ferrodiorite (plagioclase–augite–olivine–magnetite cumulate) of UZa (5181). Main House area. ×15.

F

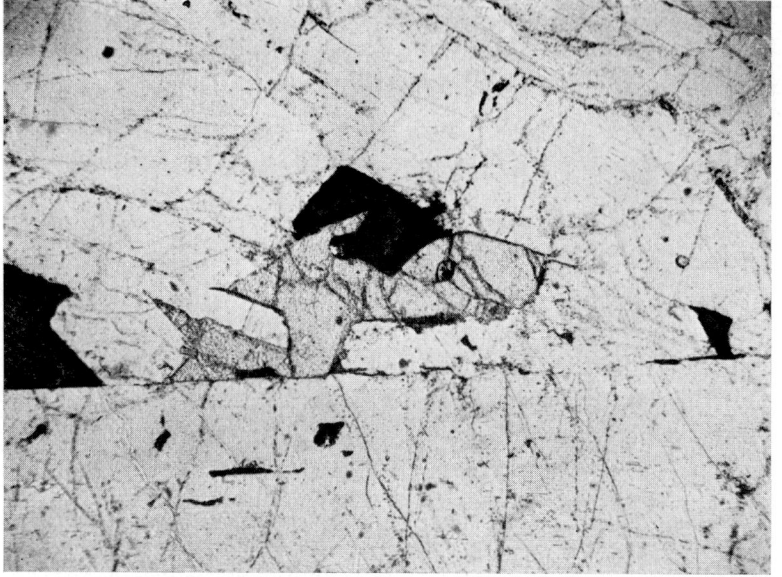

FIG. 50a.

Intercumulus apatite in a leucocratic layer of UZa (5321) from trough band, Main House area. ×15.

Intercumulus quartz in same UZa rock as in Fig. 50a. The patches are seen as separated in the thin section, but are in optical continuity. The marginal zoning of the plagioclases is only slight (cf. Fig. 17a). Cross-nicols. ×15.

FIG. 50b.

structureless ferrodiorite is found (Fig. 55 & 56). The sides of the troughs were some-times unstable and slumped inwards (Fig. 57).

The position and dip of the axial lines of the troughs, and the general maximum dip on the flanks, were put down on a topographical map on the scale of 12 inches to the mile (Fig. 89), made from a plane-table survey. It is evident that to the northwest the troughs have been removed by erosion and that towards the southeast the pitch carries them below higher horizons of the layered series. On the reasonable assumption that the structures persist for a few hundred metres in the direction of the regional dip, it is possible to draw a cross-section of the trough banding and adjacent horizons (Fig. 59). Persistent planar layers, below some of the troughs, show clearly that the troughs are not the result of folding but are primary structures formed during deposition of the layered series. Because of the later tilting of the intrusion, the troughs are not now symmetrical about a vertical axis (cf. Fig. 58). When corrected for the tilt produced by the coastal flexure, the troughs become roughly radial and are symmetrical about a vertical axial plane.

FIG. 51. Unusual coarse layers associated with gravity-stratified layers. UZa north of Main House area.

FIG. 52. Trough banding F, north of the Main House (see Fig. 58). The structure is symmetrical about an axial plane dipping at 70-80°. (From Wager & Deer, 1939.)

FIG. 53. The same trough banding as in Fig. 52, but a closer view showing the gravity-stratified character of the thicker bands. (From Wager & Deer, 1939.)

FIG. 54. Trough banding E (see Fig. 58), showing its persistence upwards, and the uniform rock on both sides. The bands extend varying distances into the uniform rock at the sides. (From Wager & Deer, 1939.)

FIG. 55. Flank of trough banding N, on the neck of the Skaergaard Peninsula. The banding is on a fine scale. Again, as in Fig. 54, the banding extends a variable distance into the adjacent uniform rock. (From Wager & Deer, 1939.)

Outside the limits of the map (Fig. 58), trough banding has been found on Pukugaqryggen of Gabbro Mountain and also on the Skaergaard Peninsula. The general direction of the axial planes, after correcting for the tilting, is shown in Fig. 60, from which it is clear that each of the troughs is at right angles to the nearest margin of the layered series, and is directed towards a point a little south of the centre of the present outcrop of the intrusion.

The trough bands show strong rhythmic layering, sometimes consisting of average rock alternating with gravity-stratified units and sometimes, simply a succession of gravity-stratified units. An extreme leucocratic layer (a plagioclase–magnetite cumulate with only a little of the latter mineral (Fig. 61)) and an extreme melanocratic layer (an olivine–augite–magnetite–ilmenite cumulate (Fig. 62)) exhibit the same textural features described from strongly contrasted layers in LZa (p. 58). The average rock at this horizon has also already been figured (Fig. 49).

FIG. 56. Uniform rock between two troughs, northeast of the trough band group A-D (Fig. 58). Some of the bands turn almost horizontal on the flanks of the troughs.

FIG. 57. Slumped flanks of a trough band, NE of the A–D group (see Fig. 58).

The origin of the remarkable trough banding structures is discussed in the last part of this chapter. When the original account was written the phenomenon was thought to be restricted to a thickness of only a few hundred metres of the layered series. Trough banding is certainly developed best in the upper part of the UZa and lower part of the UZb, but shallow troughs have been seen in the lower part of UZa (Fig. 63) and incipient troughing, detected by the form of magnetite-rich layers, occurs in MZ (Fig. 44) and LZc (Pl. VII). In addition, a rather abrupt kind of trough is seen on a rock face in LZa shown in Fig. 31.

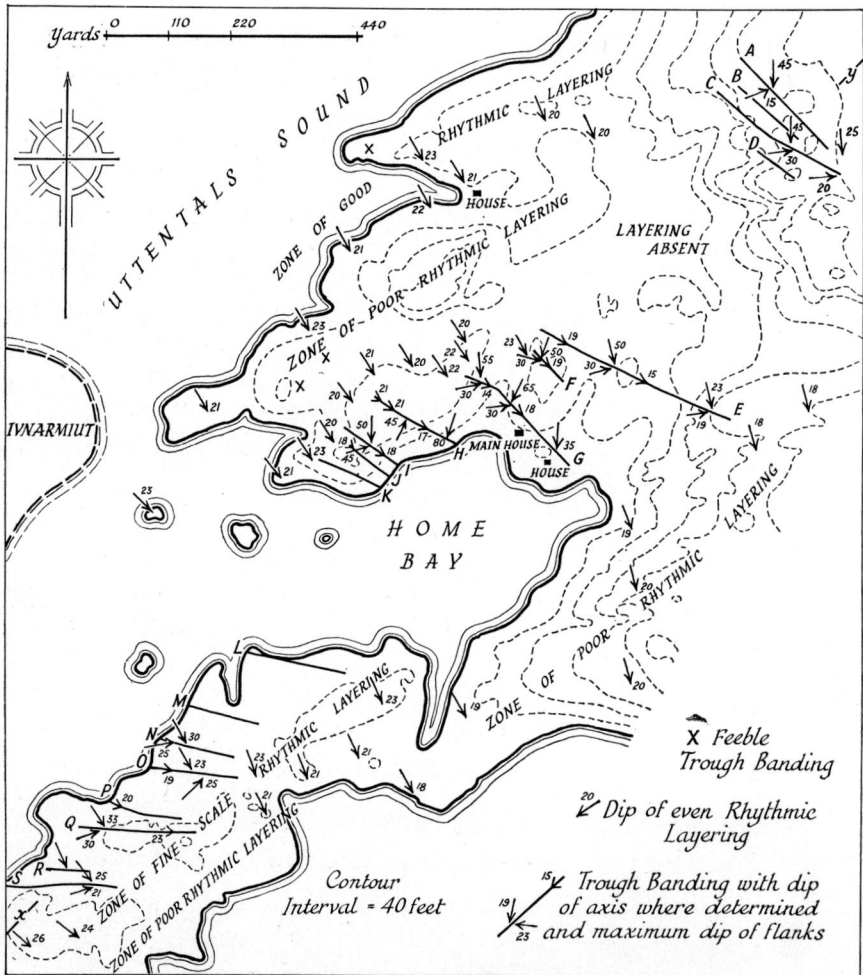

FIG. 58. Map of the area near the Base Houses, showing trough banding structures (lettered A to S) and the direction and quality of the layering. (xy marks the position of the section given in Fig. 59.) (From Wager & Deer, 1939.)

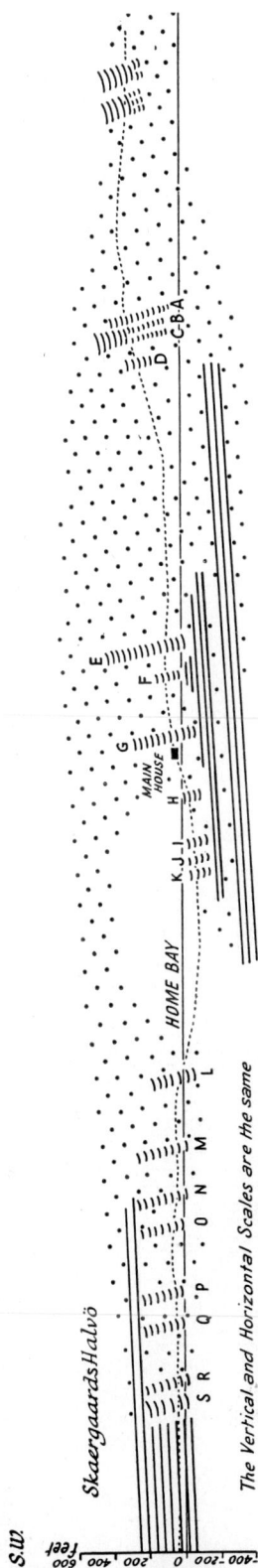

FIG. 59. Section across the Skaergaard region (along xy of Fig. 58) showing diagrammatically the trough banding (small arcs), good layering (continuous lines) and uniform rock (dots). (From Wager & Deer, 1939.)

The Vertical and Horizontal Scales are the same

UZb. The incoming of apatite as a cumulus mineral is used to define the lower limit of UZb. The rocks usually have abundant, well-shaped apatite (Fig. 64 & 65), the habit of which should be contrasted with that in the typical rock from UZa (Fig. 50a). The apatite line, marking the incoming of cumulus apatite, can be mapped approximately in the field if a hand lens is used to examine the rock surfaces, and this means was employed around the neck of the Skaergaard Peninsula. Near the margin in the Skaergaarden area, however, the occasional incoming of cumulus apatite was found to occur 200 m below the general level at which cumulus apatite appears. This is the greatest deviation from uniform cryptic layering so far recorded in the Skaergaard intrusion.

Layers, one foot thick, of olivine–plagioclase–magnetite–apatite cumulates (Fig. 65), that persist over a lateral distance of several hundred yards, are found in the lower part of UZb in the Skaergaarden area (shown on map, Fig. 89). They occur immediately above the level at which the trough banding has ceased. The highest, well gravity-stratified unit in the layered series was found near the top of UZb, at 2100 m and is just the same in appearance as one of the early gravity-stratified layers of LZb (Fig. 38).

UZc. The rocks of this subzone contain a green, iron-rich, ferrohedenbergite pyroxene which in the lower rocks occurs as clusters of interlocking grains and in the upper rocks, as extensive, poikilitic crystals (Fig. 20 & 66). Both these textures have been ascribed to sub-solidus inversion of a ferriferous β-wollastonite to ferrohedenbergite (Brown & Vincent, 1963). As a result of the crystal fractionation it is believed that the magma became so iron-rich that an iron wollastonite, although not found in the uninverted state was, temporarily, precipitated directly from the magma. With further fractionation the magma precipitated a brown ferrohedenbergite, either as rims to the green inverted pyroxenes or, in the higher rocks of the sandwich horizon, as a separate phase. These pyroxene relationships have been discussed in Chapter III (p. 41).

When the sulphides of the Skaergaard intrusion were being investigated (Wager *et al.* 1957), what is now called UZc was described as the Purple Band, as in the original memoir, and this was further subdivided on the basis of the type of sulphides present. A lower part containing only copper sulphides was called the lower Purple Band, while an upper part, characterized by about 2% of pyrrhotite mostly inverted to marcasite, was called the upper Purple Band.

The rocks of UZc are rather melanocratic meso- or adcumulates. One layer of exceptionally coarse rock was found (Fig. 66) and is believed to be an example of a

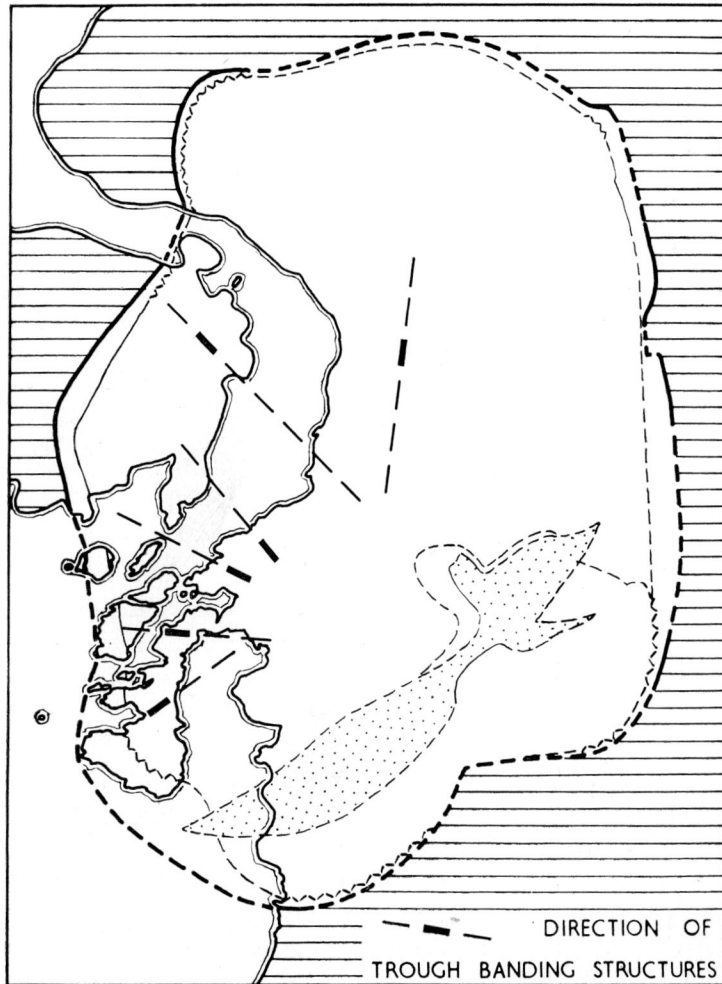

FIG. 60. The general direction and position of all the trough banding so far found in relation to the shape, in plan view, of the intrusion. Country rocks, horizontal lines; Basistoppen sheet, stippled.

PLATE VII

Layering in LZc, picked out by richness in iron ore. Some incipient trough banding is developed. Kraemers Island (position within the LZc shown on map). Gabbrofjaeld and Pukugaqryggen in the background.

FIG. 61. Leucocratic layer (plagioclase–magnetite cumulate) of UZa (5321).
Trough Band H, House Area. ×15.

FIG. 62. Melanocratic layer (olivine–augite–magnetite cumulate) of UZa (5322),
adjacent to specimen shown in Fig. 61. ×15.

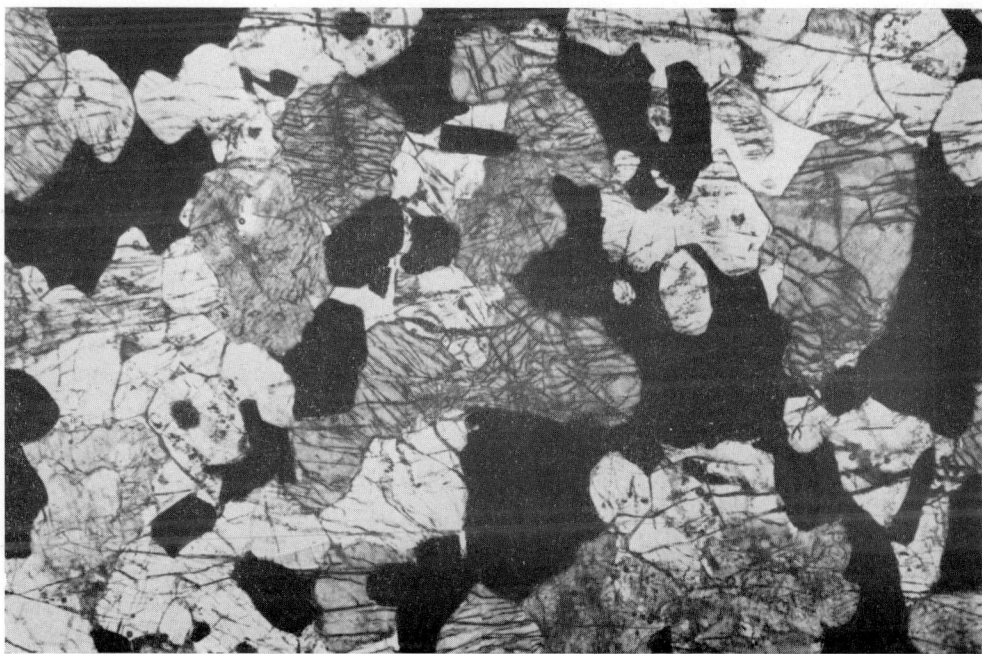

crescumulate resulting from upward growth of pyroxene, olivine, and apatite crystals. In the UZc rocks there is little or no rhythmic layering to be seen, and igneous lamination is weakly developed in the lower rocks and absent from the upper rocks. All the UZc rocks are very iron-rich, with 25–30% FeO, and might have made a useful iron ore if in a more accessible place.

THE SANDWICH HORIZON ROCKS (SHR)

The early investigations suggested that 200 m of rock called the Unlaminated Layered Series (Wager & Deer, 1939, pp. 112–7) represented a final sheet of liquid which differentiated by the sinking or floating of crystals without the intervention of convection currents. It was considered that a lower layer of fayalite ferrogabbro had

FIG. 63. Single gravity-stratified layer in UZa with a slight tendency towards trough form, just ENE of North House (cf. similar example from LZb, Fig. 38).

Fig. 64. Average layer in UZb (5197). The rock is a plagioclase–olivine–augite–magnetite–apatite cumulate. ×15.

Fig. 65. Melanocratic layer (5196), 1½ m below rock illustrated in Fig. 64. This rock is an olivine–plagioclase–magnetite–apatite cumulate. Augite was at this time a primocryst mineral but is absent from this particular layer. Note abundance of cumulus apatite crystals. ×15.

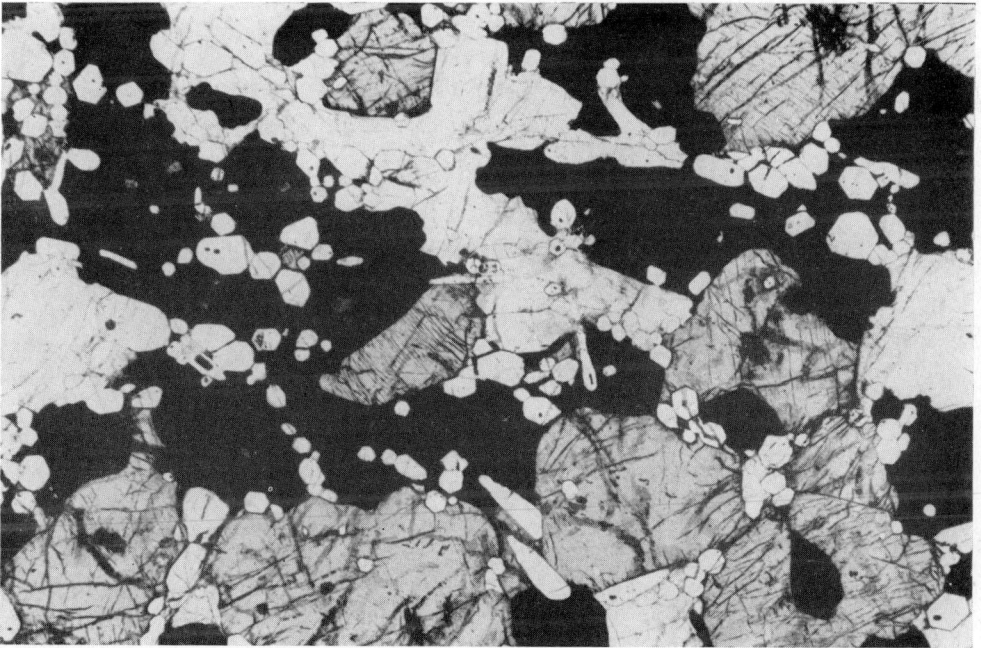

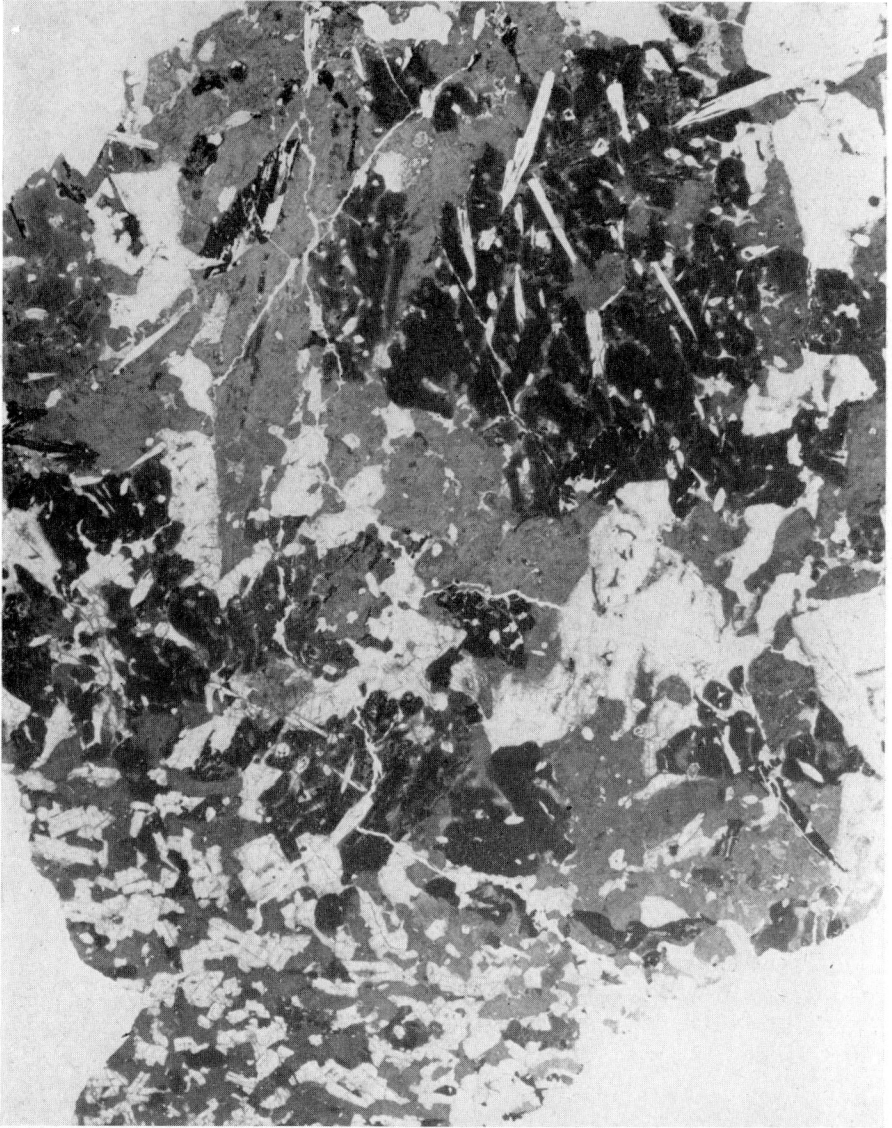

FIG. 66. Coarse crescumulate from UZc (4325). Normal ferrodiorite of UZc
type is seen at the bottom of the photograph. In the crescumulate part the
extensive, dark areas are fayalitic olivines, dense with inclusions and not easily
distinguishable from small patches of iron ore; the grey areas are of a green
ferrohedenbergite inverted from β ferriferous wollastonite. Upward-grown
apatites are conspicuous towards the top of the photograph. $\times 2 \cdot 5$.

formed by sinking of olivine, and an upper layer of andesinite, by flotation of plagioclase. Between these there seems to have been left a layer of liquid which ultimately crystallized without fractionation, giving fayalite–hedenbergite granophyre. The lower part of the Unlaminated Layered Series has since been classified with UZc, the upper part mostly as UBGβ or UBGγ, and the fayalite–hedenbergite granophyre, now called melano-granophyre, is regarded as a mildly intrusive, pene-contemporaneous sheet. However, there still remains a thin layer of rock, of distinctive composition, which is believed to have formed from a residual liquid sandwiched between the upper border group and the layered series, and which cannot satisfactorily be classified with either of the adjacent groups.

The sandwich horizon rock, SHR, on the west side of Basistoppen, is badly exposed

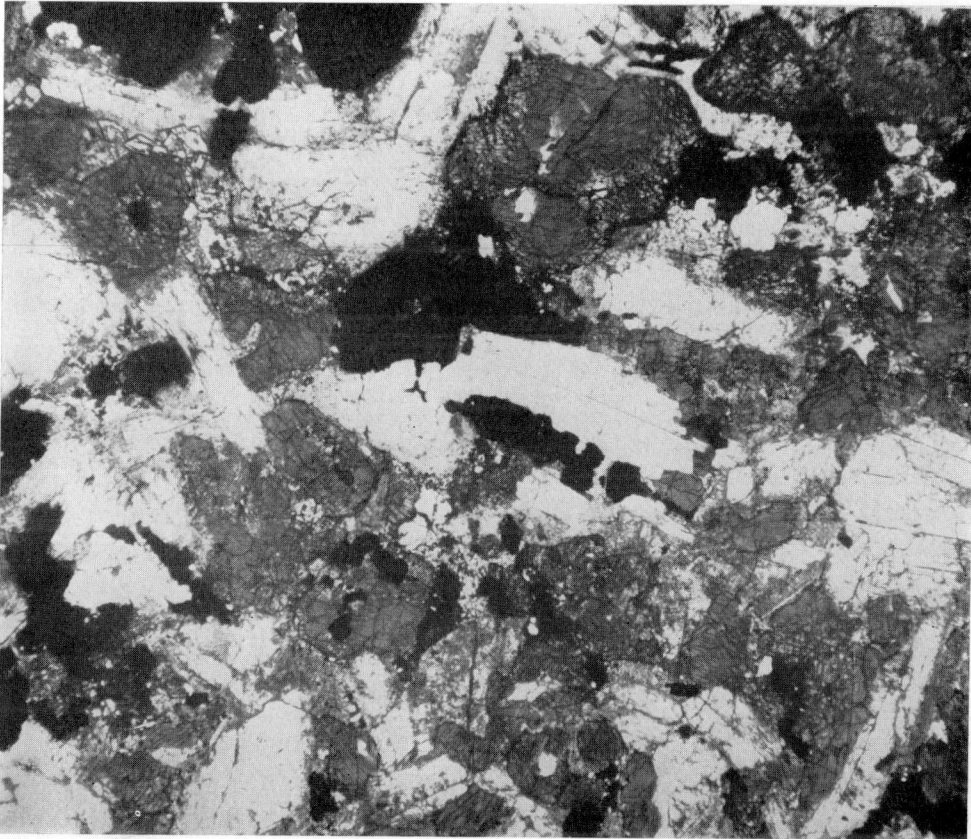

FIG. 67. Ferrodiorite of the Sandwich Horizon of the Layered Series (1330; analyzed). Plagioclase (considerably zoned), brown ferrohedenbergite (Fig. 19, No. 15), fayalite, and iron ore are the principal constituents. The fayalite (Fo$_0$) is much darkened by iron ore (e.g., just above centre). Quartz and alkali felspar, with chlorite, etc., form an abundant mesostasis. ×15.

and only a few metres are seen; similar rocks have not yet been found elsewhere, but this may be due to lack of exposure. The sandwich horizon rock is characterized by zoned plagioclases, the cores of which are An_{30}, olivine which is nearly pure fayalite, a ferrohedenbergite (Brown & Vincent, 1963, pp. 184 *et seq.*) lying at the end of the Skaergaard pyroxene trend (Fig. 19), and about 5% of micropegmatite (Fig. 67). The cores of the plagioclases are more sodic than any found in the layered series proper, and the more sodic outer zones are extensive. While the olivine is almost pure fayalite, and the pyroxene almost devoid of magnesium, the plagioclase is mostly only a sodic andesine. There is no clear evidence in this rock of any accumulation of crystals, and the rock is thought to be largely the result of direct crystallization of the magma available at this late stage. The composition of the whole rock, as judged by the albite and iron ratio (see p. 174), is more extreme than any of the layered series or upper border group and it is also extreme in many trace elements. Although, in a sense, it belongs as much to the UBG as to the LS, it is for convenience grouped with the layered series.

The Cross-bedded Belt, and Changes in the Layered Series related to Distance from the Margins

The layered rocks within about 100 m or so of the contact with the marginal border group, regardless of their horizon in the layered series, show differences in their style of layering and are richer in melanocratic minerals. Individual layers extend for only about a metre and then are cut off by an overlying layer, giving the appearance shown by some false- or cross-bedded sandstones (Fig. 68 & Pl. VIII). The individual layers are often gravity stratified, passing upwards from a melanocratic rock at the base to an average rock but not, usually, to a leucocratic one. The contrast in the general appearance of the layering between the central part of the layered series and the marginal belt is shown diagrammatically in Fig. 69, taken from the original memoir. The general decrease in the amount of felspar in the layered rocks as the inner contact is approached is clearly seen in the field, and a few specific gravity measurements have been made which indicate that the rocks, on average, are denser near the contact than further inwards (Wager & Deer, 1939, p. 85). The belt of abnormal layered rocks, stretching from the inner contact inward for about 100 m, is termed the cross-bedded belt and its observed occurrence is shown diagrammatically on the map (Fig. 89). The cross-bedded belt occurs at all horizons in the layered series, at least up to UZb, and it has been found at the north, west, and east sides of the intrusion. It is likely that it is also developed along the south side but the layered series there is below the level of exposure. It is impossible to see the rocks of the cross-bedded belt without regarding them as the result of deposition of crystals from an irregularly, and vigorously moving liquid. The cross-bedded belt coincides in position with the steeper-dipping layers which consitute the outer part of the saucer-shaped layering. Small, contemporary, normal faults, striking parallel to the nearest margin of the layered series, are a frequent feature

PLATE VIII

The cross-bedded belt as seen on a surface parallel to the dip. The whole mass of rock is relatively rich in dark minerals. Gravity stratification is obvious in some of the layers. Cross-bedding is visible in several places and, here and there, rather similar features are due to slumping of the crystal mush.

[face p. 96]

of the cross-bedded belt (Fig. 70, and Wager & Deer, 1939, pl. 5, fig. 2). The faults downthrow inwards, and are considered due to the sediment of crystals being so steeply banked up against the margin that it became unstable, resulting in movements giving structures resembling land slips, and occasional slumping (Fig. 71).

Another change taking place in relation to the margin is in the average size of the cumulus plagioclase crystals. If the lengths of the larger plagioclases, in a series of rocks belonging approximately to the same horizon in the layered series, are measured roughly in the hand specimen with an ordinary ruler, there is found to be a general increase in length in proceeding from the margin towards the centre of the intrusion. For the MZ rocks the average size of the felspars changes from about 0·5 cm at the margin to about 1 cm near the centre of the intrusion (Wager & Deer, 1939, p. 71). The cross-bedded belt is excluded from this comparison as it seems to show occasional large felspars associated with more abundant, smaller ones. A statistical investigation

FIG. 68. The cross-bedded belt in a face almost perpendicular to the strike of the marginal banding. Uttentals Sound, near Strömstedet. A single layer can usually be traced for only a short distance before it is truncated by the overlying layer.

G

would be worthwhile but there seems to be no doubt of the broad fact that an increase in general size of the plagioclase takes place in passing from the margin towards the centre, the significance of which is discussed on p. 102.

PRELIMINARY CONSIDERATION OF THE CONDITIONS OF DEPOSITION OF THE LAYERED SERIES, AND THE PART PLAYED BY CONVECTION CURRENTS

The most significant evidence for the view that the layered series was formed by a continuous process of deposition of crystals at the bottom of the magma comes from the cryptic variation. As already stated, the changes in composition of the cumulus minerals in proceeding upwards in the layered sequence is from high to low temperature members of the solid solution series. In the case of the plagioclase it is from An_{67} in the lowest exposed layered rocks and probably from An_{77} in the earliest hidden layered series rocks. The experimental work on the two-component plagioclase system (Bowen, 1913) and later work on more complex systems involving the plagioclases, leave no doubt that the more anorthite-rich plagioclase must have been formed at the higher temperatures and the successively more albite-rich plagioclases at progressively lower temperatures. Similarly the olivines vary from Fo_{68} in the lowest exposed layered rocks, and probably from Fo_{81} in the lowest HLS rocks to Fo_0 in the highest layered rocks with, of course, a break in the MZ. The experimental work of Bowen and Schairer (1935, pp. 161–3), together with later work, makes it clear that this sequence of olivines must also imply decreasing temperature of formation. The pyroxenes vary from magnesium-rich to iron-rich members and experimental work (Bowen, Schairer & Posjnak, 1933; Yoder,

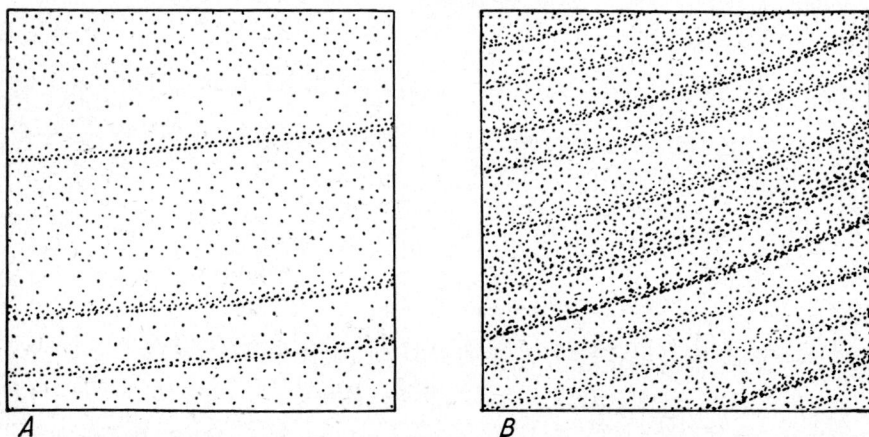

FIG. 69. *A.* Generalized appearance of the rhythmic layering found some distance away from the contact between LS and MBG.

B. The same as *A*, but close to the contact where complications (not shown) due to cross-bedding and slumping are also frequent (cf. Pl. VIII and Fig. 68, 70). (From Wager & Deer, 1939.)

FIG. 70. Small contemporary faults, striking parallel to the margin, in the cross-bedded belt of the LS and within a few metres of the contact with the MBG. Near the coast, on the S. side of Uttentals Sound, at Strömstedet.

Tilley & Schairer, 1963) again shows that there must have been the same direction of change in their temperature of formation. Thus by combining the results of experimental work with the observed cryptic variation, it is clear that when the cumulus minerals of the lower rocks crystallized, the material which ultimately formed the higher rocks must have still been liquid. The time sequence of formation of the layered rocks must be from below upwards and the cumulus minerals of the successive layers must have crystallized at steadily decreasing temperatures. The cumulus minerals are, in fact, successive crystal fractions.

Various mechanical processes may produce a separation of successive crystal fractions from a liquid but in the Skaergaard layered series, the fact that the successive fractions occur as layers resting one above the other suggests that the crystals collected as a sediment at the bottom of the liquid, primarily under the influence of gravity.

The hypothesis of the bottom accumulation of crystals is supported by the observation that any rock of the layered series can be divided, in theory at any rate, into two parts, the cumulus crystals and the intercumulus material. The cumulus crystals behave as would be expected of an accumulation of discrete entities. Thus they can vary

FIG. 71. Leucocratic and melanocratic layers disrupted and distorted by slumping. Cross-bedded belt, north side of Uttentals Plateau, near Marginal Border Group.

considerably in relative amounts, indicating that they existed as entities free to accumulate in different proportions under different conditions of sedimentation. Also the textures of the rocks, when rightly interpreted, show that at an early stage in the formation, the rocks consisted of a pile of crystals with a liquid filling the interstices.

To accumulate at the bottom of the liquid, the cumulus crystals must have sunk through the liquid in which they were formed and, therefore, must have been of greater density than the liquid. Since the heat loss from the intrusion would be mainly from the top it would be expected, as a first premise, that the crystals would form there and those denser than the liquid would then sink vertically to the floor. There is, however, also the possibility that cooling at the top of the liquid set up a convective circulation, either periodic or continuous, which would greatly modify the process of crystal deposition. In the Skaergaard intrusion, several features suggest the action of magmatic currents, and the trough banding, in particular, almost amounts to a visual demonstration. The hypothesis was therefore developed (Wager & Deer, 1939, pp. 262–77) that during cooling there were convection currents in the pool of magma, which descended at the margins, swept across the floor, and ascended at the centre (Fig. 126).

The trough banding structures have been shown to have developed during the formation of the layered series, and not as a result of subsequent folding. The troughs apparently mark the positions of vigorous lines of currents in the magma which, having descended along the sides of the intrusion, moved radially across the floor towards the centre. At the main trough-band horizon the currents must have persisted for a considerable time in one position, although, apparently, a single current sometimes split into two or more currents in the same general position, before finally dying out (cf. Fig. 59). Deposition along the line of the current gave rise to strongly gravity-stratified layering, probably because of the pulsatory character of the currents. The cumulate between the trough bands must have formed from long mounds of crystals deposited without sorting, which flanked the currents. Incipient trough banding structures have been mentioned as occurring at all heights in the outer part of the layered series, although infrequently, and this implies a tendency for the generally inward-moving convection currents to have been concentrated, at these places, into more strongly flowing streams.

The normal, planar, rhythmic layering, especially where involving gravity stratification, is also believed to be the result of currents, moving as broad sheets rather than canalized in troughs. In a general way, variation in the velocity of the currents is considered to have produced a variation in the proportion of heavy and light crystals settling out of the magma. If the current were non-turbulent and of uniform velocity (laminar flow) there would be uniform deposition of the various cumulus crystals giving average or uniform rocks. With currents of variable velocity, it is suggested that deposition would be equally variable, resulting in rhythmic layering. Further consideration is given to the origin of the various kinds of layering in Ch. VIII.

Igneous lamination, which, in varying degrees, is a common feature of the layered series except at the very top, is also ascribed to currents in the magma. Grout originally used the hypothesis of convection currents in explaining the igneous lamination of the Duluth gabbro (Grout, 1918*d*, pp. 453, 456–7). He pointed out that while thin mica flakes, sinking in still water, might tend to lie parallel to the floor without the agency of currents, because of their relatively high density and the low viscosity of the water, in the case of thick tabular felspars, sinking in tranquil basic magma, the forces tending to orientate the crystals parallel to the floor would be much less. Grout therefore suggested that the platy parallelism in igneous rocks was the result of deposition from moving magma; he considered that as the crystals touched the floor, friction between them and the flowing magma would cause them to be laid parallel to the direction of flow, that is, parallel to the floor. We accept Grout's explanation of igneous lamination, and believe that the texture is a further indication of convection currents during the formation of the Skaergaard layered series.

In rocks with igneous lamination, any acicular or bladed crystals might be expected to show a linear parallelism within the plane of layering and in the direction of flow of the currents. This is not a striking feature of the Skaergaard laminated rocks but such an orientation of elongated pyroxenes has been detected in the field in certain places in UZa, and by a petrofabric study of the orientation of elongated apatite crystals (Brothers, 1964) it has been shown to occur in certain rocks of the trough bands.

Features of the layered series which are affected by distance from the margin also suggest convection currents. Thus the layered rocks of the cross-bedded belt show structures of the kind ascribed, in deltaic sediments, to fluctuating currents. The rocks of this belt are also markedly richer in the heavier cumulus minerals, that is olivine, pyroxene and magnetite, than are rocks of the same horizon further in. If the cumulus crystals had formed at the top of the magma body and then sunk vertically, there would be no reason for the more melanocratic rocks of a layer to occur preferentially near the margins. On the other hand, assuming convection currents, the features of the cross-bedded belt may be explained as the result of especial turbulence where the vigorous downward currents, near the margin, turned to flow horizontally across the floor of the intrusion. The currents would be highly variable in velocity and might sometimes erode as well as deposit material; furthermore, the heavier constituents would be more likely to be deposited than the lighter, thus giving the more melanocratic rocks characteristic of the cross-bedded belt. Finally, the increase in size of the cumulus plagioclases of particular horizons, towards the centre of the intrusion, could not be explained by vertical sinking but is attributed to their having time to grow during transport across the floor of the intrusion (see p. 97).

In order to obtain the regular cryptic variation shown by the layered series, the cumulus crystals must have formed in a body of magma which remained essentially homogeneous. The crystallization of solid phases, having a bulk composition different from that of the magma, necessarily produces changes in the composition of the liquid. To obtain regular cryptic layering, these changes in composition of the magma must not remain localized but must be distributed evenly throughout the whole magma; in the

Skaergaard intrusion, stirring by convection currents was probably the means by which a close approach to homogeneity was attained. The departures from regular cryptic layering, which have been shown to occur sporadically (p. 30), are regarded as due to incomplete compositional or thermal homogenization of the liquid.

Many features of the rhythmic and cryptic layering of the Skaergaard intrusion suggest deposition of cumulus crystals from a convecting magma. The problems connected with the growth and deposition of crystals from such a convecting magma involve consideration of the overall rate of heat loss, the rates of sinking of the various crystals, the conditions and place of nucleation of the crystals, etc., and some attempt to consider these matters is made in Ch. VIII. At this stage it is sufficient to note that there is clear evidence for the existence of convection currents in the Skaergaard magma during most of the protracted period of its solidification.

CHAPTER V

THE SKAERGAARD MARGINAL AND UPPER BORDER GROUPS, AND THE GRANOPHYRIC DIFFERENTIATES

The chilled, olivine gabbro phase, immediately adjacent to the country rock, is characteristic of the marginal border group on all sides. Proceeding inwards from this, much variation is found which shows, however, a significant symmetry. The eastern and western border groups pass through the same series of changes, whereas the northern and southern borders are different. The similarity between eastern and western border groups is, perhaps, to be expected since the inward inclination of the contacts are the same, and the exposed parts are structurally at the same height. Some of the differences between the northern and southern border groups are due to the near-vertical disposition of the southern margin, and the gentler dip of about 45° of the northern margin, at the time of solidification of the intrusion. Further differences are due to the crustal flexure which has resulted in exposing northern, border group rocks adjacent to the lower zone of the layered series, while the exposed southern marginal border group rocks are adjacent to the upper border group and are structurally 3 km higher (Fig. 9).

For about 60 m from the contact with the country rocks, the marginal border group, on all sides, shows no banding and this part, as already mentioned, is distinguished as the tranquil division (Fig. 72, 73 & 89). Further inwards, the characteristic marginal border group banding (Fig. 74), which suggests flow of heterogeneous material at the time of consolidation, is developed; this part, described as the banded division, varies in width from nought along the northern margin to 400 m on the Skaergaard Peninsula. On proceeding inwards, the plagioclase felspars of the banded division become more albite-rich and the pyroxenes and olivines become more iron-rich—changes similar to those found in passing upwards in the layered series.

The upper border group is preserved mainly on the higher mountain summits of the southern part of the Skaergaard intrusion, and some of our lack of knowledge about it is due to the fact that two of the summits have not been climbed and many of the rock faces are inaccessible. A chilled marginal facies, against the postulated roof of agglomerates, has not been found but might exist on Tinden. The highest rocks collected are coarse felspathic gabbros, the cores of the felspar being of calcic labradorite. In descending, the plagioclase of the upper border group becomes less anorthite-rich and the compositions of the unzoned parts of the plagioclases provide a basis for the classification of the upper border group into UBGα, β & γ, subdivisions which are correlated with the lower, middle and upper zones of the layered series, respectively (Fig. 15). The downward change in composition of the plagioclases of the upper border group is similar to the inward change found in the marginal border group. From the changes in the plagioclase composition, the order of formation of the upper border group is inferred to be from above downwards, and in the marginal border group, from the outside inwards.

The sequence of the upper border rocks is complicated by the presence of an acid sheet—the Tinden granophyre—and certain associated, transitional granophyres. These may have an intrusive relationship to the layered UBG rocks but they are not chilled against them and are regarded as an integral part of the intrusion.

Among the upper rocks of the intrusion there is a further complication which has only recently been resolved by J. A. V. Douglas (1964). A differentiated sheet, having a mineralogy which excludes it from being the result of fractionation of the Skaergaard magma, was regarded in the original account (Wager & Deer, 1939) as a huge,

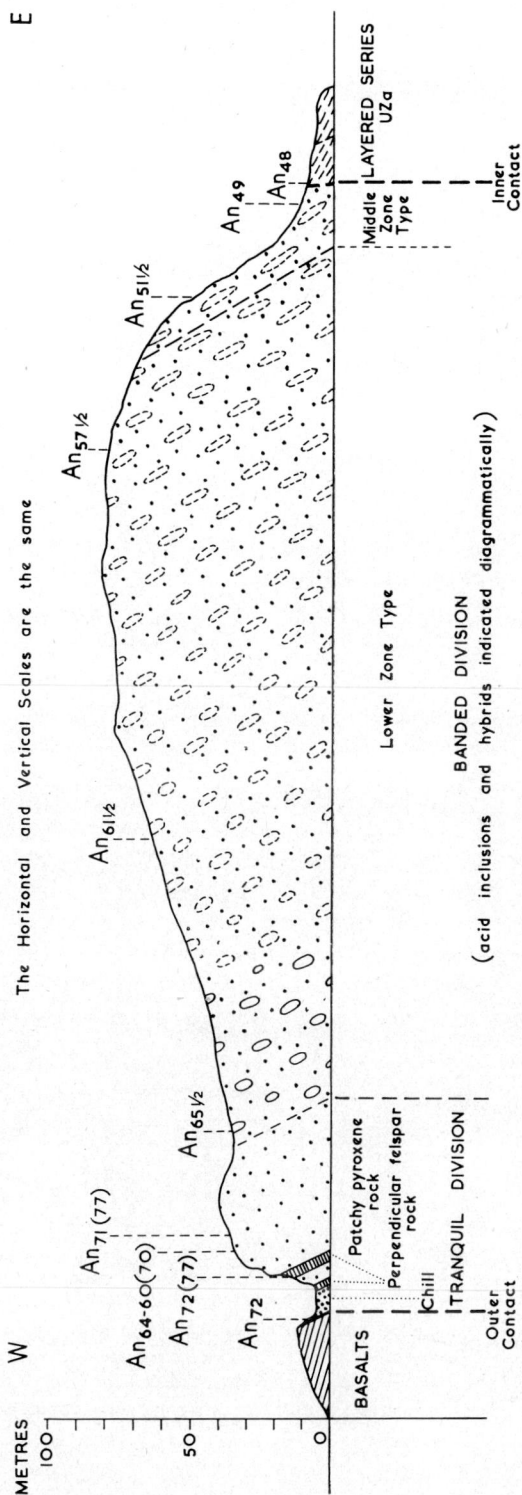

FIG. 72. Cross-section of the Marginal Border Group on Ivnarmiut (see Fig. 89). The more definite inclusions of the outer part of the Banded Division are shown in firm line. The compositions of the plagioclase cores of typical specimens are given above the appropriate point on the traverse. Note the definition of Inner and Outer Contacts, frequently referred to in the text.

raft-like inclusion which had sunk some way into the intrusion. It is now, however, considered to be the result of the injection of new magma into the Skaergaard intrusion, which was still hot at the time. Tectonically, the intrusion of the sheet is probably related to the beginning of, or to an early stage in, the crustal flexuring. Originally, these rocks were called the Basistoppen raft: they are now described as the Basistoppen sheet, and form the subject of the next chapter.

THE TRANQUIL DIVISION OF THE MARGINAL BORDER GROUP

a) *The chilled olivine gabbro at the margin, and the coarser variant further in*

The obviously chilled, olivine gabbro has been collected from within 3 m of the contact at many points around the whole intrusion. Because of its general distribution

FIG. 73. Southwest corner of Ivnarmiut showing metamorphosed basalt (left of gully), the Tranquil Division of the Marginal Border Group (centre), and the Banded Division (right). (See Fig. 72 & 89.) The Banded Division (dark) begins just to the right of the rock mound (ruins of old Eskimo house).

and uniform composition it is inferred that the Skaergaard magma was emplaced in a single episode (Ch. II). The more southerly specimens of the olivine gabbro are chilled against basaltic country rocks and are constant in mineralogical composition and texture, while the more northerly, which are adjacent to the metamorphic complex, are somewhat more variable, due to contamination with the acid gneiss.

The southern, uncontaminated type of fine-grained marginal gabbro is seen in thin section to consist of abundant plagioclase and olivine crystals surrounded in some places by poikilitic pyroxenes (both augite and inverted pigeonite) and in others by ilmenite (Fig. 75). The plagioclase is strongly zoned, the cores being An_{72}. The normative plagioclase composition, as weight per cent, is An_{62} (see Table 7). This is probably close to the average felspar composition and, combined with the observed composition of the cores, gives an indication of the extent of zoning. The olivine averages Fo_{59}

FIG. 74. Banded Division of the Marginal Border Group as seen on the south coast of Mellemö. Some of the heterogeneity is due to hydbridization with acid gneiss inclusions of the Metamorphic Complex, which give the coarser bands seen in the photograph.

from optics, and is probably also zoned, from more magnesium- to more iron-rich varieties. Augite and orthopyroxene, the latter showing textural evidence of inversion from pigeonite, have similar poikilitic habits. Contamination with acid gneiss has produced rocks with less olivine and more orthopyroxene, and a tendency for the breakdown of the poikilitic, into an equigranular texture.

By counting the number of differently orientated crystals of the different mineral species encountered during systematic traverses of thin sections, an approximate quantitative measure of the numbers of the different crystals per unit volume of rock (Table 2) has been obtained (Wager, 1961). The values imply that during the solidification process, many crystal nuclei of plagioclase and olivine, and relatively few of

TABLE 2

	Number of crystals per cm³ of rock			
	Plagioclase	Olivine	Pyroxene	Iron ore
Chilled marginal gabbro (4507, Fig. 75)	4600	1900	38	13
Patchy-pyroxene gabbro (4442, Fig. 76)	2700	75	16	22

pyroxene and iron ore, were formed. The magma, free from crystals on intrusion, was apparently chilled rapidly and nuclei of plagioclase and olivine formed abundantly in the middle labile region (Fig. 84) and then grew. The initial chilling seems to have been insufficient for the nucleation of pyroxene and magnetite, which occurred only after further cooling. The further cooling would necessarily be slow and, as a result, relatively few nuclei of pyroxene and magnetite were formed, and these were able to grow into extensive poikilitic units, since there was enough time for the necessary diffusion. The poikilitic texture of the chilled rocks is therefore viewed as the result of the pyroxenes and ilmenite nucleating relatively late, and growing into large crystals from only a few crystal nuclei. There are grounds for believing that near the contact there was considerable supercooling of the magma, of the order of 20°; the effect of this on the texture and mineral composition of the chilled olivine gabbro is considered further (p. 119).

Proceeding inwards from the obviously chilled marginal rock, the grain size is found to increase for 50 or 60 m, but, except for certain localized zones, the texture remains generally similar. The rocks are called the patchy-pyroxene gabbros because they contain large poikilitic pyroxenes (reaching 5 or 10 cm across) which surround abundant plagioclase crystals and, in contrast with the chilled gabbro, relatively few olivine crystals (Fig. 76A, B). The numbers of individual crystals per unit volume, in a sample of this rock, have also been estimated (bottom of Table 2). The total number of plagioclase crystals per cm^3 is 2700, which is about half the number found in the chilled marginal gabbro and means that the average volume of each of the individual plagioclase crystals must be about twice those in the chilled rock, since the plagioclase content is the same. There are about 160 times as many plagioclase crystals as pyroxenes—a proportion not very different from that found in the chilled olivine gabbro. However, there are only five times as many olivines as pyroxenes, contrasting with fifty times as many in the chilled rock. The estimated number of distinct ilmenite crystals is rather doubtful (the counts were not made on polished specimens and the orientation of the crystals was not determined) but it appears to be of about the same order as in the chilled rock. The composition of the central part of each of the plagioclase crystals is about An_{71}, as in the chilled olivine gabbros, but there are occasional cores, irregular patches, or reversed zones of An_{77}. The reversed zoning is similar to that shown by the perpendicular-felspar rock about to be described. In chemical composition, the coarser olivine gabbros of the tranquil division are fairly close to the chilled marginal olivine gabbro, their different texture being ascribed to slower cooling.

FIG. 75. Chilled olivine gabbro of the Tranquil Division (analyzed specimen 4507). Southern margin, on E. side of Skaergaard Bay.

b) *The perpendicular-felspar and wavy-pyroxene rocks*

In the tranquil division there are indefinite sheets, parallel to the contact, of texturally different rock types which interrupt the steady change to the coarser-textured olivine gabbros (Fig. 72). The most striking of these (Fig. 77 & 80) was described

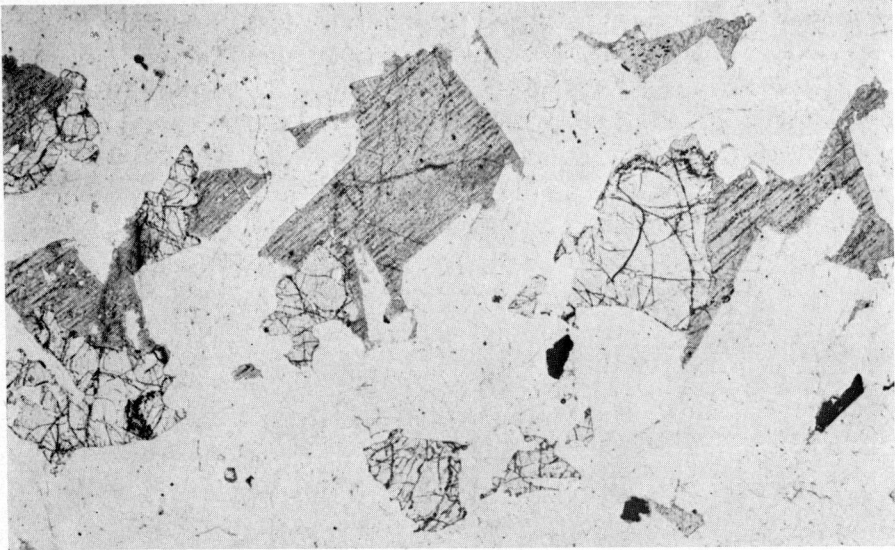

A

B

Fig. 76. Patchy-pyroxene rock (4442) of Tranquil Division, 30 m from the west margin of Ivnarmiut. × 10.

A, ordinary light.   *B*, partly crossed nicols.

originally as the perpendicular-felspar rock because of the elongated plagioclase crystals set approximately perpendicular to the margins of the intrusion (Wager & Deer, 1939). This arrangement of the bladed felspars is clearly not the result of the orientation of crystals by flow of the magma, for this would have aligned them parallel to the margins. It is, apparently, the result of plagioclase crystals growing inwards into the cooling magma, like the acicular crystals which grow inwards from the walls of a cooling crucible of molten sulphur, or like the minerals which sometimes grow inwards from the boundary walls of pegmatite bodies and mineral veins. The perpendicular-felspar rock is seen best at various points on the western margin; it has been found also on the eastern and northern margins, but not on the southern.

The other rock type, for which the field name wavy-pyroxene rock is used, occurs as thick sheets or lenses, a metre or so wide and approximately parallel to the margins. The rock is a variant of the patchy-pyroxene gabbro, the wavy, poikilitic pyroxene sheets being roughly horizontal and elongated at right angles to the margin (Fig. 78 & 79). Although the irregular pyroxene sheets tend to extend inwards at right angles to the

FIG. 77. Perpendicular-felspar rock, 20 m from Outer Contact, Mellemö. The elongated felspars (from left to right of photo) have grown roughly perpendicular (From Wager & Deer, 1939.)

contact, like the plagioclase crystals of the perpendicular-felspar rock, no constant crystallographic orientation of the pyroxene crystals within the sheets has been observed, and the pyroxenes are definitely poikilitic, not acicular in habit. Roughly horizontal, impersistent, felspar-rich or olivine-rich layers are characteristically associated with the poikilitic pyroxene sheets. The wavy-pyroxene rock has a similar distribution, in the outer part of the marginal border group, to the perpendicular-felspar rock but an incipient development of the wavy-pyroxene structure was also found in the southern border group.

The plagioclase crystals of the perpendicular-felspar rock are elongated tablets of blade-like form, measuring about $2 \cdot 5 \times 0 \cdot 3 \times 0 \cdot 1$ cm. The longest direction is parallel to the c crystallographic axis and the crystals are slightly flattened parallel to (010). The central three-quarters of each of the crystals is unzoned, and An_{70} in composition. Outside this, typically, is a narrow reversed zone of An_{77}, then follows a narrow zone of An_{72}, after which normal zoning gives, in places, a final fringe going down to about An_{50}.

FIG. 78. Wavy bands of coarse pyroxene–felspar rock in the Tranquil Division. Kraemers Island.

The perpendicular felspars form about 50% of the whole rock and between them is fine-grained, poikilitic olivine gabbro much like the chilled olivine gabbro at 1 to 3 m from the margin (Fig. 80). Some of the smaller plagioclases, between the elongated crystals, also have An_{72} cores and narrow, outer reversed zones of An_{77}. The small olivines are usually about Fo_{70} in composition. The orthopyroxene is largely free from the blebby augite inclusions considered to form by exsolution from an inverted pigeonite and is thus regarded as having crystallized direct from the magma. It is assumed to have crystallized at a higher temperature than the rims, containing augite blebs and presumed to be of inverted pigeonite.

Associated with the perpendicular-felspar rock, and sometimes forming a base from which the elongated felspars have grown, are indefinite, olivine-rich sheets with a modal olivine content of about 40% instead of the usual 20%. Some of the olivine crystals are larger than the average and skeletal in form, being composed of crystal units in sub-parallel growth (Fig. 81). Like the elongated felspars, they appear to have grown inwards into the liquid. The composition of the greater part of each of these olivines is Fo_{72}, but whether they have any marginal reversed zoning has not been established. These olivines are similar in appearance, although much smaller, to the crescumulate olivines in the layered rocks of Rhum (Ch. XI).

FIG. 79. Wavy-pyroxene rock (5076) of the Tranquil Division, collected near exposure shown in Fig. 78.

H

114

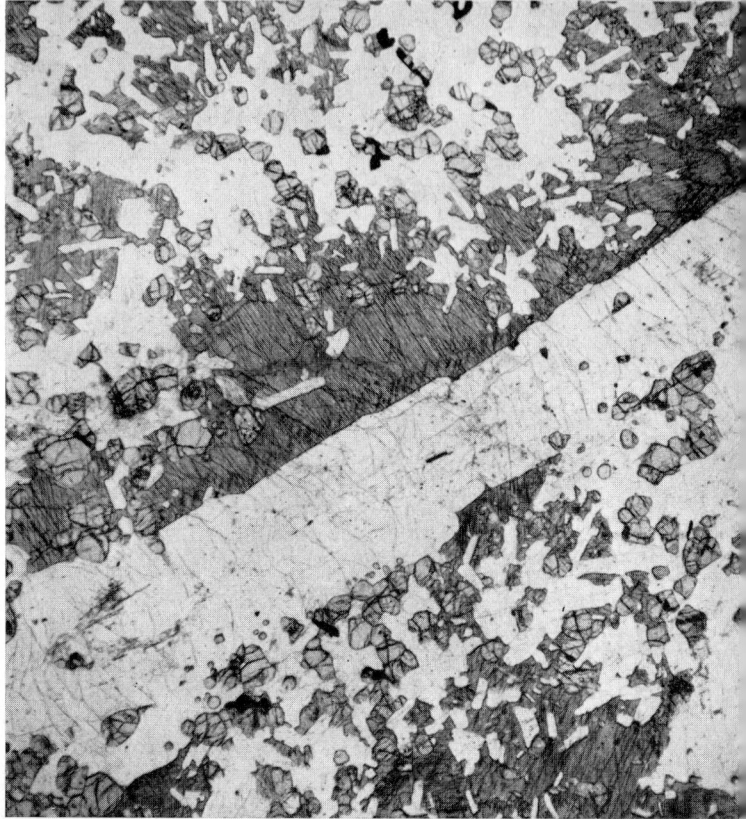

FIG. 80

Perpendicular-felspar rock (1851), 20 m from the Outer Contact, Mellemö. The section is cut parallel to the elongated felspars and shows part of one of them. The remainder of the field is essentially the same as the chilled olivine gabbro, though finer grained. ×15.

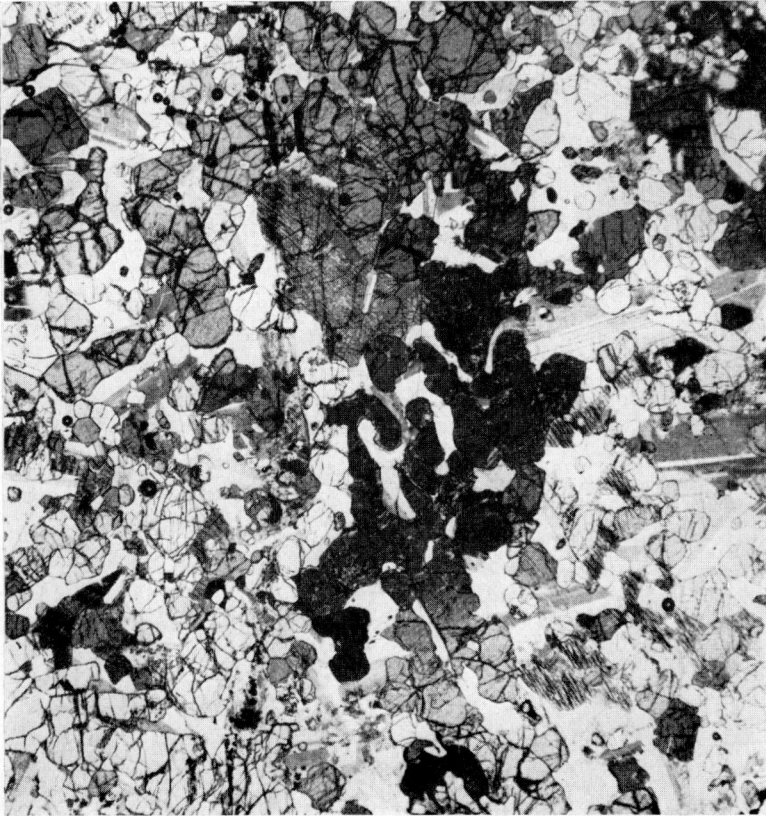

FIG. 81

Melanocratic rock (5289), 50 from Outer Contact on Mellemö an near the perpendicular-felspar rock The section shows an olivine cryst (centre, dark) with the beginnin of crescumulate texture. Partial crossed nicols. ×15.

c) *The picrite of the northern marginal border group*

Occasional examples of both the perpendicular-felspar rock and the wavy-pyroxene rock occur in the northern marginal border group, and there is also much coarse-grained, patchy-pyroxene gabbro. In addition, there is a new feature, namely a plutonic breccia consisting of rounded and sometimes broken masses of picrite (formerly called gabbro-picrite; see footnote, Ch. II, p. 31) embedded in a somewhat heterogeneous eucrite (Fig. 82). The picrite is considered an integral part of the Skaergaard intrusion since it occurs within the marginal envelope of chilled olivine gabbro. Under the microscope, it has the appearance of being an accumulation of early-crystallized olivines (Fig. 83). Occasionally, some olivines also show evidence of crescumulate growth. The northern margin dips inwards at 45°, while the other margins are much steeper or vertical, and it is believed that this is a factor leading to the formation of the picrite. It is suggested that early olivine crystals sank to form an accumulation on the gently sloping northern margin, and that periodically the partly solidified cumulate became unstable and slumped; during the slumping, the sheets of olivine-rich rock were broken up into blocks, some of which became rounded by movement. In two places, extensive rafts of picrite are seen, arranged parallel to the margin of the intrusion. These may be either cumulate preserved in its position of formation or slumped material which has not disintegrated.

The olivines of the picrite have the appearance of being a loose crystal accumulate. In composition they are Fo_{81}, while the olivines in the perpendicular-felspar rock, even where of crescumulate type, are apparently about Fo_{70} (Table 2). Some of the plagioclases in the picrite have extensive cores of An_{72}, with an outer reversed zone of An_{77} followed by normal zoning. Others have cores of An_{77} and then normal zoning. There are also magnesium-rich augites and orthopyroxenes in the picrite, which are early, high temperature crystals judged by their composition (Fig. 19); the augites, however, are small crystals compared with the accumulated olivines and would not be expected to have sunk appreciably.

d) *Explanation of the rock types of the tranquil division*

A summary of the chief mineral phases found in the tranquil division of the marginal border group, and their optically determined compositions, is given in the first half of Table 3. The rocks are of interest for the information they give on the detailed manner of congelation in the outer part of this large body of magma. The marginal olivine gabbro is also important because it is believed to be a congealed sample of the magma before any differentiation had taken place. The mineralogy and petrology of some of the other rocks of the tranquil division are also important because certain of the minerals, rightly interpreted, indicate the composition of the earliest equilibrium phases crystallizing from the original magma and give a clue to the composition of the rocks of the hidden layered series.

The olivine of the picrite, Fo_{81}, is presumed to be the olivine in equilibrium with the initial magma, as it is the most magnesian olivine encountered in the intrusion and probably formed slowly, without supercooling, in the main body of the magma. The plagioclase in the picrite, An_{77}, is equally believed to be that forming under equilibrium conditions from the initial magma. Until our recent appreciation of the effects of super-cooling, it had seemed that the plagioclase cores in the perpendicular-felspar rock, which are An_{72}, represented the composition of the plagioclase in equilibrium with the initial magma. However, it is now considered that the acicular plagioclases were formed from supercooled magma, some degrees below the equilibrium temperature, and are thus less anorthite-rich than if formed under equilibrium conditions. From the experimental investigation of the albite–anopthite system it is known that the difference in crystallization temperature between plagioclase An_{77} and An_{72} is about $30°$ C and it would be somewhat less in a multi-component system such as basic magma (see Fig. 118). The narrow, reversed zone of the acicular felspars, An_{77} in composition, was apparently formed because the latent heat of crystallization of the acicular plagioclases raised the temperature locally to the equilibrium value. The olivine of the perpendicular-felspar rock, and of the associated olivine-rich zones with crescumulate olivine, is about Fo_{66}

FIG. 82.   Rounded blocks of picrite in the northern Marginal Border Group, about 200 m north of Stromstedet.

and was, apparently, also precipitated from supercooled magma since the presumed equilibrium olivine of the picrite is Fo_{81}. The other minerals, particularly the pyroxenes of the tranquil division rocks, have been only partly investigated. However, it is clear that there are relatively high temperature, magnesian augites and orthopyroxenes present in the picrite (see p. 42) and also in the perpendicular-felspar rock and related types. The optical and textural data on the pyroxenes have provided valuable information on the early part of the trends (Fig. 19).

Search for the most calcic cores in the plagioclases of the chilled olivine gabbros failed to find any which were more calcic than An_{72}. It is considered, therefore, that crystallization of plagioclase in these rocks took place only after supercooling to the temperature appropriate for the crystallization of An_{72}. No reversed zoning, as found in the perpendicular-felspar rock, has been seen; when the chilled marginal olivine

FIG. 83. Picrite (1682) from the northern Marginal Border Group, 100 m above Uttentals Sound. The large olivine crystals are decomposed along the now conspicuous fractures. Surrounding the olivine is plagioclase, clinopyroxene in small rectangular crystals, a little iron ore, and considerable patches of poikilitic orthopyroxene. × 15.

TABLE 3
Composition of minerals in the Marginal Border Group

(From optical measurements except for the pyroxene values (in parentheses) which are from chemical analyses (Brown, 1957).)

	Tranquil division					Banded division			
	Chilled olivine gabbro 4507	Picrite N. margin 4526	Perpendicular felspar rock 1851	Wavy pyroxene rock 5290	Patchy pyroxene rock 4442	LZ type 4443	LZ type 4444	MZ type 4446	UZa type 5280
Distance from the margin	1 m from S. margin	ca. 30 m from N. margin	ca. 20 m from W. margin	ca. 30 m from W. margin	30 m from W. margin	65 m from W. margin	170 m from W. margin	350 m from W. margin	ca. 800 m from SW. margin
Plagioclase	$An_{72}\to$	$An_{72-77-70\to}$ also: $An_{77\to}$	$An_{70-77-72\to}$ (also small crystals: $An_{77-70-77}$)	*	$An_{77-70-77-70\to}$	$An_{66\to}$	$An_{62\to}$ (a few $62-46\to$)	$An_{51\to}$ (a few $51-43\to$)	$An_{45\to}$
Olivine	Fo_{59}	Fo_{81}	Usually about Fo_{70}	Fo_{67}		*	*	Abs.	*
Augite	$(Ca_{37}Mg_{40}Fe_{23})$	$(Ca_{42}Mg_{48}Fe_{10})$	*	*		*	*	*	*
Orthopyroxene	Abs.	En_{78} $(Ca_4Mg_{77}Fe_{19})$	En_{72}	Abs.	Abs.	Abs.	Abs.	Abs.	Abs.
Inverted pigeonite	En_{66}	Abs.	*	En_{70}	En_{62} (4289)	* (as intercumulus)	*	*	*
Iron ore	Ilmenite	Chrome spinel Ilmenite	*	*	*	* (as intercumulus)	* (as intercumulus)	*	*

* = Present but composition not determined.   Abs. = Absent.

gabbro was forming, the loss of heat was apparently sufficiently rapid to prevent even a local rise in temperature to give higher temperature equilibrium plagioclase such as that of the An_{77} zones of the perpendicular felspar.

The olivines of the chilled marginal gabbro have a composition of about Fo_{59}; this also implies strong supercooling from the equilibrium conditions which produced olivines of Fo_{81} composition. Zoning in the small olivines cannot satisfactorily be appreciated optically: the determined forsterite percentage is thus the average for the whole grain, and is close to the normative composition obtained from the rock analysis, which is Fo_{64}.

The orthopyroxene of the chilled olivine gabbro, deduced to have formed by inversion of pigeonite from the orientation and amount of augite exsolution lamellae, was the mineral which first suggested that there had been significant supercooling (Brown, 1957). This deduction was made because some of the marginal rocks, further in, had a more magnesian orthopyroxene which had crystallized directly from the magma at, presumably, a higher temperature. It was inferred therefore that the pigeonite of the chilled marginal gabbro had formed from a supercooled liquid. The initial Skaergaard magma was apparently such that under equilibrium conditions orthopyroxene formed, but with considerable chilling, pigeonite formed. This may be more fully explained with the help of Fig. 23. The earliest orthopyroxene, found in the picrite, is shown at B, and the corresponding liquid in equilibrium with it, at C. If there were strong supercooling of this liquid a pigeonite such as D (the composition, in terms only of En:Fs ratios, of inverted pigeonite from the chilled marginal gabbro) could be produced, the amount of supercooling being roughly of the order of 20° C (Brown, 1957, p. 530 and appendix). Greater supercooling would probably have produced an orthopyroxene, and lesser supercooling, a more magnesian pigeonite. Whether or not the actual values of the quantities shown are to be relied upon, the diagram gives a qualitative picture, at least, of how the observed distribution of orthopyroxene and pigeonite (later inverted) in the tranquil division rocks came about. It is thus estimated that the amount of supercooling in the chilled rock is of the order of 20° C and supercooling of a similar order is also indicated by the plagioclase evidence.

When the texture of the chilled marginal gabbros is considered in detail, it is realized that it must largely depend upon the frequency of nucleation under different degrees of supercooling. The conditions necessary for the formation of crystal nuclei in super-saturated solutions is by no means fully understood. Some workers accept the existence of two distinct regions of different crystallization behaviour, described long ago by Ostwald as the metastable and labile regions. In the metastable region, it is believed that crystallization does not occur at all unless seed crystals are introduced, while in the labile region, spontaneous crystallization can occur, especially when the liquid is stirred or suffers mechanical shock. Some investigators, for example Miers and his collaborators, gave evidence for a sharp distinction between the metastable and labile fields (Miers & Isaac, 1907). Others doubt the existence of a sharp boundary between the two contrasted sets of conditions; they consider that there may be no truly metastable region, but only a region where nucleation is extremely slow, followed by the labile region where spontaneous nucleation is at first infrequent, but abundant further within the region (see Buckley, 1951, pp. 7–42 for a summary).

To simplify the discussion of the crystallization history of the chilled marginal gabbro, four regions will be postulated (Fig. 84) that probably pass gradually into each other:

1. A narrow metastable region.
2. A region called here the *early labile* region where nuclei may spontaneously develop, given enough time.
3. A *middle labile* region where abundant spontaneous crystallization readily occurs.
4. A *late labile* region, accepted by all workers, where again there is very slow, spontaneous nucleation.

The arguments developed below, to explain the textures of the chilled marginal gabbro, would not be directly applicable if what is here called the early labile region did not exist, but it may well be in a region which only becomes significant when cooling is slow.

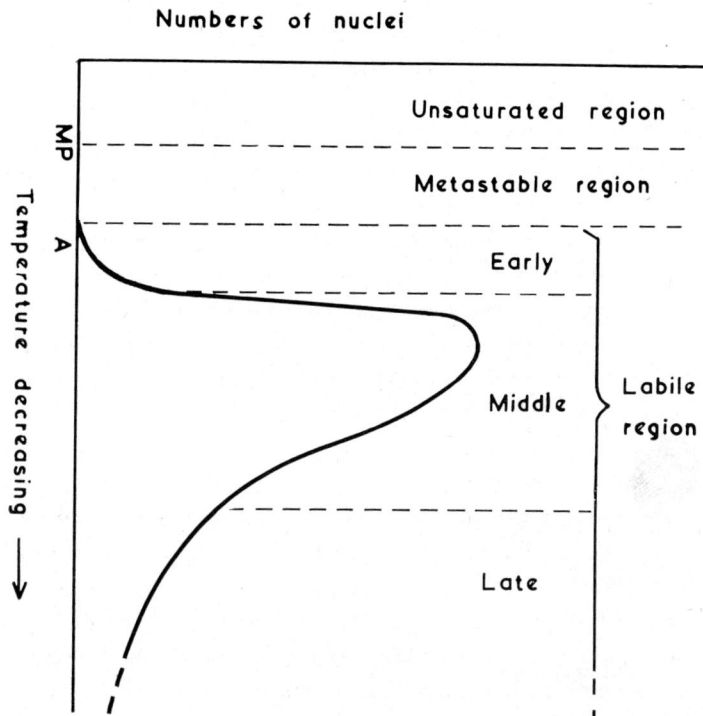

FIG. 84. Diagrammatic representation of the relationship between crystal nucleation and the degree of supersaturation. The suggested subdivision of the Labile region into Early, Middle, and Late stages is indicated.

The abundant, early nucleation of plagioclase and olivine, relative to pyroxene and ilmenite, is the cause of the poikilitic textures of the two latter minerals, seen in the rapidly chilled marginal rock. The plagioclase and olivine presumably nucleated abundantly, because supercooling for these minerals reached into the middle labile region. By the time the temperature was low enough for pyroxene to be a possible phase, the temperature was falling more slowly, partly because of the latent heat supplied by the crystallization of the plagioclase and olivine, and partly because the immediate country rock would have begun to warm up. Thus, the conditions were appropriate for the formation of a few nuclei of pyroxene in the early labile region. The amount of the appropriate substances diffusing to the pyroxene centres, in the considerable time available under the conditions of slower cooling, allowed the few pyroxenes to develop into large, poikilitic crystals. Similar arguments may be used to explain the poikilitic texture of the ilmenite.

The textures and relative abundance of the crystals in the coarse, patchy-pyroxene rocks may be interpreted by using the same hypothesis but assuming slower cooling and thus, less supercooling. The abundant plagioclase crystals suggest that super-cooling for this mineral reached the middle labile region, while the much less abundant olivine crystals suggest that supercooling was not sufficient to reach the middle labile region for this mineral but only the early labile region, where relatively few crystal nuclei of olivine formed. It was believed that this was due to plagioclase having crystallized under equilibrium conditions, from this particular magma, at a somewhat higher temperature than olivine, and this has been confirmed experimentally (see addendum to Wager, 1961, pp. 364–5, and Yoder, Tilley & Schairer, 1963, pp. 80–1). This sequence of events is not seen in the strongly supercooled gabbro at the margin because the temperature of the liquid was carried rapidly down to a point where both plagioclase and olivine were roughly in their respective, middle labile, nucleation regions. It is interesting to note that in the coarser olivine gabbros, where the number of olivine nuclei was not very different from the number of pyroxene nuclei, the olivine crystals tend to grow into the spaces between the plagioclases and to exhibit what may be described as sub-ophitic texture (Fig. 76A).

The manner of formation of the wavy-pyroxene rock remained unsolved until the meaning of the texture of the chilled marginal olivine gabbro was understood. The pyroxenes in this rock are not inward-growing, acicular crystals, comparable with the plagioclases of the perpendicular-felspar rock, but inward-extending patches of poikilitic pyroxene, the crystallographic orientation of which has no particular relation to the elongation of the patches. They are, none the less, believed similar to the elongated felspars in being due to growth from a liquid supersaturated for pyroxene, which occupies the interstices of the earlier plagioclase and olivine crystals. The texture of this rock indicates that plagioclase and olivine nucleated abundantly and grew to a considerable size before any pyroxene formed; then, on further loss of heat, a few pyroxene nuclei developed which grew inwards within the interstitial liquid and formed extensive, poikilitic crystals having whatever orientations the original nuclei happened to have. Thus they tended to become indefinite, poikilitic sheets, set perpendicular to the contact because the approximately vertical front of crystallization moved inwards as heat was lost to the country rocks across the nearby, almost vertical contact.

Some igneous pyroxenes have acicular form—for instance, those in the pegmatitic schlieren of many dolerite sills—and in these cases it is clear that the rate of growth of pyroxene crystals is at a maximum along the c crystallographic axis. It would be possible, therefore, to have acicular pyroxenes like the bladed felspars of the perpendicular-felspar rock, but they do not occur in the Skaergaard marginal rocks. The difference in the habit of the pyroxenes in the two cases is due to the order of crystallization and not to intrinsic differences in the growth of the crystals. The bladed felspars of the perpendicular-felspar rock were probably formed like the elongated olivines of the Rhum crescumulates (Wadsworth, 1961) as a result of abundant, small crystals, with all orientations, forming at the interfaces between the solidified magma and liquid; these grew into the supercooled magma, but at different rates depending on their orientation; those crystals having the fastest direction of growth perpendicular to the interface grew into long, acicular crystals and inhibited the growth of the unfavourably orientated ones. In the case of the wavy pyroxenes, nucleation occurred late and relatively few nuclei developed, so that the kind of competition suffered by the felspars did not occur.

The texture of the wavy-pyroxene rock also seems to indicate some disentangling of the olivine from the plagioclase crystals, during sinking over small distances. Sinking of early olivines seems to have been responsible for the picrite, and some sinking of both olivine and plagioclase, but at different rates, is probably responsible for the small-scale heterogeneity shown by the wavy-pyroxene rock (Fig. 78 & 79). Although slight sinking of crystals occurred during the formation of the tranquil division of the marginal border group, the wavy-pyroxene and the perpendicular-felspar rocks indicate that the magma was essentially stationary during the period of their formation; convective movement began only with the succeeding banded division.

The Banded Division of the Marginal Border Group

The tranquil, outer division of the marginal border group gives place, at about 70 m from the margin on all sides, to rocks showing fluxion banding parallel to the nearby margin (Fig. 72). In rock faces which are approximately vertical, radial sections, the banding is seen to have a constant dip (Fig. 74), but on horizontal surfaces the outcrop of the banding tends to be wavy, with only a general parallelism to the contact (Fig. 85). In three dimensions the banding would seem to be corrugated. Much of the heterogeneity which is seen as banding is due to patches or streaks (Fig. 86 & 87) of coarse acid gabbros, sometimes with centres of granophyre, that have resulted from melting and assimilation of inclusions of the acid gneiss of the metamorphic complex. Other banding is due to slight variation in the relative abundance of the dark and light constituents of the rock or to variation in the grain size (Fig. 85). The banding implies flow, up or down, parallel to the nearby margin. The heterogeneity resulting from contamination by the acid gneiss has made the study of these rocks a difficult and rather uninviting problem. On the other hand, the existence of the gneiss inclusions, in the position in which they are now found, is of importance as they provide evidence for the velocity of the downward currents which are believed to have carried them downwards to their present position (p. 213).

Typical of the banded division along the western margin are the rocks exposed on Ivnarmiut (Fig. 89). The coarse, patchy-pyroxene gabbro of the tranquil division gives place inwards to rocks with rather larger felspars and olivines, and pyroxenes which are more granular than poikilitic. The cores of the plagioclases are An_{66}, while they are An_{77} in the nearby tranquil zone (see Table 3). From this position inwards, there are a series of changes corresponding approximately to those found in proceeding upwards in the layered series. Thus the next specimen examined has plagioclase with cores of An_{62}, and pyroxene occurs as non-poikilitic crystals, roughly similar in optical characteristics,

FIG. 85. The Banded Division of the Marginal Border Group on Mellemö, near centre of island. On the more level surfaces the banding is tortuous while on the steeper faces it is more even. (From Wager & Deer, 1939.)

and presumably in composition, to those of the middle of LZb. Later rocks in the banded division have no olivine, plagioclase of about An_{51}, and abundant magnetite; these rocks thus correspond to the MZ of the layered series. A little further in, on Ivnarmiut, the inner contact is reached; this is a rather indefinite junction which is steep or vertical, and is the result of the layered series being banked against the marginal border rocks (Fig. 88). The layered rocks belong to the cross-bedded belt and are decidedly more melanocratic than those further in. On Ivnarmiut, the mineral

FIG. 86. Acid inclusion in the olivine gabbro of the Marginal Border Group, 30 m from the Outer Contact, Mellemö. Large plagioclase and pyroxene crystals are developed marginally. (The small elongated white patch, 20 cm above the hammer, is lichen.) (From Wager & Deer, 1939.)

assemblage of the layered rocks near the inner contact is anomalous: olivine is present as cumulus crystals and for that reason the rock is regarded as belonging to UZa, but the plagioclase, which is little zoned, is An_{48}, the composition of the plagioclase in the upper part of the MZ. Judging by the felspar composition, the layered rocks which are banked against the border group are only just later in the fractionation series than the innermost border rock. Traced south, the banded division of the marginal border group widens to include not only LZ and MZ types but a narrow, innermost zone in which the plagioclase is An_{45} and iron-rich olivine is abundant, hence corresponding to UZa.

FIG. 87. Elongated streak of coarse plagioclase–pyroxene rock with granophyre patch (top right). All derived, presumably, from an acid-gneiss inclusion.

The layered rocks banked against this rock have cumulus apatite and belong to UZb. The relationships between the banded division types on the west side of the intrusion are shown on the detailed map (Fig. 89 & 90).

The evidence from the eastern margin is, at present, fragmentary. Rocks of the tranquil division give place to the banded division rocks of LZ type and in one place, olivine-free, MZ type rocks were found. The width of the border group on the east is probably much like that on the west.

Along the northern border there is a rather complex tranquil division, including the slumped picrite and occasional examples of patchy-pyroxene gabbro, perpendicular-felspar rock, and wavy-pyroxene rock. The felspars are zoned, usually with cores of about An_{73}, and the olivine is about Fo_{59}. Inwards are layered rocks of the LZa cross-bedded belt, dipping at $30°$ and containing plagioclase with An_{67} cores. There seem to

FIG. 88. The Inner Contact, almost vertical, between the Banded Division of the Marginal Border Group (right) and the UZa layered series (middle and left), at SE corner of Ivnarmiut (see Fig. 89). The dip of the layering can be seen and is roughly parallel to the boards of the boat. Fluxion structures in the border group are parallel to the inner contact with the LS. The layered rocks are more melanocratic than further in, and the layering is rather uneven. (From Wager & Deer, 1939.)

the inclination of the walls. The evidence from the northern border, where the lowest exposed part of the marginal border group is to be seen, indicates that there are no banded division rocks: had they been present they would have been expected to be of

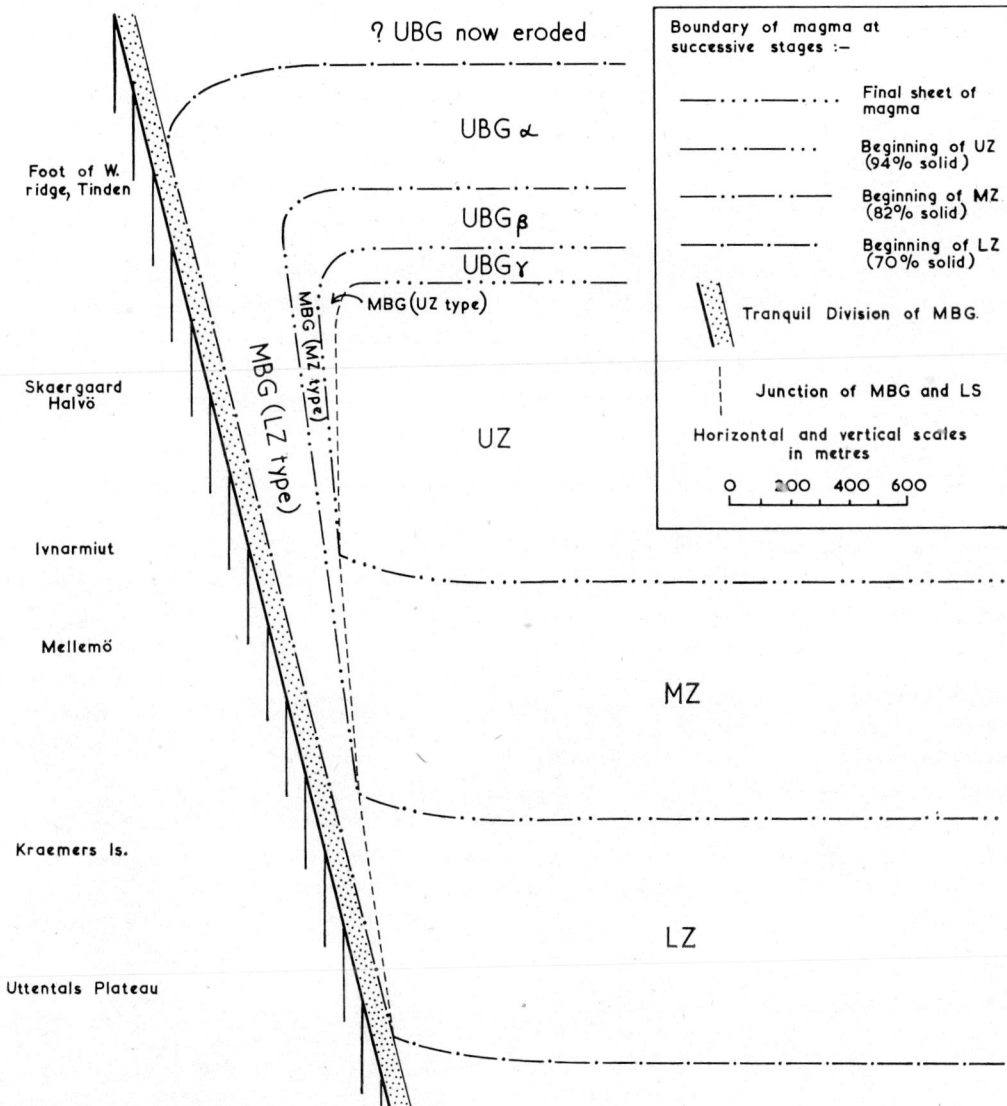

FIG. 90. Deduced disposition of Marginal Border Group types in relation to the Upper Border Group and the Layered Series. Localities providing the evidence for the nature of the Marginal Border Group are given at appropriate levels.

I

hidden zone type. None of MZ or UZ type could exist in this position because the middle and upper zone stages in the evolution of the magma had not been reached, but rocks of HZ type could have been present, and are not. Their absence implies that there was no formation of banded division, marginal border group material when the HZ rocks were forming. Proceeding in a southwesterly direction from the northern contact, to Kraemers Island, banded division rocks of LZ type are found. On passing further southwards, which is equivalent to moving upwards in the intrusion, the LZ types apparently increase in width and the MZ and UZa types of the banded division are successively developed. Finally there is the evidence from the southern margin, at the foot of Tinden, that suggests that there is here a gradual passage from a marginal border group of LZ type to upper border group rocks (UBGα), equivalent in time of formation.

The banded division of the marginal border group is considered to be the result of congelation of successive samples of the residual magma together with some suspended primocrysts. The suspended crystals, which were caught by the congelation process, are similar to those called cumulus crystals when they have settled to the floor. While suspended in the magma they can scarcely be described as cumulus crystals, since they have not accumulated, and it is for this reason that the term primocrysts has been applied to them.

When the changing compositions of the successive Skaergaard residual magmas are estimated in Ch. VII, it will be shown that the marginal border rocks contain a greater amount of crystallized contemporary magma than the layered rocks. The P_2O_5 content of the marginal border rocks may also be used as an indication of the amount of liquid relative to primocrysts. Thus, two banded division rocks (see Fig. 110) have 0·1 and 0·3% P_2O_5, respectively, and must originally have had a greater proportion of liquid to primocrysts than the contemporary average layered rocks, with only about 0·05 to 0·07% P_2O_5. Although the banded division rocks of the marginal border group are not typical cumulates, they have not simply the composition of the contemporaneous liquid because they contain varying amounts of primocrysts. The banded marginal border group rocks approach cumulates in their make-up and composition and they are described, for the present, as *congelation cumulates*. In such rocks, the crystal to liquid ratio is not a function of packing or the amount of adcumulus growth, but is complexly related to the rate of heat loss from the convecting magma.

PLATE IX

The upper 500 metres of Sydtoppen from the west. Dark Upper Border Group rocks show some layering, especially to the left. Beneath these a light horizon of rock consists of an upper 20 m of the Tinden Granophyre Sill and 10 m of the Sydtoppen Transitional Granophyre below.

[PLATE IX to face p. 130]

The Upper Border Group

There are just enough large-scale structures in the upper group for it to be considered as a series of layers lying fairly evenly, one above the other. Layers can be traced laterally by eye, for 2 km in the rock face above Brödretoppen glacier (Pl. IX), and a general dip of 25–30° southward is inferred. This indicates that the layering in the upper border group must have been approximately horizontal before the flexuring.

The structurally highest rocks examined in any detail are on the summits of Brödretoppen and Östoppen. They are coarse gabbros, and the most conspicuous constituents are large, tabular plagioclase crystals, considered to be primocrysts; these are often half a centimetre or more in length, and form 55 to 60% of the rocks. The other constituents are usually sub-poikilitic augite and magnetite, and interstitial quartz or micropegmatite (Fig. 91). Rare layers also contain altered, sub-poikilitic olivine. The inner half of the plagioclase crystals is essentially unzoned and is An_{66} in the highest rock. In most of the rocks there is an irregular arrangement of the tabular felspars but in some there is a marked igneous lamination (Fig. 93) indicating orientation, by flow, parallel to the general direction of the roof. The pyroxene, magnetite, and rare olivine of these rocks are usually interstitial and must have crystallized after the primocryst plagioclase had been orientated.

The most calcic plagioclase in the highest upper border rocks, so far collected, is An_{66} which, by coincidence, is the same as that in the lowest exposed layered rocks. In descending about 150 m on Brödretoppen, the anorthite content decreases to An_{60}: then, after an interval of about 150 m for which no specimens are available, rocks containing plagioclase with An_{40} cores were collected. On the neighbouring peak, Östoppen, 250 m of rocks with plagioclase varying downwards from An_{60} to An_{54} are found. Rather similar, coarse gabbros are encountered along the coast at the foot of Tinden. Here, in moving inwards from the southern margin, which is the same as descending in the sequence of upper border group rocks because of the dip, the plagioclase cores vary from An_{62} to An_{58}, and the rocks broadly resemble those on Brödretoppen except that some have abundant, large grains of olivine and augite which must be regarded as having been primocrysts in the congealing magma (Fig. 92). Interstitially, there is inverted pigeonite, iron ore, and micropegmatite. All those rocks which have a plagioclase composition between An_{66} and An_{54} are classified as UBGα and are considered to have formed simultaneously with the LZ of the layered series.

The typical UBGα rocks contain plagioclase crystals similar to, but larger than those deposited to form the layered rocks; they are also similar to those caught by congelation in the LZ type of the banded division of the marginal border group. At the time of congelation, the primocryst plagioclase crystals were, apparently, surrounded by a considerable amount of magma which crystallized to give the outer zones of the plagioclase and the intercumulus pyroxene, iron ore, micropegmatite and, occasionally, olivine. In a few cases, besides the abundant primocryst plagioclase, there were also primocryst olivines and augites as judged by the textural relations. The composition of the intercumulus augite, which is not conspicuously zoned, has sometimes been shown to be more iron-rich than the cumulus pyroxene of the equivalent layered series rocks and so is roughly similar in composition to the intercumulus pyroxenes of the equivalent layered rocks. Micropegmatite is relatively abundant in these upper border group rocks, whereas micropegmatite is only found in the plagioclase cumulates of the equivalent lower zone of the layered series. Douglas (1961) has shown that throughout the upper border rocks, there is a tendency for the composition of the presumed intercumulus minerals to belong to a more advanced fractionation stage than the primocryst plagioclases or other primocryst minerals of the same rocks. The commonest UBGα rocks are interpreted as consisting of primocryst plagioclases surrounded by the products of crystallization of the contemporary magma, while occasional rocks, especially from the UBGα rocks of the west foot of Tinden, also have olivine and augite as primocrysts. The UBGα rocks differ from the contemporaneous layered series in being usually free from the denser primocrysts, such as olivine and pyroxene, and in having a greater proportion of material derived from the contemporary magma.

Rocks of the upper border group with plagioclase primocrysts ranging from An_{54} to An_{43}, are grouped as UBGβ. They form a zone about 150 m thick on Basistoppen and they also occur on Kilen and Brödretoppen (Pl. XI). The rocks are usually rich in andesine plagioclase (Fig. 94) and can be described as andesinites, as was done originally (Wager & Deer, 1939). The intercumulus material consists of iron-rich augite and 5–10% of quartz and micropegmatite; in some cases there is also a little altered olivine. The primocryst plagioclase corresponds in composition to that of the MZ of the layered series, while the iron-rich, intercumulus augite, and the absence of inverted pigeonite, clearly indicate a more advanced fractionation stage. Near the contact with the Basistoppen sheet, the rocks show evidence of having been altered by the later intrusion. There is some particularly iron-rich, intercumulus augite which may be the result of metamorphism, and the tridymite, now paramorphed by quartz, may be

FIG. 91 (*opposite*)

Coarse felspathic gabbro, typical of the structurally highest UBGα as found on the summit of Brödretoppen (5269≡analyzed 3052). Tabular plagioclase primocrysts tend to be poikilitically enclosed by augite (top). Olivine, partly altered to serpentine, shows a subpoikilitic relationship towards plagioclase (left) but may have been primocrystal. The amount of olivine is, on average, less abundant than shown, and in the structurally lower UBGα rocks has not been found. A mesostasis (lower centre) consists of quartz, alkali feldspar, chlorite, and some apatite; the quartz has an acicular habit typical of paramorphs of tridymite. ×15.

FIG. 92 (*opposite*)

The type of UBGα rock with olivine, from coast at west foot of Tinden (5271, analyzed). Plagioclase, some showing an extreme platy habit, and olivine crystals, partially replaced by iron ore, are apparently primocrysts, while augite and iron ore crystallized interstitially. ×15.

FIG. 91

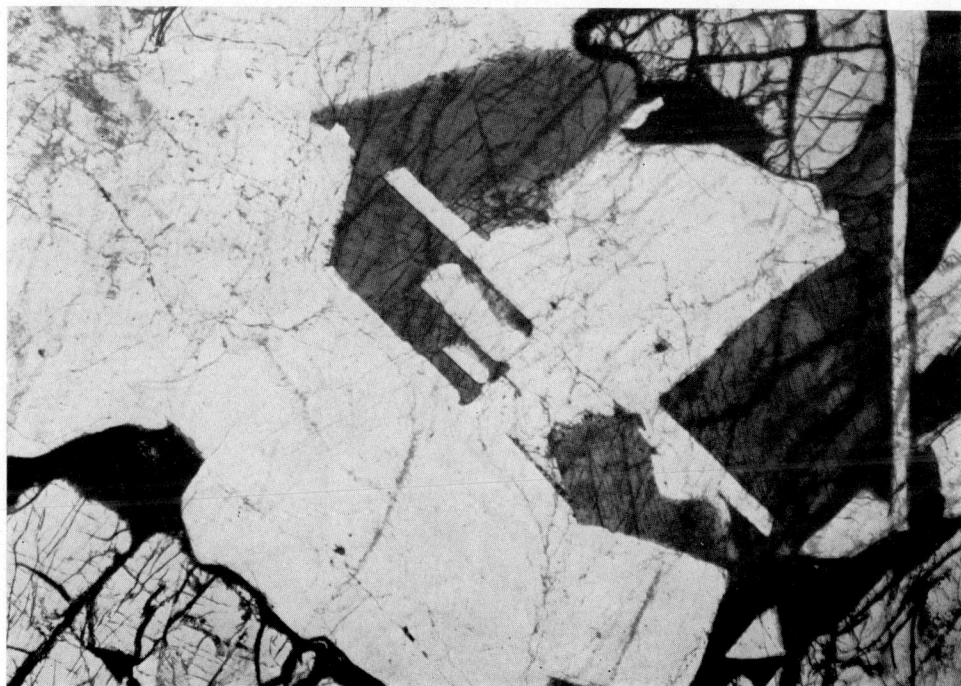

FIG. 92

of similar origin. The rocks are somewhat granulitized, which is perhaps due to the mechanical effects of the intrusion of the sheet.

Below the UBGβ on the west side of Basistoppen, there are about 100 m of rocks containing plagioclases with cores ranging from An_{43} to An_{37}, to thus corresponding to those of the UZ layered rocks; these rocks are grouped as UBGγ. Many of the rocks of UBGγ are best described as pyroxene ferrodiorites (see Fig. 95); others are felspar-rich ferrodiorites. Intimately associated with these ferrodiorites are other rocks richer in quartz and micropegmatite which we here call melanogranophyres. In the ferrodiorites, the amount of quartz and micropegmatite is about 5%, while it is up to 45% in the melanogranophyres where, also, the iron-rich pyroxenes are long and feathery in form (Fig. 96). The field relationship between these rather gradational types is not certain, but we believe that the melanogranophyre magma was produced by gentle filter-press action and that it was injected as irregular, contemporaneous sheets into the UBG rocks which were still at a high temperature. The melanogranophyres

FIG. 93. Strongly laminated gabbro of UBGα (4158) from Östtoppen. At various levels in UBGα, the tabular plagioclase primocrysts appear to have been orientated by convection currents parallel to the roof. Augite, iron ore, and quartz with alkali felspar (bottom left) crystallized interstitially. ×15.

are not necessarily considered to belong to UBGγ but may be related to the granophyres of Tinden, with which they are described in the last section of this chapter.

The pyroxene ferrodiorite contains about 5% of quartz and micropegmatite, occasional fayalitic olivine, ferrohedenbergite, iron ore, apatite and zircon. The texture is variable: the plagioclase occurs generally as rather large, discrete crystals with cores of fairly constant composition, outside which is strong normal zoning; the pyroxene is subophitic to granular; olivine, when present, varies between Fo_{18} and Fo_5; white quartz is mainly intergrown with alkali felspar but it also occurs with a form indicating that it is paramorphic after tridymite. These rocks are iron-rich, and intermediate in relation to silica content (see Table 8 and Fig. 111). On the basis of their primocryst plagioclase compositions, they correspond with the pyroxene ferrodiorites of the UZ of the layered series but, in general, they are considerably richer in interstitial quartz and alkali felspar, and their pyroxenes are rather different (Brown & Vincent, 1963, fig. 2 & p. 185).

Although the collection from Brödretoppen is scanty, and many of the steeper rock

FIG. 94   →

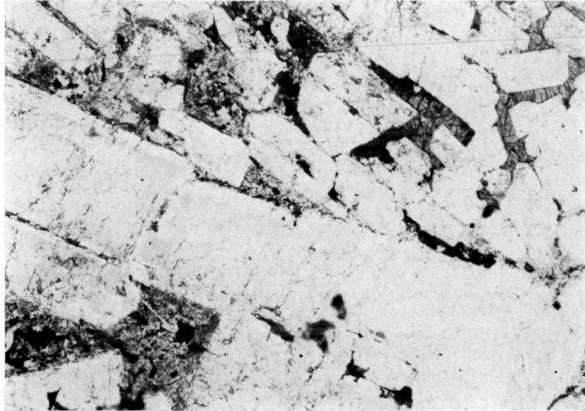

Plagioclase-rich diorite, of UBGβ (5146≡analyzed 4136) from Western Basistoppen.   Plagioclase primocrysts, partially clouded, are surrounded by ferrohedenbergite (top right) and a mesostasis of alkali felspar and quartz, darkened by decomposition products; quartz, in part, has an acicular habit indicating inversion from tridymite. × 15.

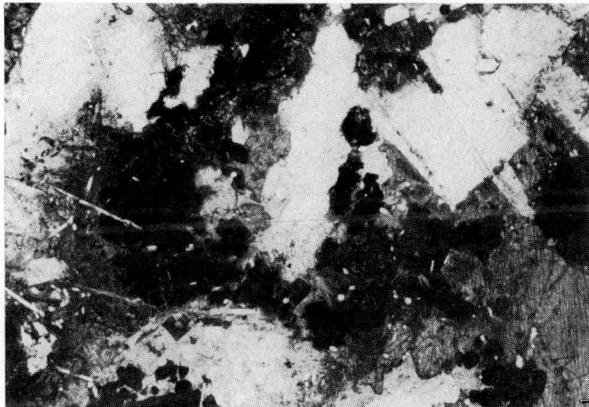

←   FIG. 95

Ferrodiorite of UBGγ from Western Basistoppen (5150, analyzed). Plagioclase primocrysts are seen partly enclosed by ferroaugite (lower right). The fayalitic olivine is clouded with iron ore. The mesostasis of quartz (after tridymite) and alkali felspar (especially lower left) is much darkened by decomposition products. × 15.

faces have not been visited, the evidence is sufficient to indicate that the lower part of UBGα and all of UBGβ are absent on the west ridge by which the mountain has been climbed. Possible explanations are that the missing UBG rocks were not developed here or, if they were, they subsequently broke away and sank in the magma. There is evidence of the latter process, at an earlier stage, in the blocks of UBGα type which are found as inclusions in the middle zone of the layered series on Kraemers Island.

Among the rocks of UBGα, occasional blocky masses of acid rock occur which, by analogy with similar masses in the marginal border group and layered series (Wager & Deer, 1939 are interpreted as altered inclusions of grey gneiss from the metamorphic complex. Besides the blocks which are obvious inclusions, because of their form, there are rafts of intermediate to acid material that are considered to be remelted and hybridized grey gneiss, streaked out parallel to the layering. The mineral phases in these altered inclusions are the same as in the surrounding UBGα gabbros, but the proportions are different. It was originally thought probable that the acid inclusions had considerably modified the adjacent UBG rocks by hybridization. As yet, there are no satisfactory criteria for deciding how much contamination there may have been, but from the general chemistry and petrology of the upper border group rocks (Douglas, 1961) it is thought to have been considerable in the UBGα, and progressively less in UBGβ and UBGγ. Recent work on the isotopic composition of strontium in Skaergaard rocks by Hamilton (1963) also gives evidence of some contamination with the grey gneiss in a specimen of UBGα. Convection currents carried down acid inclusions to contribute to the formation of the banded division, marginal border group rocks of LZ type, and to a lesser extent of MZ type. Gneiss inclusions are also found in the UBGα and to a lesser extent in UBGβ rocks, while by UZ times, when UBGγ rocks were forming, there were, apparently, few distinct acid masses left to be incorporated in the liquid.

For the most part, the UBG rocks are to be regarded as an accumulation of plagioclase primocrysts surrounded by the products of crystallization of the contemporary magma. The amount of contemporary magma varied from layer to layer, but is probably least in the UBGα rocks. The UBG rocks also include, to an unknown extent, the products of melting of acid gneiss blocks; on the whole, however, fractionation rather than contamination is thought to be responsible for their general richness in interstitial micropegmatite.

In a preliminary consideration of the origin of the layered series (p. 98), it has been assumed that plagioclase crystals sank through the magma during the whole period of formation of the layered series. From a consideration of the UBG rocks, it is believed that plagioclase primocrysts contributed largely to their make-up, and the question arises as to how primocryst plagioclase, which sank to give the layered series, could also reach the top of the liquid. The answer seems to be that the plagioclase primocrysts were carried upwards to the top of the liquid by the central ascending convection currents, when the latter had a sufficiently high velocity to carry plagioclase crystals upwards, against their tendency to sink in a slightly less dense liquid. Having reached the top of the liquid, primocryst plagioclases apparently became stuck in the congealing liquid, giving the felspar-rich gabbros of the UBG. The olivine, pyroxene, and iron ore primocrysts, which are only rarely present in the UBG, are considered generally to have sunk through the upward moving current because of their greater sinking velocity. At one time it was thought that the upper border rocks might be true flotation cumulates, the plagioclase being lighter than the basic magma when, in the later stages of fractionation, it had become iron-rich and dense, but this hypothesis runs counter to the evidence of the plagioclase having sunk to the floor throughout the whole period of formation of the layered series.

Granophyric Differentiates—the Melanogranophyres, the Sydtoppen Transitional Granophyres, and the Tinden Acid Granophyres

The most acid of the granophyres, that which forms the Tinden Sill, is a conspicuous, mappable unit emplaced mainly within the lower part of UBGα (Pl. XI). The sill is at its thickest, about 30 m, under Sydtoppen where it is seen to have an unchilled upper contact with UBGα rocks and to pass downwards, without a clear break, into the darker and more basic-looking transitional granophyre of Sydtoppen. The sill thins rapidly northwards, while southward it thins a little and transgresses into the higher parts of the UBGα, where it ends rather abruptly as it cannot be seen extending into the marginal border group.

The Sydtoppen transitional granophyre is exposed below the Tinden acid granophyre on the west face of Sydtoppen (Pl. IX). The contact between the two is not chilled but was not examined in detail. In the field, it was thought that the lower rock might be a basal facies of the Tinden acid granophyre, but it is now considered to represent a different magma injected just prior to the Tinden sill.

The field relations of the melanogranophyres, which largely correspond to the rocks called hedenbergite granophyres in the original paper on the intrusion (Wager & Deer, 1939, pp. 209–13), are even less definite than those of the more acid granophyres. They occur, without chilled contacts, among the UBG rocks as high as the lower part of UBGα on the west face of Brödretoppen, and they are found immediately above the sandwich horizon rocks on Basistoppen. In the field, they were not always distinguished with certainty from the UBGβ and γ rocks, but in chemical and mineralogical composition they seem essentially distinct, and by analogy with the Tinden Sill they are considered to have formed from contemporaneous liquid which was injected here and there into the UBGβ and γ, and on Brödretoppen, into the lower part of UBGα, above the Tinden Sill.

The melanogranophyres, a hand specimen of which is illustrated in Fig. 96, are presumably the earliest rocks of the granophyric differentiates because of their higher temperature minerals. In thin section (Fig. 97 & 98) the plagioclases have cores from An_{38} to An_{36} and strong zoning so that margins down to about An_{15}, or even lower, are conspicuous. Alkali felspar, turbid with decomposition products, occurs as an outer fringe to the plagioclases and has been shown by X-ray methods to be unmixed, and to consist of an albite-twinned sodium-rich, and a potassium-rich phase. The pyroxenes, usually pale green, are elongated and branching. Three chemical analyses (Brown & Vincent, 1963) show that they are ferrohedenbergite but are more calcium rich than the layered series pyroxenes at comparable magnesium–iron ratios (Fig. 24, Ch. III). Iron-rich olivine is present in most of the rocks and is between Fo_9 and Fo_5. Quartz varies in abundance and frequently has a criss-cross, acicular form, indicating that it has inverted from tridymite (Fig. 98).

The Sydtoppen granophyre is transitional in many of its features between the melanogranophyres and the Tinden acid granophyre (Fig. 99). The plagioclase occurs as prismatic crystals containing cores of An_{32}, although most of each crystal is about An_8, probably produced by late stage reactions. The pyroxene is the most iron-rich of any found in the granophyres and is less elongated than that of the melanogranophyres. The groundmass is largely an intergrowth of quartz and alkali felspar. In composition the rock is more acid than the melanogranophyres and more extreme in iron to magnesium ratio (Ch. VII).

The Tinden Sill, in its upper part, is a rather uniform acid granophyre with about 10% of albite phenocrysts (An_{10-5}). The groundmass is a fine-scale intergrowth of quartz and felspar, the latter, from X-ray diffraction measurements, consisting of both potassic and sodic phases. The other constituents, amounting to about 8%, are

Fig. 96 (*opposite*)
Basistoppen-type melanogranophyre (3021) from northwest side of Basistoppen. Ferrohedenbergite and fayalite form elongate, feathery crystals. About natural size.

Fig. 97 (*opposite*)
Melanogranophyre (4322, analyzed) from the west face of Basistoppen, just above the Sandwich Horizon, consisting of sodic plagioclase, ferrohedenbergite, fayalite, apatite, and an abundant mesostasis of quartz and alkali felspar. $\times 15$.

Fig. 98 (*opposite*)
Another melanogranophyre (5152) from west face of Basistoppen, containing a considerable amount of granophyric intergrowth in which a criss-cross pattern of quartz paramorphs after acicular tridymite is dominant. $\times 15$.

Fig. 99 (*opposite*)
Sydtoppen transitional granophyre (4489). Plagioclase occurs in elongated, prismatic crystals which are cloudy with decomposition products (bottom right). The plagioclase is generally An_8 in composition but some crystals contain cores of An_{32}. Ferrohedenbergite occurs in prismatic crystals, fayalite is represented by dark decomposition products, and there is some iron ore. The groundmass is a granophyric intergrowth of quartz and alkali felspar. $\times 15$.

FIG. 96

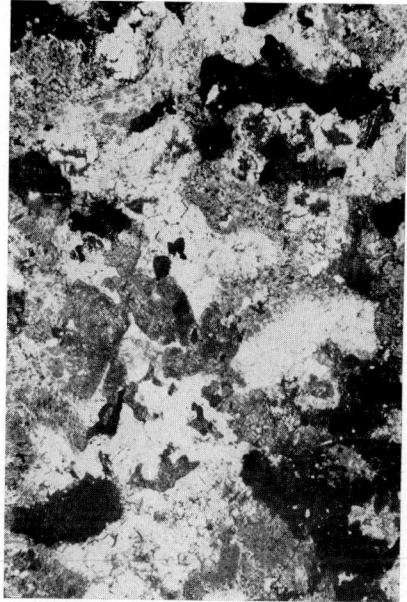

FIG. 97

FIG. 98   FIG. 99

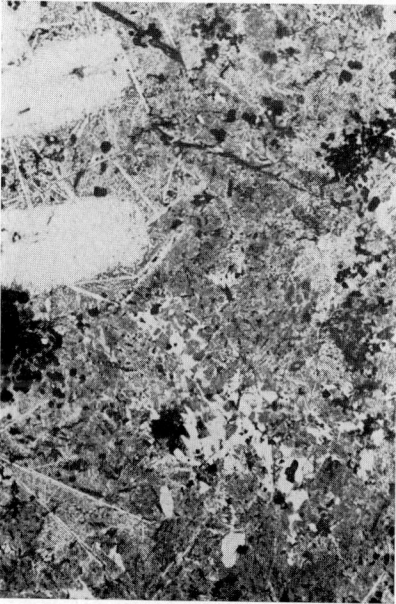

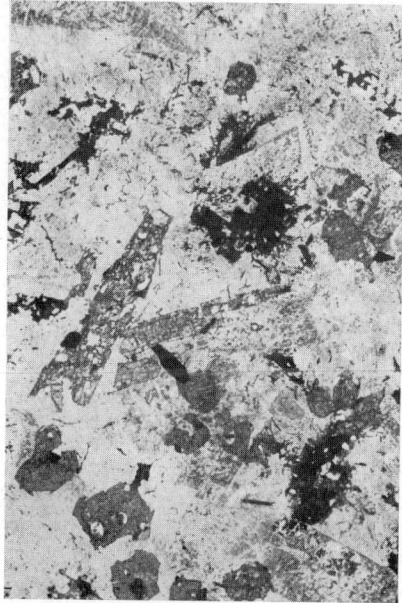

chlorite, calcite, ilmenite, zircon, and sulphides.   Close to the upper contact the
texture is felsitic, but elsewhere the groundmass consists of a granophyric intergrowth,
appearing as polygonal patches in thin section, and having the albite phenocrysts
embedded in it (Fig. 100A, B).   Towards the base of the sheet, the granophyre contains
more plagioclase phenocrysts and more chloritic material.   Chemically, the rock is a
sodic granite in composition.   It is interesting in containing considerable amounts
of iron sulphide, up to 5% in some specimens.   The sulphide is either pyrrhotite in
irregular patches, half a centimetre across, or pyrite in well-shaped crystals (Wager,
Vincent & Smales, 1957).

The acid granophyre of the Tinden Sill, the transitional granophyre of Sydtoppen,
and the melanogranophyres are believed to have originated within the intrusion because
they show no chilling at their margins, and no granophyres are found in the surrounding
country rocks.   It seems likely, therefore, that these rocks were produced by filter-press
action acting on the partly solidified UBG rocks and the partly melted gneiss inclusions,
and, possibly, on the upper part of the layered series which contains some micropegmatite of presumably low crystallizing-temperature.   In several of the specimens of
melanogranophyre, and in the Sydtoppen transitional granophyre there are contemporaneous acid veins, produced by local filter-press action, that approach the acid

A   B

FIG. 100.   Typical transgressive leucogranophyre from the Tinden Sill (A,
ordinary light;   B, partly-crossed nicols).   Extensive, somewhat decomposed
plagioclases, about An_5, are set in a granophyric groundmass.   Ilmenite and
brown, semi-opaque clots of late-stage chlorite are the chief dark minerals. $\times$ 15.

granophyre of the Tinden Sill in mineralogical composition. It is suggested that some coastal flexuring, occurring at the appropriate time, produced a filter-press liquid which was emplaced in a region of lower compressive stress. The existence of some coastal flexuring around this time is indicated by the somewhat phacolithic form of the Tinden Sill and the even more definitely phacolithic form of the Basistoppen sheet, which was injected immediately after the Tinden Sill (see next chapter).

Transgressive acid veins with unchilled margins, from a few inches to many yards

FIG. 101. Unchilled granophyre veins cutting MZ gabbro, west side of Sound on Kraemers Island. The rock face is about 50 m high.

in thickness, are also found, sporadically, in other parts of the intrusion, especially in the MZ and UZa of Kraemers Island and on the western flanks of Pukugaqryggen and Basistoppen (Fig. 101). Petrologically and chemically, these granophyre veins are variable but some resemble the granophyre of the Tinden Sill and a transgressive, 2-ft vein in the marginal border group of the Skaergaard Peninsula resembles the melano-granophyres. It may be that the flexuring, which is believed to have produced the granophyric liquids by filter-press action on the still partly liquid UZc and UBG rocks, also produced fractures in the largely consolidated MZ and UZa rocks, which then became filled with any granophyric liquid available at the time.

The ultimate origin of the material forming the granophyres is still an open question. Although the original Skaergaard magma had abundant normative olivine, it yielded, under conditions of strong fractionation, a micropegmatitic residuum which occurs as a normal interstitial constituent of the ferrodiorites, and even of cumulates as low as those of the LZ when these happen to be free from cumulus forsteritic olivine (p. 60). The granophyre magmas are, therefore, regarded as possible products of the crystal fractionation of the Skaergaard magma.

Micropegmatite is more abundant in the UBG rocks than in the layered series, however, and whereas some of this is likely to be due to fractionation of the original magma, some may be the result of melting of the intermediate and acid gneiss inclusions. The more standard petrological and geochemical methods have provided no clear evidence on the question of how much of the micropegmatite of the upper layered series, the UBG, and the granophyres is derived from melting of inclusions of the metamorphic complex and how much is from crystal fractionation of the original basic magma. However, when the normative amounts of quartz, albite and orthoclase of the granophyric differentiates are plotted on Tuttle and Bowen's experimental $Or–Ab–SiO_2–H_2O$ diagram (1958) for an isobaric section at 1000 bars water pressure (Fig. 102), the points for the Skaergaard rocks all lie in the felspar field and on the albite side of the line defining the felspar thermal valley (see plot by Carmichael, 1963, fig. 9). The granophyres contain albite phenocrysts, which is in harmony with their plotting in this part of the felspar field. The fact that the points lie on the albite side of the felspar thermal valley further suggests that they were derived from basic magma by crystal fractionation at a relatively high level. In this connection, it should be noted that the one available analysis of the acid gneiss of the metamorphic complex lies on the orthoclase side of the felspar thermal-valley line. The trend of points on Carmichael's plot suggests a fractionation sequence, roughly from the sandwich horizon, through the melanogranophyres, to the Sydtoppen transitional granophyre and the Tinden acid granophyre, which is the sequence suggested by their mineralogy.

Oxygen isotope studies by Taylor and Epstein (1963), summarized on p. 185 indicate first that the Sydtoppen transitional granophyre is likely to be mainly the result of fractionation of the original Skaergaard magma, and secondly that the Tinden acid granophyre could be accounted for, in the present state of our knowledge of oxygen isotope fractionation processes in magma, either by crystal fractionation of the original Skaergaard magma or by contamination with the more acid part of the metamorphic complex. On the other hand, the isotopic composition of strontium from Skaergaard rocks (Hamilton, 1963), also discussed later (p. 187), shows that, despite various unexplained fluctuations, there are high values of the Sr^{87}/Sr^{86} ratio among the Tinden acid granophyres which suggest that at least some of the strontium has been derived from the intermediate and acid parts of the ancient metamorphic complex of the region in which Sr^{87}, produced by the decay of Rb^{87}, is relatively abundant.   Thus the strontium isotope data seems to suggest that an appreciable part of the acid granophyre material has originated by remelting of acid gneiss inclusions, while the oxygen isotope data indicate that the transitional granophyre is probably a product of crystal fractionation of the original basic magma and the acid granophyre is probably similar, although some contamination by acid gneiss is not excluded.

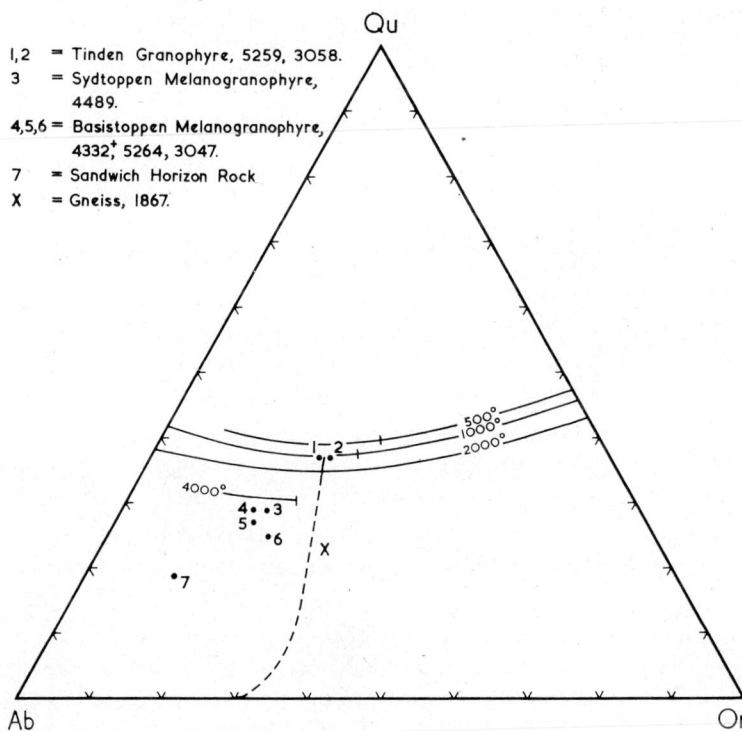

FIG. 102. Normative amounts of quartz, orthoclase and albite of the analyzed granophyres and Sandwich Horizon rocks, plotted on the Qu–Or–Ab triangular diagram. On the diagram the projection of the quartz–felspar boundary at various pressures (kg/cm²) of H_2O, and the felspar thermal valley at 1000 kg/cm² are shown from Tuttle and Bowen (1958, fig. 30). The positions of the plotted points suggest that the granophyres are the result of fractionation of basic magma (cf. Carmichael, 1963).

CHAPTER VI

THE BASISTOPPEN SHEET

An extensive slab of basic rocks, at least 4 km long and 500 m thick, is found within the upper border group (Pl. XI), and from the beginning it has been considered that it could not be part of the Skaergaard intrusion proper because of the rock types present (Fig. 103). In the original memoir (Wager & Deer, 1939) it was explained as a sunken raft of extraneous rock which pre-dated the intrusion. This view was put forward because veins of andesinite of UBGβ type were found penetrating the lower part of it. However, it was difficult to conceive how this relatively thin sheet, presumably part of a differentiated sill because it closely resembles certain sills found cutting the basalt outside the intrusion, became detached from its surroundings, and sank into the intrusion without breaking up. In an essay review of the original Skaergaard memoir, Professor L. Hawkes (1940) accepted most of the general findings on the intrusion but regarded the idea of the raft inclusion as an unlikely story.

Further petrographic work by C. J. Hughes (1956) on a series of rocks from the supposed raft showed that there was strong cryptic variation in the plagioclases and pyroxenes and some discontinuous, or phase-change cryptic layering. These features suggested that the rocks are part of a gravity-differentiated sheet but not part of the Skaergaard intrusion because of the significantly different compositions of the associated mineral phases. J. A. V. Douglas has recently re-examined all available rocks, and re-visited a critical area in the field in 1965. He has arrived at the conclusion that what had previously been considered to be a gigantic, raft-like inclusion is, in fact, a later intrusion, into the Skaergaard intrusion and differentiated in position (Douglas, 1961 & 1964). Marginal fine-grained rocks, previously considered due to granulitization as a result of metamorphism, are now interpreted as having been produced by some degree of chilling, and the andesinite veins intruded into the sheet are considered as back-veining, the result of rheomorphism of the andesinites of the UBGβ and γ into which the sheet is locally injected. Some of the unusual phenomena which confused the original interpretation of these rocks are now realized to be due to the sheet having been injected into the Skaergaard intrusion when it was solid but still at a high temperature.

Evidence of cryptic variation in the Basistoppen sheet has been obtained from various series of rocks collected at widely separated localities and is summarized in Fig. 103 (after Douglas, 1961). Excluding the border facies, the lowest rocks are picrites which are, however, impersistent in their occurrence, being absent, for instance, on the higher part of Basistoppen. The picrites are olivine–chromite cumulates (Fig. 104A), the olivine being about Fo_{80}; the intercumulus, poikilitic minerals are plagioclase (about An_{75}), augite, and bronzite.

The picrite passes up into a bronzite-gabbro zone, in places 125 m thick. Olivine is no longer present, and cumulus bronzite (En_{75}) has appeared in its place; plagioclase, An_{73}, first appears as a cumulus mineral a little after the incoming of cumulus augite and bronzite; and cumulus iron ore comes in just below the top of the bronzite-gabbro zone. The rocks are rhythmically layered, the more average rock being shown in Fig. 104D; a layer rich in cumulus augite in Fig. 104B; and a thin layer with large

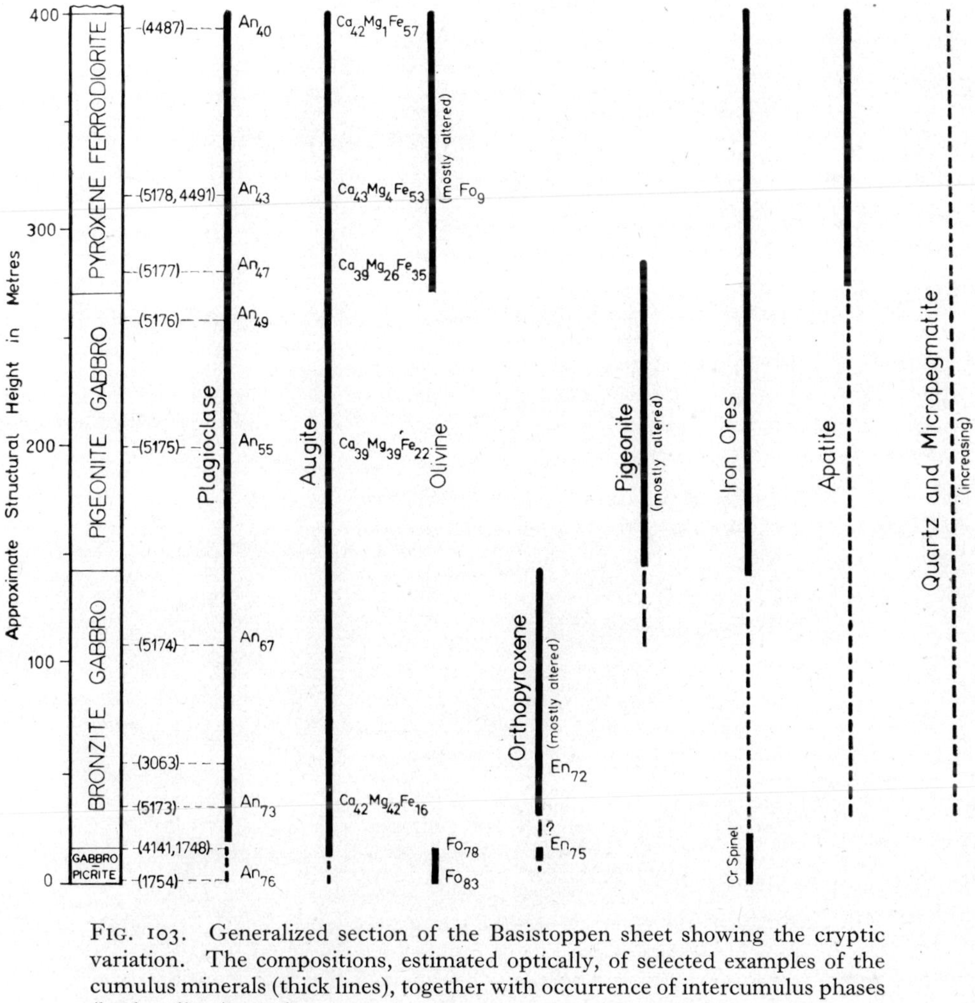

FIG. 103. Generalized section of the Basistoppen sheet showing the cryptic variation. The compositions, estimated optically, of selected examples of the cumulus minerals (thick lines), together with occurrence of intercumulus phases (broken lines) are shown.

K

cumulus bronzites in Fig. 104*C*. Igneous lamination is a feature of some of the rocks, particularly after the incoming of cumulus plagioclase.

Above the bronzite-gabbro zone there is a pigeonite-gabbro zone. Here, cumulus pigeonite has crystallized in place of the cumulus bronzite. Although both minerals are often completely altered, they can still be identified by augite exsolution and by alteration characteristics. In this zone the plagioclase ranges from An_{60} to An_{50} and iron ore is very clearly a cumulus phase (Fig. 104*E*). Interstitial micropegmatite, first found near the top of the preceding zone, is here common.

The uppermost zone is of pyroxene-ferrodiorite and has, as cumulus phases, plagioclase ($An_{49\cdot40}$), olivine (mostly altered, but fresh in two rocks where it is Fo_9), clinopyroxene ($Ca_{40}Mg_{21}Fe_{39}$ to $Ca_{42}Mg_1Fe_{57}$, from optics), iron ore, and apatite. A typical pyroxene-ferrodiorite is illustrated in Fig. 104*F*. At times, rhythmic layering was seen in the field (Fig. 105), and light coloured, contemporaneous acid veins of granophyre occur, not uncommonly, in the ferrodiorite (also Fig. 105).

The top junction between the sheet and Skaergaard rocks has not so far been examined in the field but must be present on the steep, northwest face of Brödretoppen. The lower contact, however, is seen on the west and east faces of Basistoppen and on

FIG. 104 (*opposite*)

Examples of main rock types from the Basistoppen sheet. All × 15. The approximate structural heights in the sheet of most of these specimens are given in Fig. 103.

(*A*) Picrite (1511) from Basistoppen col. Cumulus olivine and chrome-spinel. Intercumulus minerals are sub-poikilitic orthopyroxene and augite, and poikilitic plagioclase (all the plagioclase shown in the photograph forms one crystal).

(*B*) Pyroxene-rich layer (1749) overlying the picrite, from south side of upper part of Basisgletcher. The cumulus minerals are augite (partially schillerized) and chrome-spinel. The chief intercumulus minerals are plagioclase and orthopyroxene (partly replaced by bastite).

(*C*) Bronzite gabbro (3063) from near base of Skillenunatak.

(*D*) Bronzite gabbro (5174) from south side of Skillenunatak. Cumulus minerals are plagioclase, augite, and orthopyroxene. A large, prismatic, bastite pseudomorph (lower left) is mantled by another alteration product, possibly after pigeonite. The iron ore is intercumulus.

(*E*) Pigeonite gabbro (4474) from south side of Kobbernunatak. The cumulus minerals are plagioclase, augite, pigeonite (largely altered to chlorite), and iron ore. The mesostasis consists of cloudy micropegmatite, chlorite, and small apatites.

(*F*) Ferrodiorite (5178) from near top of Skillenunatak. The cumulus minerals are plagioclase, iron-rich augite and olivine (largely altered to chlorite and iron ore), iron ore, and apatite. The interstitial micropegmatite is cloudy with decomposition products.

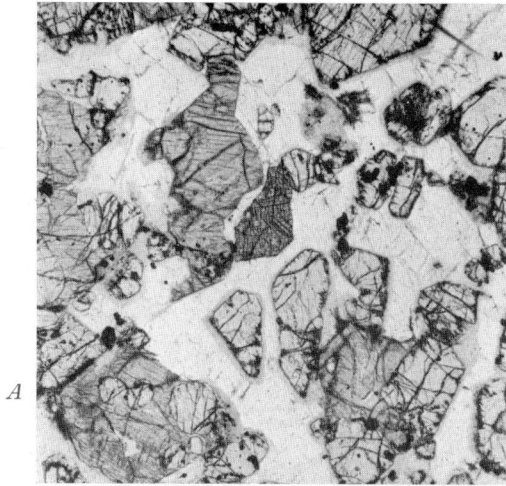

A

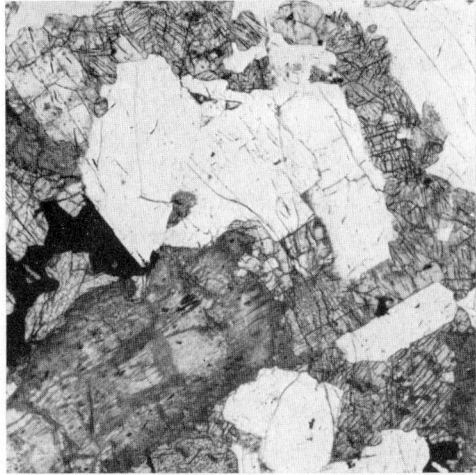

D

B

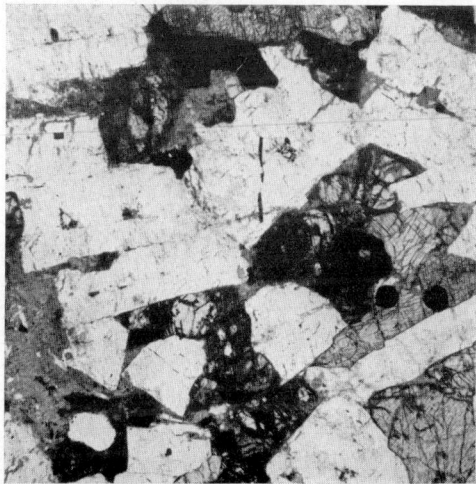

E

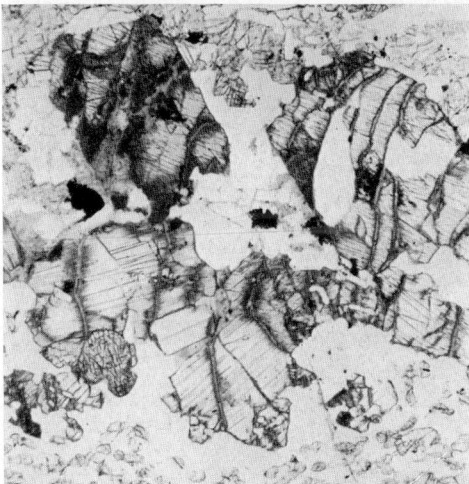

C

F

FIG. 104

Nunatak I. The contact rock is fine-grained, the felspar averaging 0·5 mm in length. It consists mainly of plagioclase (about An_{46}) and augite. The fine-grained rocks are here only a few tens of centimetres thick and they are seen passing into medium-grained, bronzite-gabbro. Cutting these rocks and the overlying bronzite-gabbro are small, irregular veins, without chilled margins (Fig. 106), having relatively large plagioclase crystals and sometimes much iron ore. The plagioclase crystals of the veins are ten times the size of those in the fine-grained border rocks, and are about An_{55} in composition. The large plagioclases are similar in composition and size to those of adjacent andesinites of UBGβ and γ, and the veins are regarded as rheomorphosed UBGβ and γ rocks which have been injected into, and in places mixed with, the fine- and medium-grained marginal rocks. The rheomorphosed UBG material was only partly remelted; in particular the large andesines are, apparently, the unmelted plagioclases of the UBG rocks. In some cases, the plagioclase-rich veins can be seen to change into more magnetite-rich veins towards their extremities; the ore-rich terminations, sometimes also rich in zircon, are considered to have been derived from the felspar-rich parts by filter-press action contemporaneous with their injection.

The full extent of the Basistoppen sheet cannot be seen because it is covered by ice to the northeast, and by sea to the southwest. On Basistoppen, the sheet is transgressive to the layering of the Skaergaard rocks and this may be generally the case. It is pictured

FIG. 105. Slight rhythmic layering in a rock face of ferrodiorite at the top of Kobbernunatak ridge. One of the not uncommon, contemporaneous, transgressive veins of granophyre is seen to the right.

as injected, somewhat transgressively, in response to stresses set up by the continuing earth-flexuring process (cf. Fig. 6, Ch. II). The thickness of the sheet and of the various zones probably varies from place to place; in particular, the thickness of the pyroxene-ferrodiorite relative to the lower zones may have been enhanced considerably in the parts which have been examined, otherwise the overall chemistry cannot be properly understood. It is probable that the northeast-trending macrodyke, a part of which is shown on the map (Pl. XI), represents an extension of the sheet.

The sheet is remarkable in showing marked cryptic layering, the rocks varying from a picrite having Fo_{80} and An_{76} in the lower part, to a pyroxene-ferrodiorite with an olivine at least as low as Fo_9 and a plagioclase of An_{43} in the upper part. The strong iron enrichment suggests that the initial magma had a low oxidation state and, therefore, was similar to, but certainly not identical with, the Skaergaard initial magma. Such strong differentiation in a sheet only about 500 m thick has not, to our knowledge, been found so well developed elsewhere. The reason for the strong fractionation of the Basistoppen sheet may be that the magma was injected into the still-hot Skaergaard intrusion and, as a result, underwent exceptionally slow cooling which provided enough time for fractionation by crystal settling.

FIG. 106

Photomicrograph of a large thin-section of the fine-grained marginal gabbro (4126) of the Basistoppen sheet on Nunatak *I*. The veins of plagioclase and iron ore are believed to be derived by rheomorphism of the Skaergaard ferrodiorite into which the sheet is here injected. × 1.6.

CHAPTER VII

THE CHEMISTRY OF THE SKAERGAARD INTRUSION

If cumulus crystals are separated from the rocks and have little, or a roughly constant amount of the lower temperature zones adhering to them, the chemical analyses give smooth changes in composition when plotted against height, as has been particularly well shown for the augites (e.g. Fig. 19). On the other hand, chemical analyses of whole rocks cannot be expected to show any such simple relationship to the stage of fractionation because the rocks contain variable proportions of the available cumulus minerals, and there is a variable amount of pore material, from trapped contemporary magma. Analyses of the Skaergaard rocks, therefore, have little value unless they are interpreted in terms of (*a*) the composition of the cumulus crystals, enlarged sometimes by adcumulus growth, and (*b*) the composition of the contemporary magma.

To understand the chemistry of the intrusion it is therefore necessary to know as much as possible about the composition of the continually changing residual magma as well as the composition of the primocrysts. Rocks representing quenched samples of the successive residual liquids might have been available if, for instance, the magma had been periodically ejected from an overlying volcano, or injected into the country rocks to form dykes or sills; unfortunately, no rocks in the area have yet been shown to have this kind of connection with the intrusion. Apart from the composition of the initial Skaergaard magma, obtained by the analysis of the chilled marginal olivine gabbro, the composition of the successive residual magmas, which had only an ephemeral existence, can be estimated only by indirect means.

COMPOSITION OF THE INITIAL SKAERGAARD MAGMA

The chilled marginal olivine gabbro, already described (p. 106), is regarded as the quenched, initial magma. Against the metamorphic complex on the northern and western sides, there has apparently been contamination and the chilled rock, where it occurs adjacent to the basalts at the southern margin, has, therefore, been preferred for analysis. The basalts of the country rock have been thermally metamorphosed to granulites, and the opportunities for transfer of material into, and out of, the chilling Skaergaard magma, either mechanically or by diffusion, seem to have been small except, perhaps, in the case of water which is discussed below. On the other hand, contamination clearly occurred when the melting temperature of the country rock was lower than the temperature of the initial Skaergaard magma, as is the case where the intrusion cuts through the more acid parts of the metamorphic basement complex. Although, at the present level of erosion, the chilled olivine gabbro of the southern margin of the intrusion is against basalts, some 1000 m below this level the intrusion must be cutting through the metamorphic complex and some contamination may have taken place there. It does not seem likely, however, that contamination by acid gneiss at the lower levels would affect the composition of the magma chilled against the basalts at higher levels, because of the large bulk of the intrusion and the turbulent movements in the magma which must have accompanied the sudden emplacement; thus the chilled rock, where against basalt along the southern margin, is believed to be a satisfactory representative, in most respects, of the composition of the original magma.

One specimen of the chilled olivine gabbro (EG.4507), one metre from the basaltic country rock, has been subjected to a very full analytical study (Table 4). To represent the initial composition of the Skaergaard magma, the latest and more complete analysis of the marginal rock, EG.4507, is preferred to the average of the three analyses in Table 7. To provide a standard of comparison for the trace element data, the values for the standard diabase (W1 from New Jersey) are also given. Because the standard diabase is considerably less basic than the Skaergaard chilled gabbro the trace elements, in many cases, differ considerably.

In general composition, the Skaergaard initial magma is basaltic, with relatively low contents of silica and alkalis, relatively high alumina, and a strongly reduced state. Classification of primary basaltic magmas is a subject for much controversy nowadays but the original division of the Hebridean basalts into two main types by Bailey *et al.* (1924) has continued to receive support from recent studies such as those of Yoder and Tilley (1962). According to the latter the tholeiitic (Non-Porphyritic Central Type) and alkali (Plateau Magma Type) basalts can be distinguished by the SiO_2 relative to the total alkalis ($Na_2O + K_2O$), by the normative compositions, and by the fractionation products. In addition, the Hebridean Porphyritic Central Type basalt of Bailey *et al.*, with a high Al_2O_3 content, and called High-alumina basalt by Kuno (1960), may exist as a fairly distinct type of which the Warner basalt of the Medicine Lake Highlands is an aphyric example. The Skaergaard initial magma has the general chemistry of a tholeiitic basalt when plotted on a SiO_2–alkalis diagram (Tilley, 1950). In terms of normative constituents, it contains neither nepheline nor quartz, but both olivine and hypersthene (Table 7). In this respect, according to Yoder and Tilley's classification (1962, p. 352) it would be termed an olivine tholeiite. Significantly, both a Ca-rich and a Ca-poor pyroxene are present in the chilled border rock and in the layered series cumulates, rather than the single diopsidic pyroxene characteristic of alkali basalt derivatives. A wide variety of basalt magmas have tholeiitic affinities, including some with a high content of Al_2O_3 resulting in high normative plagioclase, and the Skaergaard magma, which is relatively rich in Al_2O_3, should, perhaps, be termed a high-alumina, olivine tholeiite. However, the classification of basalt magmas seems to us to be still problematical, and from the point of view of the behaviour of the Skaergaard magma during fractionation, the high ratio of $Fe^{++}/Fe^{+++}+Fe^{++}$, indicating a highly reduced state, is another very significant feature of its composition.

TABLE 4

Analysis of chilled marginal gabbro (E.G.4507), 1 m from southern margin, Skaergaard intrusion

Major element analyses by E. A. Vincent; trace elements, by various other workers, referred to where the elements are dealt with systematically (pp. 190–203). For comparison of the trace elements, the values for W1, the Triassic diabase, Fairfax County, Va., U.S.A. (Fleischer & Stevens, 1962) are given.

Major elements as percentages of oxides	Skaergaard chilled gabbro	W1
SiO_2	48·08	52·46
Al_2O_3	17·22	15·03
Fe_2O_3	1·32	1·41
FeO	8·44	8·74
MgO	8·62	6·62
CaO	11·38	10·96
Na_2O	2·37	2·07
K_2O	0·25	0·64
H_2O^+	1·01	0·53
H_2O^-	0·05	0·16
P_2O_5	0·10	0·14

Trace elements as parts per million listed in order of increasing atomic number, with values for W1 for comparison.

	Skaergaard	W1		Skaergaard	W1
S	55	n.d.	Pd	0·018	0·019
Sc	26	33	Ag	0·11	0·06
Ti	7020	6400	Cd	0·13	0·3
V	190	240	In	0·054	0·064
Mn	1240	1320	Sb	0·13	1·1
Co	56	38 or 51	Cs	0·10	1·08
Ni	193	82	Ta	0·46	0·7
Cu	126	110	W	0·22	0·45
Zn	132	82	Au	0·0046	0·007
As	0·30	2·2	Tl	0·047	0·13
Rb	5·8	2·2	U	0·2–0·3	0·52
Sr	267	220 or 175			

Major Element Variation

It is probably not possible to make a logical distinction between major and trace elements for the whole range of igneous rocks. The major elements are here taken as those which, on the whole, are more abundant than 0·1% but it is unsatisfactory to define major elements as those more abundant than a specific value, as it is unsatisfactory to distinguish between a mountain and a hill by saying that the former is higher than a thousand feet, without taking into consideration the general level of the region. In a particular intrusion, it may be convenient to consider that the major elements are those essential for the building up of the crystal phases found in the rocks, while elements which substitute for the major elements, and do not cause a separate crystal phase to form, are the trace elements. Using these criteria for the Skaergaard intrusion, O, H, Si, Al, Fe, Mg, Ca, Na, K, Ti, P, S, Cu, Cr, and Zr would be major elements. We do not propose, however, to use this grouping in its entirety but to consider S, Cu, Cr, and Zr as trace elements although they are essential for the production of the rare pyrrhotite, chalcopyrite, bornite, chrome-spinel and zircon, which occur as distinct crystal phases in the Skaergaard rocks. Hydrogen is also grouped with the trace elements. The variation in the major elements tends to be interpreted in terms of the composition, temperature and pressure of phase equilibrium assemblages, while the variation of the trace elements is interpreted by means of their ionic properties in relation to the structures of the crystalline phases formed from the major elements.

a) *Amounts of the major elements in the exposed layered rocks*

All satisfactory analyses of the layered and sandwich horizon rocks, together with their modes and norms, are given in Table 5 and are shown, plotted against height, in Fig. 107, 108 & 109. The points for the layered rocks which, as judged in the field, are approximately average, are indicated separately from those for leucocratic and melanocratic layered rocks. Two sets of curves have been drawn; one for the average cumulates and the other for the plagioclase-rich cumulates. The three analyzed melanocratic cumulates do not form a coherent group; one of them (5108) is an olivine cumulate with a moderate amount of pore material, another (2308) is a magnetite-rich mesocumulate with only small amounts of pore material, and the third (5322) is an olivine–augite–magnetite mesocumulate, also with little pore material. No useful curves can be put through points for such diverse cumulates.

Even for average rocks, the curves cannot be smooth because of the sudden beginning or cessation of crystallization of certain primocryst minerals during the fractionation. When magnetite, for example, becomes a primocryst, the amounts of FeO, Fe_2O_3 and TiO_2 must increase abruptly, and in Fig. 108 the curves for these elements have been broken at the appropriate level, as also has the curve for P_2O_5 (Fig. 109) at the level where apatite becomes a primocryst. There should certainly be breaks in the curves of the other elements and there should also be sudden changes of slope (cf. Bowen's

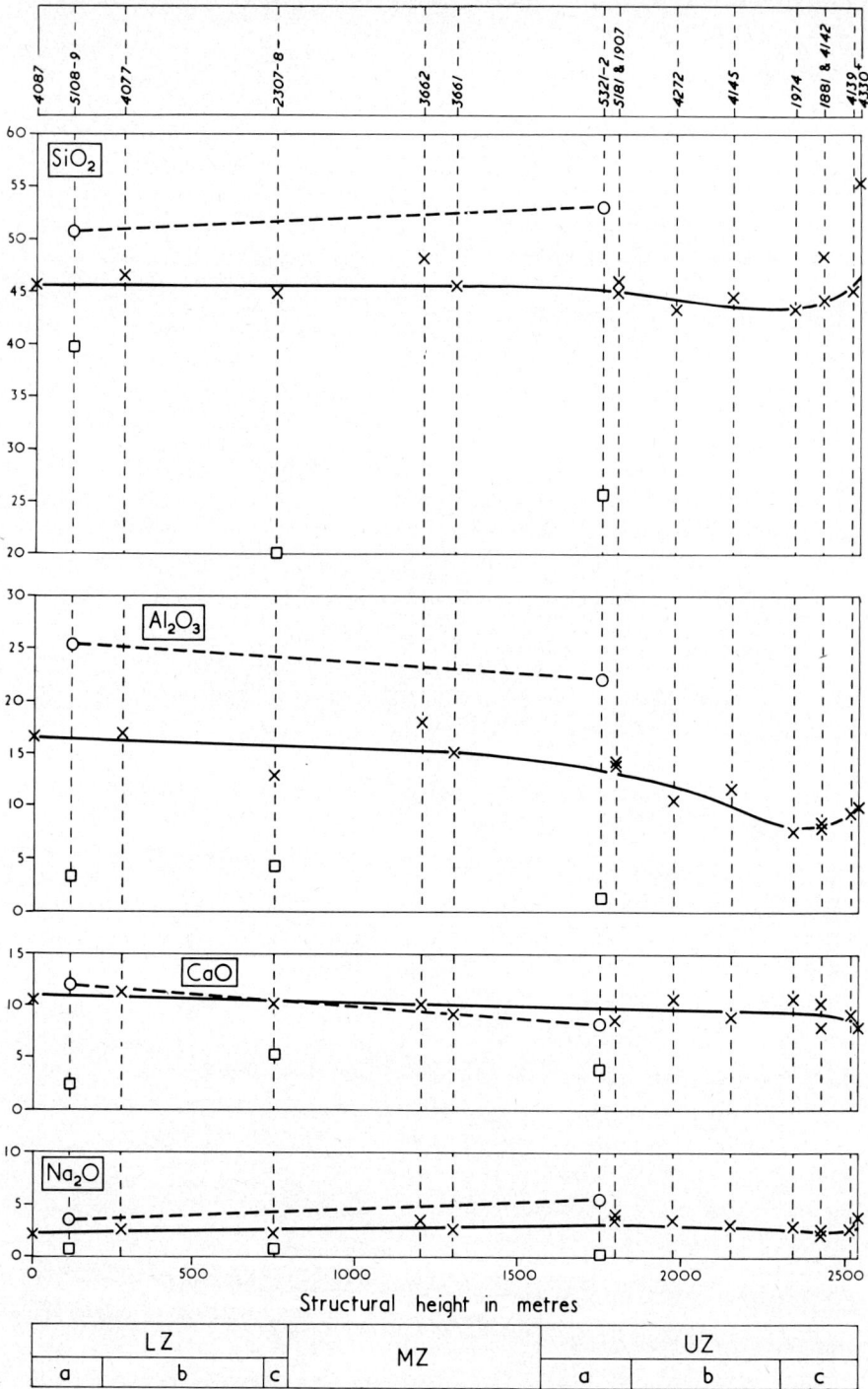

FIG. 107. Chemical variation of the observable layered series plotted against height for : SiO₂, Al₂O₃, CaO, and Na₂O. (From Wager, 1960.)

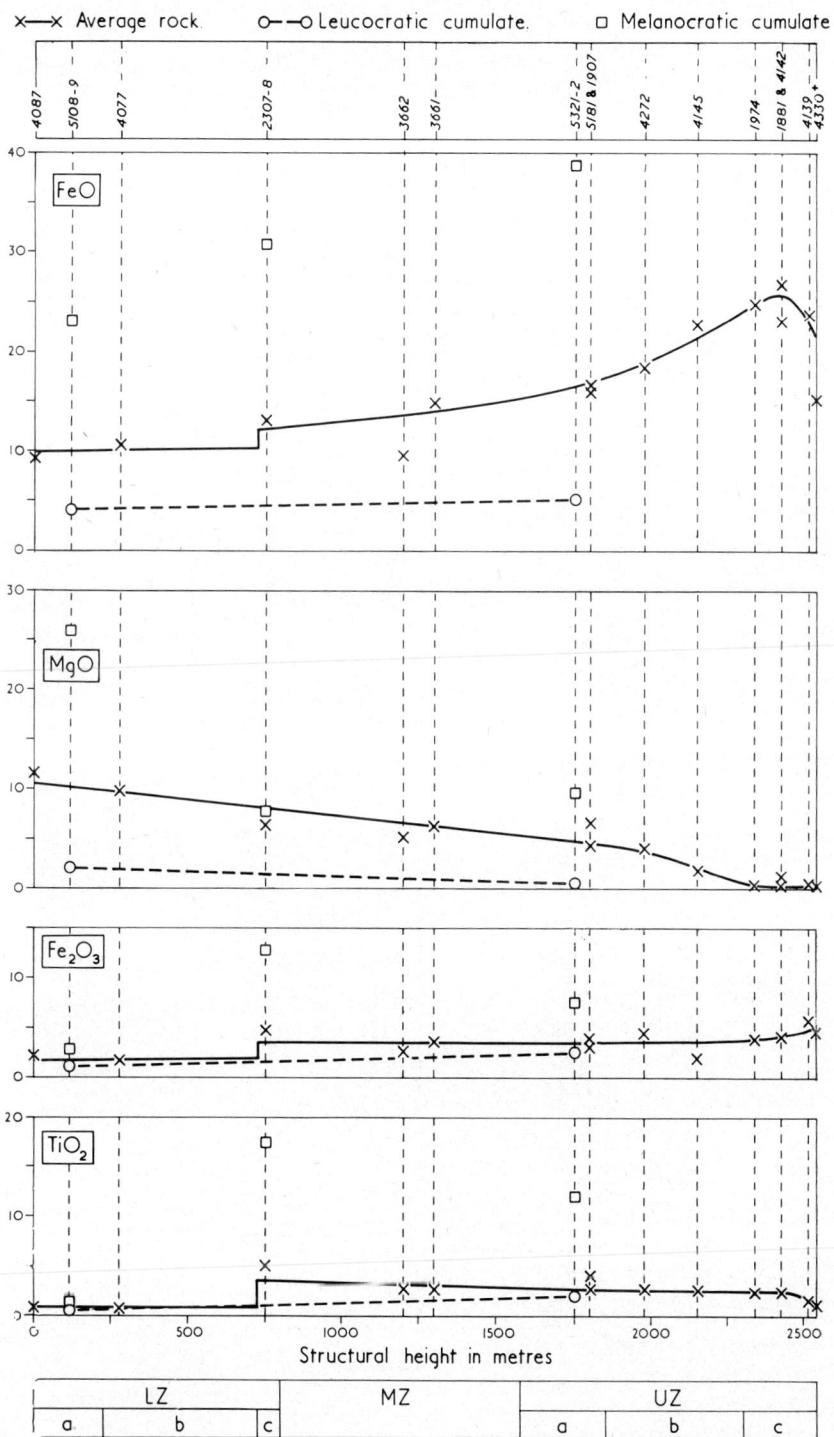

FIG. 108. Chemical variation of the observable layered series plotted against height for: FeO, MgO, Fe₂O₃, and TiO₂. (From Wager, 1960.)

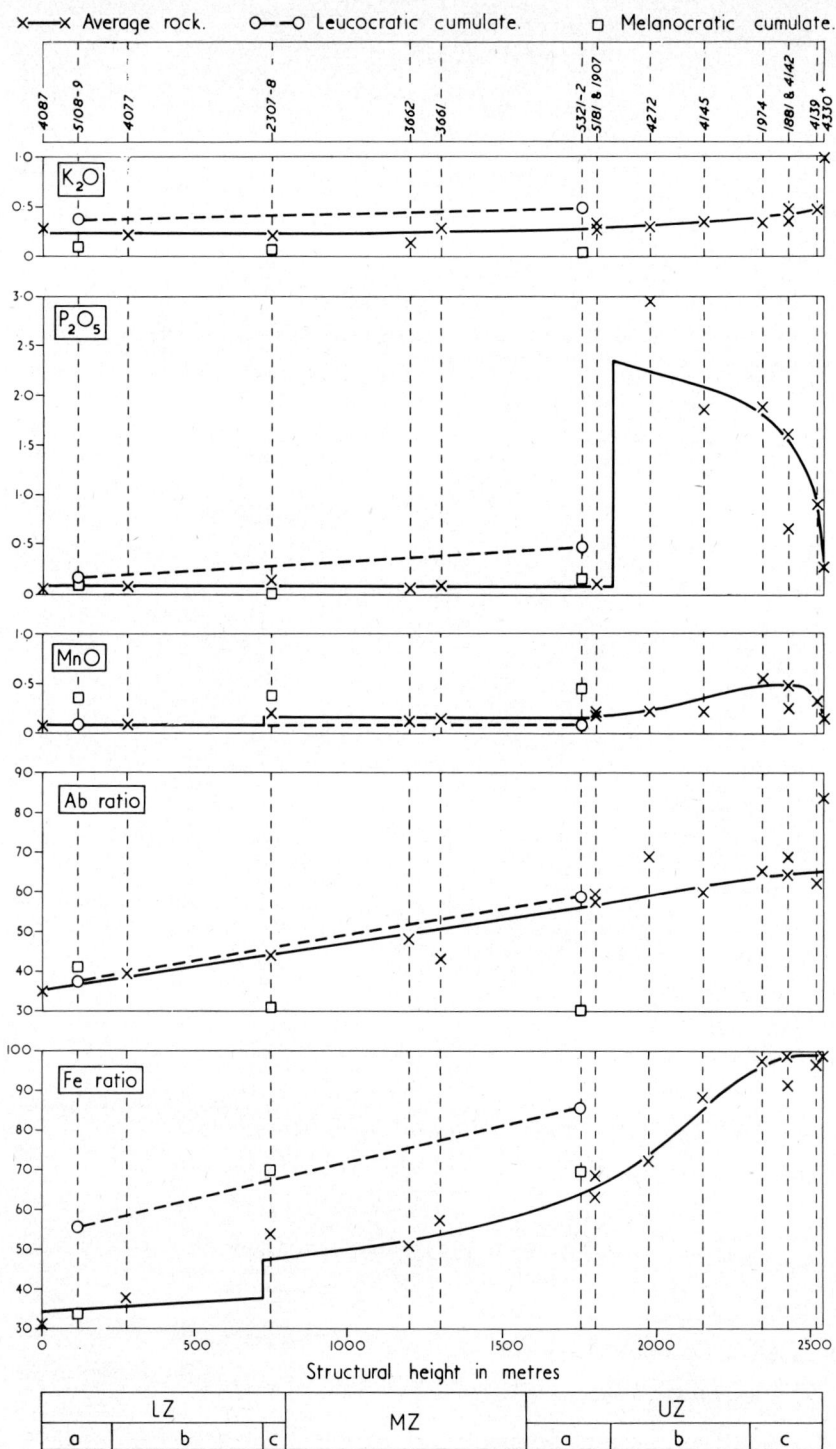

FIG. 109. Chemical variation of the observable layered series plotted against height for: K₂O, P₂O₅, MnO, and the albite and iron ratios. (From Wager, 1960.)

theoretical discussion based on Fenner's argument; 1928, pp. 96–9), but these would only become apparent if we could select truly average rocks or had a sufficiently large number of analyses to average out the irregularities resulting from rhythmic layering.

A precise meaning may be attached to the idea of an average rock, so far as the cumulus part of the rock is concerned. Thus a rock with cumulus crystals corresponding in proportions to those of the total primocryst assemblage precipitating from the liquid at that particular time is to be regarded as giving the average rock. Besides the cumulus part of any layered rock there is also the pore material formed from the trapped contemporary liquid; variations in the amount of this will affect the composition of the average rock as just defined, but generally less strongly than the variations in the proportions of the cumulus crystals. Because of the rhythmic layering and variation in amount of the pore material, only extensive and elaborate sampling would make it possible to obtain the real composition of the average rock at a particular horizon. An approach to it has, however, been obtained by plotting the analyses of rocks which were judged to be average in the field, then putting the best possible curves through the plotted points, and reading off from the curves the average composition at any desired level in the layered series. The compositions obtained in this way, for average rocks of the middle of zones and certain subzones, are given in Table 6. The following significant features may be noted: (1) Surprisingly, the amount of SiO_2 falls slightly during fractionation, until a very late stage is reached. (2) The FeO content shows an increase from 4·5 to 25%, while MgO decreases even more sharply, from about 9 to 0·3%. (3) The average rock contains about 0·05% P_2O_5 until apatite becomes a cumulus mineral, at the UZb stage, when the amount rises suddenly to about 2·2%. (4) As a result of rhythmic layering, the spread of points representing the compositions of contrasted layers at the same general horizon is liable to be greater than the total variation resulting from the cryptic layering.

b) *Composition of the sandwich horizon rocks*

The sandwich horizon rocks (EG4330 and 4331), collected at the junction of the layered series and upper border group on Basistoppen, have been briefly described already (p. 92); in composition, they are believed to represent the liquid from which the latest, UZc layered rocks were formed. Chemically, the analyzed rock (4330) is of an extreme composition (see Table 5). The MgO content is the lowest found in any of the analyzed rocks, being only 0·13%, while FeO is as high as 15·1%. The iron ratio (see Fig. 121) is nearly 99%, and is higher than in any other layered or border group rock except the related 4331 and 5166. The albite ratio (see Fig. 121) is 84%, and is only exceeded by some of the granophyres which are considered to be the result of filter-press action. From its mineralogy (Ch. III) and chemistry the SHR is clearly later than any of the rocks we have classed as layered series. As discussed earlier, it is considered to result from crystallization of the latest liquid evolved directly from the crystal accumulation process.

TABLE 6

Composition of average layered rocks at the middle point of certain zones and subzones deduced from graphs (Fig. 115)

	1st layered rock	Rock HZ	Rock LZ	Rock MZ	Rock UZa	Rock UZb	Rock UZc
SiO_2	50.8	48.7	45.7	45.7	45.2	43.8	44.2
Al_2O_3	19.5	18.0	16.0	15.3	13.3	11.0	8.1
Fe_2O_3	...	0.7	1.8	3.3	3.5	3.6	4.2
FeO	4.5	6.8	10.2	13.3	16.4	20.2	25.4
MgO	9.3	9.5	9.3	6.7	4.8	2.8	0.3
CaO	13.0	12.0	10.7	10.2	9.7	9.3	8.8
Na_2O	2.12	2.20	2.43	2.85	3.17	3.08	2.62
K_2O	0.21	0.23	0.23	0.25	0.28	0.32	0.40
TiO_2	0.55	0.71	0.88	2.95	2.67	2.50	2.28
P_2O_5	0.04	0.03	0.07	0.08	0.08	2.17	1.54
MnO	0.16	0.13	0.08	0.16	0.16	0.27	0.48
Total	100.18	99.00	97.39	100.79	99.26	99.04	98.32
Albite ratio	30.49	35.68	40.31	47.15	57.14	64.07	71.42
Iron ratio	21.70	29.22	38.23	52.97	65.91	80.50	98.09

C.I.P.W. Norm

	1st layered rock	Rock HZ	Rock LZ	Rock MZ	Rock UZa	Rock UZb	Rock UZc
Q	...	...	...	...	...	...	...
Or	1.11	1.11	1.11	1.67	1.67	1.67	2.22
Ab	17.82	18.34	19.70	23.58	24.49	26.20	22.01
An	43.10	38.62	32.28	28.08	21.41	15.29	9.45
Ne	...	...	0.39	0.28	1.21	...	...
(feldspar + Ne Σ)	62.03	58.07	53.48	53.61	48.78	43.16	33.68
Di — Wo	8.93	8.70	8.70	9.40	11.14	7.66	10.44
Di — En	6.20	5.50	4.90	4.50	3.80	1.50	0.20
Di — Fs	1.98	2.64	3.43	4.75	7.66	6.73	11.62
(Di Σ)	17.11	16.84	17.03	18.65	22.60	15.89	22.26
Hy — En	9.70	5.80	...	...	...	2.40	0.40
Hy — Fs	3.17	2.77	...	...	...	10.56	26.14
(Hy Σ)	12.87	8.57	...	...	...	12.96	26.54
Ol — Fo	5.04	8.47	12.74	8.47	5.67	2.10	0.03
Ol — Fa	1.84	4.38	9.69	9.48	11.93	10.20	2.00
(Ol Σ)	6.88	12.85	22.43	17.95	17.60	12.30	2.03
Mt	...	1.16	2.55	4.87	5.10	5.34	6.03
Ilm	1.06	1.37	1.67	5.62	5.02	4.71	4.41
(Mt + Ilm Σ)	...	2.53	4.22	10.49	10.12	10.05	10.44
Ap	...	2.53	...	...	...	5.04	3.70

TUNGSTEN

p.p.m. Tungsten

Percentage solidified

URANIUM

p.p.m. Uranium.

Percentage Solidified

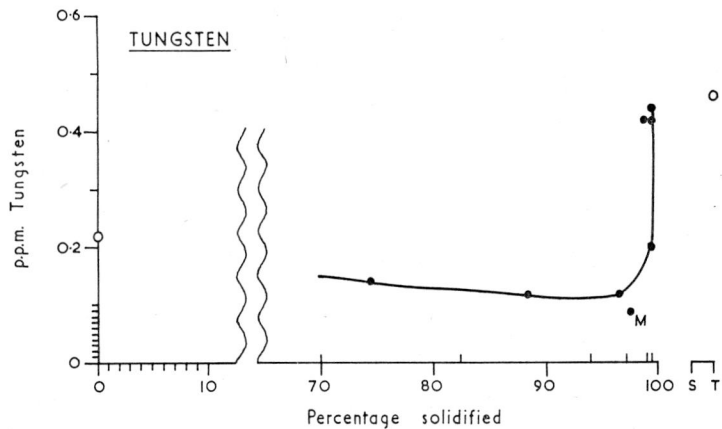

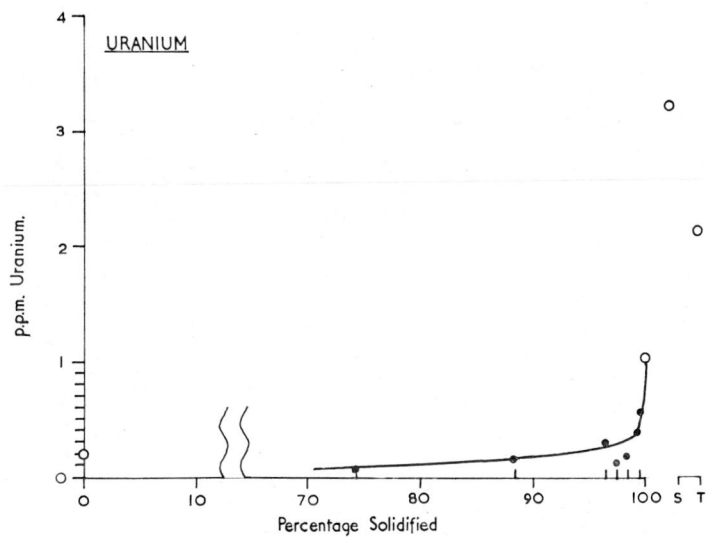

- •   Layered Series Rocks.
- ○   Rocks representative of liquids.
- — — — Layered Series trend.
- .... Liquid trend.
- F   Felspar-rich rock } where
- M   Melanocratic rock } distinguished.

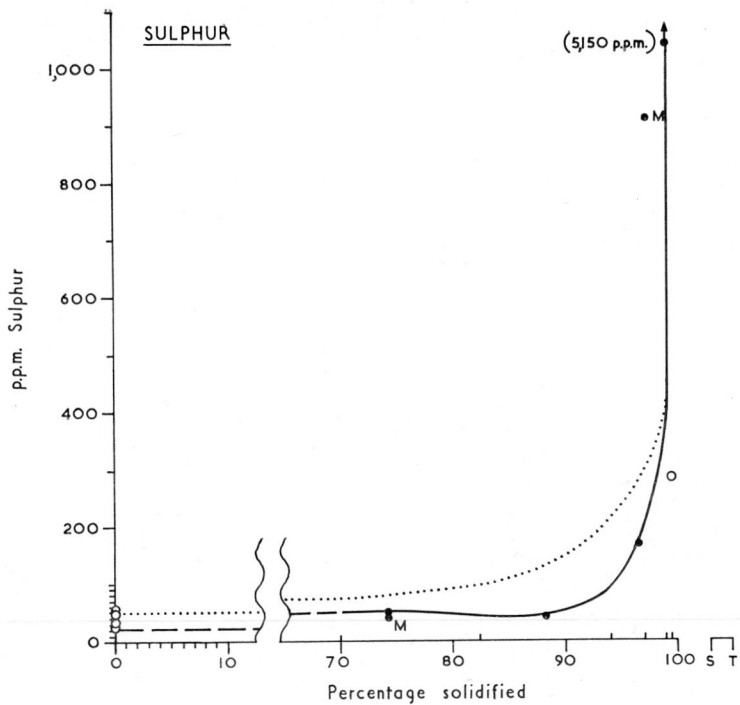

SULPHUR

(5,150 p.p.m.)

● M

p.p.m. Sulphur

1,000 —
800 —
600 —
400 —
200 —
0 —

0   10   70   80   90   100 S T

Percentage solidified

● M

●   Layered Series Rocks.
o   Rocks representative of liquids.
— — — Layered Series trend.
··········· Liquid trend.
F   Felspar-rich rock } where
M   Melanocratic rock } distinguished.

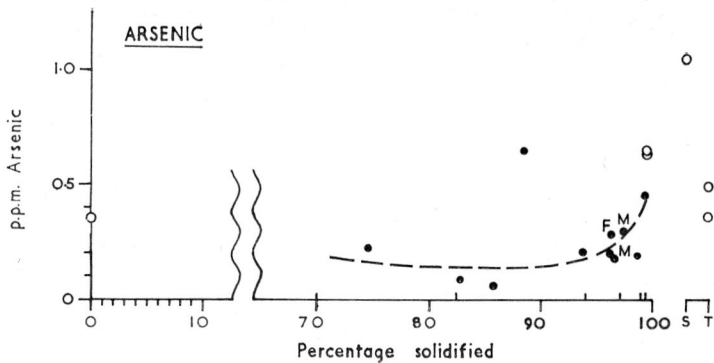

ARSENIC

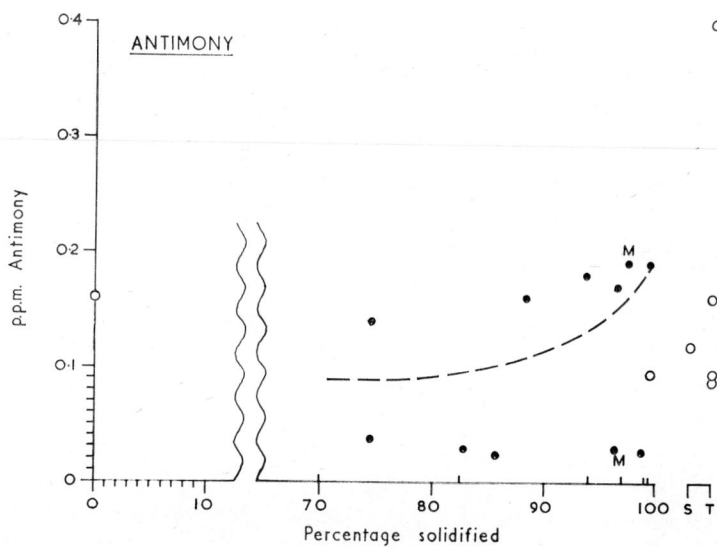

ANTIMONY

SILVER

GOLD

(0.07 p.p.m.)

p.p.m. Cobalt

p.p.m. Silver

Percentage solidified

p.p.m. Nickel

p.p.m. Gold

Percentage solidified

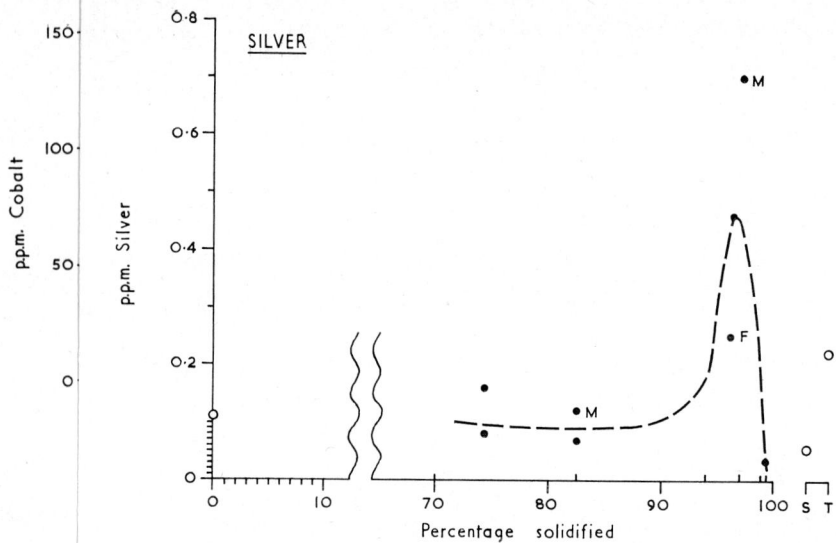

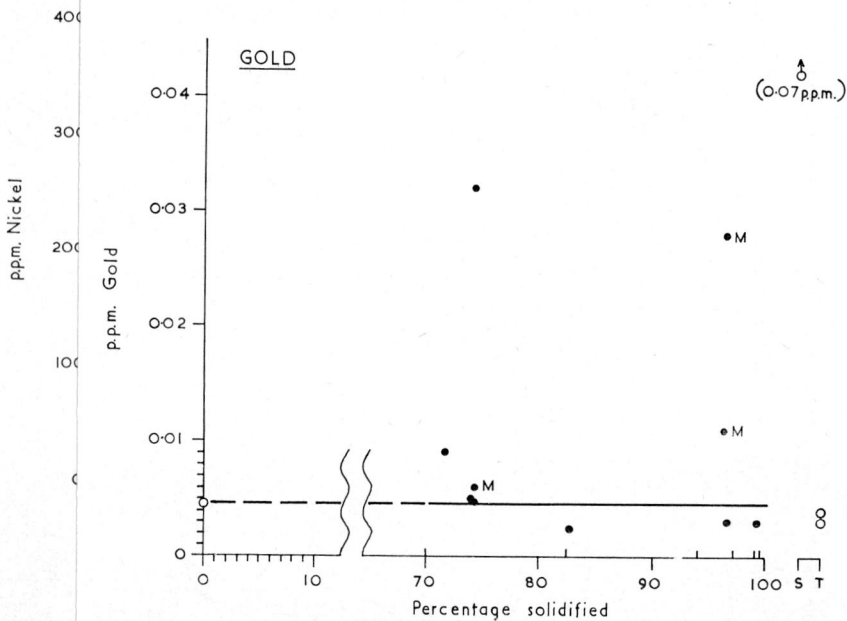

•   Layered Series Rocks.
○   Rocks representative of liquids.
— — —   Layered Series trend.
··········   Liquid trend.
F   Felspar-rich rock } where
M   Melanocratic rock } distinguished.

	U.B.G.β		U.B.G.γ	
	4163	4136	5150	4122
	An$_{54}$	An$_{49}$	An$_{42}$	An$_{40}$
	57·00	55·30	50·32	48·90
	13·02	18·52	10·89	11·97
	2·15	2·18	5·78	4·12
	9·25	7·47	15·55	17·18
	2·64	0·21	1·50	1·46
	5·75	8·20	6·63	8·24
	3·50	4·51	3·13	3·18
	0·96	1·30	1·07	0·62
3052. Quartz- Doug	0·97	0·41	1·42	0·67
*3050. Quartz- B. A.	0·41	0·06	0·14	0·09
	1·78	0·94	2·38	2·48
2275. Olivine Anal.	2·11	0·42	0·90	0·81
*5271. Olivine foot o	0·12	0·09	0·33	0·34
	99·66	99·61	100·04	100·06

4163. Contam...ms Anal.

4136. Andesi... Deer.

*5150. Melano... E. A.

*4122. Ferrodi...

	4163	4136	5150	4122
	16·98	5·07	7·93	3·35
	5·67 } 50·02	7·68 } 72·30	6·32 } 45·31	3·66 } 47·13
	29·61	38·16	26·48	26·91
	14·74	26·46	12·51	16·56
	0·82	...	...	...
	...	4·79	6·06	7·95
	...	0·25 } 10·15	1·01 } 12·62	1·11 } 16·62
	...	5·11	5·55	7·56
	6·58 } 19·07	0·27 } 5·69	2·72 } 17·63	2·52 } 19·64
	12·49	5·42	14·91	17·12
	...	...	...	...
	...	...	...	...
	3·12	3·16	8·38	5·97
	3·38	1·79	4·52	4·71
	...	...	...	...
	4·98	0·99	2·12	1·91
	...	...	...	...
	1·29	0·45	1·52	0·73
	64·9	60·3	68·2	63·4
	66·5	95·4	85·6	87·1

(%)

	4163	4136	5150	4122
	21	26	10	9
	42	56	45	52
	...	...	3	11
	28	15	30	21
	9	2	10	5
	...	1	2	2

SrO, 0·12; BaO, 0·04; NiO, 0·01.

		Transitional granophyre, Sydtoppen	Acid granophyres, Tinden Sill	
	4332	4489	5259	3058
	An_{38}	ca. An_{36}	ca. An_{10-0}	ca. An_{10-0}

		4332	4489	5259	3058
		60·23	64·39	72·69	75·03
		11·19	12·37	13·15	13·17
		5·52	3·41	0·92	1·56
		9·11	6·74	2·80	0·58
		0·51	0·31	0·16	0·15
		5·11	3·67	1·80	0·69
4330.	Ferrodi	3·92	4·30	4·02	4·24
	4139	1·94	2·33	3·26	3·85
*5264.	Melano	0·80	1·12	0·66	0·28
	granc	0·10	0·24	0·09	0·13
3047.	Melano	1·18	0·92	0·46	0·31
	Deer	0·27	0·086	0·032	0·034
*4332.	Melano	0·24	0·21	0·04	0·01
	Butcl				
		100·12	100·10	100·08	100·03
*4489.	Sydtop				
	Sydt				
*5259.	Acid gs				
	(Alka				
3058.	Acid g				

ridge 7·74 } 20·71 } 31·25 } 34·03 }
1·47 13·77 19·27 22·76
3·17 }51·85 36·38 }57·72 34·01 }61·49 35·87 }61·83
7·21 7·57 8·21 3·20

...   .

5·83 } 4·21 } 0·21 } ...
0·80 }14·34 0·39 } 8·86 0·02 } 0·44 ...
5·71 4·26 0·21 ...

0·47 } 0·38 } 0·38 } 0·37 }
3·96 } 4·43 4·17 } 4·55 3·48 } 3·86 ... } 0·37

...   .
...   .

3·00   4·94   1·33   1·00
...   .   0·87
2·24   1·75   0·87   0·59
...   .
0·64   0·20   0·08   0·08
...   .
0·89   1·36   0·75   0·41

	82·3	84·3	81·5	87·8
	91·7	93·3	90·7	72·8

	33		93	
	40		4	
	5		...	
	14		...	
	7		1	
	1		...	

c) *Composition of the marginal border group rocks*

The available analyses of the marginal border group rocks are presented in Table 7, and all except the picrite analysis are plotted in Fig. 110 against the distance from the western margin, except that rocks which are from margins other than the western are assigned positions in the western marginal border group sequence based on a combination of field and petrological evidence.

The points for the chilled marginal olivine gabbro, and the patchy-pyroxene gabbro of the tranquil division, cluster together, confirming that these are produced essentially by crystallization of the unmodified original magma. The points for the perpendicular-felspar rock diverge from those for the initial magma in the ways which would be anticipated from the greater amounts of relatively calcic plagioclase in these rocks. The picrite analysis is not plotted because it is so different from the rest of the rocks that to do so would greatly enlarge the graph.

From what has already been said about the nature of the marginal border group rocks (Ch. V), the chemical variation of the banded division would be expected to resemble broadly that of the layered series, and the three analyzed rocks of this division, ranging from LZ to MZ types, show this to some extent. Unfortunately, there are not enough analyses of marginal border group rocks to warrant detailed discussion. The amount of contemporary magma relative to primocrysts is believed to be greater in the marginal border group than in the layered series and this is indicated by the higher amount of phosphorus in the border rocks (see Fig. 37) compared with that in the corresponding layered rocks. The reefs of acid and hybrid rocks, derived from inclusions of the metamorphic complex, have probably contaminated the border rocks (other than the chilled facies) to some extent and caused some erratic variations.

d) *Composition of the upper border group rocks*

The analyses of the upper border group rocks are given in Table 8. They cannot be plotted satisfactorily against structural height as this is not always significant because of non-sequences. The analyses, therefore, have been plotted against decreasing anorthite percentage of the plagioclase cores, which best corresponds to the fractionation sequence (Fig. 111, left-hand side). Curves have been drawn, ignoring some of the irregularities which may be due to variation in the amount of primocrysts or to contamination from inclusions of acid gneiss. The curves, as would be anticipated, resemble those for the layered series but are less accentuated.

e) *Composition of the granophyric differentiates*

The melanogranophyres, the Sydtoppen transitional granophyre, and the Tinden acid granophyre, form a chemically related series and the analyses are presented in Table 9. As explained earlier, only the Tinden acid granophyre is a clearly transgressive sheet, formed from a late liquid, but the others are also believed to have formed from liquid pressed out of earlier rocks and intruded contemporaneously at about the sandwich horizon or a little above. For the granophyres, the rock analyses are believed to represent the compositions of the liquids from which the rocks were formed.

The melanogranophyres, with silica percentages from 57 to 60, albite ratios from 78 to 82, and iron ratios from 89 to 92, contain zoned plagioclase, the cores of which are between An_{40} to An_{38}. Judging by the plagioclase composition, and the albite and iron ratios, the melanogranophyres are not so far advanced in crystal fractionation as the sandwich horizon rocks although they are richer in SiO_2.

The Sydtoppen transitional granophyre differs in composition from the melano-granophyres in being much richer in SiO_2, poorer in FeO and MgO, and richer in alkalis. The rock is also extreme in its trace element composition, and in its exceptionally high O^{16}/O^{18} isotope ratio (Taylor & Epstein, 1963). The oxygen isotopes suggest that the liquid was formed by continued crystal fractionation of the original basic magma (p. 184).

The acid granophyre of the Tinden sill seems to be an even later product of filter-press action. The SiO_2 and K_2O amounts are comparable with those of common granitic rocks, and so are most of the other constituents. The iron ratio, however, is far less extreme than that of the transitional granophyre. This may be because the acid granophyre includes material pressed out of melted gneiss inclusions, a hypothesis supported by the strontium isotope ratios (p. 187).

The granophyric differentiates have been plotted (Fig. 111, right-hand side) against the compositions of their plagioclase cores, with one exception (see explanation of figures). The order is essentially the same as would be obtained by using the albite ratios but the iron ratios would place them somewhat differently. The Tinden grano-phyre is placed at an arbitrary position because the plagioclase composition has probably been altered by late-stage reactions.

f) *Estimated composition of the hidden layered rocks*

If the total weight of an element present in the exposed rocks be subtracted from the total weight originally present in the whole mass of magma forming the intrusion, then the total weight in the hidden rocks is obtained, and this amount will be distributed in some systematic way throughout the hidden rocks. Since the amount of an element separating from the magma changed during fractionation, it is convenient to use a graphical method for estimating the amounts of oxides or elements over any particular range of solidification. The case of vanadium may be used to explain the method (Fig. 112). The curve *APBCD* represents the percentage amount of the element separating at successive stages from 0 to 100% solidified. The amount of the element in the first 20% of rocks solidified is proportional to the area *APQO* and the amount in the whole intrusion is also proportional to the rectangular area *LMEO* when *LO* is the percentage of the element assumed to be present in the initial magma, i.e. the amount determined in the chilled marginal gabbro. Thus the shaded area must equal the area *LMEO*. Of the curve *APBCD*, the part *BCD* is known from analyses of the exposed layered

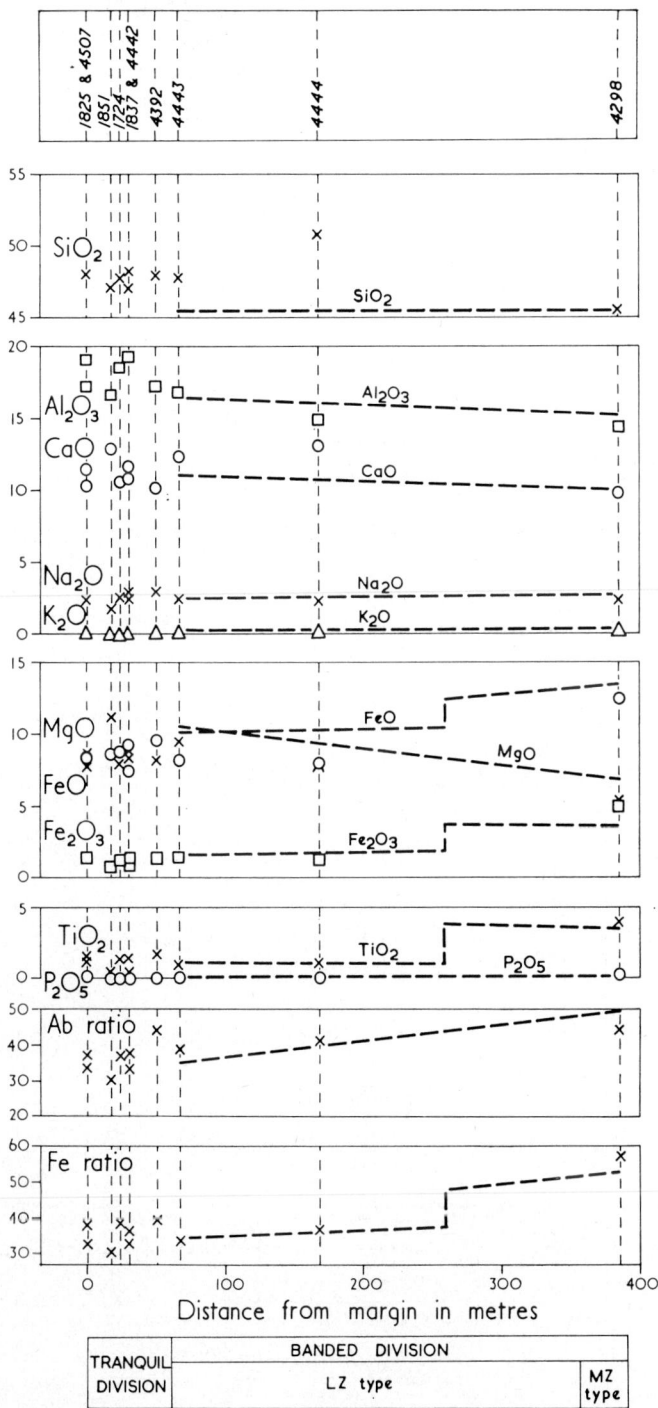

FIGURE 110

Chemical variation of the Marginal Border Group magma. There are too few analyses of the Banded rocks plotted against distance from margin. The Division rocks to permit a significant curve to be put points for the picrite 1682 (Table 7), are omitted in through them. The broken line curves are those for order to keep the graphs of manageable size. The the layered series between LZ and MZ and may be points for the Tranquil Division form clusters around compared with the actual points for the Banded Division what is considered the composition of the original rocks.

162

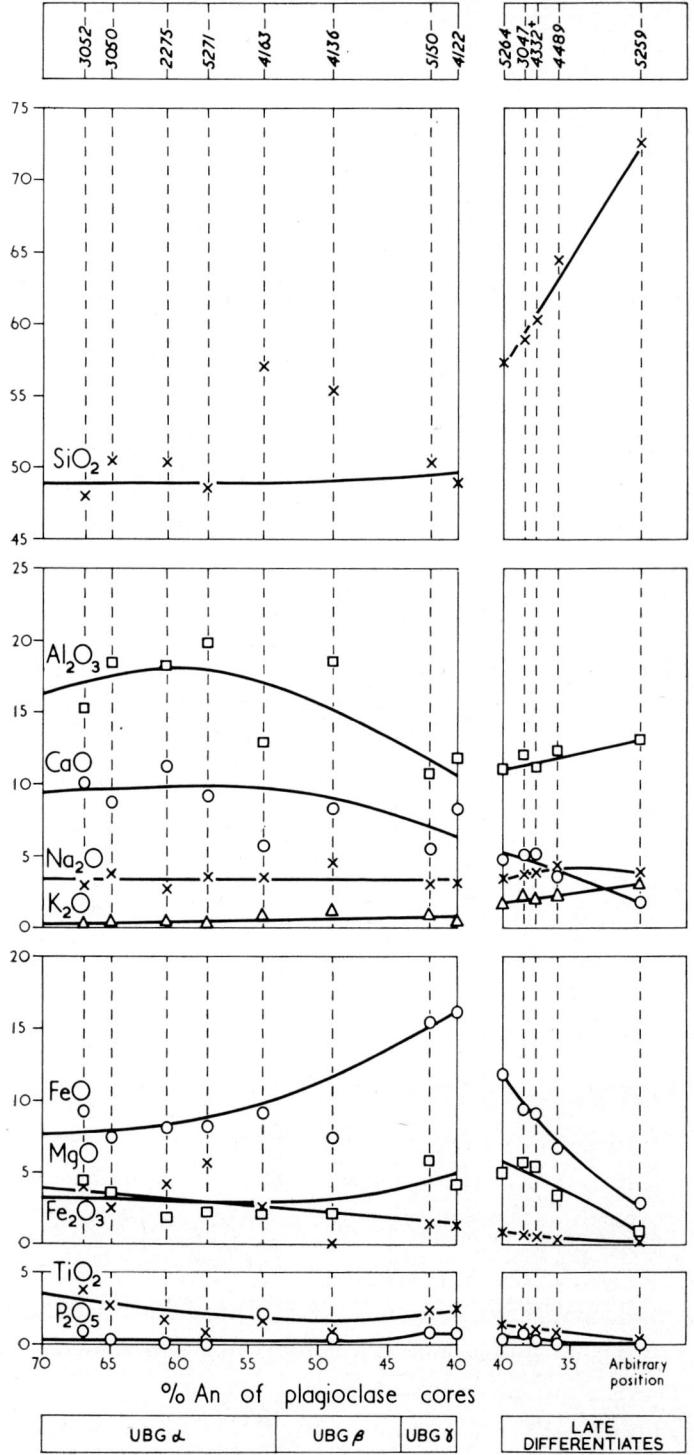

series rocks, while the first part of the curve, *APB*, is drawn so as to fit the requirement that the shaded area shall equal the rectangle *LMEO*. In this way, the total amount of the element in the hidden zone, as a proportion of the total weight of the intrusion, is proportional to the area *OAPBR*.* The curve *APB* shows a possible way in which the amount of the element could vary with the amount solidified; it is clear that the form of the curve *APB* can vary within certain limits.

In a recent paper giving revised estimates of the overall major-element chemistry of the intrusion (Wager, 1960), the effect of assuming that the exposed part of the intrusion formed either 40, 30 or 20% of the whole was tested. It was found that the assumption of 30% is reasonable for some oxides (e.g. TiO_2; Fig. 113B), but is impossible for P_2O_5 and sulphur because, with this assumption, there is more P_2O_5 and S

* The actual weight of the element is this proportion multiplied by the total weight of the intrusion which is estimated later as $1 \cdot 5 \times 10^{12}$ metric tons.

← FIGURE 111 (*facing and below*)
↓

Chemical variation of the Upper Border Group rocks and the Late Differentiate granophyres, plotted against the percentage of anorthite in the plagioclase cores (except for the Tinden granophyre, 5259, which is at an arbitrary position). The points for 4136 and 4163 are often far from the curve because the first is plagioclase-rich rock and the second, contaminated with acid gneiss.

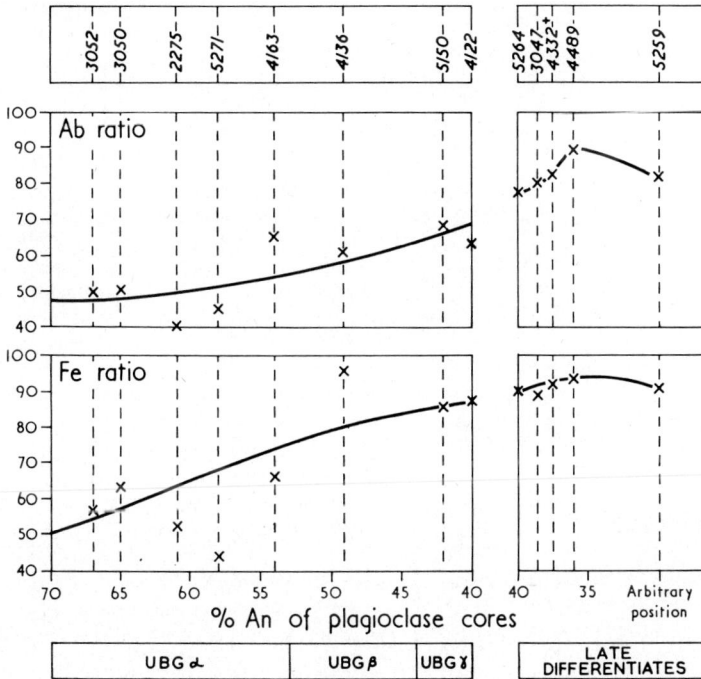

estimated to be present in the exposed rocks than in the whole of the intrusion (assuming the chilled gabbro correctly gives the initial magma composition).

To make the overall chemical picture for P_2O_5 and sulphur reasonable, it is necessary to assume that the volume of the UZ rocks is considerably less than would be the case if their amount were directly proportional to their thickness; this is not an unreasonable assumption, and suggests that the areal extent of the upper zone layered rocks at the time of their formation was less than the area of the intrusion at that level, perhaps due to the continued flexuring. In the two graphs (Fig. 114) showing P_2O_5 plotted against percentage solidified, the assumption in both is that the exposed rocks form 30% of the whole intrusion. In Fig. 114A the relative abundance of the various kinds of exposed layered rocks is taken as proportional to their thickness and in this case we have the impossible result that they contain more P_2O_5 than was present in the whole of the original magma. In the second graph, the assumption is made that the volume of the successive UZ rocks is progressively less than is implied by direct proportionality to thickness; the overall volume of UZ rocks, in fact, has been assumed to be half the amount it would be if it were proportional to their thickness. This assumption is equivalent to saying that the area originally covered by the successive UZ rocks decreased steadily so that it became, on the average, only half that of the LZ and MZ rocks. No field evidence is opposed to this assumption, and certain general arguments can be used to support it. With the assumption of the smaller volume of the UZ rocks (Fig. 114B), the amount of P_2O_5 in the exposed rocks is reduced and the estimated average amount in the HZ rocks becomes 0·03%. Such an amount seems

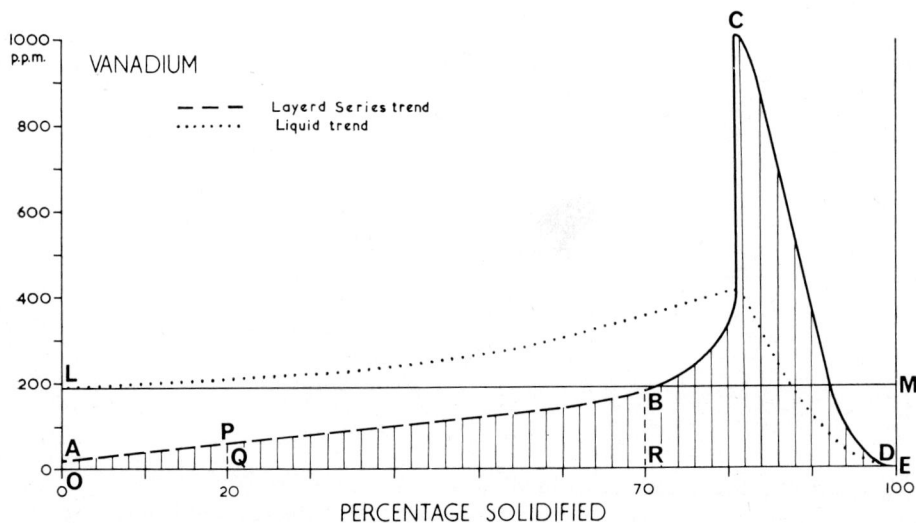

FIG. 112. Vanadium—an example of a graph to show the method used to calculate the composition of the hidden zone.

reasonable in relation to that present in the lowest observed rocks, and accords with the assumption, found satisfactory also for TiO_2, that the earliest layered rocks contained about 35% of trapped liquid of the composition of the original magma (Wager, 1960, p. 379).

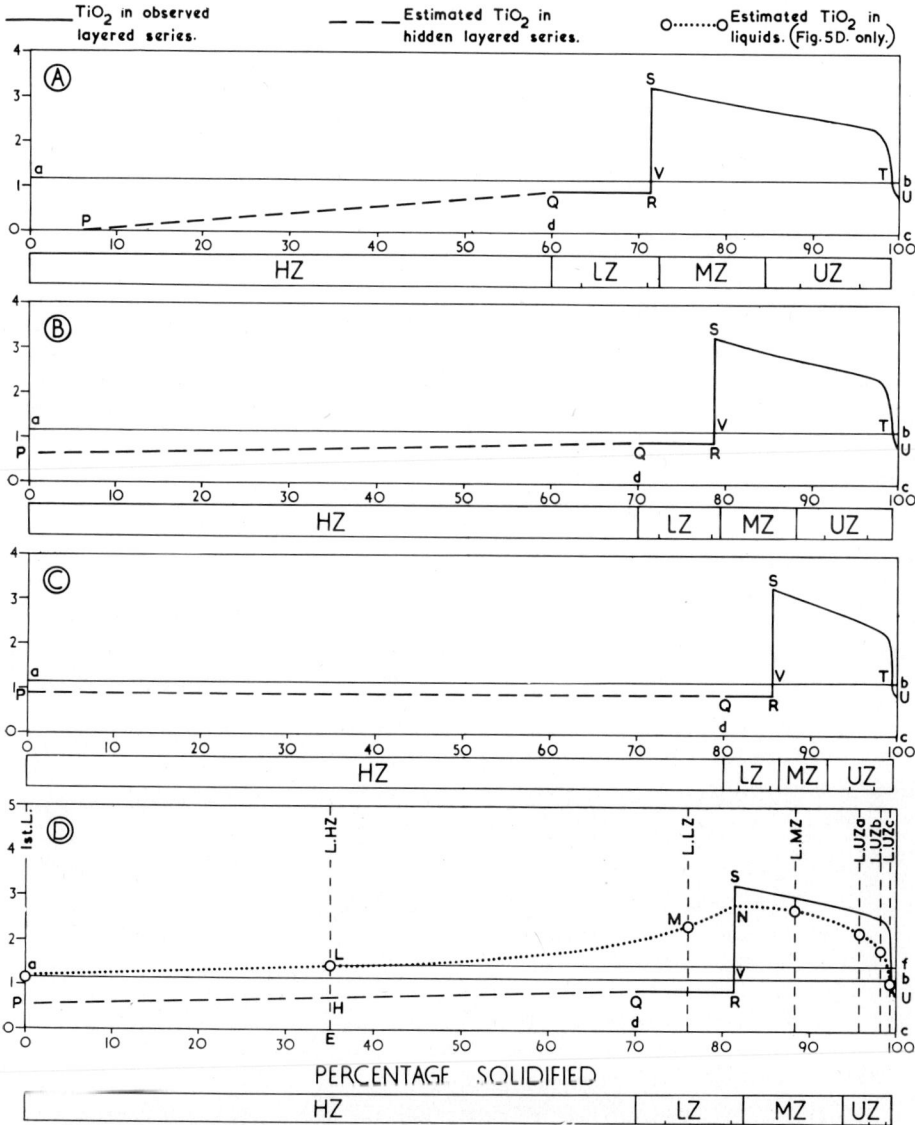

FIG. 113. The TiO_2 content of layered rocks, plotted against percentage solidified, with various assumptions about the relative volumes of the observed (60, 70, and 80%) and hidden layered series (see text). Also the deduced TiO_2 composition of HZ rocks, and of successive liquids (bottom graph only). (From Wager, 1960.)

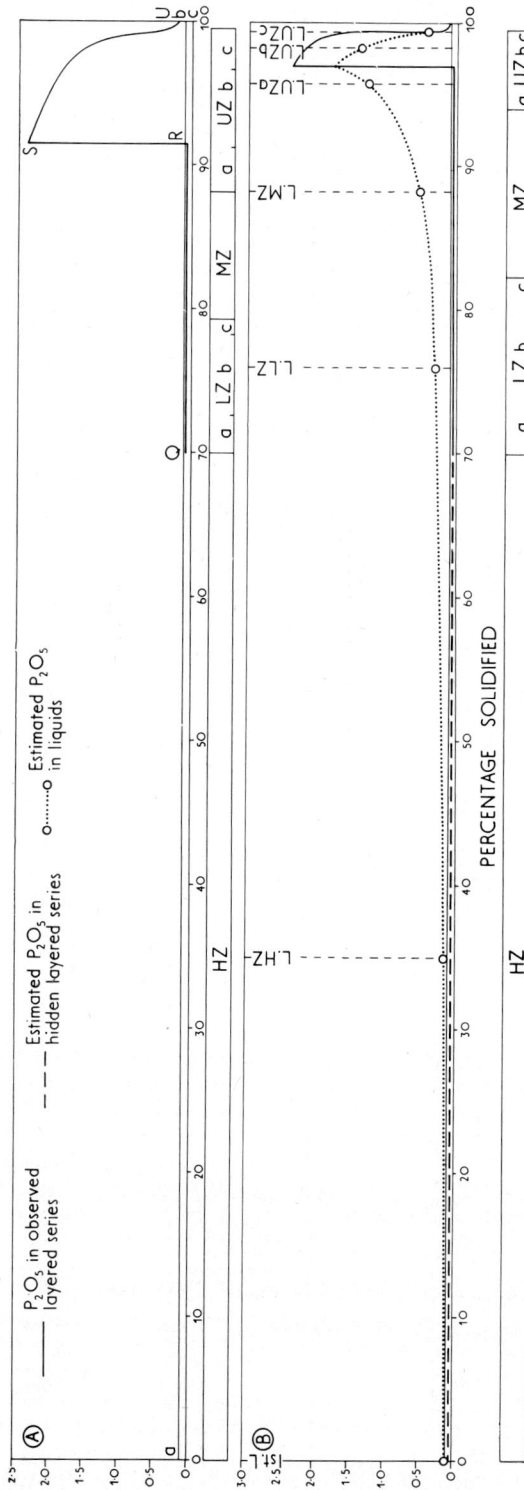

FIG. 114. The P$_2$O$_5$ content of layered rocks, plotted against percentage solidified, and deduced P$_2$O$_5$ composition, with various assumptions (see text), of the HZ and of successive liquids. (From Wager, 1960.)

In estimating the amounts of TiO_2 and P_2O_5 in the HZ rocks, it has been assumed that the curves showing the amounts of these two oxides in the exposed layered rocks represent satisfactorily the whole of the TiO_2 and P_2O_5 separating from the magma at each successive stage. This is true only if the amounts of these oxides in the simultaneously forming marginal and upper border groups is the same as in the contemporaneous layered rocks. The bulk of the marginal border group is small and may be disregarded for the present purposes; the bulk of the upper border group, however, is perhaps one-third the amount of the contemporary layered series rocks (cf. Fig. 90) and these rocks are more plagioclase-rich than most of the latter. To allow for the differences in composition between the UBG and the layered series rocks, minor adjustments are necessary in the case of certain elements (see Wager, 1960, p. 381). Thus the curve for the average amount of an oxide in the total material separating may sometimes be a certain amount above or below the curve for the layered series. In estimating the composition of the HZ and the successive liquids, these adjustments have been worth making in a rough way for certain oxides.

The assumptions best fitting the known facts for P_2O_5 and TiO_2, are (1) that the first observable layered rocks formed at 70% solidified, and (2) that the area of the UZ rocks became gradually reduced so that on average it was half the area of the LZ and MZ rocks. With these assumptions and the adjustments required by the composition of the upper border group rocks, the composition of the HZ rocks is worked out afresh and given in Fig. 113D 114B. The overall picture of the chemistry of the intrusion, so obtained, is an improvement upon that originally put forward (Wager & Deer, 1939, p. 220) but is not greatly different. The effects of the somewhat arbitrary assumptions made to allow for the composition of the UBG, the volume and composition of which are not well known, are slight and do not affect the results significantly, except in the case of Na_2O which is considered specifically below.

The composition of the average rock of the HZ, taken as the amounts of oxides at 35% solidified, may be read from the graphs (Fig. 115), and is given in Table 6 along with the CIPW norm. Rocks having this composition could well be plagioclase–pyroxene–olivine cumulates, rich in a calcic plagioclase. In the original paper, it was surmised that the hidden layered rocks would be, on average, plagioclase-rich olivine eucrites (Wager & Deer, 1939, p. 220) and the newly estimated composition also indicates a rock which could be described in these terms.

g) *The changing composition of the successive residual magmas*

Whereas the composition of the hidden layered series might be obtained if a very deep borehole were drilled, the composition of the successive residual magmas, which only existed ephemerally, can only be estimated by calculation. From the compositions and relative amounts of the various parts of the intrusion a graphical method has been used to estimate the compositions of the successive liquids, rather in the way it has been used in the previous section for estimating the composition of the hidden layered series rocks.

The method may be illustrated by using TiO_2 as an example (Fig. 113D). The amount of TiO_2 in the initial magma, from the chilled gabbro analysis, is 1·17% (point a). The amount of TiO_2 in the liquid at 35% solidified, the mean for the HZ, is shown by point L which is fixed by means of the following considerations: the amount of TiO_2 separated in the rocks from 35 to 100% solidified is proportional to the area $HQRSUcE$, PHQ having been established from the calculation described in the previous section. All this TiO_2 was present in the liquid at 35% solidified, and the point L is fixed, by trial and error, so that the area of the rectangle $LfcE$ is equal to the area $HQRSUcE$. Sufficient points (LMN, etc.) have been fixed, by this method, to draw the general curve for TiO_2 in the liquid. The curve is concave upwards to N where the maximum content of TiO_2 in the liquid, 2·80%, is reached. At this stage, which is the beginning of the LZc, a cumulus titaniferous magnetite was precipitated, and from then onwards the amount of TiO_2 in the liquid began to fall, at first gently, but in the later stages, rapidly. The curve for P_2O_5 in the liquid has been drawn in a similar way (Fig. 114B).

The variation of all the major oxides in the successive residual liquids, determined as for TiO_2, is presented in Fig. 115. In drawing the graphs, the estimated composition of the total material separating from the successive liquids is accepted, in some cases, as being the same as given by the curve for the analyzed average rocks of the layered series (continuous line), while in other cases (SiO_2, Al_2O_3, FeO, MgO, Na_2O, K_2O) the curve has been modified (broken line) from the curve for the analyzed layered series rocks, for reasons already mentioned (p. 153). For most oxides, the estimated changes in composition of the liquid during the last few per cent of the solidification process are rapid and highly sensitive to the particular assumptions made about the amount and composition of the late-stage, intermediate and acid rocks. The graphs given here are based on the assumption that the volume of the intermediate and acid differentiates is 0·5% of the volume of the whole intrusion. At 99% solidified, the curves turn off towards the average composition of the intermediate and acid rocks but the rapidly changing compositions cannot adequately be shown at the scale used in Fig. 115 and are further illustrated later (Fig. 121). The estimated compositions of the liquids at selected stages are given in Table 10.

The chief features of the variation in the major elements in the successive residual liquids (Fig. 115; Table 10), are as follows:

SiO_2. Silica in the liquid, from o to 95% solidified, decreases slightly. This surprising result, commented upon in the original paper (Wager & Deer, 1939, p. 297, and pl. 27), is now believed to be caused partly by the large amount of plagioclase which was probably precipitated in the early and middle stages of the HZ, but chiefly because of the low partial pressure of oxygen in the liquid, residual liquids *decreasing* in silica in the system $MgO–FeO–SiO_2$ (cf. Osborn's explanation given later, p. 243). In the later stages of fractionation, the amount of SiO_2 being precipitated in all the various rocks approximately equals the amount in the liquid, and the SiO_2 content of the latter does not change appreciably until the latest stage of differentiation when the amount increases rapidly.

Unless the amount of the late, silica-rich rocks has been grossly underestimated (which seems improbable), or silica has, in some unexpected way, been lost from the system, it seems clear that the silica content of the HZ rocks must be relatively high. The composition of the hidden layered rocks, calculated on the assumption that

TABLE 10

The compositions of the liquids at the middle points of the various zones as read from the graph (Fig. 115)

	First liquid (= 4507)	L.HZ	L.LZ	L.MZ	L.UZa	L.UZb	L.UZc
Percentage solidified	0	35	76	88·2	95·7	98·2	99·3
SiO$_2$	48·1	47·3	46·8	46·9	47·5	49·8	55·0
Al$_2$O$_3$	17·2	16·4	15·2	14·1	12·5	11·8	11·8
Fe$_2$O$_3$	1·3	2·0	3·3	3·6	3·8	4·3	4·5
FeO	8·4	10·0	13·0	15·0	17·3	17·8	14·5
MgO	8·6	8·1	5·8	4·0	2·5	1·3	0·8
CaO	11·4	10·9	10·0	9·5	9·5	8·0	7·0
Na$_2$O	2·37	2·45	2·80	3·00	3·20	3·40	3·65
K$_2$O	0·25	0·29	0·33	0·40	0·65	1·00	1·40
TiO$_2$	1·17	1·42	2·37	2·65 (2·80*)	2·20	1·80	1·05
P$_2$O$_5$	0·10	0·13	0·28	0·50	1·25	1·35 (1·75*)	0·40
MnO	0·16	0·16	0·16	0·20	0·28	0·35	0·25
Albite ratio (atomic)	37·2	39·6	47·2	52·8	62·0	68·3	73·8
Iron ratio (atomic)	35·8	41·2	55·9	68·1	79·6	88·8	91·2

* Maximum reached at an intermediate stage.

plagioclase and olivine are the only cumulus crystals (Wager, 1960, table III), has 3% less silica than has been estimated graphically (Fig. 115). The only way in which the early HZ rocks could have a silica content approaching that estimated by the graphical method, is for them to have orthopyroxene instead of olivine as a cumulus mineral. Thus the earliest rock, assuming it to have 15% bronzite, 50% plagioclase, and 35% of the initial magma, would have 49·5% SiO$_2$ (Wager, 1960, table 3, column B) which is closer to the estimated figure of 51% than if the rock contained cumulus olivine and no bronzite, when the SiO$_2$ would be 47·0%. Bronzite, directly precipitated from the magma, occurs in the outer part of the marginal border group and occasionally, tiny grains of bronzite are found in the lowest exposed layered series. These facts suggest the possibility that orthopyroxene may have been a cumulus mineral in the earlier, unexposed layered rocks (Brown, 1957). The chilled marginal gabbro shows clearly that at the higher levels, especially where there was supercooling, one of the early phases to separate was olivine, but it is likely (Wager, 1960, p. 384), that orthopyroxene, in place of olivine, may have formed as an early phase in the deeper levels where the hydrostatic pressure was high. This has since been made more probable from certain experimental evidence (Davis & England, 1963).

Al_2O_3. The content of alumina in the successive liquids falls slightly. Over the greater part of the cooling history it thus follows silica but at the late stage, when silica rises, alumina remains low. The fall in alumina, we believe, is real, and may be a feature of fractionation of high-alumina basalts.

Fe_2O_3. The amount of Fe_2O_3 in the exposed layered series is such that, with the postulated relative volumes, the amount in the earliest hidden layered rocks is reduced to zero. This cannot be the case in reality, the amount of Fe_2O_3 in the earliest rock being probably about 0·5% (cf. Table 6). To fit the assumptions which prove satisfactory for the other major oxides, the amount of Fe_2O_3 in the exposed layered series should be two-thirds of that actually found. It seems probable that the relatively high content of Fe_2O_3 in the observable rocks is due to oxidation of some of the ferrous to ferric iron during fractionation (see below, p. 187). The curve for the successive liquids shows a gradual, small rise, which is presumably a combination of oxidation and the normal fractionation effects.

FeO. The FeO content of the magma apparently increased until it reached 18% in the UZb. After this the iron ratio continued to increase (Table 10) and iron-wollastonite crystallized temporarily in place of an iron-rich augite, although the estimated FeO content of the liquid was by then beginning to fall. When titaniferous magnetite first began to occur as a cumulus mineral at the beginning of LZc, there should have been a slowing up of the rate of increase of FeO in the liquid but the graph of FeO in the liquid is not sufficiently refined to show this.

MgO. From the beginning, the amount of MgO separating in the ferromagnesian minerals exceeds the amount in the liquid. Thus the curve for MgO in the liquid falls, slowly at first and then rapidly, until the amount is less than 0·8%.

CaO. The first liquid contains 11% of CaO and the amount falls slowly so that there is still 7% in the UZc liquid.

Na_2O. There must have been a slight but definite increase in the Na_2O content of the liquid (2·4–3·1%) until the beginning of the UZ; after this stage, the effect of the composition of the upper border group rocks begins to be important. As already mentioned, it is only in the case of Na_2O that the assumed effect of the composition of the upper border rocks is such as to affect significantly our estimate of the general trend of the liquid compositions. The soda content of the analyzed rocks of the UBGβ and γ

is between 0·5 and 1% greater than for the corresponding average layered rocks. The total volume of these rocks is, however, presumed to be considerably less than that of the corresponding layered rocks. The graph showing the total Na_2O separating has, therefore, been raised by a few tenths of a percent from that derived from the layered series analyses. This is sufficient to make the Na_2O content of the liquid increase throughout the differentiation, which is normally expected to happen in the fractionation of basic magma. In the original memoir, where the effect of the UBG rocks was not considered, the Na_2O content of the liquid was shown to have fallen slightly at a late stage (Wager & Deer, 1939, p. 225).

K_2O. In the later stages, as for Na_2O, the graph for K_2O separating in the rocks should be drawn higher than that for the layered rocks, because of the UBG compositions, but the difference is small and has been ignored. Whatever reasonable assumptions are made, the rate of increase of K_2O in the liquid must have been very slight over the first 98% solidified.

TiO_2. On slender evidence it has been suggested that ilmenite was a cumulus mineral from the beginning of crystallization, or soon after. Whether this be so or not, the amount of TiO_2 precipitated in the HZ was less than in the liquid and therefore the amount of TiO_2 in the liquid increased upwards, with an increasing rate until titaniferous magnetite separated as a cumulus phase. This occurred when the TiO_2 content of the magma was about 2·8% and the FeO and Fe_2O_3 contents were 14 and 3·5%, respectively. These amounts are, therefore, the percentages of TiO_2, FeO and Fe_2O_3 for this particular magma when it became saturated with titaniferous magnetite. The amount of TiO_2 in the liquid falls from 2·8% to about 1% with increasing fractionation. It is interesting to note (Fig. 115) the big difference between the TiO_2 content of the average layered rock and the liquid just before titaniferous magnetite began to precipitate as a cumulus mineral in the LZc.

P_2O_5. This increased in the liquid in the same way as TiO_2 until apatite was precipitated as cumulus crystals, at 97% solidified. To precipitate apatite, the amount of P_2O_5 necessary in the liquid was apparently about 1·75%. The unexpectedly low amount of P_2O_5 in the average layered rocks up to the end of UZa, compared with the increasing amount in the contemporaneous liquid, has been interpreted as meaning that the amount of trapped liquid in the layered rocks, on average, became less with ascent in the layered series, up to the time when apatite appeared as a cumulus mineral. The P_2O_5 content of the layered rocks is used, in fact, as a means of estimating the proportion of trapped liquid (see earlier, p. 66).

MnO. The amount of manganese estimated to be present in the hidden zone is higher than might have been expected. In the liquid, the amount of MnO starts at 0·16% and rises slightly to a maximum of 0·35% at about 98% solidified.

The calculated iron and albite ratios for the successive Skaergaard liquids are also given in Table 10. While the iron ratio reaches 91% in the UZc liquid at 99% solidified and 99% in the SHR (Table 5), at which stages a fayalite occurs in the rocks, the albite ratio has only reached 74% and 84% respectively, and the plagioclase is sodic andesine.

The methods used, so far, give the composition of successive liquids from o to about 99% solidified; the later changes in composition of the liquid are believed to be represented directly by the analyses of the intermediate and acid rocks of the intrusion (Table 9). The Skaergaard intrusion provides an example of the formation of intermediate* and acid rocks by the crystal fractionation of a basic magma, which was originally rich in the constituents of olivine. The mechanism of the process was first discussed by Bowen in 1915 and since in various places (1915, pp. 25–7; 1928, pp. 70–8). The rhythmic layers of the LZ which happen to be free from olivine, contain 1% or so of quartz as a result of crystal fractionation of the trapped liquid (see Ch. IV, p. 60). Quartz does not, however, develop in the interstices of those rocks of the LZ which contain olivine as cumulus crystals, apparently because the trapped liquid reacted with the olivine to produce pyroxene and then, at a later stage, hornblende, serpentine, etc., so desilicifying itself. The granophyric liquids which were eventually produced in the Skaergaard intrusion by filter-press action were formed after the ferrodiorite stage and were, of course, far removed from any contact with the magnesian olivines of the LZ, or HZ, with which they might have reacted. The original Skaergaard magma was an olivine tholeiite as defined by Yoder and Tilley (1962, p. 352), and probably rocks from such a magma will not normally produce evidence, from their interstitial minerals or segregation veins, of their capacity to give a quartz-bearing residuum by strong fractionation, because the small amount of potential quartz reacts with the magnesian olivine.

h) *The major element chemistry, illustrated graphically*

In order to discuss the petrogenesis of a complex of igneous rocks it is necessary to know the order in which the rocks were developed, and it is desirable that this order should be used in the graphical presentation of the chemistry. It is often not possible to establish the order with any certainty, but for the layered series, an order is firmly established from the order of superposition, and for the marginal and upper border groups, an order is established by correlation with the layered series by means of the felspar compositions. Thus the layered series forms the basis for defining the successive stages in the solidification process, and by means of the arguments already given, the stages can be stated approximately in terms of the percentage of the whole intrusion which has, by any particular stage, solidified.

A way of illustrating certain significant aspects of the fractionation of basic magma is by plotting the percentage of Mg, Fe^{++}+Mn, and Na+K as atoms or oxides at the corners of a triangular diagram (Fig. 116). This method was first used in the Skaergaard memoir (1939, pp. 231–2 & 313–6), it being argued that the early, middle and late fractionation stages were picked out by this grouping of elements. The trend in

---

* Preferred by one of us (L.R.W.) in referring to the melanogranophyres, but not to the ferrodiorites.

composition of both average rocks and the liquids is from a point near the Mg corner towards the $Fe^{++}+Mn$ corner, and then towards the $Na+K$ corner (Fig. 116). On the diagram, some examples of rocks from extreme rhythmic layers have also been plotted; they lie very far from the curves for both the liquids and the average rocks.

Other chemical methods of defining fractionation stages for series of liquids can also be used. One such method is based on plotting the so-called albite and iron ratios* derived from the analyses (Wager, 1956). The albite and iron ratio plot should give the order in which the rocks were formed when a single fractionation series is being considered. In the Skaergaard case, the method is particularly useful for the later

* The albite ratio is defined as the molecular percentage of albite in the normative plagioclase. If nepheline is present in the norm this is converted to albite and, in deriving the Ab ratio, is included with it. The iron ratio is the atomic ratio $\dfrac{(Fe^{++}+Mn)\times 100}{Fe^{++}+Mn+Mg}$.

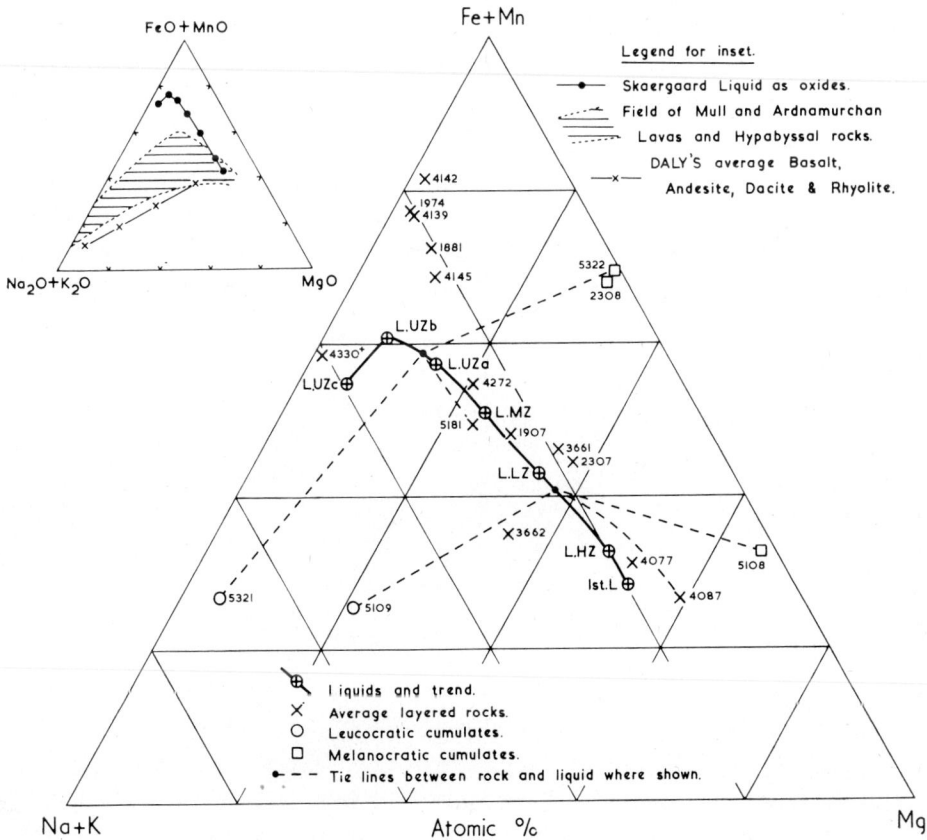

FIG. 116. Successive Skaergaard liquids and selected layered rocks plotted on a triangular diagram showing at the three corners the chief elements of high-, medium-, and low-temperature minerals.

rocks of the intrusion, from about 98 to 100% solidified, where the order of formation cannot be obtained with certainty from field evidence.

The fractionation stages must primarily be defined from the composition of the liquid and not from the composition of rocks representing crystal accumulation. Thus the concept of fractionation stages is particularly valuable for rocks which can be assumed to lie on what Bowen called the liquid line of descent—for example, a series of non-porphyritic lavas. For the Skaergaard intrusion, the albite and iron ratios for some of the average rocks and for the estimated liquids are plotted in Fig. 117, the points for the liquids being tied to those for the corresponding rocks. The points for the various average rocks, except for LZc, are below and to the left of the corresponding contemporary liquid. This is to be expected, since the rocks contain an accumulation

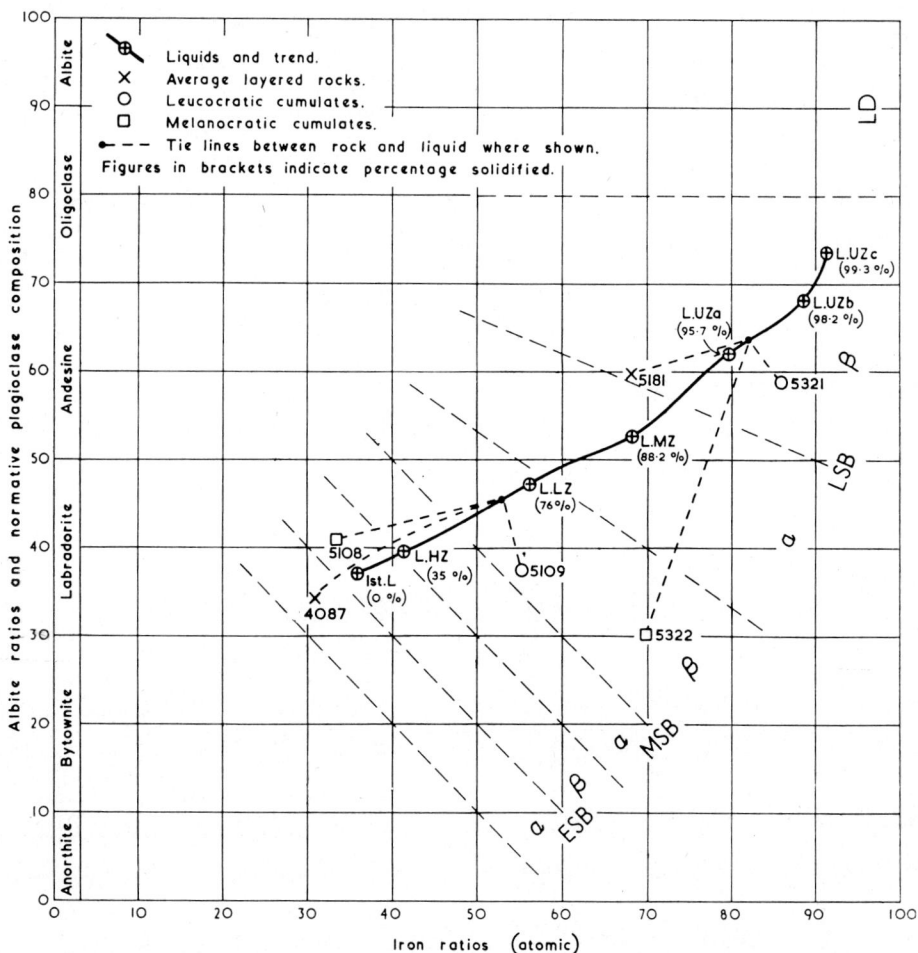

FIG. 117. Skaergaard liquid compositions plotted to show fractionation stages. Points for certain layered cumulates are shown tied to the corresponding liquid.

of early crystals having lower albite and iron ratios than the liquids from which the crystals formed. Whether the curves should be smoother is not known; it is possible that some of the apparent irregularities are due to difficulty in selecting a specimen to represent the average rock. On the graph, the percentages solidified, at the stages of the selected liquids, are given in brackets. In the inset to the figure, the position, on the trend line, of the liquid at every successive 10% solidified is indicated and shows that the rate of change increases with fractionation. The positions of the early, middle, and late-stage basalts, as defined chemically in the previous paper (Wager, 1956), are also indicated on the inset.

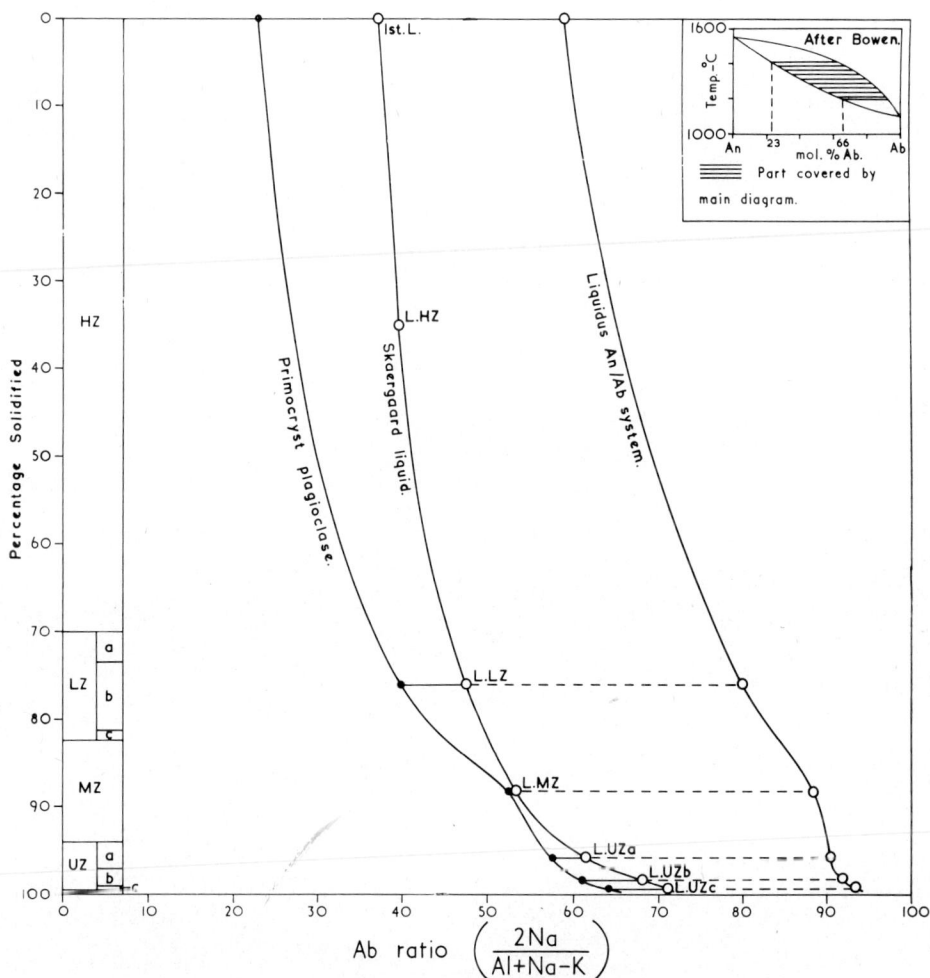

FIG. 118. Plot of the albite ratios of the Skaergaard liquids; the corresponding primocryst plagioclases; and the liquids in equilibrium with these plagioclases in the synthetic An–Ab system. (From Wager, 1960.)

The relationship between the composition of the cumulus plagioclase in the layered rocks, as determined optically, and the albite ratio of the liquid (which is the same as the normative plagioclase composition with the nepheline adjustment) is shown in Fig. 118. On the diagram, the albite ratio of the melt for the two-component, anorthite–albite system (Bowen, 1913) with which the various cumulus plagioclases would be in equilibrium, is indicated also. In the natural system there is much less difference in composition between the precipitated plagioclase and normative plagioclase in the liquid than there is between the plagioclase and the liquid in the synthetic, two-component system. An essentially similar result was found by Matthews (1957) when

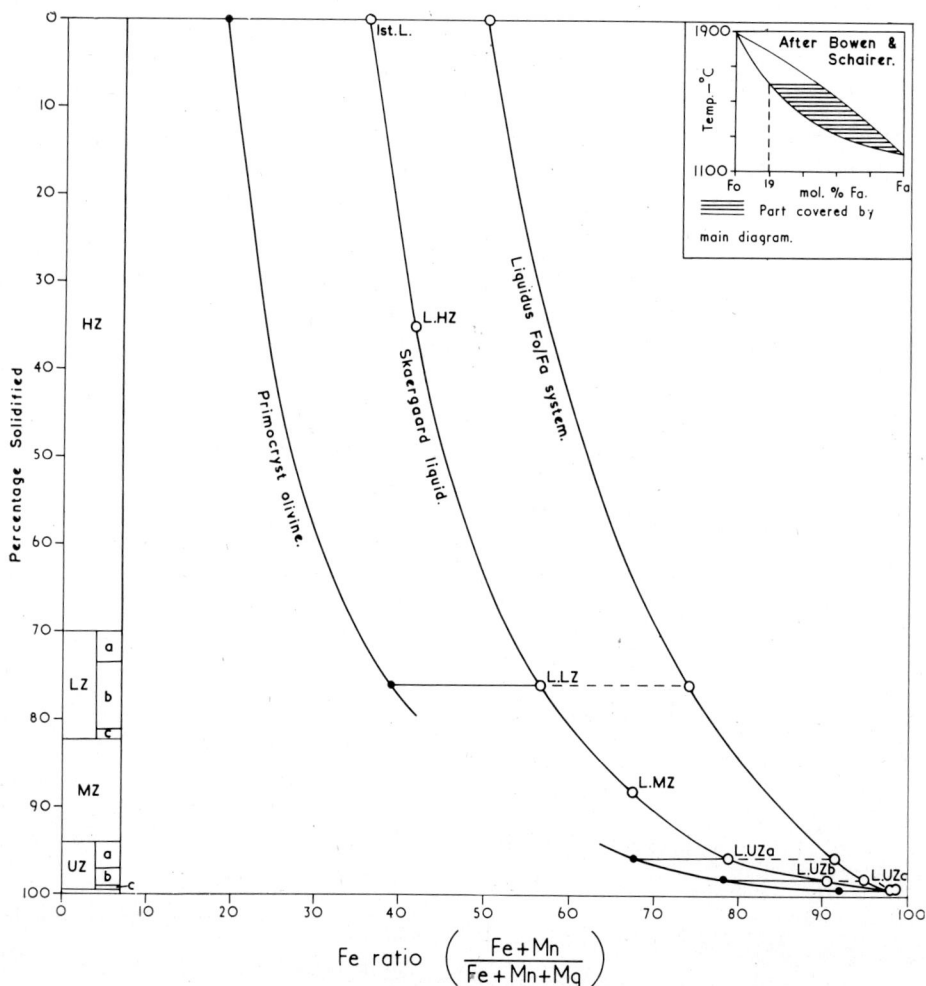

FIG. 119. Plot of the iron ratios of the Skaergaard liquids; the corresponding primocryst olivines; and the liquids in equilibrium with these olivines in the synthetic Fo–Fa system. (From Wager, 1960.)

comparing the compositions of plagioclase phenocrysts and the normative plagioclases of a series of Quaternary lavas from Mount Garibaldi, British Columbia.

The relationship between the iron ratio of the successive liquids and the same ratio for the cumulus olivines in equilibrium with them are shown in Fig. 119 and compared with the iron ratio of the liquid in the Mg_2SiO_4–Fe_2SiO_4 system (Bowen & Schairer, 1935, fig. 7). As for plagioclase, there is less difference in the natural system between the composition of the precipitated olivine and the normative olivine as calculated for the liquid, than in the synthetic, two-component system.

The stage of fractionation of the border group rocks relative to the layered rocks can be established by using the Ab and Fe ratios plot, an average position being obtained by dropping perpendiculars from the established points onto the diagonal from the origin, as was first done in a thesis on upper border group rocks by J. A. V. Douglas (1961). The positions on the Ab and Fe ratio diagram, of analyzed marginal border group rocks, are shown in Fig. 120, and compared with the curve for the liquids and average layered rocks. The three chilled rocks and the two coarser, average rocks of the

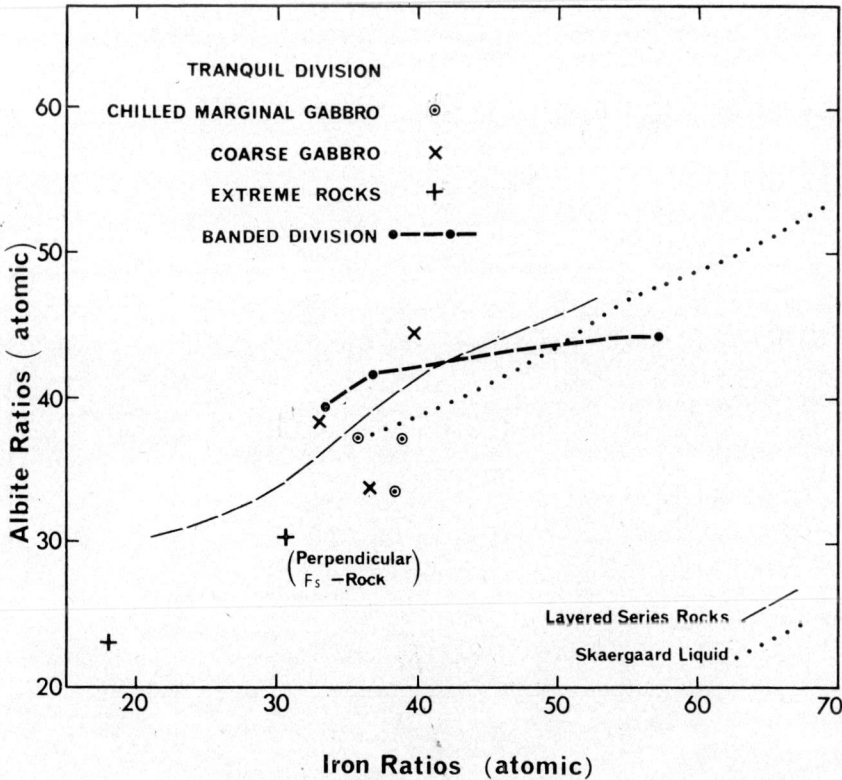

FIG. 120. Albite and iron ratios plot of Marginal Border Group rocks.

M

tranquil division all plot fairly close together, as expected, while the rocks of the banded division trail off, also in the expected direction. The outer rocks of the banded division (4443, 4444 and 4298) correspond in plotted position to LZa rocks, and the innermost to MZ rocks. The banded division rocks probably have 30 to 50% of primocrysts, and the points should lie closer to the average-rock curve than to the liquid curve but this is not shown by this small number of analyses.

The analyzed perpendicular-felspar rock (1851) has the same albite ratio as the estimated earliest layered rock (cf. Tables 7 & 6), although the crescumulate plagioclase is of a relatively albitic composition because formed from a supercooled magma. The iron ratio, on the other hand, is nearer to that of the first liquid, as would be anticipated since there is little accumulation of early ferromagnesian phases in this rock. The other analyzed extreme rock of the tranquil division, the picrite (1682), has a somewhat lower iron ratio than the estimated earliest layered rock of the HZ (cf. Tables 7 & 6), and this also is to be expected as it contains a larger amount of the earliest olivine, which should have the lowest possible iron ratios for ferromagnesian minerals from this magma. The albite ratio for the picrite is not a significant value, since there is so little plagioclase that

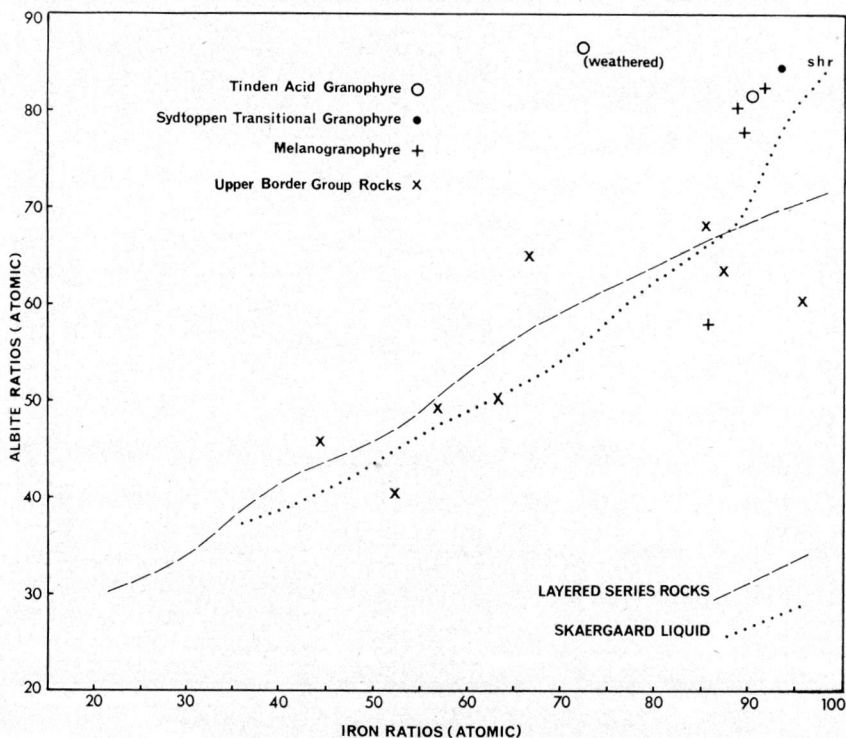

FIG. 121. Albite and iron ratios plot of Upper Border Group rocks and Granophyric Differentiates. Curves for Layered Series shown for comparison (liquids curve ending with Sandwich Horizon rock).

its normative composition is liable to have been affected by the amounts of Al and Na in minerals other than plagioclase, which is not taken into consideration in the calculating of the ratio.

The upper border group rocks are plotted in the same way, on an Ab and Fe ratios plot, in Fig. 121, and compared with the liquids and average layered rocks. The five rocks classed as UBGα, on the basis of the primocryst plagioclase, have the lower ratios, as would be expected (except 4163 which is probably contaminated). All the UBG rocks (Table 8) should have a higher iron ratio than the contemporaneous layered rocks because they have little or no primocryst ferromagnesian mineral and this is true in the majority of cases. The rock classed as UBGβ has an unexpectedly high iron ratio, probably because of metasomatic effects. The melanogranophyres, the Sydtoppen transitional granophyre, and the Tinden acid granophyre, which are believed, from their general mineralogy and chemistry, (Table 9) to represent successive late liquids, are, in fact, rather clustered together on the albite and iron ratio graphs (Fig. 121), their Ab ratio being between 78 and 84 and the Fe ratio between 89 and 93. The plot does not show their relationships satisfactorily because the effects of the incoming of the low-temperature orthoclase and quartz is not taken into account, and there has probably been some oxidation which affected the iron ratio. The rather magnesian pyroxenes of these rocks (cf. Ch. III) is further evidence of oxidation. Despite some anomalies, however, the graph shows that the early UBG rocks have low albite and iron ratios and that the successively later rocks have increasingly higher ratios. The graphs of the albite and iron ratios should be compared with the plots of these ratios in Fig. 117.

TRACE ELEMENT VARIATION IN THE MINERALS, ROCKS, AND RESIDUAL LIQUIDS

The first systematic investigation of the abundance of the trace elements in the Skaergaard intrusion was carried out by a semi-quantitative spectrographic method (Wager & Mitchell, 1951). The results gave the relative amounts in the different minerals and the general trend of variation in the rocks, resulting from crystal fractionation. This early work still gives the best available, overall picture of the Skaergaard trace element variation, especially for the separated minerals. Since then, however, new analytical methods have been used by E. A. Vincent, A. A. Smales and others, which have produced data for elements not previously considered and, also, more reliable figures, in some cases, than had previously been obtained by emission spectrography. A summary of the best trace element data now available on the rocks is presented graphically in a later section of this chapter. For the actual analytical data on the rocks and minerals, the original papers should be consulted.

a) *Trace elements in the minerals*

The trace element amounts, determined spectrographically on separated cumulus minerals, are presented graphically in Fig. 122. The minerals were separated from the

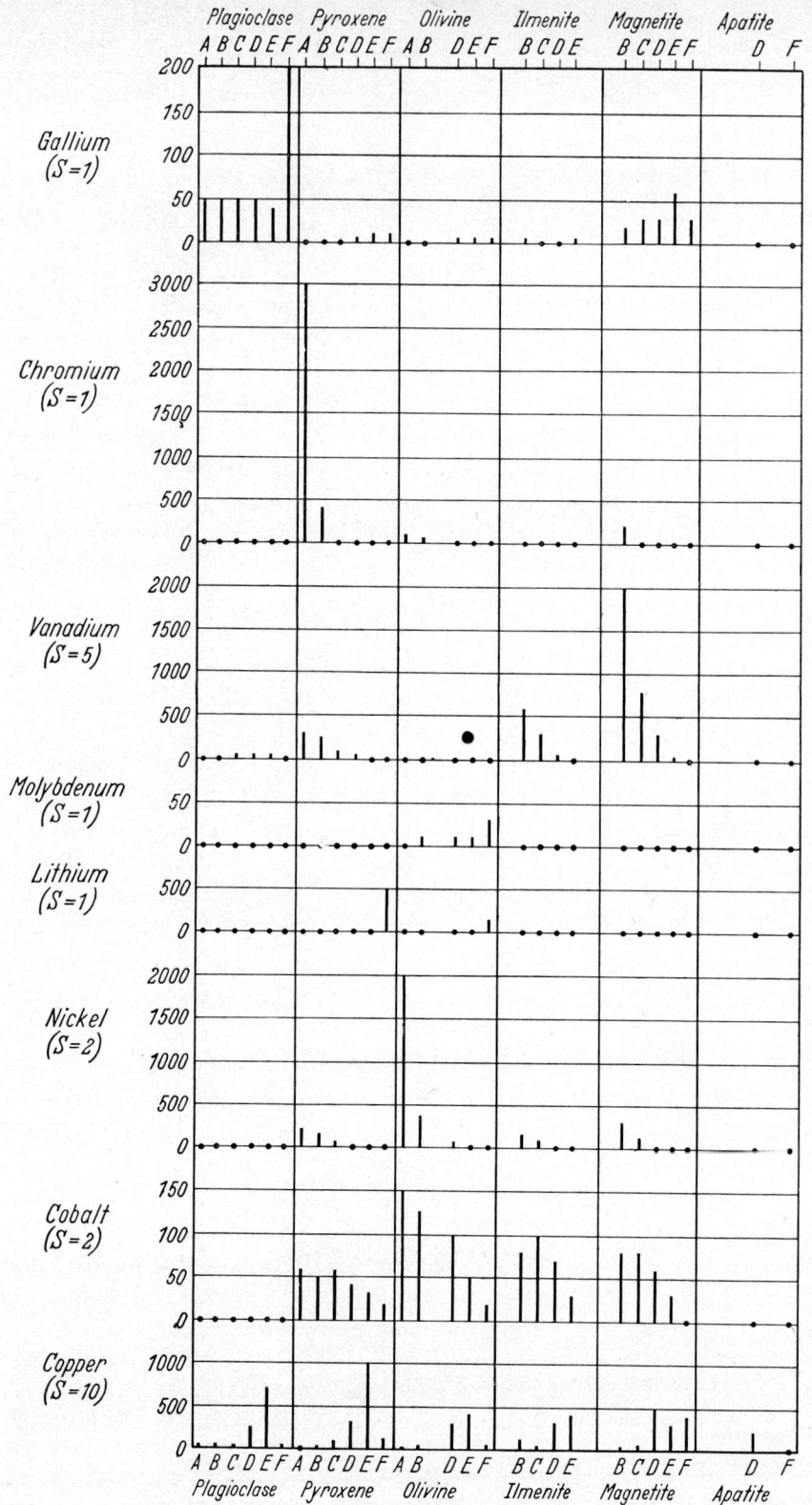

FIG. 122, Gallium—Copper.

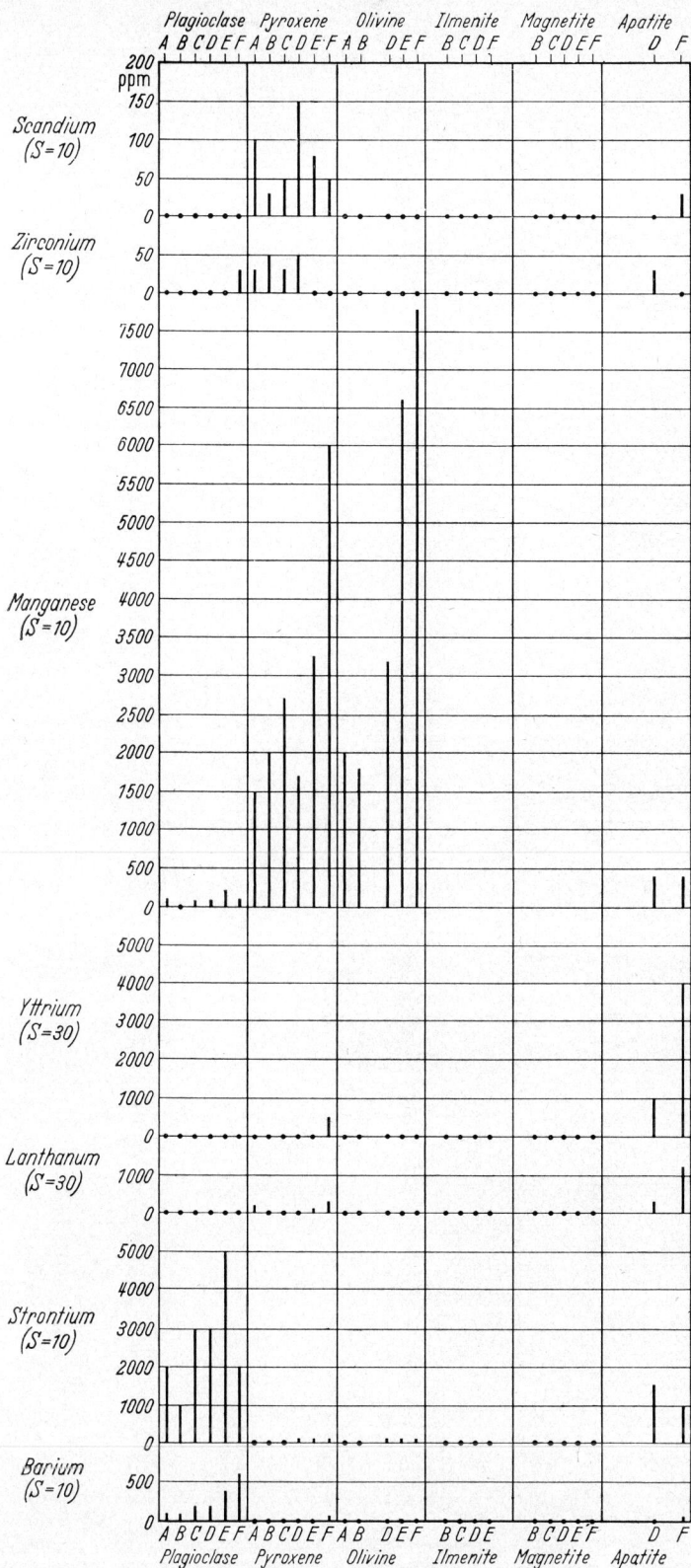

N.B. For caption
see overleaf on
p. 182

NOTE : • = 5 ppm or less

FIG. 122, Scandium—Barium.

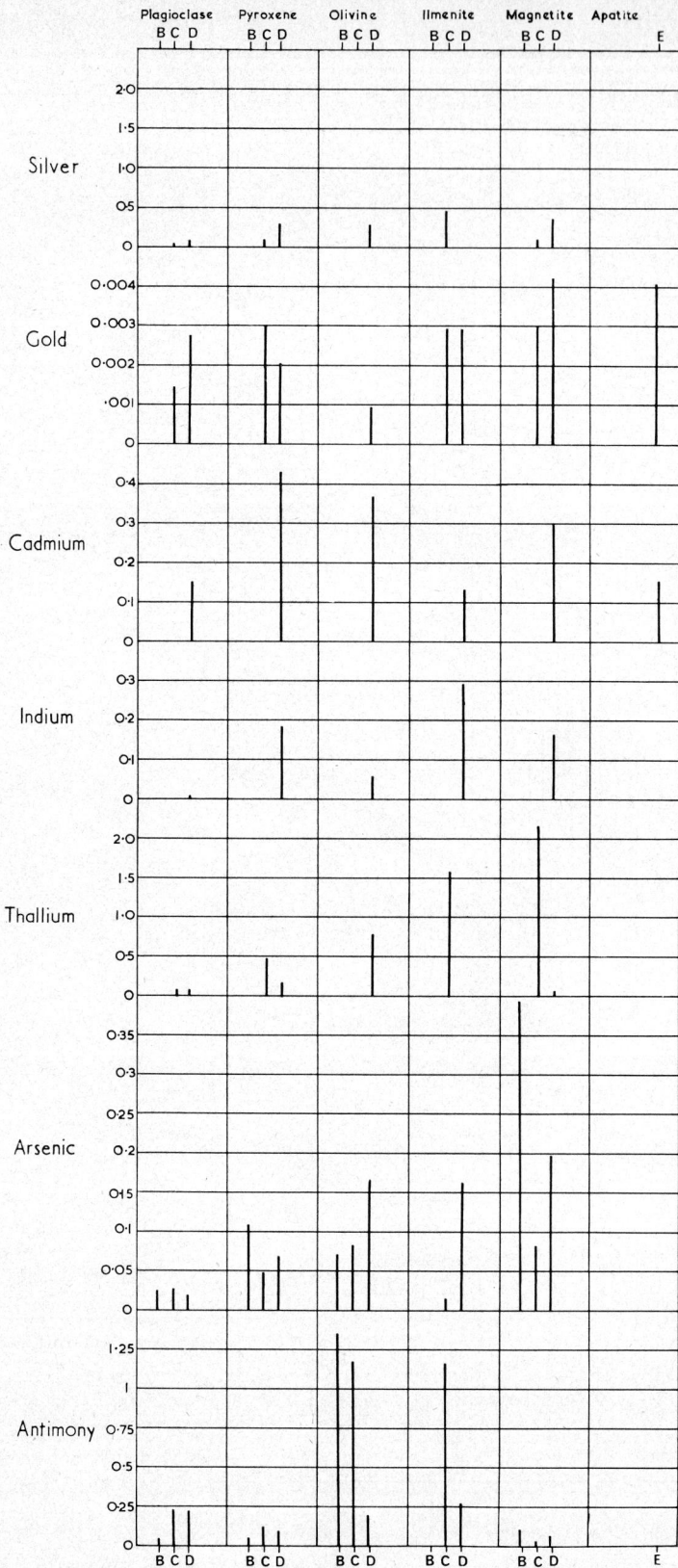

Fɪɢ. 122

Trace elements (p.p.m.) in the minerals of the layered series. Dots in any column indicate amounts below the sensitivity, S. MZ rocks; D, from UZa rocks; E from UZb rocks; F, from UZc rocks. (Ga to Ba after Wager & Mitchell, 1951; Ag to Sb from recent data discussed in text.)

A, from MBG rocks; B, from LZ rocks; C, from

Fɪɢ. 122, Silver—Antimony.

relatively coarse rocks in sufficient quantities for spectrographic analysis, by hand picking or simple magnetic methods, thus avoiding the danger of contamination by heavy liquids (except that a bromoform separation was used for apatite). Where possible, the minerals were separated from rocks belonging to six different fractionation stages: A, the tranquil division of MBG; B, the LZ stage; C, the MZ stage; D, the UZa stage; E, the UZb stage; and F, the UZc stage. For three elements, i.e. indium (Wager, Smit & Irving, 1958), gold (Vincent & Crocket, 1960) and cadmium (Vincent & Bilefield, 1960), there are now more reliable data obtained by radioactivation analysis, but the original diagram has not been modified as it is only intended to show in a general way the distribution of the elements in the various minerals.

The diagrams of Fig. 122 show that, usually, an element exhibits a marked preference for certain minerals. They show also that the variation in the amount of an element in the same mineral phase at different fractionation stages is generally in conformity with Goldschmidt's rules. On the other hand, gold has been shown by Vincent and Crocket (1960) to be present in nearly equal amounts in the different crystal phases (Fig. 122). Specific comments on the distribution of the various trace elements in the minerals so far investigated are made in the concluding section of this chapter.

b) *Trace elements in the rocks and estimates of the amounts in the successive residual liquids*

The variation in the trace element composition of selected rocks has been established in a broad way by emission spectrography (Wager & Mitchell, 1951) and with greater precision by radioactivity methods, largely the work of A. A. Smales and A. E. Vincent. The data are extensive and here are only summarized graphically (Pl. X): for references, the reader is referred to the specific papers quoted below in the sections dealing separately with each element.

The results of the analyses for trace elements of the chilled marginal olivine-gabbro, sometimes supported by data from the other rocks of the tranquil division, are assumed to give the trace element abundances of the original Skaergaard magma. Using the same graphical method as for the major elements, the average composition of the hidden zone rocks and the composition of the successive residual liquids are given on the graphs, when there are sufficient data to justify handling them in this way. The trace element contents of the sandwich horizon rock, the Sydtoppen transitional granophyre, and the Tinden acid granophyre, which represent three of the later liquids produced by fractionation, are shown on the right-hand side of the graphs, where they are plotted against arbitrary abscissae.

The trace elements, like the major elements, fall into three groups which, in summary, are:

1. Elements which enter, readily, the early-stage cumulus minerals; for these, the amounts in the rocks and liquid decrease with fractionation, e.g. Cr and Ni.

2. Elements which enter, most easily, the cumulus minerals forming in the middle and rather late stages of the fractionation; for these, the amounts in the rocks and residual liquid rise to a maximum, and then fall, e.g. Co and Sr.

3. Elements which tend to remain in the residual liquid and are, therefore, most abundant in the latest rocks, especially the intermediate and acid rocks produced by filter-press action, e.g. Rb and Ba.

It is interesting to note that the trace elements often show a relatively greater change with fractionation than the major elements. Data on trace elements may, therefore, provide a better index of the fractionation stage reached by a particular rock than do the major elements, as argued by Robson and Spector (1962).

ISOTOPIC COMPOSITIONS OF OXYGEN AND STRONTIUM FROM THE SKAERGAARD INTRUSION

a) *Variation of O^{18}/O^{16} in the rocks and separated minerals*

Schwander (1953) investigated the ratio of O^{16} to O^{18} in three layered rocks of the Skaergaard intrusion and showed: (1) that strong differences exist and (2) that one of the rocks was exceptionally low in O^{18}. Since then, determinations of the O^{18}/O^{16} ratio on a number of Skaergaard rocks have been made by Taylor and Epstein (1963) and they have discussed in detail the meaning of the unexpectedly large variations found.

The values, i.e. the deviation, in parts per thousand, in the O^{18}/O^{16} ratio from that of standard Hawaiian ocean water,* are shown for a number of Skaergaard rocks and separated minerals in Fig. 123, which is based on a diagram by Taylor and Epstein (1963, fig. 2). Olivines and pyroxenes have lower δ values, while plagioclases have higher values, than the whole rocks from which they come, with the exception of plagioclase and pyroxene in the chilled marginal gabbro. Taking into consideration the relative amounts of these three dominant minerals, the summation of the δ values of the minerals is reasonably close to the value obtained for the rock. There is little change in the δ values of the plagioclase, pyroxene and ilmenite in passing from low to high (UZa) in the exposed layered series, but in the next higher rock analyzed, the fayalite ferrodiorite of the UZc, there is a marked fall in the δ value of the plagioclase and pyroxene and a corresponding fall in that for the rock. The formation of the fayalite ferrodiorite was the result of strong crystal fractionation at the stage when the amount of residual liquid was only a few per cent of the whole. The two melanogranophyres, regarded as produced by filter-press action, continue the trend to lower δ values and the Sydtoppen transitional granophyre, resulting from more vigorous filter-pressing, has the lowest δ value of any rock so far investigated by Taylor and Epstein. This trend in δ values is the opposite

---

$$* \quad \delta \; = \; \left(\frac{R \text{ sample}}{R \text{ standard}} - 1 \right) \; \times \; 1000, \quad \text{where R is the ratio of } O^{18}/O^{16}.$$

from that shown in passing from basic, through intermediate, to acid rocks in the Southern California Batholith (Taylor & Epstein, 1962, pp. 680–2).

The latest rock from filter-pressing, the Tinden acid granophyre, shows a reversal of the trend. It has been thought possible that remelted acid gneiss inclusions may have contributed to the material of the melanogranophyres and the transitional and acid granophyres. Taylor and Epstein, therefore, determined the δ values for two specimens of acid gneiss from the country rocks and for an inclusion, in the marginal border group, believed to have been originally acid gneiss. All these were found to have high δ values, like those of certain intermediate and acid rocks from California and elsewhere, but very different from the granophyric rocks of the Skaergaard. The oxygen isotope investigations thus support the view (see p. 142) that assimilation of acid gneiss cannot be the main source of the material of the melanogranophyres and the transitional granophyre. The higher δ value of the acid granophyre, however, is a change in the

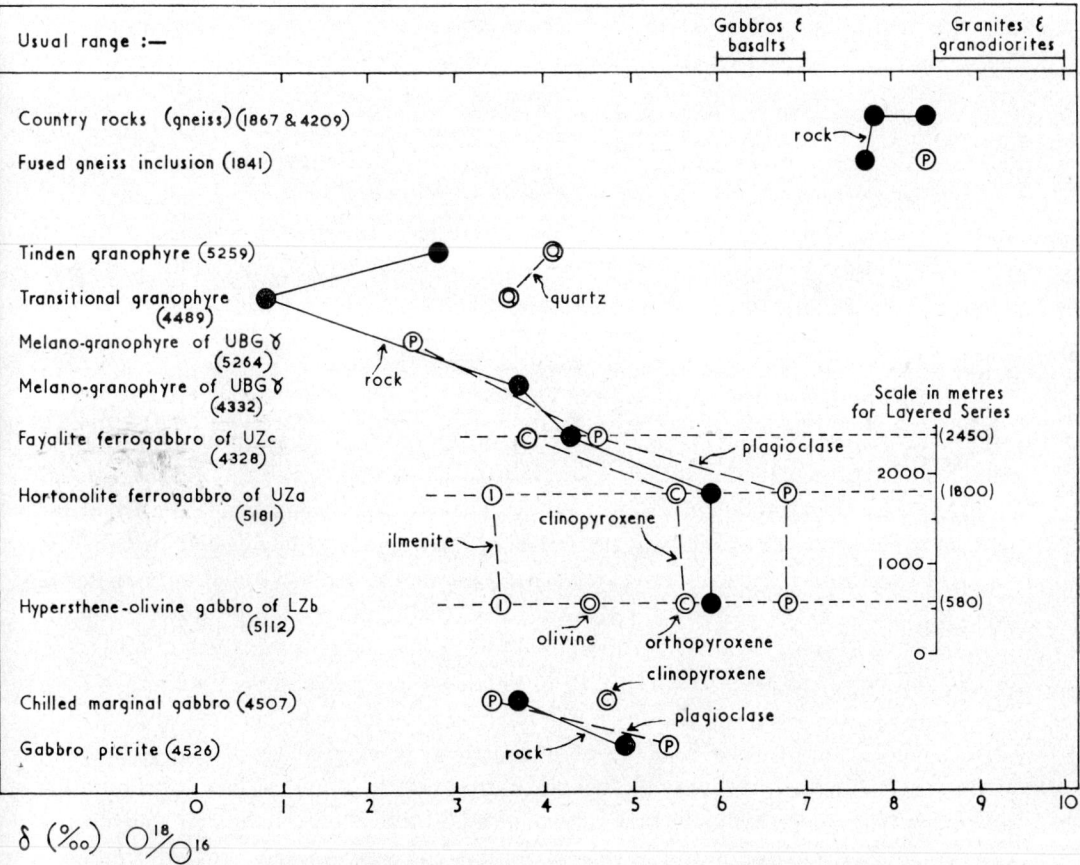

FIG. 123. A graphical representation of the oxygen-isotopic analyses of minerals and rocks of the Skaergaard intrusion. (After Taylor & Epstein, 1963.)

direction which would be caused by contamination by a small amount of acid gneiss, but Taylor and Epstein (1963) consider that the higher δ values may equally be due to the fractionation which, by this stage, was perhaps of the same kind as occurred in the Southern California Batholith (Taylor & Epstein, 1962).

The constancy in the δ values of the rocks and individual minerals from LZb to UZa makes it likely that the hidden layered series would have similar values. If the intrusion is behaving as a closed system with respect to oxygen, presumably 95% of the layered rocks must have had a δ value of about 5·9 and this must be very close to the value for the initial magma, since the effect of the lower δ values for the few per cent of UZb and UZc rocks, and the granophyres, must be slight. The value of 5·9 is just within the range of δ values (see top line of Fig. 123) for the basalts and gabbros from other regions so far determined by Taylor and Epstein (1962, p. 470). The chilled marginal olivine gabbro, regarded as approximating to the composition of the original Skaergaard liquid for most elements, has a δ value of 3·6; so far as the oxygen isotopes are concerned it cannot, therefore, be representative of the initial Skaergaard magma. The δ value for the pyroxene from this rock is greater than that for the plagioclase, which is the reverse of what has been found in the case of the other three pairs of analyses of these minerals. Taylor and Epstein (1963) consider that the data from the marginal border rock indicate extreme oxygen-isotope disequilibrium and they suggest that the minerals of this rock have undergone oxygen-isotope exchange with meteoric water, which is very low in O^{18} having, in fact, a negative δ value.

The cause of the trend to lower δ values in the later layered rocks and the later, filter-pressed fractions as far as the transitional granophyre, is still obscure. Some process has caused the Skaergaard liquids to become increasingly depleted in O^{18} relative to O^{16}, which is the opposite of the changes found in passing from gabbro to tonalite, granodiorite, and leucogranite in the Southern California Batholith (Taylor & Epstein, 1962, pp. 680–3). From LZ to UZa, the δ values of the liquid probably remained at about 5·9, like that of the rocks. The melanogranophyres and the transitional granophyre are themselves considered to represent late liquids and hence the trend for the δ value of the liquid throughout the whole sequence must follow closely that for the rocks. If this trend is caused by the separation of crystals then, up to UZa, the material separating must have averaged the same δ value as the liquid, while in the later stages it must have had a higher δ value in order to push the liquid to the lower δ values observed. The separation of plagioclase in especial abundance during UZb and UZc might have contributed to this latter change in trend but it is not clear whether it is sufficient to account wholly for it; the separation of quartz (as suggested by Taylor & Epstein, 1963, p. 61) cannot contribute to the change in trend of the liquid as it is not a cumulus mineral. Other processes are tentatively suggested by Taylor and Epstein but on the evidence so far available it is difficult to understand why the Skaergaard trend, so definitely due to crystal fractionation, is in the opposite direction from that of the Southern California Batholith rocks, unless the latter are not due to fractionation.

If oxygen from meteoric water, with its low δ value, were absorbed into the Skaergaard intrusion in large enough quantities, it could cause the observed trend. The Skaergaard initial magma was probably relatively dry and it may have been able to absorb some water from the surrounding rocks until it became saturated. It was also a fairly reduced magma and absorbed water would react with the ferrous iron to produce ferric, provided hydrogen could escape. If water were capable of diffusing into the magma, then hydrogen should be capable of diffusing out, and so long as Fe^{++} was available, the extent of this reaction would be dependent upon the effectiveness of the diffusion process. The increase of Fe^{+++} at the expense of Fe^{++} by this process has been suggested as the reason for the greater amount of ferric iron in the exposed rocks of the Skaergaard intrusion than can reasonably be accounted for on the basis of the amount in the chilled marginal gabbro, assuming it to have the composition of the initial liquid (Wager, 1960, p. 384). If there has been oxidation of the chilled contact rock after solidification, as suggested by Taylor and Epstein (1963, pp. 65–7), then the real value of Fe^{+++} in the original magma is lower than previously assumed, and there is a somewhat greater amount of Fe^{+++} unaccounted for, unless the oxidation of Fe^{++} to Fe^{+++} in the main body of magma, as postulated here, has occurred. Some absorption of water from the surroundings, therefore, seems likely to have occurred and if this were meteoric water with the usual low δ value, it would undoubtedly contribute to the low δ values of the later Skaergaard rocks. However, the amount of water which would have to be absorbed, to produce the excess of ferric iron found, is an order of magnitude too low to account for the change in oxygen-isotope ratios by this means. Thus the exceptional trend of δ values seems to be the result of crystal fractionation, as Taylor and Epstein believe. Like the very strong iron enrichment, the trend to lower δ values seems to be another example of the characteristic course of fractional crystallization in the Skaergaard intrusion.

b) *Variations in the Sr^{87}/Sr^{86} ratio of various rock types*

Determinations of the isotopic composition of strontium in various rocks of the Skaergaard intrusion have recently been made by Hamilton (1963). Because of the small percentage mass-difference between Sr^{87} and Sr^{86}, it must be assumed that no significant differences in the relative amounts of these isotopes will develop as a result of fractional crystallization and Hamilton, in fact, finds that the Sr^{87}/Sr^{86} ratio in the chilled marginal gabbro and in the layered series is almost constant, and similar to that of common basalts (Faure & Hurley, 1963). Significantly higher values of this ratio, however, are found in the Sydtoppen transitional granophyre and the Tinden acid granophyres, suggesting that these rocks have incorporated some material relatively rich in Sr^{87}. The age of the metamorphic complex is around 1800 million years (Wager & Hamilton, 1964) and the acid complex must contain appreciably more radiogenic Sr^{87} than basalt. From the strontium-isotope data it would appear that the

Sydtoppen and Tinden granophyres contain a significant amount of material derived, ultimately, from the acid gneisses of the metamorphic complex. At present this hypothesis, for the Sydtoppen transitional granophyre, is at variance with the oxygen-isotope data.

EVIDENCE FOR THE DEVELOPMENT OF AN IMMISCIBLE SULPHIDE LIQUID AND ITS COMPOSITION

Small patches of sulphide minerals, containing a little magnetite and an acicular silicate, probably amphibole, can be found in very small amounts in all the fresh rocks of the intrusion (see p. 55). Sometimes, the composite sulphide units have a form and texture suggesting that they are the result of crystallization of an immiscible globule of liquid (Ch. III, Fig. 28); more usually their shape is irregular, probably due to modification of an original globular form by the growth of the minerals from the main silicate liquid while the sulphide portion was still liquid.

Experimental evidence, provided long ago by Vogt (1916–8), has shown that sulphur, even when present in only very small quantities in silicate melts, forms an immiscible sulphide liquid, of low crystallization temperature. It is therefore suggested that the sulphides of the Skaergaard rocks were at one time immiscible droplets in the silicate magma. Evidence for this view comes not only from the globular form of some of the sulphide patches but also from their complex mineral composition, and the fact that the amounts of sulphide in adjacent layers may sometimes vary by a factor of 10 or 20. The high concentration of sulphides is by no means always in the melanocratic layers, but is more often there than in the leucocratic layers. The fact that the amount of sulphide in adjacent layers can vary greatly, indicates that the sulphide material, at one stage, occurred in discrete units capable of independent variation like the cumulus minerals of the layered series. At the time of deposition of the cumulus crystals, the temperature would certainly have been too high for crystals of chalcopyrite or pyrrhotite to have existed; thus the discrete sulphide units which collected together with the cumulus crystals must have been immiscible sulphide droplets rather than crystals.

The sulphide-rich patches, consisting of a special assemblage of minerals, can be regarded as a sulphide-rich rock within the main body of silicate–oxide rock. The separation of a fair sample of the small, composite units of sulphide rock for analysis has proved impossible, and the approximate chemical composition of the sulphide liquids has only been obtained by indirect means, except for the large pyrrhotite patches in the Tinden acid granophyre, which were successfully separated and analyzed (Table 11). Microscopic examination of the sulphide patches shows that up to the UZc stage, the immiscible liquid was dominantly a copper sulphide and that it then, abruptly, became dominantly an iron sulphide. Using the standard composition of the sulphide minerals and the rough modal estimates of the amounts of the different minerals in the sulphide patches, the general composition of the immiscible liquids at different stages has been roughly deduced (Table 11). The amounts of nickel and cobalt in the sulphides, have also been obtained by an indirect method. Precise values for Ni and Co, in various rocks which have a considerable range in the amount of the sulphides, were obtained by radioactivation methods, and by correlating the observed Ni and Co quantities in the various whole rocks with the different amounts of the sulphides in the rocks, a rough figure for the amount of these elements in the sulphide part of the rock

TABLE 11

Estimated composition (wt. per cent) of the immiscible sulphide liquid at various fractionation stages (Wager, Vincent & Smales, 1957, table VIII)

	I	II	III
S	23	34	38
Fe^{++}	10	61	62
Cu	53	2·3	0·015
Mn	...	...	0·03
Ni	<0·025	0·06	0·0001
Co	2	0·02	0·02
Fe_3O_4	10	2	...
Silicate	ca. 2	ca. 1	...

    I. Copper sulphide liquid at UZa levels.
   II. Iron sulphide liquid at UZc levels.
  III. Iron sulphide liquid at the Tinden acid granophyre stage (actual analysis).

has been established (see Wager, Vincent & Smales, 1957, pp. 873–81). Combining the results from the modal and radioactivation analyses, some features of the compositions of successive sulphide liquids can be seen (Table 11). The estimated partition ratio between sulphide and silicate liquids for Cu, Ni and Co are given in Table 12. Apparently, during 98% of the solidification process, the sulphide liquid was copper-rich, and the concentration of copper in the sulphide liquid was approximately 1000 times higher than in the contemporary silicate liquid, while the concentrations of Co and Ni in the sulphide liquid were about 700 and 100 times higher, respectively. The later, iron-rich sulphide liquid of the UZc was only about 120 times richer in Cu, 20 times richer in Co, and 50 times richer in Ni, than the silicate–oxide liquid with which it was in equilibrium. The latest iron sulphide, analyzed from the acid granophyre, has a remarkably low nickel content, as already remarked (p. 57), but it is not certain whether the pyrrhotite liquid was in equilibrium with the acid magma or with some earlier magma poorer in Ni, from which it became separated by a filter-press process.

In attempting to generalize on the conditions for the separation of an immiscible sulphide magma from a silicate–oxide magma, the system may be treated as binary and the compositions plotted in terms of the proportion of the total silicate and oxide to total sulphide (Fig. 30). From the Skaergaard data, it is estimated that the silicate–oxide liquid in equilibrium with a sulphide liquid contained only about 0·05% of sulphides dissolved in it, while the sulphide liquid included about 3 to 10% of iron oxides and silicates.

TABLE 12

Partition ratios (sulphide liquid/silicate liquid) for copper, nickel and cobalt (Wager, Vincent & Smales, 1957, table X)

	I	II	III
Cu	1000	120	1
Ni	< 100	50	0·04
Co	700	20	30

I. At the bornite-rich stage, around UZa.
II. At the pyrrhotite stage, UZc.
III. At the Tinden acid granophyre stage (but see text).

SORTING OF THE ELEMENTS DURING FRACTIONATION

The elements present in the Skaergaard magma when it was originally intruded were, of course, uniformly distributed throughout the whole liquid.* On the basis of the analytical data available for the major, and many trace elements, the way in which the elements sorted themselves into the various mineral phases and rocks during the cooling and differentiation of the Skaergaard magma can be evaluated. It is now proposed to review this data systematically, considering each element in the order in which it occurs in the Periodic Table which, for the present purposes, is better than the order of increasing size of the ions used in the earlier discussion (Wager & Mitchell, 1951, pp. 177–96). The elements for which there are satisfactory data are indicated in thick print (apart from Pb, since determined) in the Periodic Table given in Fig. 124.

The overall picture of the distribution of each element in the different parts of the

* The size of the Skaergaard intrusion, before erosion, is estimated later (p. 204) to have been approximately 500 cubic kilometres. Assuming the rocks to have an average density of 3·0, the weight of the whole intrusion was $1·5 \times 10^{12}$ metric tons. The amount of any element in metric tons in the initial magma can be obtained from the analysis of the chilled olivine gabbro (Table 4) by multiplying the parts per million of the element by $1·5 \times 10^{6}$; the amount of copper in the intrusion, for example, could be 190 000 000 metric tons, and the amount of gold, 7000 metric tons.

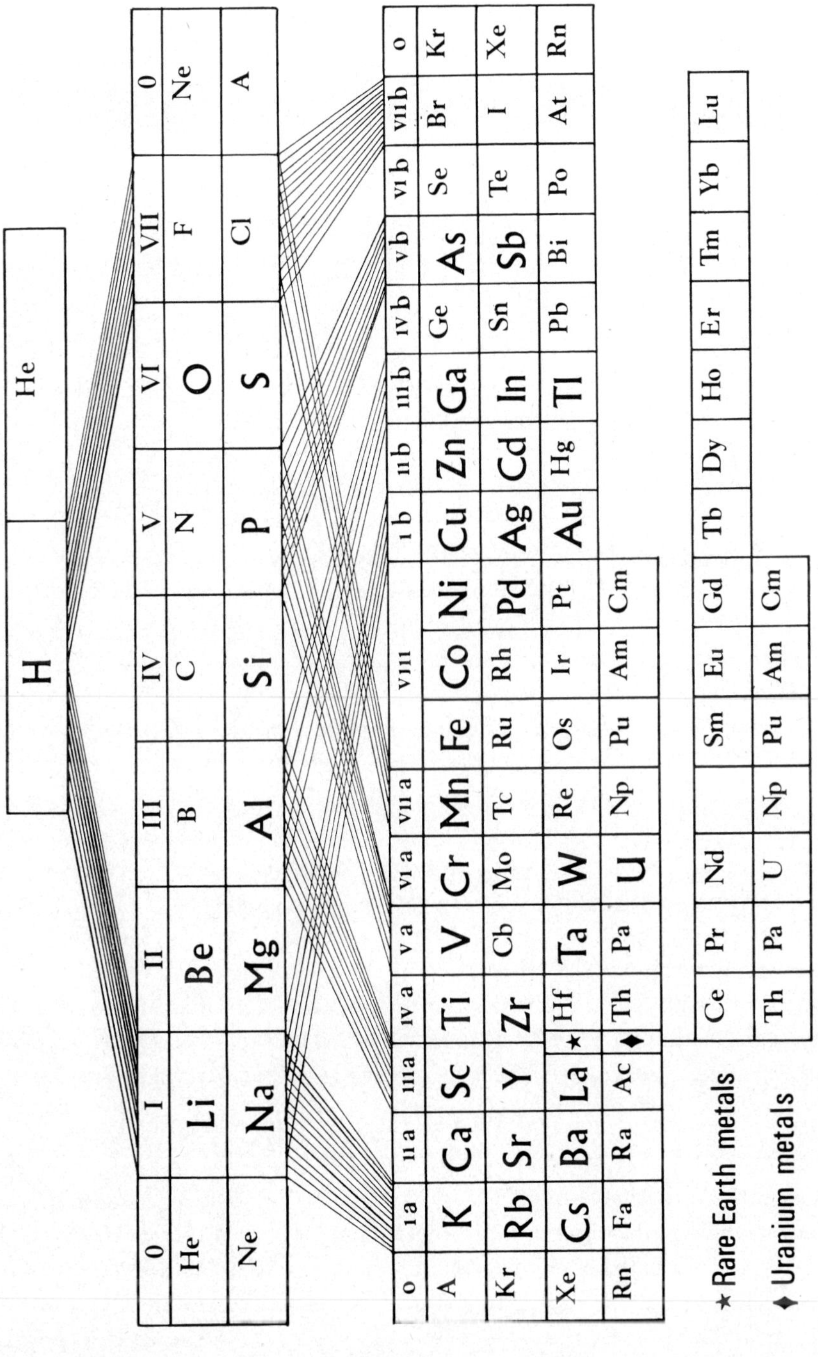

Fig. 124. The Periodic System of the Elements. The elements determined so far in the Skaergaard intrusion are in bold type. [(Pb determined since this diagram was prepared).

intrusion is presented graphically, the data being too voluminous to be easily tabulated In some cases, the early trace-element data obtained by emission spectroscopy (Wager & Mitchell, 1951) have been discarded, since there are better data available from radio-activation or other methods of analysis. The curves for the rocks are drawn for the average rocks; the points for rocks in which sorting of the cumulus crystals has produced extreme types, felspathic or melanocratic, may fall very far from the average-rock curve. Certain rocks represent the compositions of former liquids; other liquid compositions have been estimated graphically by the method described earlier (p. 168) and in the case of many elements, a curve for the liquid has been constructed.

a) *H*

The amount of *hydrogen* in the original magma can only be estimated from the value for H_2O^+ in the analyses of the chilled marginal border rock and the other rocks of the tranquil division of the marginal border group. The values range from 0·24 to 1·01%, the latter being for the rock one metre from the margin which may have absorbed water from the country rock, as suggested by some of the oxygen-isotope data (Taylor & Epstein, 1963, p. 65). The mineralogy of the whole intrusion suggests a dry magma, and perhaps a value of 0·25% H_2O, although little more than a guess, is the best estimate at present.

The cumulus minerals separating from the magma are all anhydrous and the rocks should, therefore, contain only the water present in the trapped magma, ranging from about 40% downwards to 10 or 5%. The amount of H_2O in the LZ rocks, obtained by analyses, varies from 0·21 to 0·81%. The amount of water in these rocks probably averages about 0·4%; if this is in say 30% trapped liquid, the magma at this stage must have contained about 1·2% H_2O, which is appreciably more than in the initial magma. For the later rocks, the values for H_2O^+ in the analyses are even more variable and seem to lead to no useful generalizations.

The data on ferric iron (p. 198), and possibly that on the oxygen isotopes, (p. 187) suggest that some meteoric water was absorbed by the Skaergaard magma, resulting in the oxidation of ferrous iron and the liberation of hydrogen which, it is suggested, diffused out of the system. Although of great interest, the question of the hydrogen content of the rocks and magma at successive stages of cooling is, at present, highly speculative.

b) *Group I: Li, Na. Group I(a): K, Rb, Cs*

Considering the cumulus minerals, *Lithium* (Pl. X) is most abundant in the late pyroxenes and fayalites, presumably replacing Mg and Fe^{++} (Wager & Mitchell, 1951). It was found to be below the sensitivity of the spectrographic method (2 ppm) in the plagioclase and iron ores. In the early rocks, it is less abundant than in the magma and the amount in the magma must have built up slightly. In the rocks representing the late liquids, the amount reaches at least 15 ppm (Douglas, 1961). The ratio Li/Mg is about 0·0004 in the original magma and in the LZ rocks, while it is between 0·005 and 0·015 in the rocks derived from the late liquids.

Sodium (Pl. X), being a major constituent of plagioclase which is present in all the rocks, only changes by about a factor of 2 during the whole process of fractionation. The range in Na_2O content of the liquid is from 2·4% (Na = 18 000 ppm) in the original magma to 4·3% (Na = 32 000 ppm) in the acid granophyre.

Potassium (Pl. X), over the greater part of the differentiation series, is present as a trace element mainly replacing Na and Ca in the plagioclase. It also enters into brown mica which crystallized in very small amounts from the trapped liquid, in the lower zones of the layered series. In the initial magma, K_2O is present to the extent of 0·25% (K = 2200 ppm) and there is less than this in the average rocks of the LZ. In the latest rocks, K_2O reaches 3·8% (K = 32 000 ppm) and forms potassic felspar as a separate phase.

Rubidium (Pl. X) distribution in the rocks is now known from a few radioactivation analyses by Cabell and Smales (1957) and from X-ray fluorescence analyses by Hamilton (1963) and Douglas (1961); the earlier spectrographic analyses have been rejected. The amount in the original magma was about 4 ppm, and in the LZ, MZ and UZa rocks it is only about 2 ppm but it may have been slightly more abundant in the HZ as judged from the graphs. In the UZb there are 13 ppm, and in the sandwich horizon rock, 23 ppm. In the transitional and acid granophyres formed from filter-pressed liquids, the amounts found were 47 and 96 ppm, respectively. Rb is here presumably entering late orthoclase and, no doubt, the hydrothermal minerals. From the LZ rocks to the latest granophyre fractions there is an increase in the amount of Rb by a factor of about 50.

Caesium (Pl. X), determined by radioactivation analysis (Cabell & Smales, 1957) is about 50 times less abundant, on the average, than rubidium. In the initial magma it is present to the extent of 0·1 ppm; in the LZ, MZ and UZa it is less than this, while in the later rocks it is erratically present in amounts from 0·2 to 0·5 ppm. It may be slightly concentrated in early rocks with traces of mica, as indicated by some unpublished spectrographic data obtained by C. J. Liebenberg and S. R. Taylor.

c) *Group II: Be, Mg. Group II(a): Ca, Sr, Ba*

Beryllium (Pl. X) in all the rocks is apparently less than 10 ppm (Wager & Mitchell, 1951).

Magnesium (Pl. X) falls from the usual values for basic rocks to surprisingly low amounts, while the rock remains basic or intermediate in silica percentage, and contains abundant iron-rich pyroxene and olivine. This is one of the striking features of the Skaergaard differentiation, and the reason is discussed in Ch. IX.

Calcium (Pl. X) only varies from 12% down to 9% CaO (92 000 to 15 000 ppm of Ca) during 98% of the solidification process. In the later stages it falls rapidly to a value characteristic of alkali granites.

N

Strontium (Pl. X), as can be seen from the mineral data summarized in Fig. 122, enters plagioclase most readily, and apatite to a moderate extent, while little is replacing calcium in the augites. The Sr content of plagioclase, and also the ratio of Sr/Ca, rises from that in the early plagioclases (about An_{60}) to a maximum in the plagioclase of UZa (about An_{40}), after which it falls. The amount in the chilled original magma, determined spectrographically (Wager & Mitchell, 1951) is 350 ppm, while in a related rock, 4507, determined by radioactivation analysis (Loveridge, Webster, Morgan, Thomas & Smales, 1959) it is 267 ppm. Probably all the early spectrographic data are high, especially those given for the separated felspars in Fig. 122; more, good Sr data on the cumulus minerals are needed.

The data on Sr given in Pl. X are from radioactivation and X-ray fluorescence (McReath, 1963) analysis only; they show a considerable spread, and no curve for the layered series rocks can reasonably be drawn. Some of the spread is clearly related to the amount of felspar in the rock; the felspathic rocks (indicated by F) have consistently high Sr, and the two melanocratic rocks (M) have very small amounts. In detail, however, we cannot offer an explanation for the spread; all that can be said at present is that the average content in the whole of the exposed layered series is about 200 ppm of Sr, an amount which is about the same as that in the original magma. Thus there is little change in the Sr content of the liquid over this fractionation range. The sandwich horizon rock, considering that it is not rich in plagioclase, has a high Sr content (450 ppm), while the amounts in the granophyres fall, reaching 125 ppm in the acid granophyre.

The slight increase in Sr^{87} relative to Sr^{86} in the granophyric differentiates, shown by Hamilton (1963), is considered due to contamination with the acid gneiss of the metamorphic complex (p. 187).

Barium (Pl. X), like strontium, is present mainly in the felspars, presumably occupying the Ca positions but, unlike Sr, it remains low in apatite (Fig. 122). The spectrographic data for the rocks were originally interpreted as indicating that the content of Ba in the initial magma was 40 ppm. This was based on an average of four rocks, one of which has a high value now thought to be due to contamination with acid gneiss. Probably a better value would be 25 ppm, which is about one-tenth the amount of the Sr (compare the rather similar case of Rb and Cs). The amount of Ba in the HZ rocks was presumably a little less than that in the original magma. The Ba content of the residual liquids apparently rose slowly until the filter-press granophyre stages, when the rise became rapid, ultimately giving rocks with about 1000 ppm Ba.

The graphs for the distribution of Mg, Ca, Sr and Ba, shown together on Pl. X, show systematic changes, presumably related to the differing sizes of the ions.

d) *Group III: Al. Group III(a): Sc, Y, La*

Aluminium. The amount of aluminium (Pl. X) does not change much in the average rocks as a result of crystal fractionation, because felspar is an important constituent of all of them except the melanocratic layers, where Al is low.

Scandium (Pl. X) enters pyroxene and apatite rather than plagioclase, olivine, ilmenite or magnetite (Fig. 122). The amount in the original Skaergaard magma is 26 ppm (Kemp & Smales, 1960*b*). The amount in the HZ is probably a little less and it rises to 50 ppm in the MZ and then falls slowly to 7 ppm, and less, in the transitional and acid granophyres Goldsbrough (1962).

Yttrium and Lanthanum have only been determined spectrographically, when the lower limit of detection was 30 ppm (Wager & Mitchell, 1951). Both elements are most abundant in apatite, especially in the later one (Fig. 122). The next most favourable mineral is pyroxene. In the rocks, both elements are below the sensitivity (30 ppm) until UZc, and both are most abundant in the melanogranophyres and acid granophyres where they reach about 300 ppm.

e) *Group IV: Si. Group IV(a): Ti, Zr*

Silicon (Pl. X) changes little in amount during 97% of the fractionation process, and the slight change is, on the whole, a decrease. After 97% solidified, silicon rapidly increases, reaching 70% SiO_2 (327 000 ppm Si) in the acid granophyre.

Titanium (Pl. X) separated early as ilmenite, but the precise stage at which ilmenite became a cumulus mineral has not been established. It also enters pyroxene in appreciable quantities. Nevertheless, titanium accumulated in the magma until the LZc stage when it had reached about 3% TiO_2 (18 500 ppm Ti) and titaniferous magnetite formed as a cumulus mineral. After this the amount in the liquid fell slowly.

Zirconium (Pl. X) enters pyroxene and apatite in appreciable amounts, but is below the sensitivity in the other minerals so far analyzed (Fig. 122). The amount in the original magma was about 50 ppm, and towards the end of the fractionation it increased rapidly to reach 2800 ppm in the transitional granophyre, the amount decreasing in the latest acid granophyre stage.

f) *Group V(a): V, Ta*

Vanadium (Pl. X). The amounts estimated spectrographically (Wager & Mitchell, 1951) are, probably, systematically low from the evidence of the radioactivation analyses by Kemp and Smales (1960*a*, p. 407). However, combining data obtained by both methods, the pattern of behaviour of the element is clear. Vanadium enters early pyroxene, ilmenite, and magnetite (Fig. 122), the amount being greatest in the early, cumulus titaniferous magnetite where it reaches at least 1·7% V_2O_3 (Vincent & Phillips, 1954, p. 12). The amount in all of these minerals falls with fractionation.

The initial magma contained 190 ppm of vanadium. The amount separating in the LZa and LZb rocks is about the same, while probably the amount in the HZ rocks is appreciably less. Where magnetite becomes a cumulus mineral, the amount in the rocks increases sharply; an analyzed average rock of the MZ has 1700 ppm and the amount would probably be considerably greater in the LZc rocks where magnetite first becomes a cumulus mineral. The position of the maximum on the graph is given at the

LZc stage where magnetite first becomes a cumulus mineral, although there is at present no analytical data to show this. The general form of the graph is based on the approximate, but more abundant data of Wager and Mitchell (1951) although these data are otherwise excluded from Pl. X. The amount in the liquid may have reached 300 or 400 ppm at the LZc stage, after which it must have fallen rapidly, reaching, ultimately, very low amounts. In the reduced Skaergaard magma, magnetite became a cumulus mineral at a relatively late stage, by which time the vanadium had become strongly concentrated in the magma, so that the early cumulus magnetite is, no doubt, much richer in vanadium than would be the case in a more oxidized magma.

Tantalum (Pl. X) in the chilled original magma has been determined by radio-activation analysis (Atkins & Smales, 1960, p. 472) to be about 0·5 ppm. In the HZ and LZ it is probably lower than this and then it increases steadily to 1·5 ppm in the UZc and to 3·5 ppm in a melanogranophyre (not shown on the graph).

g) *Group VI(a): Cr, W, U*

Chromium (Pl. X). The amount of chromium in the initial magma seems to have been 170 ppm, using an average of four rather dissimilar spectrographic determinations on four different samples (Wager & Mitchell, 1951). Although the amount of chromium present in the magma was small, a chrome spinel was precipitated for a brief time in the earliest stages, from the evidence of chromite grains within the olivines of the marginal border-group picrites. This indicates the very low solubility of chromite in basic magma, precipitation taking place when Cr_2O_3 was only about 0·025%; this is about the same as the solubility of zircon in granite magma.

Chromium readily enters the early pyroxenes (Fig. 122). However, when magnetite became a cumulus mineral, chromium entered this phase even more readily: one of the early cumulus magnetites has about 1000 ppm of chromium, while the contemporaneous augite has only 350 ppm, and the ilmenite <2 ppm. Later, both magnetite and pyroxene have less than 1 ppm chromium; this is not considered to be the result of the chromium ion being unable to enter the magnetite or iron-rich augite, but the result of extreme depletion of the magma in chromium by this stage.

Chromium in the magma clearly falls rapidly from about 170 ppm in the initial magma down to a very low amount (<1 ppm) at the UZ stage. There is, however, a return to 10 or 15 ppm Cr in the granophyres. It is difficult to understand this unless it be due to oxidizing conditions changing the chromium from a trivalent to a hexavalent state, resulting in some accumulation of hexavalent chromium in the residual liquid because none of the precipitating minerals were able to incorporate it.

Tungsten (Pl. X) was present to the extent of about 0·2 ppm in the initial magma, as determined by Atkins and Smales (1960). There is less in the layered series rocks until the UZa, where the amount is about 0·4 ppm. The amount of tungsten in the rocks so far analyzed reaches its maximum of 1·1 ppm in the melanogranophyre 4332, after which it falls to 0·5 ppm in the acid granophyre of the Tinden Sill.

Uranium (Pl. X). The amount of uranium in the chilled marginal gabbro (the initial magma) is estimated as 0·2 ppm by Hamilton (1959, pp. 8–12) who used neutron activation and fluorimetric methods of analysis. The amount in the exposed layered rocks up to the base of UZc varies from about 0·1 to 0·2 ppm. Although some of the low values may be due to weathering, it is likely that the main cause is the relatively small amount of trapped liquid in the higher rocks. While the UZc ferrodiorites contain 0·3 to 0·5 ppm, the SHR is estimated to contain 1·0 ppm. The Sydtoppen transitional granophyre is the richest in uranium of the Skaergaard differentiates, containing 3·5 ppm; the later Tinden Sill acid granophyre contains only 1·5 to 2·5 ppm. The distribution of α-activity in the various minerals of the layered series, determined by Hamilton using an emulsion method, indicates that there is about the same amount of U and Th in the plagioclases and pyroxenes, rather less in the olivines and iron ores, and considerably more in the apatites.

The amount of uranium separating in the layered rocks was, apparently, less than the amount in the contemporaneous magma, with the result that it accumulated to some extent in the residual magma. It is estimated that the U content of the liquid rises steadily from 0·2 to 3 or 4 ppm in the filter-pressed differentiates.

It had been hoped that the U content or the α-activity might be a guide, in the same way as P_2O_5, to the proportion of trapped liquid, but technical difficulties with the analyses and the possible effects of slight weathering have so far prevented its use in this way.

h) *Group VII(a): Mn*

The greatest amount of *manganese* in the minerals is found in ilmenite, and the amount decreases in the order magnetite, olivine, augite. For each individual mineral, the amount of MnO rises with fractionation, reaching 0·93% in the UZc inverted iron wollastonite (Brown & Vincent, 1963) and 1·44 MnO (11 000 ppm Mn) in ilmenite from the same horizon. The amount in the average rocks reaches a maximum of about 0·5% MnO (4300 ppm Mn) in the UZc ferrodiorite; it then falls rapidly in the filter-pressed differentiates.

i) *Group VIII: Fe, Co, Ni, Pd*

Iron. This element is the only one for which the values for the amounts present in two valency states are available (Pl. X). The initial magma contained an average of 1·23% Fe_2O_3 (9200 ppm Fe^{+++}) and 8·58% FeO (66 000 ppm Fe^{++}). The magma was reduced in character but not uniquely so. In the average material separating, the amount of both FeO and Fe_2O_3 increased until, at the UZc stage, it was 25% and 4%, respectively (Table 6). The total amount of $FeO + Fe_2O_3$ in the magma is estimated to have risen from the initial 9·8% until it reached 22·1% at 98% solidified, after which it fell rapidly (Table 10). Possible reasons for the development of such an extremely iron-rich liquid are considered in Ch. IX.

It has been pointed out above (p. 170) that the amount of Fe_2O_3 in the observed rocks is such that, with the assumptions which give a reasonable, overall, chemical picture for all the other elements, there is $1\frac{1}{2}$ times too much Fe_2O_3 in the observable rocks. It has therefore been suggested (Wager & Deer, 1939, p. 183) that some oxidation took place by means of the reaction $2FeO + H_2O \rightarrow Fe_2O_3 + H_2$, and that the hydrogen diffused out of the intrusion. The total increase in Fe_2O_3 is estimated to be about 0·12% of the whole mass of the intrusion and to produce this, about 0·02% of water would need to have been absorbed. If the absorbed water were meteoric, with the usual O^{18}/O^{16} ratio, this process should affect the oxygen-isotope ratios, but the change in oxygen-isotope ratios shown by Taylor and Epstein (see p. 185) indicates that the process is probably not significant. The limited oxidation of the iron may be due to the slowness with which water could diffuse into the magma, or hydrogen could be removed either by diffusion or by the steaming off of hydrogen gas.

Cobalt and *Nickel* will be considered together. Both enter pyroxene, olivine, ilmenite and magnetite readily, and they are more abundant in the early-formed crystals than in the later (Fig. 122). The best figures for cobalt and nickel in the original magma, determined by radioactivation analysis (Wager, Vincent & Smales, 1957, pp. 875–6) are 55 ppm and 180 ppm, respectively. In the layered rocks, cobalt apparently rises to a maximum of about 100 ppm at the MZ stage and falls gradually to amounts between 2 and 10 ppm in later rocks. Nickel, on the other hand, is about 250 ppm in the LZ rocks and falls steadily to 1 ppm in the UZb rocks. Some olivine-rich rocks of the tranquil division of the marginal border group have 430 ppm Ni. It is none the less likely, as shown by the graph, that the average HZ rocks, because they are felspar rich, have only about 200 ppm Ni. There is an unexpected return to greater amounts of Ni in the late melanogranophyres and acid granophyres. Since these rocks are not thought to be made up of considerable proportions of acid gneiss of the metamorphic complex, the reason for the return of Ni, and to some extent Co, in the latest rocks is still not satisfactorily understood, but may be due to a valency change as has been suggested for chromium.

Co is strongly concentrated in the sulphides formed from the immiscible sulphide liquid (see previous section). However, the amount of sulphide in the layered rocks is only about 0·025% by volume, and such small amounts of sulphide, even though cobalt-rich, would contribute only a small amount of the total Co occurring in the rock, the greater part of which is in the ferromagnesian minerals. However, in sulphide-rich layers, for example the rock 5196 in the UZb, the amount of sulphide is estimated to be 0·4% by volume. Of the 147 ppm cobalt in this rock, 95 ppm is estimated to be present in the non-sulphide minerals, and thus about 50 ppm is present in the sulphides. Similarly the sudden rise from 1·5 ppm Ni in the early UZc to 14 ppm in the upper part of UZc is ascribed to the sudden precipitation of abundant, iron-rich sulphide droplets which crystallized largely to pyrrhotite.

The ratio of the Ni to Co in the Skaergaard rocks produced by crystal fractionation is not significantly affected by the small amounts of sulphide present, except in UZc, and it decreases regularly, as shown in Table 13.

From the Skaergaard data, the partition between the liquid and various crystal phases, for Co and Ni at various stages of the fractionation process, can also be approximately determined (see Wager & Mitchell, 1951, pp. 154–5).

Palladium is the only other Group VIII element so far successfully determined. In the chilled marginal gabbro, the amount present is 0·018 ppm and there is about the same amount in the ferrodiorites (Vincent & Smales, 1956; Wager, Vincent & Smales, 1957, pp. 876 & 886). The rocks with greater than average amounts of sulphides show no greater amount of Pd.

TABLE 13

Ni/Co ratios at various fractionation stages

	Original magma	HZ	LZ	MZ	UZa	UZb	UZc*
Ni/Co in rocks	3	5	2·5	1	0·1	0·03	...

* Figures not comparable because of the relatively large amount of sulphide.

j) Group I(b): Cu, Ag, Au

Copper (Pl. X) has been determined by radioactivation analysis (Wager, Vincent & Smales, 1957). It is rather variable in the rocks, probably as a result of the irregularities in the amount of sulphides. In the initial magma, 130 ppm of copper were present. In the average ferrodiorite of UZa and UZb there are 600–800 ppm, while in a melanocratic layer with abundant sulphides, there are 2000 ppm. From about the middle of the UZb, the copper content of the average rock begins to fall and by the UZc stage, when the immiscible liquid droplets consisted dominantly of iron sulphides, the copper content of the rocks continues to fall, reaching about 200 ppm or less. When the element sulphur is considered (p. 202), evidence is given that a copper sulphide-rich liquid separated as an immiscible phase at about 50% solidified, at which stage the copper and sulphur contents of the magma were both about 100 ppm, and it seems likely that immiscibility relationships played a dominant part in controlling the amount of copper in the rocks.

Silver (Pl. X) has been determined by radioactivation analysis (Adams, 1961). In the chilled olivine gabbro, the amount is 0·1 ppm, a thousand times less than the amount of copper. The amount in the LZ rocks is also about 0·1 ppm and this increases to about 0·4 ppm in average rock of the UZa, after which it falls. There is a close parallelism in the behaviour of copper and silver but the latter is 1000 times less abundant.

Silver apparently enters readily into the copper-rich sulphide liquid but not into the iron-rich sulphide liquid, since the rock with 2% of pyrrhotite, or marcasite replacing it, has the least silver of any analyzed rock.

Gold (Pl. X) has been carefully studied in the Skaergaard rocks and minerals by Vincent and Crocket (1960). Unlike most elements, gold enters plagioclase, pyroxene, olivine, ilmenite and magnetite in closely similar amounts, the total range being only from 0·001 to 0·004 ppm. The initial magma contains 0·0046 ppm Au, and all the other rocks analyzed throughout the differentiation series, including the acid granophyres, are within a factor of 2 of this amount. From the nearly constant amounts of gold in the rocks and minerals of the Skaergaard intrusion, combined with a consideration of the chemical behaviour of the element, Vincent and Crocket conclude that the gold was present in the magma in the form of uncharged atoms, and that these became incorporated in the various minerals, apart from the sulphides, in a random manner. The amount of gold in most of the minerals is so small that in the case of a pyroxene crystal, for instance, there is only one atom of gold to every hundred million unit cells.

The ferrodiorite (5196) of UZb, which has a relatively high copper sulphide content, has ten times the gold content of the average rock, but no such impressive increase in gold occurs in the UZc ferrodiorites, containing mainly iron sulphide. Vincent and Crocket conclude (1960, pp. 135 & 141–2) that gold was strongly concentrated in the dominantly copper sulphide liquid but had little tendency to concentrate in the iron sulphide liquid.

k) *Group II(b): Zn, Cd*

Zinc. Some tentative data (Woode, 1961) indicate that Zn (132 ppm in the initial magma) is concentrated to some extent in magnetite, as might be expected. It is concentrated to some limited extent in the late granophyres and apparently not concentrated to any marked extent in the immiscible sulphide liquids.

Cadmium (Pl. X) has been more successfully determined by radioactivation analysis (Vincent & Bilefield, 1960). It is present in the initial magma to the extent of 0·13 ppm. The analysis of separated minerals shows that it is concentrated to some extent in iron-rich pyroxene, iron-rich olivine, and in magnetite but not in ilmenite. On the whole, cadmium tends to remain in the liquid, becoming a little more abundant in the later rocks and reaching 0·6 ppm in the acid granophyres. It does not show any tendency to be concentrated in the immiscible sulphide liquids.

l) *Group III(b): Ga, In, Tl*

Gallium (Pl. X) has been determined spectrographically and more recently by a radioactivation method (Nightingale, 1962). It enters plagioclase and magnetite fairly readily (Fig. 122). There is a decided increase in the amount with fractionation, except that in the late acid granophyre the abundance of quartz relative to felspar results in lower gallium contents.

Indium (Pl. X), determined by a radioactivation method (Wager, Smit & Irving, 1958), enters preferentially into pyroxene and ilmenite (Fig. 122), presumably in the trivalent condition. The amount in the initial magma was 0·05 ppm. The rocks with the highest indium content are the ferrodiorites of UZc which contain 0·18 ppm, while the late acid granophyre, with little pyroxene and magnetite, has only 0·09 ppm.

Thallium (Pl. X), determined by radioactivation analysis (Adams, 1961), is present in the chilled marginal gabbro to the extent of 0·05 ppm. In the LZ rocks it is less abundant, and upwards the amount falls slightly until UZc is reached, where a rapid increase occurs which is continued in the filter-pressed differentiates. Thallium, being a large monovalent ion, tends to accumulate in the residual liquid along with K and Rb. The slight fall in the MZ, UZa, and UZb rocks is, no doubt, due to the small amount of trapped liquid in these rocks.

m) *Group IV(b): Pb*

Some unpublished analyses by E. I. Hamilton (personal communication) show that there is 30 ppm of *lead* in the Sydtoppen transitional granophyre, and rather less in one melanogranophyre (4332). The acid granophyre of Tinden gives variable results between 10 and 40 ppm. Attempts have been made to extract lead from the plagioclases of the layered series for the determination of the isotope ratios, but so far unsuccessfully, as the amounts present are so small.

n) *Group V: P. Group V(b): As, Sb*

Phosphorus (Pl. X). So far as is known, phosphorus does not enter appreciably the early cumulus minerals but even with a sensitive analytical method (Curren, 1958), this would be difficult to prove because in separating the minerals for analysis there would always be the possibility of small grains of apatite, which had crystallized from the intercumulus liquid, being present as an impurity. (For the igneous minerals the recent results of Koritnig (1965) are inconclusive.)

In the initial liquid, 0·1% P_2O_5 (400 ppm P) is present and it is estimated that the amount rose gradually in the liquid to 1·75% (7500 ppm P) at the beginning of UZb where apatite began to precipitate as a cumulus phase. Previous to the precipitation of apatite as a cumulus mineral, the phosphorus content of the rocks is due to trapped liquid, and the fall in phosphorus content has been interpreted as due to increasing adcumulus growth, which reduced the amount of pore material. From the stage when apatite becomes a cumulus phase, the liquid presumably remained saturated with respect to phosphorus. Thus the fall in the amount of phosphorus in the liquid during the subsequent fractionation is considered to be due to the decreasing solubility of apatite in the successive residual liquids, especially as they become more silica-rich.

Arsenic (Pl. X) has been determined by Esson, Stevens and Vincent, using a neutron activation method (1965). The amount in the chilled marginal gabbro is 0·3 ppm. In the lower and middle zones and in UZa and UZb rocks, it is on the whole less than this, but rises to 0·6 ppm in the sandwich horizon rock, and to 1·0 ppm in the Syd-toppen transitional granophyre, after which it falls to about 0·4 ppm in the acid grano-phyre. The entry of As into the different minerals at the various stages of fractionation is shown in Fig. 122, from Esson *et al*. The latter suggest that As^{+++} is accepted into those octahedral lattice sites which are usually occupied by Fe^{+++}, Mg^{++}, Ti^{++++}, and Al^{+++}, and also shows preference for the sulphides.

Antimony has also been determined by neutron activation analysis, by Esson *et al*. (1965). The amount in one specimen of chilled marginal gabbro was found to be 0·13 ppm, and the amount in the layered rocks shows an apparently unsystematic and considerable spread from about 0·2 to about 0·02 ppm. There is no particular concen-tration in the granophyric differentiates. Antimony enters the early olivine and ilmenite more readily than the plagioclase, pyroxene or magnetite (Fig. 122); it decreases in amount in the olivine and ilmenite with fractionation. Esson *et al*. conclude that Sb^{+++} probably substitutes for Fe^{++} in both silicate and oxide minerals.

o) *Group VI: O, S*

Oxygen, of course, is the most abundant element in the magma, entering into all the minerals except the rare sulphides. The chief geochemical interest of oxygen in the intrusion is in relation to the oxidation–reduction potential (discussed briefly under iron, pp. 197–8) and the isotope ratio O^{18}/O^{16} (discussed on pp. 184–7).

Sulphur is not an easy element to determine with precision at the low levels in which it usually occurs in the Skaergaard rocks but it has been done, not entirely to his satisfaction, by E. A. Vincent (Wager, Vincent & Smales, 1957). Apparently, the amount of sulphur varies considerably from rock to rock, and the number of samples so far investigated does not adequately represent the sulphur content for this reason, also. However, the amount of sulphur in the initial magma was apparently about 50 ppm and it is believed to have increased in amount until an immiscible, copper-rich sulphide liquid began to separate at about 60% solidified (Wager *et al*., 1957, p. 883) when the concentration of S was about 100 ppm and that of copper about 200 ppm. From this stage (and possibly earlier), droplets of an immiscible sulphide liquid formed, and accumulated with the sediment of cumulus minerals in irregular amounts. The amount of copper-sulphide liquid forming was presumably such that the silicate liquid just remained saturated with copper sulphide. The rising graph for the amounts of copper and sulphur in the liquid is interpreted as indicating increased solubility of the copper sulphide with the changing, overall composition of the silicate magma. At about the beginning of UZb, when 500 ppm of copper sulphide were being precipitated in the average rocks, the amount of copper separating, as immiscible copper-sulphide liquid, exceeded that present in the silicate liquid. The curve for sulphur in the liquid continues to rise, presumably because the total amount of copper-sulphide liquid separating was small and did not remove enough sulphur from the silicate magma to prevent it increasing there. For a short time, the rise in sulphur content may have increased the amount of copper-sulphide liquid separating, since this amount would depend upon the product of the concentration of both copper and sulphur. However, the analyses of the rocks show that the curve for copper eventually falls steeply. This is considered to be the result of the silicate liquid having increasingly reduced amounts of copper due to the decreasing solubility of copper-sulphide liquid as the sulphur content of the silicate magma rose. With the fall in the quantity of copper-sulphide liquid separating, the sulphur content of the silicate magma increased rapidly until, at about the middle of UZc, an immiscible iron-sulphide liquid began to form. Immiscible iron-sulphide droplets first developed when the sulphur content of the silicate liquid had risen to about 400 ppm and the concentration of ferrous iron was also very high. The beginning of the formation of the iron-sulphide liquid coincided with a rapid change in composition of the silicate liquid towards intermediate and acid, siliceous compositions, and from this time onwards, the sulphur content of the silicate magma fell steeply,* implying a decreasing solubility of iron-sulphide liquid in the silicate liquid. Thus, from a stage probably corresponding to about 60% crystallized, the silicate liquid remained saturated with respect to an immiscible sulphide liquid, which at first was largely copper sulphide and later, largely iron sulphide.

* The iron sulphide-rich rocks of the upper part of UZc contain around 5000 ppm. of sulphur and are off the scale of the graph (Pl. X). No points are given corresponding to P and Q on the graphs as the sulphur contents of these rocks are highly variable. In the case of the Tinden granophyre, as previously mentioned, some specimens contain considerable masses of sulphide amounting to 4 or 5% of the whole rock.

CHAPTER VIII

THE MECHANISM OF THE SKAERGAARD DIFFERENTIATION

In this chapter it is proposed to accept that the layered series formed from the magma as a sediment of crystals, and to go on from there to consider the processes by which the crystals accumulated to give the observed cryptic and rhythmic layering. This involves discussion of the extent to which the crystals sank directly through the magma to the floor or were carried by convection currents, the problem of where and under what conditions the crystals nucleated, and a consideration of how the heat was removed from the intrusion to allow adcumulus growth and the crystallization of the trapped liquid. First, however, an attempt will be made to construct a picture of the initial act of intrusion and the stages of solidification before bottom accumulation began.

THE INITIAL ACT OF INTRUSION

It is believed that injection of the magma to form the Skaergaard intrusion took place as a single, violent episode which filled the space, that is now the intrusion with, a uniform magma (Ch. II). The observed boundaries of the intrusion mainly dip inwards, defining an inverted cone shape; the material removed is, therefore, considered to have been ejected upwards. Apart from the shape of the intrusion, which seems to exclude downward removal of the material by any stoping mechanism, there is the general difficulty that sialic crustal material is unlikely to have sunk into basic magma. Blocks of metamorphosed and melted acid gneiss occur at low levels in the intrusion but their position is explained by the action of convection currents which dragged the light material downwards against its tendency to float. The large mass of gneiss on Tinden (see map, Pl. XI) may be one of a great pile of gneiss blocks which were blown out by the intrusion, and once formed a capping to it.

A re-estimation of the volume, before any erosion of the exposed layered series, together with the contemporaneous marginal and upper border groups gives a figure of 150 km³. Assuming that the observed rocks form 30% of the whole intrusion, as is suggested by considerations of the overall chemistry (Ch. VII), then the total volume of the intrusion was 500 km³. An earlier estimate of the total volume (Wager & Mitchell, 1951, p. 132) was 300 km³, the lower figure being due mainly to the assumption that the exposed layered rocks formed 40% of the whole.

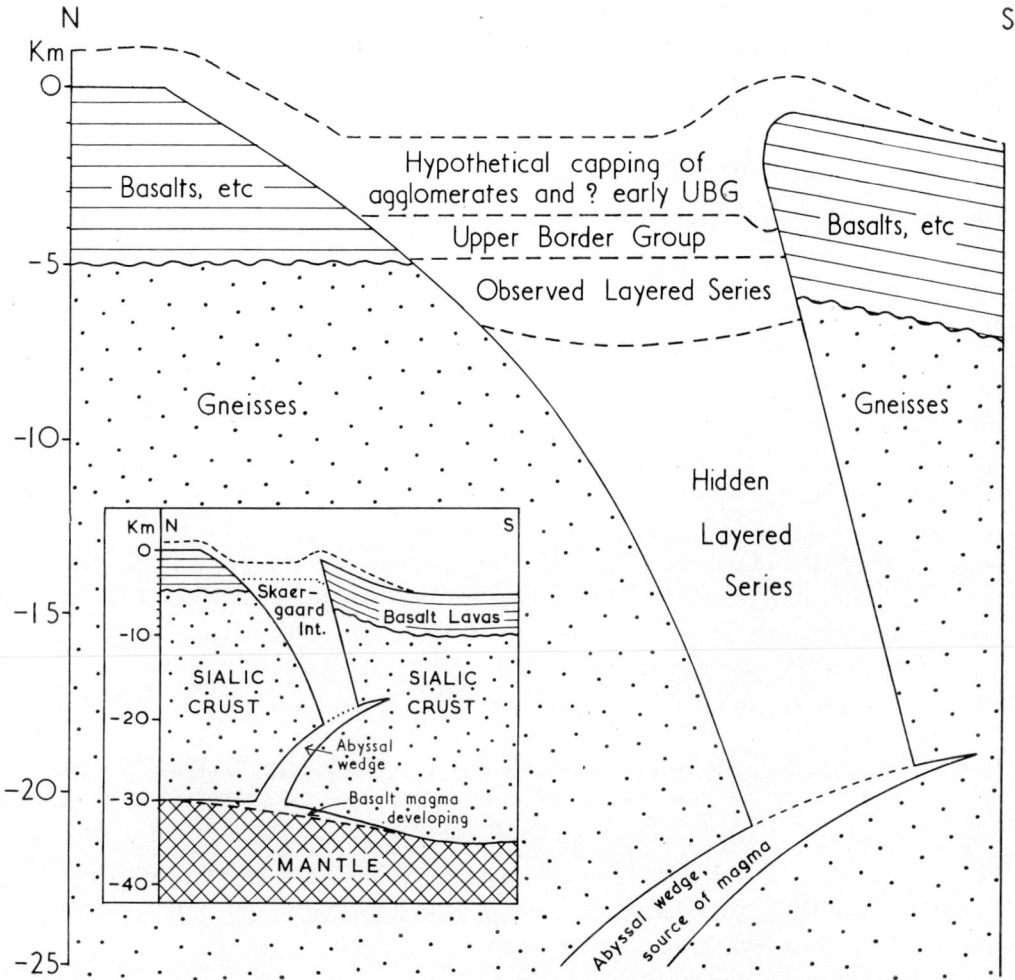

FIG. 125

Hypothetical N-S section of the Skaergaard intrusion at the time of intrusion. The section is drawn assuming that the intrusion reached down to a magma reservoir at a depth of 20 km, and that the volume of the hidden layered series, 70% of the whole, is 350 km³.

Some of the crustal flexuring is assumed to have taken place before injection, so that the section differs from Fig. 8 which assumed all flexuring after injection. The inset shows a possible relationship between the Skaergaard intrusion and an abyssal wedge injected as a result of the beginning of the crustal flexure. An explosion from the upper side of the abyssal wedge is believed to have blown out the crustal rocks to make room for the intrusion.

Various possible forms for the intrusion may be obtained by extrapolating downwards, with different assumptions about the inclination of the walls and the depth to which the intrusion reaches. One possible form is shown in Fig. 125; in its construction, it has been assumed that the crustal flexure had begun before the intrusion was formed, and that it had already produced a dip of 10° in the basalts now lying near the southern boundary. This assumption results in some differences from the section previously given (Fig. 8) in which the flexuring was considered to have happened entirely after the intrusion. In drawing the section in Fig. 125, the depth of the intrusion has been taken as 25 km, and this means that it is still 5 km in diameter at the bottom. Perhaps the assumption of a greater depth, resulting in a narrower intrusion, would be equally reasonable but we prefer to take as our working model the form shown in Fig. 125.

The form of the intrusion is that of an asymmetrical, inverted cone, the vertical axis being tilted northwards. Like the cone sheets of the British Tertiary igneous province, it must have been formed as a result of high magmatic pressures. Cone sheets, however, are symmetrical about a vertical axis, and to account for the inclined form of the Skaergaard intrusion it was suggested in the original paper (Wager & Deer, 1939, p. 56) that the intrusion was due to magmatic pressure from the upper surface of an inclined wedge of magma. This hypothesis still seems the best available, and is illustrated in the inset to Fig. 125. As a result of the downward movement of the crust to the south-west, giving the flexure, an abyssal wedge of magma is considered to have risen from the level of the Mohorovičić discontinuity but it did not reach the surface because of the stresses set up by the flexuring. The basic magma is pictured as a partial melt derived from the mantle, which before being intruded had accumulated in a low-pressure region produced by the crustal flexuring. High magmatic pressure exerted over the top surface of the wedge eventually blew out the cone-shaped mass of crustal rocks and, simultaneously, the space was occupied by magma. The abyssal wedge is considered to have been a linear structure, parallel to the flexure, and so could have been sufficiently extensive to have contained the large volume of basic magma required to fill completely, and in one act, the space which is now the intrusion.

Evidence from the chilled marginal gabbro indicates that the magma was completely liquid when the intrusion formed. The intruding magma presumably moved upwards rapidly from a region of high pressure to one of lower, and hence it is possible that it became somewhat superheated, although no effects which might be ascribed to this have been noted.

FORMATION OF THE TRANQUIL DIVISION OF THE MARGINAL BORDER GROUP

After an initial turbulence which must have existed at the time of injection, the magma must have become essentially stationary, during the formation of the tranquil division of the marginal border group. At the time of formation of the perpendicular-felspar rock of the marginal border group, there could have been no appreciable movement of the magma for, otherwise the inward projecting felspar crystals would have been broken off and become aligned parallel to the walls. Equally, but less dramatically, some of the other textures of the tranquil division can be seen to have required tranquil conditions for their formation. The width of this division in the upper part of the intrusion, where it can be observed, is everywhere about 70 m. Any comparable upper border group rocks, if they ever existed at the top of the intrusion, have been eroded away, except possibly on the upper part of Tinden which has not yet been investigated. In the formation of the tranquil division rocks, we may picture rather steady conditions in which a solid–liquid interface moved slowly inwards as heat was conducted outwards into the country rock. The conditions must have been essentially the same as occur at the margins of a dyke or a sill when the magma is stationary. The rate of inward movement of such a solid–liquid boundary must vary in a complicated way with time, but using graphs made from calculations by Shimazu (1959a, pp. 32–3) or Jaeger (1959), the amount of time required to produce the tranquil division, 70 m wide, is apparently of the order of 1500 years.

The details of the solidification process in the tranquil division have been established to some extent. To produce the special mineral compositions and textures of the chilled margin, supercooling of the magma by perhaps 20° C, and some delay in nucleation, seem to be required. The crescumulate plagioclases, and the less well-developed crescumulate olivines, also suggest growth into a supercooled liquid from the wall of already formed crystals. The rate of the inward growth was apparently not sufficient to keep pace with the cooling, and after 10–15 cm of crescumulate plagioclase had formed, the inward growth was brought to an end by plagioclase nucleating ahead of the growing tips of the crescumulate crystals, and many disorientated, smaller plagioclase crystals developed; then, after an interval, crescumulate growth began again. In a general way, the rate of outward diffusion of heat was such that the velocity of inward crystal growth of the plagioclase crystals could not quite keep pace with it, except for the short periods when the crescumulates formed.

The picrite of the northern marginal border group was also formed during the tranquil stage, and indicates appreciable sinking of olivines when they had attained a few millimetres in diameter. Along the northern margin of the intrusion, where the inward inclination is about 45°, the olivines apparently accumulated to give an olivine-rich mush which at times slumped, giving the pseudo-agglomerate of rounded picrite masses. On the east and west margins, some sinking of olivines may be responsible for rather olivine-rich bands in the tranquil division. The wavy-pyroxene rock also has a texture suggesting slight sinking of olivines and plagioclases to give indefinite sheets rich in one or other of these minerals. The extent of vertical sinking suggested by the wavy-pyroxene rock is perhaps only a few centimetres, while in the case of the picrite it was considerably more. In the tranquil division, sinking of crystals was relatively un-important; the rocks are essentially the result of congelation of a stationary liquid without any important change of bulk composition due to relative movements of crystals and magma.

The Mechanism of the Main Solidification

The tranquil congelation of stationary magma against the margins of the intrusion eventually gave place to very different conditions, involving flow of the liquid. Evidence for what happened is, of course, available only from the exposed layered series and border groups, which represent only the last third of the solidification process. However, some extrapolation, backwards in time to the conditions of formation of the hidden layered series, can be made.

a) *The extent to which crystals sank or remained in suspension*

The observed layered series gives clear evidence of the bottom accumulation of crystals of plagioclase, olivine, pyroxene, and, later, of magnetite and apatite. It appears that these primocrysts nucleated in the body of the magma, grew to a fairly constant size for each mineral species, and then collected as cumulus crystals on the temporary floor formed by the previous sediment of crystals. The banded division of the marginal border group was growing simultaneously inwards, due to the loss of heat outwards into the country rock which produced slow congelation of the magma with its suspended primocrysts. By a similar process of congelation, the upper border group apparently formed from above downwards; large primocrysts of plagioclase and, more rarely, other minerals, in suspension in the magma, were caught by the congelation process, giving the felspathic gabbros of the upper border group. This process of congelation would not be expected at the bottom of the diminishing pool of magma because the loss of heat, through the already formed and still hot sediment of crystals, would be relatively small.

It is easy to accept that crystals of olivine, pyroxene, and iron ore, which can confidently be assumed to be denser than the basic magma from which they were forming, could collect as a precipitate at the bottom of the liquid. On the other hand, less dense plagioclase crystals could, apparently, either accumulate on the floor along with the denser cumulus crystals, or remain in suspension and be incorporated in the congealing magma forming the upper and marginal border groups. Both experimental evidence (cf. Bowen, 1915a) and observational evidence (e.g. Fermor, 1925) suggest that the difference in density between calcic plagioclase crystals and the usual basic magma from which they might form is small. During all the changes in composition of the Skaergaard liquid, plagioclase probably just remained denser than the liquid, because plagioclase crystals are amongst the other cumulus minerals of the layered rocks throughout the whole thickness. At the same time, from the evidence of the upper border group, it seems that the turbulence of the liquid could keep abundant plagioclase crystals, and sometimes other minerals, in suspension, so that they were caught in the congealing magma.

The rate of sinking of solids in liquids depends upon the density difference between the two, on the size and shape of the solid particles, and on the viscosity of the liquid. The rate of sinking of spheres in a liquid, if the spheres are not above a certain size, is satisfactorily given by Stokes's Law:

$$V = \frac{2}{9} g r^2 \frac{d_s - d_l}{\eta},$$

where r is the radius of the sinking sphere; g, the acceleration due to gravity; d_s, the specific gravity of the solid and d_l, of the liquid; and η the viscosity of the liquid in poises. This equation can be used to give an approximate rate of sinking of the variously-shaped crystals formed in the Skaergaard magma. Using Stokes's Law and Hess's estimates of the viscosity of typical basic magmas and of the densities of the various crystals forming from it (Hess, 1960) a table of approximate sinking velocities of different crystals of various sizes can be drawn up (Table 14). The table illustrates,

TABLE 14

Settling velocities (metres per year) of crystals in basic magma (density of magma, $d_l = 2\cdot58$; viscosity $= 3000$ poises)

	Plagioclase				Olivine			Augite			Titaniferous Magnetite		
d_s	2·68				3·70			3·28			4·92		
$d_s - d_l$	0·10				1·12			0·70			2·34		
Radius (mm)	0·5	1·0	2·0	4·0	0·5	1·0	2·0	0·5	1·0	2·0	0·25	0·5	1·0
Sinking velocity (m per year)	5·7	23	92	368	64	256	1024	40	160	640	3	134	535

Velocities calculated from Stokes's Law

$$V = \frac{2}{9} g r^2 \frac{d_s - d_l}{\eta}.$$

generally, how slowly crystals sink in magma, and how important is the size of the particles in controlling the rate of sinking. A plagioclase crystal, 2 mm in diameter, settles in basic magma at about the same rate as a crystal of chromite that is $\frac{1}{2}$ mm in diameter, despite the much higher density of the latter mineral. In bottom accumulation, size sorting of the crystals would, therefore, be expected to be an obvious feature. In sedimentary rocks, graded bedding due mainly to size, rather than to density differences, is common. In the layered intrusions studied so far, size sorting is not conspicuous, as Hess has pointed out (1960), and a possible reason is mentioned later. On the other hand, gravity stratification, apparently due to density differences, is not infrequent, and its origin is discussed below.

O

b) *Rate of accumulation of the layered series*

If the total amount of heat lost from an intrusion, in unit time, can be estimated, then, using the latent heat of crystallization of the mineral phases, the amount of crystals formed in that time can be deduced. Hess (1960, pp. 144–6) has made an estimate of this kind for the Stillwater layered series and arrives at the highly interesting result that only 10 cm of crystal sediment will accumulate per year (i.e. about 1 ft in 3 years). In the case of the Skaergaard intrusion, the crystals formed, when heat is lost, go mainly to the building up of the layered series but some contribute to the border groups, especially the upper border group. The rate of heat loss from an intrusion must be heavily dependent upon the shape of the mass of magma and upon the nature and extent of any convection occurring within it. In estimating the heat loss, Hess was only able to allow for the effect of convection by making the calculation for a stationary magma and then modifying it on the assumption of an increase in the diffusivity of heat, and in order to produce some sort of model, he assumed an increase of 50% in diffusivity above the normal figure. The formidable task of calculating the rate of heat loss in a convecting magma has not been attempted for any intrusion, so far as the authors know. For the Skaergaard intrusion, the shape and the vigorous convection of the magma would probably increase the overall rate of heat loss compared with that from an extensive sheet such as the Stillwater intrusion; other factors, such as the greater depth of the Skaergaard intrusion, which reached into hotter regions of the crust, would work in the reverse direction. It is likely that the rate of heat loss from the Skaergaard intrusion was somewhat greater than that from the Stillwater, and therefore, that the rate of accumulation of crystals on the floor was greater. In developing a rough quantitative picture of the conditions for the Skaergaard magma during cooling, the rate of accumulation of the layered series will be assumed to be about twice that estimated for the Stillwater, i.e. 20 cm a year or 0·6 mm per day. With this assumption, the time for development of the 2·5 km of exposed layered series is about 12 000 years. The time required for the solidification of the whole Skaergaard intrusion, including the time for cooling down to normal temperatures, is likely to be of the order of 40 000 years.

c) *Evidence for two types of thermal convective movement in the*
    Skaergaard magma

The evidence for the existence of thermal convection currents in the Skaergaard magma has been given already (end of Ch. IV). At the trough-banding horizon there is visual evidence of a former single system of currents which flowed over the floor of accumulated sediments, from the walls to the centre. The currents indicated by the trough banding persisted in essentially the same place, during the time taken to accumulate two or three hundred metres of the layered series. This must have been for a considerable period, probably to be measured in thousands of years. However, the currents in the troughs presumably fluctuated violently in strength, because gravity stratification is characteristic of the layers, but the position of the currents remained constant.

The banded division of the marginal border group provides evidence of magmatic currents descending parallel to the walls. The inclusions, formed from the acid gneiss of the metamorphic complex, must have been considerably less dense than the basic magma, especially by the time they had been fused to masses of viscous liquid (see Wager & Deer, 1939). They cannot be considered to have sunk in the magma but must have been carried down by descending currents against their tendency to rise; from the evidence of the acid inclusions an estimate is made, later, of the velocity of the descending marginal currents. Although often of high velocity, the currents were probably inter-mittent since they were responsible for the cross-bedded belt of sedimentation. During the formation of the LZ type of the banded division of the marginal border group, the acid inclusions were abundant and provide evidence for vigorous, but not necessarily continuous, descending marginal currents. When the MZ type of the banded division of the marginal border group was formed, the acid inclusions were not so abundant and were more hybridized, but clearly the convection currents had still the necessary velocity to carry down relatively light acid material. On the other hand, there is little UZ type of marginal border group available for investigation and thus the evidence for marginal descending currents at that time is indefinite.

To account for the different kinds of rhythmic layering found in the layered series, it is suggested that two essentially different kinds of convection current were present in the magma. The first is considered similar to that postulated by Hess (1960) for the Stillwater, and described by him in the following terms:

> . . . a layer of liquid might develop below the roof which, by loss of heat and impregnation by crystals, became denser than the underlying liquid. Although mechanically unstable it might develop and remain in this position for a time, just as in winter a layer of colder water sometimes exists metastably for a short time at the surface of the sea. Such a condition would be more readily imaginable in a viscous liquid than in water. Eventually overturn of the unstable mass of liquid will occur. One might expect that some inhomogeneity in the layer would cause it to get started at one point first. A downward bulge would form, the nose of which would accelerate rapidly. The velocity would be enormously greater than the settling of crystals. A single descending column of denser liquid might draw off the whole of the layer from below the roof. It would spread rapidly over the floor and come to rest.

Each overturn of this type would give rise to a single gravity-stratified layer. Because of the inclination of the central axis of the Skaergaard intrusion, the downward plunging currents in the early stages would strike the northern wall and the return current might well be near or against the southern wall (Fig. 126A), while in the later stages the return current would, presumably, be roughly central (Fig. 126C).

In the Skaergaard intrusion, a common feature of the rhythmic layering is the alternation of layers of uniform, average rock with the thin, gravity-stratified layers

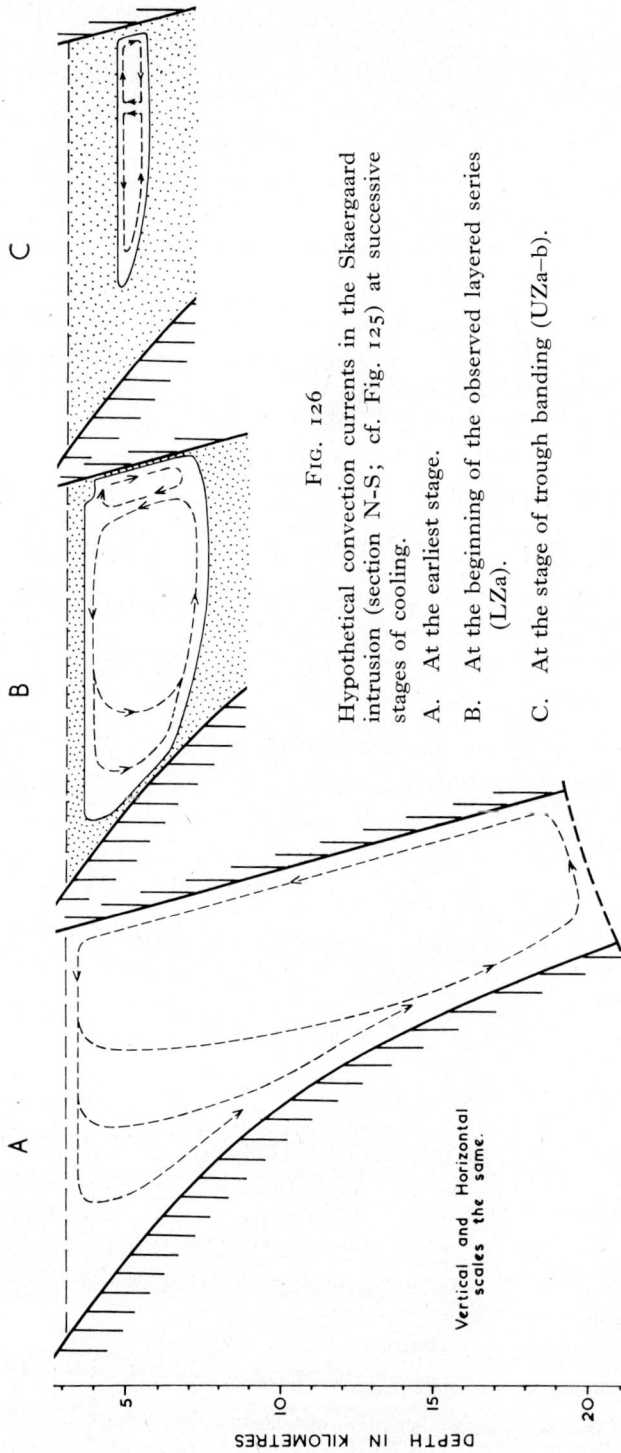

FIG. 126

Hypothetical convection currents in the Skaergaard intrusion (section N-S; cf. Fig. 125) at successive stages of cooling.

A. At the earliest stage.

B. At the beginning of the observed layered series (LZa).

C. At the stage of trough banding (UZa–b).

The thickness of the line bounding the intrusion represents approximately the 70 m of the Tranquil Division of the Marginal Border Group.

At the two later stages considered, the dotted areas represent the Layered Series and Upper Border Group

rocks already formed.

The HZ rocks of the early stages are believed to be mainly orthocumulates, while at the middle stage they are mainly mesocumulates and in the late stage mainly adcumulates.

(Fig. 10). To account for such sequences of rhythmic layers, the hypothesis previously suggested (Wager & Deer, 1939), of changes in current velocities, seems inadequate (cf. also Wager, 1953) and the hypothesis that there were two distinct kinds of convection currents has been developed: (1) a violent, downward-plunging type as described above, and (2) a slow, steady, laminar-flow type of current through which crystals settled uniformly to give the uniform rocks. The fast, turbulent, density currents of the first type are considered to have pushed out some way under the slow, laminar-flow currents, getting gradually slower and, finally, coming virtually to rest; they are considered to have been intermittent and to have operated independently of the steady current. When the intermittent currents came to rest, their load of crystals would have been deposited as a gravity-stratified layer. Apparently such currents sometimes followed one another in quick succession, giving a series of gravity-stratified layers, but sometimes layers of uniform rock, formed from the steady laminar-flow current, are interbedded with them. In the next section, an attempt is made to set up a model to explain the observed layering assuming that these two main kinds of convection currents existed. The fast, but intermittent density currents, believed responsible for depositing the gravity-stratified layers, will be called, briefly, the *intermittent currents* and the slow, more continuous current will be called the *steady current*.

d) *A semi-quantitative model of the convection currents in the*
   Skaergaard magma

The general velocity of circulation of convection currents in an intrusion of basic magma, under conditions perhaps close to those of the Skaergaard intrusion, have been roughly evaluated by Shimazu (1959a, b). The velocities will vary greatly in the different parts of the current but Shimazu's results indicate a general velocity of the order of 10 m per day.

Evidence for the velocity of the descending currents near the margin of the Skaergaard intrusion is provided by the gneiss inclusions of the marginal border group that were dragged downwards by the currents, against their tendency to float. Dr E. R. Oxburgh (personal communication) has calculated that blocks of up to about 50 cm in diameter, in average basic magma, should obey Stokes's Law closely. Assuming the viscosity of the magma to be 3000 poises, the density of the liquid 2·7 g/cc, and the density of the blocks 2·6 g/cc, then the upward velocity of blocks 50 cm in diameter would be 160 m per hour or 4 km per day. Accepting this figure, a descending current of about this velocity (in fact, slightly greater) must have existed to carry down the gneiss inclusions, against their tendency to rise. Such vigorous currents—very different from the 10 m per day currents estimated by Shimazu—probably happened only periodically and locally, but they seem to be of the appropriate kind to have produced the gravity-stratified layers of the layered series.

Evidence on the velocity of the postulated steady currents, from which the uniform layered rocks are believed to have been deposited, is to be obtained from the fact that the large felspars of the upper border group were carried up to the top of the liquid by the ascending, central convection current, while olivine and pyroxene crystals were usually not carried up to this position. The felspars, often 4 mm in average diameter, would sink in stationary magma at a rate of about 1 m per day. If they were carried upwards, therefore, the ascending current must have exceeded this and in drawing up the model, it is proposed to assume a velocity of 2 m per day for the upward currents and a similar velocity for the steady current which crossed the floor of the intrusion from the walls to the centre.

A simplified diagram of the two contrasted types of currents is given in Fig. 130 for the stage when the pool of magma was $3\frac{1}{2}$ km thick and 5 km wide. The steady current,

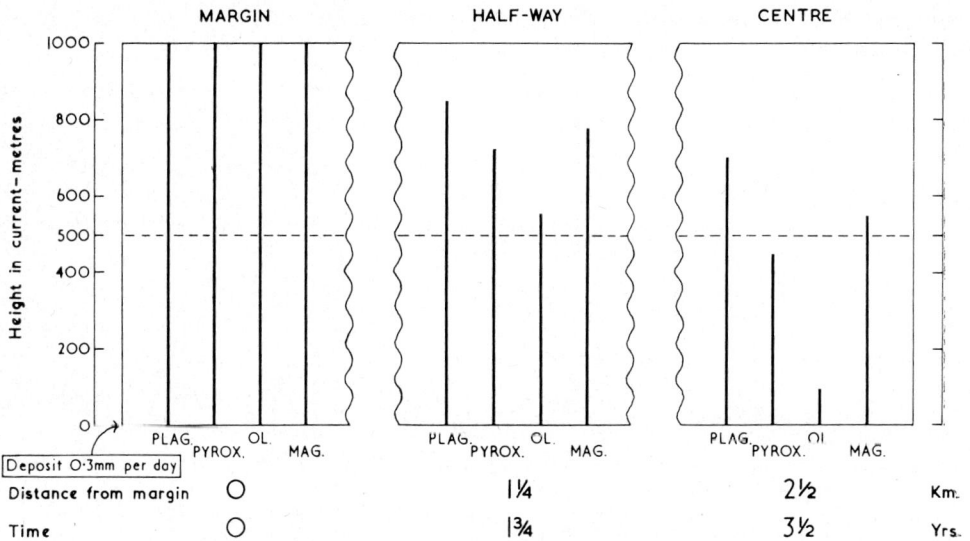

FIG. 127. Distribution of the larger crystals in the steady current as it moves from margin to centre of the magma pool. The velocity of current is assumed to be 2 m per day. The assumed size and velocity of sinking of the different crystals is as given in Table 14.

Near the margin the crystals are assumed to be distributed throughout the whole current, 1000 m thick. All the crystals sink, but at different rates, and when halfway to the centre, after $1\frac{3}{4}$ years, the crystals have sunk leaving various thicknesses of the upper part of the current free from crystals. By the time the centre is reached ($3\frac{1}{2}$ years) the upper 500 m of the current is free from pyroxene and olivine but contains some plagioclase and magnetite. (The reason that magnetite has sunk slowly is due to its smaller average size; see Tables 14 and 15.)

The horizontal broken line is inserted to show the situation if the current had been 500 m thick. In this case all the pyroxene and olivine would have been deposited before the centre was reached. The cumulus crystals in the central region, with this assumption, would only be plagioclase and iron ore. Locally some rocks of this kind have been formed, but whether they are abundant in the central region is not known.

where crossing the floor, is shown with a velocity of 2 m per day and its thickness is considered to have been considerable, say about a kilometre. The intermittent current, undercutting the steady current, is pictured as reaching, at times, a velocity two thousand times greater, and to have been much thinner, than that of the steady current. Although shown here as working independently of each other, it is clear that the currents must have interacted on one another, and probably currents of intermediate kinds existed.

The amount of material precipitated, per day, as layered series rock, has been

A. Early stage—high velocity and turbulence

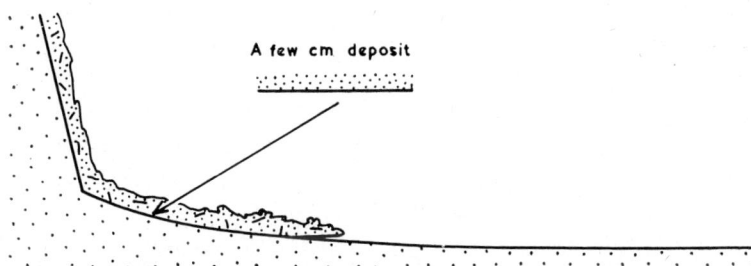

A few cm deposit

B. Later stage—turbidity current movement ceased

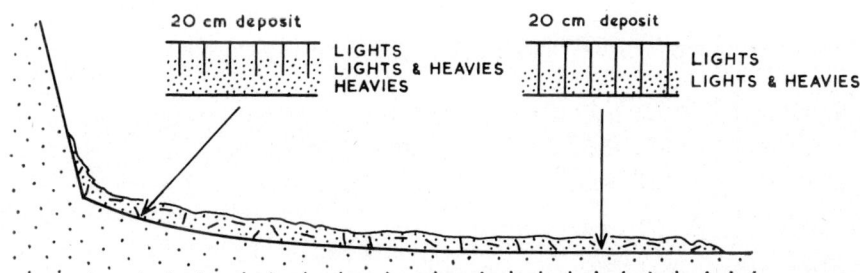

20 cm deposit   20 cm deposit

LIGHTS   LIGHTS
LIGHTS & HEAVIES   LIGHTS & HEAVIES
HEAVIES

FIG. 128. Suggested manner of formation of a gravity-stratified layer from a dense turbidity current. The current is assumed to be 2 m thick and has 10% of crystals in suspension, giving 20 cm of deposit. After the slowing of the current the time of sinking of the standard-sized crystal through the 2 m layer would be 8 days.

At the early stage (A) only the large, dense crystals, or no crystals at all, would be deposited. The current, in fact, would be often eroding, giving the cross-bedded belt structures.

At the later stage (B) the turbidity current, as such, is assumed to have ceased and the material formerly involved has only the very small velocity of the steady current. From the layer of essentially stationary liquid, a gravity-stratified layer of sediment would form. The nature of this sediment should differ between the margin and centre, as shown.

estimated as 0·6 mm (see p. 210). This material must have been partly deposited from the intermittent currents and partly from the steady currents, but on average the thickness of the deposit over the whole area of the intrusion must have been the same, otherwise the general saucer-shape of the layering would not have persisted; no doubt if the thickness of the deposit became excessive at any place, then temporary erosion, slumping, or less deposition would have occurred there to restore the saucer shape. The deposition of the cumulus crystals forming the cross-bedded belt, near the margin, is pictured as having taken place largely from the intermittent currents because of the gravity-stratified character of the deposit, and the evidence of periodic erosion as well as deposition. On the other hand, the layered rocks inwards from the cross-bedded belt tend to be largely uniform, and are considered to have formed mainly from the steady current with only periodic incursions of deposits from the intermittent currents.

To obtain a more detailed picture of the manner of deposition of the layered rocks, let us consider, first, the deposit from the steady current. Of the 0·6 mm of sediment deposited per day, let it be assumed that 0·3 mm was from the steady current and 0·3 mm from the intermittent current. Since the distance from margin to centre is 2500 m and the postulated velocity of the steady current is 2 m per day, 1200 days ($3\frac{1}{2}$ years) are required for this current to cross the floor. The total thickness of deposit from any unit of the current is 1200 × 0·3 mm, i.e. about 0·4 m. If this 0·4 m of material were dispersed evenly in a current 1 km thick then the percentage by volume of crystals would be 0·04%; if the current was only 0·1 km thick then the percentage of crystals in it would be 0·4%. These considerations suggest that the proportion of crystals to liquid in the steady current is certainly small.

Let us continue with the assumption that the current was 1 km thick and had 0·04% of crystals in it. The conditions must have been such that there was enough time for the crystals to sink through the current to the floor, where they formed, on average, 0·3 mm of precipitate per day. The size of a cumulus crystal in the LZ rocks is fairly constant for each mineral species, and the rates of sinking of such crystals have already been estimated (Table 14). Using these rates of sinking, the distances which the various crystals would sink in the $3\frac{1}{2}$ years, the time taken for the steady current to move from the margin to the centre of the intrusion, have been estimated (Table 15). The figures obtained imply that the plagioclase must have sunk only out of the lower 300 m of the current, because plagioclase crystals above this level would not have had time to reach the floor. Similarly, the olivines must have sunk out of the lower 900 m and so on. If the current is 1000 m thick as here postulated, then, by the time it reached the centre, plagioclase of the size here considered would still be present in the lower 700 m, olivine in the lower 100 m, pyroxene in the lower 440 m and iron ore in the lower 530 m. Smaller crystals of these minerals would, of course, be present above these heights. Assuming a current 1 km thick, the circumstances are indicated graphically in Fig. 127. If the current had been only 500 m thick the conditions are represented by the upper half of the figure, i.e. above the broken line. In this case, by the time a unit of magma had reached the central region, all the standard-size olivines and pyroxenes would have sunk out and only plagioclase and iron ore of the standard size would be accumulating, presumably with a sprinkling of smaller olivines and pyroxenes which would still be in suspension because of their lower sinking rate. The rather constant size of the cumulus crystals of each mineral species and the absence of conspicuous size grading in the Skaergaard layered series may be because the currents were of such a thickness, and were moving with such a velocity that, on the whole, there was only time for crystals of a certain size or above to sink out, although a few small crystals from nearer the floor

TABLE 15

Estimated distance sunk in $3\frac{1}{2}$ years for cumulus phases of common sizes of the Lower Zone of the Layered Series

Cumulus phase	Plagioclase	Olivine	Augite	Titaniferous Magnetite
Radius (mm)	2	1	1	0·5
Velocity of sinking (m per year)	92	256	160	134
Distance (m) sunk in $3\frac{1}{2}$ years	322	910	560	470
Percentage by volume of the cumulus crystals in the average rock	55	15	25	5

would occur along with the abundant large ones. This preliminary analysis of the way crystals might sink out of the steady current shows the complexity of the process; so far, no serious attempt has been made to investigate whether the sizes of the various crystals in rocks at any particular position in the layered series conform to this sort of hypothesis or not.

Because of the different rates of sinking, the proportions of the different crystals in the current will differ from the proportion in the deposit formed by their accumulation. The estimate of the proportion of crystals to liquid given earlier (p. 216) was based on the assumption that all the crystals carried by the current were deposited. If there be crystals in the liquid which do not have time to settle but are carried round for a second time, the estimates will be too low. However, it is likely that the crystals would not be carried up more than once for if they were, they would grow to such a size that they would sink rapidly and certainly reach the floor. Thus the estimate of the proportion of crystals in suspension may be on the low side but only slightly.

So far we have considered the slow, steady currents, the deposition from which has been arbitrarily assumed to account for half the layered series. The other type, believed responsible for the gravity-stratified layers, are considered to have been intermittent, small in volume, and of relatively high density. The velocity of these currents, when descending, has been estimated to have exceeded sometimes 3000 m per day from the presence of the acid gneiss inclusions. The high velocity is, presumably, due to a relatively high density, resulting from abundant primocrysts in suspension. On turning horizontally across the floor, the dense, high velocity currents are considered to have undercut the slow, steady current, and being of small volume and relatively thin, they are pictured as being quickly slowed down by contact with the floor and the overlying slow current. After a variable distance, they apparently came virtually to rest, and from the almost stationary, thin sheet of liquid, rich in suspended crystals, a gravity-stratified layer is considered to have been produced (Fig. 128).

The gravity-stratified layers away from the marginal cross-bedded belt are usually between 5 and 40 cm thick (excluding the Triple Group which is not considered here). To obtain some idea of the possible conditions of their formation, let it be assumed that a 20 cm layer was formed from a dense current having 10% of primocrysts instead of the 0·04 to 0·4% of the steady current. To produce the postulated 20 cm of deposit from a liquid with 10% of primocrysts, the current would only be 2 m thick. Near the margin, such thin, dense currents, moving with high velocity and presumably turbulent, might sometimes deposit crystals and at other times, erode the already deposited material, giving layering of the kind found in the cross-bedded belt. Further inwards, as the currents slowed down, deposition would be dominant. Eventually, when movement

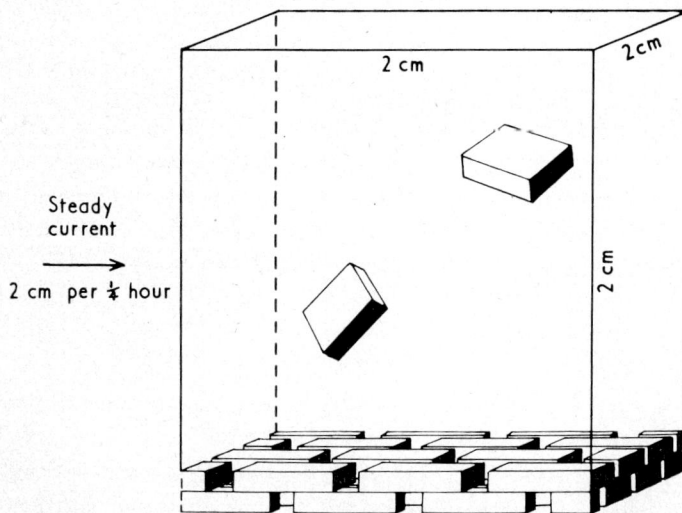

FIG. 129. Suggested conditions during the formation of a felspar adcumulate. The cumulus is considered to consist of felspar crystals, 0·5 × 0·5 × 0·1 cm in size. In a 2 cm-cube of liquid it is estimated (see text) that there will be two such felspar crystals, as shown. A new layer would take 3 days to form and 300 of the 2 cm-cubes of liquid, with sporadic crystals, would have passed over the area. (The cumulate of felspar is shown with a perfect lineation to simplify the drawing; in the actual rocks, even slight lineation is rare.)

virtually ceased, the heavy and light crystals, still in suspension, would sink out. The first material to be deposited would largely be the more melanocratic; this would be followed by more average material and, finally, by mainly leucocratic, felspar-rich material that would take the longest time to sink out. The intermittent currents would not come to rest completely but would take on the velocity of the steady current, i.e. 2 m per day, and this, it is believed, would be enough to orientate the tabular felspars, giving the igneous lamination which is commonly observed. With the above assumptions, the time taken for standard-sized plagioclase crystals to sink (Table 15) through the postulated density current, 2 m thick, would be only 8 days; thus the production of a gravity-stratified layer may well have been a relatively rapid process.

The intermittent currents were, presumably, frequent and vigorous in the marginal cross-bedded belt and became less common further in. The deposits from the

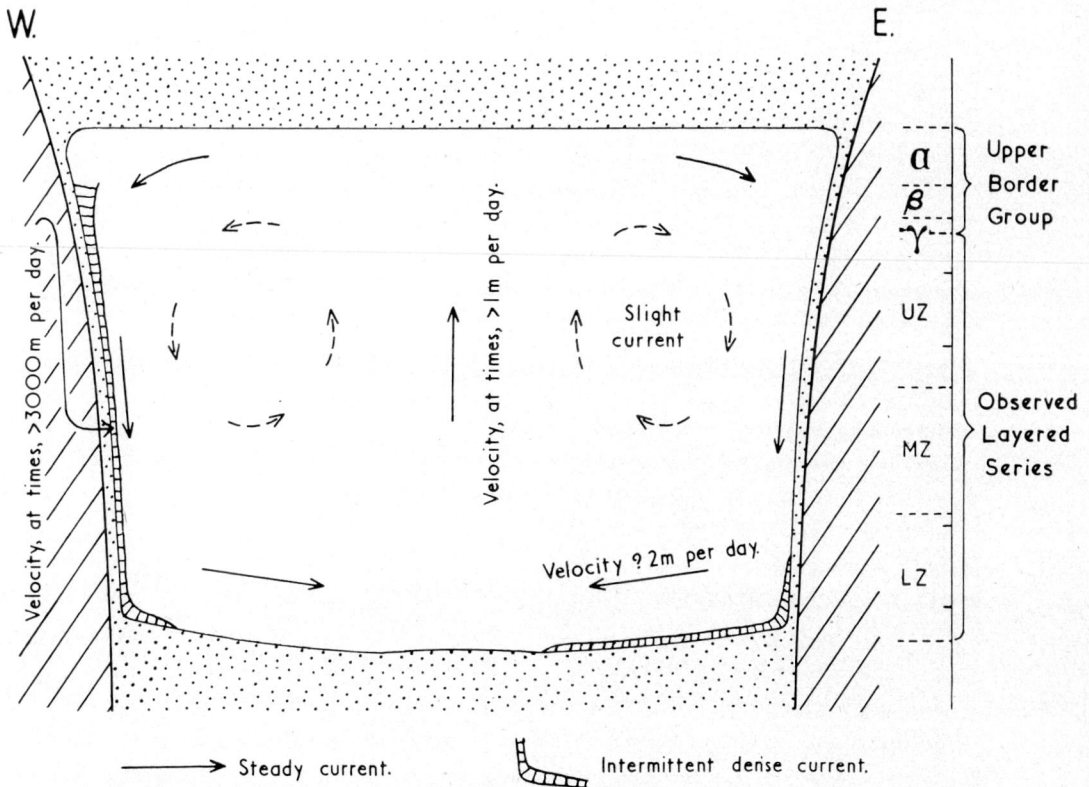

FIG. 130. Suggested conditions in the Skaergaard liquid at the beginning of the observed layered series, showing: (1) the slow, steady, convective circulation, with a maximum velocity of the order of 2 m per day, and (2) the intermittent, dense, convection currents descending along the walls and having a high velocity, at times > 3000 m per day. The intermittent currents are shown undercutting the steady current, and coming to rest over part of the floor.

intermittent current would thus be expected to be thicker near the margin. However, if the deposits near the margin became unduly thick, the slope of the top surfaces would be greater, and lead to higher current velocities and a tendency to erode the material already laid down, redepositing it further in. The erosion effect is, in fact, the characteristic feature of the cross-bedded belt to which it owes its name. In addition these outer layered rocks show the results of slumping which, also, resulted in material being carried inwards. Any deposit formed near the margin from the steady currents, would also be liable to suffer redistribution in the same way. It has been assumed above that the deposit from the steady current would be laid down equally from the margin to centre. However, while tending to form uniformity over the whole distance, the bulk of the more marginal layered rocks must have been deposited under the influence of the intermittent currents, and the bulk of the more central layered rocks must have been deposited from the steady currents.

This picture of the formation of the gravity-stratified layers is, no doubt, greatly oversimplified, and some day it will be profitable to make detailed field observations, allowing a more rigorous treatment of the problem. One point, however, may be noted now; during the sinking of the crystals from the 2 m sheet produced by an undercutting, intermittent current, there would be some addition of crystals from the overlying, steadily flowing magma. If the amount from the latter is 0·3 mm per day, then the amount sinking in 8 days is only 2·4 mm and this would spread thinly through a gravity-stratified layer 20 cm thick, thus forming only about 1% of the layer. The appearance of the gravity-stratified layer would not, therefore, be appreciably altered by the small proportion of crystals from the steady current above, but the crystals may have had a significant effect on the texture of the rock by providing seed crystals to produce poikilitic growth of minerals from the intercumulus liquid, in cases where the particular mineral species happened to be missing from among the cumulus crystals deposited from the intermittent current.

The conditions in a convecting liquid are complex enough, but if, at the same time, the liquid is crystallizing, the complexity is greatly increased. The fact that we have found it difficult to picture the details of the processes of solidification in the Skaergaard layered series is, therefore, not surprising. We have not been able to give any mathematical treatment of the problems, and in our tentative model of the convection currents the distinction between the steady current and the undercutting, intermittent currents is likely to be too clear cut. It is based on the characteristics of the strongly contrasted rhythmic layers such as shown in Fig. 12, but less extreme differences in the kinds of currents involved may be responsible for some of the other types of layering which have been observed. The relative importance of the intermittent density currents and the steady current may well have changed during the gradual reduction in size, and change in shape, of the body of magma. During movement of the liquid from one environment to another, the suspended crystals may have grown larger at times, and at other times have been partly dissolved. Differences in hydrostatic pressure would particularly affect growth or solution. Thus the discussion of the rates of sinking of the crystals is necessarily crude, not only because of the variable sizes, but also because changes in size may have occurred. Despite the tentative nature of the proposed model it will be shown later (pp. 223–27) that it gives a qualitative picture of how some of the detailed features of the layering may have developed.

e) *Supersaturation, nucleation, and crystal growth, under conditions of*
   thermal convection

The effect of thermal convection currents is to reduce greatly the general thermal gradient from what it would be if loss of heat were by conduction, without bodily movement of the liquid. For basic magma, Jeffreys (1959, pp. 287–8) calculated that the temperature gradient, as a result of convection, might be as low as 0·3° C per km. If this were so for the convecting Skaergaard magma, when $3\frac{1}{2}$ km thick as at the beginning of LZ, the temperature would be only about 1° C lower at the top of the liquid column than at the bottom.

As magma descends in a convection current, the hydrostatic pressure to which it is subjected must necessarily increase, and this will result in a rise of the melting temperature of the ordinary silicate minerals. For diopside, Yoder (1952) has determined the relationship between melting point and pressure, and gives the following equation: $t_m = 1391\cdot5 + 0\cdot01297P$, where $t_m =$ melting temperature in °C, at hydrostatic pressure P in bars. For each kilometre of descent in the liquid (S.G. = 2·7) there will be an increase in the melting point of about 3° C, and data now available for other minerals show effects of a similar order of magnitude. Even if convection in a magma does not reduce the temperature gradient to as low a figure as estimated by Jeffreys, there will certainly be a strong tendency for increasing crystallization as the magma descends, because it is subjected to increasing hydrostatic pressure.

Increase of hydrostatic pressure as liquid moves downwards in a convection current should also affect the composition of the minerals which are solid solutions. If plagioclase of a certain anorthite percentage was forming at the top of the magma, and this part of the magma with its enclosed plagioclase crystals be carried down, without appreciable change in temperature, to regions of higher pressures, then the experimental evidence suggests that the plagioclase precipitating at the lower level will be richer in anorthite. Slight, irregular, reversed zoning, which is common in the plagioclase primocrysts deposited in the LZ, may perhaps have this origin. Repeated reversed zoning has been found in certain rare crystals in the UZ, and this is thought to be the result of crystals having been carried round more than once in the circulating magma before finally coming to rest on the floor. In a descending column of magma, it seems possible that in certain circumstances the temperature of the magma and enclosed crystals might increase slightly, due to the latent heat of any crystallization which took place under the higher hydrostatic pressure, and a higher anorthite content of the plagioclase may result from the higher temperature. There is, of course, a severe limitation on this effect because any considerable heating would stop the convection process.

If the conditions of a unit of magma lying near the top of the intrusion is such that crystals are just beginning to form, they will tend to grow as the magma moves downwards into regions of higher hydrostatic pressure, even if the temperature remains the same or rises slightly. On the other hand, if a unit of magma at the top of the intrusion were just saturated but free from crystal phases, the magma, on moving downwards, might not nucleate but become supersaturated, and this effect is perhaps the means by which adcumulus growth takes place.

To allow adcumulus growth, the latent heat of crystallization has to be removed rapidly from the site, before the crystals are covered by the next layer of deposit. In a sill, loss of heat by conduction will take place into the underlying, cooler, country rock, as well as into the overlying rocks, but in a steep-sided, deep intrusion like the Skaergaard, the deposit of sediment is underlain, except at the beginning, by a thickness of hot cumulus crystals, surrounded by intercumulus liquid which crystallizes only as the temperature slowly falls. In the Skaergaard case there would have been some loss of heat downwards and outwards, as the intrusion lay within cooler crustal rocks, but it seems clear, in a general way, that the rate of heat loss downwards, through the already deposited material, must have been far too small to produce adcumulus crystallization.

The explanation of adcumulus growth which has been proposed (Wager, 1963) depends on a moderate degree of supersaturation in the magma. Evidence that supersaturation occurred in the Rhum layered intrusion is provided by Wadsworth (1961, p. 61) who has shown that the large, elongated olivines of the harrisites of Rhum grew into an overlying, supercooled liquid; and evidence for supercooling in the marginal border group of the Skaergaard intrusion is provided by certain mineralogical and textural features (see Ch. V). In adcumulus growth, the crystals grow while forming the top surface of the crystal pile, at a temperature close to that of the main body of the liquid, since the composition of the adcumulus material and the cumulus crystals is essentially the same. For any crystal growth, heat has to be removed and in ordinary circumstances, the heat is lost into the surrounding, lower-temperature rocks. It has already been suggested above that the conduction of heat from the top layer of accumulated crystals into the underlying crystal mush would not be sufficiently rapid for adcumulus growth to take place, and it is believed, therefore, that the loss of heat must be into an overlying, supercooled magma. It is envisaged that as a result of convection, fresh supplies of supercooled magma are brought successively into contact with the top surface of the cumulus crystals, allowing them to grow so long as heat can be transferred into the supercooled magma in this way, and providing that there is no fresh nucleation.

In origin the crescumulates are probably closely similar to adcumulates, both requiring the existence of a supersaturated magma for their development. In the case of the crescumulates the supersaturated liquid was probably stationary and completely free from crystals. The crystals that were favourably orientated at the top surface of the sediment grew upward into this liquid with acicular habit. With any appreciable flow of the magma the crystals could not have this growth habit; thus the movement would tend to push over and break off the crystals, and as the maximum rate of growth of the crystals would be on those parts of the crystal which first came into contact with the supercooled liquid, this would not lead to vertical growth. On the other hand where there was adcumulus growth without the production of any conspicuous crescumulate structure, as is nearly always the case in the Skaergaard, Bushveld and many other intrusions, it is considered that growth took place from a supercooled magma which was flowing over the crystals forming the top surface of the deposit.

The cause of the distinctive textures of adcumulates and crescumulates may be a simple consequence of whether or not the magma was in movement at the time of crystallization, a point returned to below. During the formation of the layered series of the Skaergaard intrusion, the magma could apparently deposit crystals on the floor, and yet be sufficiently supercooled to allow adcumulus growth. The magma was probably able to perform these apparently contradictory functions because in some places it was saturated, and in others, supersaturated. We believe that the steady-current magma only had widely spaced primocrysts suspended in it (probably only 0·4% or less) and as a result, parts of the liquid, distant from the primocrysts, were supersaturated and promoted adcumulus growth, while other parts near the suspended primocrysts were necessarily at, or close to, saturation.

f) *Synthesis of the conditions during the formation of the layered series*

Only a marginal part of the lower zone of the layered series is exposed, the more central being hidden by the middle and upper zones. Near the margin, the characteristic cross-bedded belt (Pl. VIII) gives evidence of fast intermittent, density currents which apparently only deposited the heavier crystals, usually melanocratic minerals of medium size but sometimes, large plagioclases. At times, the currents were powerful enough to erode material already deposited, giving the characteristic cross-bedded layering. It is also inferred that the surface of the sediment of cumulus crystal had rather a steep dip, for inward slips occurred, seen as small, normal faults roughly parallel to the margin (Fig. 70). A strike section in the cross-bedded belt sometimes shows the former existence of broad V-shaped valleys, now filled by later layered rock; these also seem to be the result of inward slumping of unstable parts of the sediment (Fig. 71). Some way from the margin, the sediment is of more uniform composition, interrupted at intervals by gravity-stratified layers (e.g. Fig. 69), presumably deposited from the intermittent currents. The gravity-stratified layers extend inwards for at least $1\frac{1}{2}$ km but at this distance they are infrequent; it is unlikely that they extended much further but evidence is lacking because the LZ rocks are covered by the MZ.

At irregular intervals in the vertical sequences of the LZ, about $\frac{1}{2}$ to $1\frac{1}{2}$ km from the margin, slumped rafts of rock are abundant. They are well seen in the northwest face of Gabbro Mountain (Fig. 31) where they are usually felspathic, and have apparently slid centrewards; on the whole, they get larger but rarer as the distance from the margin increases. It is believed that if more central rocks of the LZ were exposed they would be composed of more uniform material, showing perhaps zebra-type banding as occurs in the more central rocks of the middle zone.

The middle zone of the layered series also shows well-developed cross-bedding (Fig. 44) at the margin, as a result of strong descending currents. For some distance inwards, the rocks mainly show gravity stratification after which more uniform layers, presumably from the steady current, become more abundant. In the central half of the MZ, gravity-stratified layers are rare and it is inferred that undercutting density currents did not usually reach so far from the margin. In this inner part, zebra-type layering is characteristic and is considered the result of slight variations of some kind, in the steady current. The region where gravity layering peters out and zebra layering becomes common, about one kilometre from the margin, is well exposed in the case of the middle zone but many details of the layering remain unexplained.

Inclusions, which are sometimes 50 m thick, are common in a belt 1–2 km from the margin. They are of variable rock, some apparently belonging to the UBG. They seem to have slithered into place, for they have rucked-up the layered series (Fig. 46). Large inclusions are not found near the margin, perhaps because the slope of the bottom deposits was here too great for them to come to rest.

The Triple Group, occurring towards the top of the middle zone, consists of three gravity-stratified layers, ten or twenty times thicker than those ascribed to undercutting density currents. The group occupies an extensive central region and implies three distinct events, affecting at least half the area of the intrusion. While the thin gravity-stratified layers, found throughout the layered series except for the topmost UZc, are pictured as produced by undercutting density currents of small volume, pushing out periodically from the margins towards the centre, each member of the Triple Group is considered the result of a half overturn of the greater part, or the whole, of the liquid existing at the time, followed by quiescence during which a thick, gravity-stratified deposit formed. The two lower units of the Triple Group are each about 10 m thick, and if they settled out of layers of liquid 500 m thick, which is about half the total thickness of the liquid at this stage, then the liquid must have had 2% of crystals in it, which seems reasonable. The upper unit, which is considered to be 50 m thick, is perhaps a complex of two or three ill-defined units as it is unlikely that there would be enough crystals in suspension at one time to form 50 m of rock.

The trough banding structures, which occur in the UZa, imply a regular system of currents which seem to have persisted along certain radial lines for a considerable period of time (see map, Fig. 58). In the troughs, layers showing intense gravity stratification, and indicating intermittent currents, usually follow one after the other but, at times, layers of uniform rock are found separating the gravity-stratified layers, the whole being conveniently described as a *trough banding* sequence. Between the various trough banding sequences deposition of uniform material took place, presumably from non-pulsatory currents. The intermittent density currents along the troughs must have resembled, in intensity and character, the currents of the marginal cross-bedded belt since the kind of gravity-stratified layering is similar. From evidence on the neck of the Skaergaard Peninsula (cf. right-hand side of Fig. 130) it seems that the vigorous currents of the cross-bedded belt became gradually canalized as they pushed centrewards and produced intermittent currents persistent along certain radial lines, that are now made evident by the troughs. At times, the trough currents may have been able to erode, though cross-bedding is not commonly seen in longitudinal sections where it might have been expected (cf. Wager & Deer, 1939, pl. 12, fig. 1). The mounds of uniform rock which must have occurred along both sides of the troughs may have formed, partly, as levees, from material that spilled over the edge into a more tranquil environment. The troughs were sometimes scoured so deep that the sides slumped inwards (Fig. 57). Sometimes, two adjacent troughs gave place upwards to one, indicating a general broadening of the current (Fig. 56). The margins of the trough-structures are of interest in showing how the trough current varied in width (Pl. VI). In the House area, below the best development of troughs, there are wide, gravity-stratified layers with slight trough form indicating formation from a broad density current (Fig. 63). Other gravity-stratified layers below the trough structures are essentially planar, but limited in extent laterally, suggesting broad, density currents pushing out over level surfaces. Throughout the observed layered series, but particularly at the UZa horizon, the details of layering provide evidence of the varying current patterns, and would undoubtedly be worth a much fuller sedimentological study than they have so far received.

Above the trough banding horizon, fairly persistent, melanocratic layers, half a metre or so thick and succeeded by leucocratic material usually two or three metres thick, have been traced on the Skaergaard peninsula and in the House area (see Fig. 89). They reach $1\frac{1}{2}$ km from the margin and their occurrence is limited to about 100 m of structural height. They are first detectable at about 200 m from the inner contact, that is, just within the cross-bedded belt; here they are about 300 m wide but they grow wider inwards. These extensive gravity-stratified units differ from the usual gravity-stratified layers in being on a larger scale, and they differ from the Triple Group in being marginal, rather than central. They seem to be the result of an unusually thick and extensive density current, but why they should only be developed at this particular stage is not understood.

Above the level of the thick gravity-stratified layers there is still some 500 m of the layered series in which rhythmic layering persists, although gradually becoming less common upwards. The last layer showing good gravity stratification between uniform material was noted about the middle of the UZb and has very much the appearance of that shown in Fig. 38, from early in the ELS. Igneous lamination also becomes less common upwards, and is not conspicuous in the rocks forming the uppermost 200 m of

P

the layered series. The absence of igneous lamination suggests that in the final, thin sheet of magma, currents were feeble or non-existent.

The form of a liquid mass most favourable to the development of a single thermal convection system has a rather greater diameter than height. The residual Skaergaard liquid had this form at about the LZ stage (cf. Fig. 126B, bearing in mind that it shows a N–S section and that the E–W section is only about two-thirds as wide). During the period of formation of the lower zone, there was probably a regular system of steady currents and there is no evidence suggesting more than one convection cell. The steady current system is postulated to have persisted until near the end of the layered series deposition. The intermittent, convection currents were superimposed on the single, steady convection system, apparently from LZ to UZb times. The Triple Group conditions, involving three or more half overturns of the magma, seem to have applied only during MZ times. At a late stage, when the liquid had become a relatively thin sheet, it would be expected that the single, steady-current convection cell would have broken down into two or more cells, approximating to the optimum proportions for convection. However, no certain evidence for this has yet been found, although it might be obtained by detailed petrofabric studies on the uppermost layered rocks.

There are abundant field indications of downward currents along the walls of the Skaergaard intrusion and of quasi-horizontal currents crossing the floor, but we have found only one place where there is any evidence for the upward, return current. This is perhaps not surprising, for the upward current would have occurred at a unique local point, whereas the floor currents were spread over 30 km² and the chances of observing their effects are correspondingly greater. The place where the uprising current may have occurred is on the south side of Connecting Glacier, where two dip arrows of conflicting direction are marked (map, Fig. 7). This position is roughly the centre towards which the trough banding structures are directed (Fig. 60). The rocks here show poor igneous lamination and only rare, irregular layering and it seems that these features could be due to the central, ascending convection current. (East of this exposure of the supposed centre, there might be further useful evidence on a steep rock face except that stone-fall from the cliffs makes detailed investigation inadvisable.)

The conditions of accumulation of the layered series have so far been deduced from the varying pattern and distribution of the rhythmic layering. Some further light is thrown on the conditions existing during deposition of the layered series by the irregularities in the cryptic variation that have recently come to light. Magnetite and apatite, for instance, enter as primocrysts in certain thin layers well below the level at which they are continuously present, and sporadic olivines are found in the middle zone where olivine is not a usual cumulus mineral. The sporadic appearance of cumulus minerals of slightly anomalous composition must be due to a local cooling of a small amount of magma below the temperature of the main mass of the liquid, to which temperature the bulk of the cumulus minerals correspond in composition. It is suggested that some of the magma under the roof was cooled, locally, a degree or so below the average temperature, and that it then descended as a fast, intermittent current, spreading primocrysts of slightly lower temperature compositions than the average for this particular stage. Although minor irregularities occur, the overall regular cryptic variation in the Skaergaard layered series implies that, on the whole, the convection currents were able to keep the magma well stirred.

The bulk of the evidence for the transport of crystals by convection currents in the slowly solidifying Skaergaard magma comes from the layered series, and has now been presented. Consideration of the origin of the border groups, however, provides some additional evidence which has been mentioned in Chapter V, and is considered further below (pp. 231–33).

g) *Post-deposition crystallization*

In any rock which has been formed by the sedimentation of crystals, there must have been an early stage when magma existed between the cumulus crystals, and further crystallization has to take place before the deposit is the rock as we see it. It has been shown, earlier, that the crystallization of the intercumulus liquid, where trapped between the cumulus crystals, took place over a wide range of temperatures, as proved by the strong normal zoning of certain of the crystals and by the range of minerals formed. In the more complex adcumulus crystallization process, the added material was formed at the same temperature as the cumulus crystals, since it has the same composition, and in an extreme adcumulate the solid rock must have been developed immediately after the crystals settled out.

Adcumulus growth pushes out, into the main body of the magma, some of the intercumulus liquid and thus reduces the amount of liquid which is finally trapped between the cumulus crystals. As already demonstrated, by observations of the textures and by determination of the P_2O_5 content, the rocks of the Skaergaard layered series vary from orthocumulates, showing virtually no adcumulus growth and much pore material formed from trapped liquid, to adcumulates having little or no pore material formed from trapped liquid.

i) *Adcumulus growth.* Ortho- or mesocumulates, that is, rocks containing a considerable amount of material derived from trapped liquid, are characteristic of the LZ layered rocks of the Skaergaard intrusion. With increasing height in the layered series there is an increasing amount of adcumulus growth, as judged both by the textures and the phosphorus content, so that by UZa times the rocks are chiefly adcumulates, with little or no material formed from trapped liquid. The amount of adcumulus growth must depend, among other things, on the rate at which heat can be removed to allow the crystals to grow at the top surface of the sediment, and on the overall rate of accumulation of the sediment of crystals. If the rate of accumulation is slow, then the cumulus crystals remain as the top layer for a longer time and there is a greater opportunity for further crystallization of material on them. Nevertheless, the fundamental factor controlling the amount of adcumulus growth is the effectiveness of the removal of heat from the top surface of the sediment. It has been suggested earlier that the heat is removed into adjacent, overlying supercooled liquid, and therefore the more the conditions favour the presence of supercooled liquid above the surface of the sediment, the more adcumulus growth there should be.

The crystal accumulation to give the uniform rocks of the layered series is pictured as having taken place from the steady convection currents which contained only sporadic crystals of fair size, probably making up less than half a per cent of the total of the liquid and crystals. On the other hand, the rocks of the cross-bedded belt and of the gravity-stratified layers are pictured as having formed from a denser, 10%-suspension of crystals carried by the fast, intermittent currents. These contrasted conditions of deposition probably resulted in differences in the amounts of adcumulus growth, and there is some evidence of this from the phosphorus contents of the rocks. Accepting that a rock with a relatively high phosphorus content implies that it once contained a considerable amount of trapped liquid, then the felspar cumulates (see Fig. 109) had more trapped liquid than the average rocks. On the other hand, the few cases of melanocratic cumulates so far analyzed for phosphorus indicate less trapped liquid than in the average rock. Further work may show that the rhythmically layered rocks of both the cross-bedded belt and the gravity-stratified layers, further in, differ from the uniform rocks in the amounts of trapped liquid they originally contained. So far, all that can confidently be stated is that the plagioclase cumulates have more trapped liquid, in general, than the rocks of adjacent uniform layers, and this is probably due to the gentler sinking of the low-density plagioclase crystals, so that they were not closely packed but surrounded by abundant trapped liquid.

To account for the adcumulus growth of the uniform layers, it has been suggested that the liquid of the steady current is largely supersaturated, and is largely at a uniform temperature, a little below the equilibrium temperature of the crystal phases capable of forming from it. In the immediate neighbourhood of the sporadic crystals, however, the liquid cannot be supersaturated. Either the crystals will have the composition appropriate to the lower temperature, or the temperature in their immediate vicinity will be temporarily higher. While travelling along in the current, the sporadic crystals will be growing slowly by diffusion of the appropriate material to the surface of the crystal with concomitant diffusion of the latent heat of crystallization back into the supercooled liquid. The rate of diffusion of matter as ions, or molecules, through a silicate liquid is slow (cf. Bowen, 1921, pp. 295–317), while the diffusity of heat, in circumstances of the kind considered here, is likely to be relatively greater. We are therefore assuming that the liquid of the steady current is nearly uniform in temperature but varies in degree of supersaturation from place to place. It is suggested that as the steady currents carrying sporadic primocrysts moved slowly across the floor of the intrusion, there was usually not enough time for the establishment of equilibrium conditions. Depending on a fine balance of factors, conditions could exist so that parts of the liquid, well away from the sporadic primocrysts, remained supersaturated for the limited period of time during which the currents moved across the floor. The supersaturated parts of the liquid, moving along in the convection current, would pass over the crystals forming the top of the pile, and thus provide the appropriate conditions for adcumulus growth, both in regard to precipitation of material and to transfer of heat. To obtain a quantitative picture, the interplay of the various processes concerned with the diffusion of matter, and the diffusion of heat, would have to be considered in relation to the distances apart of the crystals in the suspension, the amount of turbulent mixing, and so on. At present, it is only possible to suggest that the wide spacing of the crystals in the suspension, for which evidence has already been given, makes it reasonable to believe that a mechanism of this sort is responsible for adcumulus growth.

The gravity-stratified layers consist of a plagioclase-rich upper part with much trapped liquid, and a thin, melanocratic lower part, usually with little trapped liquid. If the gravity-stratified layers are formed from a dense suspension of crystals, as suggested earlier (p. 218), then the time taken for their formation should be less than for the uniform layers and there would be less opportunity for adcumulus growth. Also, in the dense suspension there would be a lower proportion of supersaturated liquid because none would be far from a growing crystal. For both these reasons, the gravity-stratified layers would be expected to show less adcumulus growth and more pore material from trapped liquid. The felspathic parts of the gravity-stratified layers behave as expected, at any rate in some cases, but the behaviour of the melanocratic parts requires further investigation.

In the formation of crescumulates, it appears that the favourably orientated crystal grew upwards into a supercooled liquid, free from all suspended crystals. The period of upward growth was apparently ended, as described by Wadsworth (1961, p. 60), when a shower of crystals produced, presumably, by nucleation from a more-intensely supersaturated magma, sank and covered the crescumulate olivines. In the Skaergaard intrusion, only a few cases of crescumulus growth have been identified and these are in the UZc. Conditions in the upper part of the layered series were favourable for adcumulus, but not crescumulus growth. Earlier (p. 222) we have suggested that for adcumulus growth the partly supersaturated magma must flow over the crystals on the floor, while crescumulate formation requires a stationary, supersaturated magma. There is plenty of evidence in the Skaergaard intrusion of magma flow, and the rarity of crescumulates suggests that the magma in fact seldom became stationary. In the development of crescumulates, convection is, no doubt, usually involved in bringing liquid from near the top down to the floor and so causing supersaturation, but after this it appears that the magma had to become stagnant for a long enough period to allow the crescumulus growth.

Among the layered rocks accessible to observation, adcumulus growth is generally least in the LZ rocks and increases steadily until near the top of the layered series. However, the LZ rocks available for observation are near the margin as well as low in the sequence, and so far it has not been possible to decide whether the factor responsible for the small amount of adcumulus growth is depth within the intrusion or nearness to the margin; deep drilling of the intrusion would, no doubt, decide this point. If the amount of adcumulus growth decreases with depth, as seems most probable, then the HZ rocks are probably ortho- or mesocumulates. In the early stages when the hidden layered series was forming, the convective circulation is pictured as being of intermittent type. The conditions were probably close to those for the formation of the average and leucocratic parts of the gravity-stratified layers, and like them, the hidden layered rocks perhaps had much trapped liquid and are orthocumulates. If part of the magma sank rapidly to great depths, the greatly increased hydrostatic pressure would tend to produce high supersaturation which would, presumably, cause much nucleation. In circumstances of abundant nucleation, the bulk of the liquid would not become supersaturated, because the necessary diffusion to produce equilibrium has only to operate over short distances. Rapid accumulation of crystals on the floor would also be expected to result from deposition from a thick suspension of crystals. For both these reasons, it is likely that there is not much adcumulus growth in the hidden zone. On the other hand, adcumulates in abundance, and the only crescumulates so far found in the Skaergaard intrusion, occur near the upper part of the layered series. At this stage, the liquid was only a thin layer, and supercooling due to descent of the magma in a convection current must have been slight; as a result, nucleation may sometimes have failed to occur and slightly supersaturated liquid without any crystals suspended in it may have provided the conditions for adcumulus and crescumulus growth.

ii) *Crystallization of the trapped liquid.* Adcumulus growth takes place while the crystals form the top surface of the crystal pile, and must have occurred in a very short time compared with the crystallization of the trapped liquid. A single layer of crystals is sufficient to blanket-off the underlying crystals and prevent their adcumulus growth (see Ch. 4). The average rate of deposition of cumulus crystals is about 0·6 mm per day; therefore, if individual crystals were 2 mm thick, a continuous layer of them, able to prevent adcumulus growth beneath, would have formed in about 3 days. This argument suggests that the adcumulus growth in the Skaergaard intrusion must normally have taken place in a matter of a few days from the time of deposition. On the other hand, the time for the crystallization of the trapped liquid must have been very long, for it would require a fall in temperature of the order of 100° C to crystallize the bulk of it, and an even greater fall in temperature for crystallization of the last dregs of the liquid, to give the hydrothermal stage replacement and deposition of minerals. The time required to cool the bottom accumulation of crystals through 100° C could be roughly estimated by a consideration of the thermal conductivities of the various rocks with their trapped liquids, but this has not yet been done; it is likely, however, to be of the order of a few tens of thousands of years.

In the Stillwater intrusion Hess (Ch. XIII) has shown that slumping generally affected only the top 6–10 ft of the crystal mush and he has suggested, as a reason, that at greater depths below the top surface of the sediment, there had been sufficient crystallization of the intercumulus liquid to bind the crystals together and prevent slumping. In the case of the ortho- and mesocumulates of the Skaergaard intrusion, the sediment of crystals would not be effectively bound together by crystallization of the intercumulus liquid until a much later stage. In the Stillwater case, a more rigid mass was probably produced at an earlier stage since the rocks are usually adcumulates.

The slow cooling of the sediment of cumulus crystals and the trapped liquid accounts for the development of the large, poikilitic intercumulus crystals of those minerals which are not represented amongst the cumulus crystals, because it allows plenty of time for extensive inter-crystal diffusion. It is also worth emphasizing that despite the extremely slow cooling, the plagioclase crystals are zoned. A little making over of the earlier formed plagioclase crystals to lower temperature solid solutions may have taken place, as Bowen believed possible in plutonic rocks (1928) but clearly no complete equilibrium was reached, even in the tens of thousands of years at the high temperatures that were available for solid diffusion to take place.

Conditions of Formation of the Upper Border Group and the Banded Division of the Marginal Border Group

a) *The Upper Border Group*

Characteristic of the Upper Border Group rocks, discussed in detail earlier (p. 131), are the large plagioclases with uniform, basic cores, the cores being interpreted as primocrysts from textural evidence, and from the systematic variation in their composition with structural height. The pyroxenes, olivines and iron ores are not usually found as primocrysts but apparently crystallized from the liquid in the position in which they are now found, the only conspicuous exception being certain olivines and pyroxenes of UBGα rocks from the foot of Tinden, near the southern margin. The period of formation of the UBGα rocks, judged by their plagioclase compositions, corresponds to that of the lower zone of the layered series, which contains olivine and augite, in addition to plagioclase, as cumulus phases. The presence of plagioclase as primocrysts in the upper border rocks, and the rarity of melanocratic minerals, is considered to be due to their differing densities. Using the hypothesis of a convecting magma, it is considered that the upward current had sufficient velocity to carry up the plagioclase primocrysts, while the olivine, augite and iron ore crystals usually sank out of the upward current. At times and in special places, fair-sized olivines and augites were apparently carried up, presumably by a particularly vigorous current. When the upward current turned to flow horizontally under the roof in various outward directions, the plagioclase crystals would also begin to sink away from the top contact, but some of them managed to reach the cooling surface and become incorporated in the congealing magma, perhaps being carried there by turbulence in the current. If the plagioclase cores of the typical UBG rocks were not primocrysts but large crystals grown in place, the problem would arise as to why augite, olivine, and iron ores had not formed large crystals with similar textural relations, instead of their usual ophitic habit.   A further argument for the plagioclase crystals being present as discrete crystals in the material congealing to give the usual UBG rocks comes from the laminated, plagioclase-rich gabbros (Fig. 93), in which the tabular plagioclases had clearly been orientated by flow of the magma before the crystallization, poikilitically around them, of the pyroxene and iron ore.

Many of the presumed primocryst plagioclases are about 4 mm in radius and would sink at about 1 m per day (Table 14); it is, therefore, suggested that a velocity of 1 m per day was frequently reached by the rising current, and that occasionally, the upward current had a velocity of 2 or 3 m per day (cf. p. 214) so that about equally large olivine and augite primocrysts were sometimes also carried upwards.

Smaller primocrysts of the denser minerals should have been carried up in the ascending currents along with the larger, but lighter plagioclases, but their presence has not been identified with certainty.   Perhaps the small extent of the poikilitic crystals in the UBG rocks is an indication that such small primocrysts were present in the liquid, acting as centres of growth and preventing the formation of really extensive, poikilitic, intercumulus minerals.   The current of *one metre* per day, able to carry upwards the plagioclase crystals of 4 mm radius, would be able to carry up pyroxenes, about $1\frac{1}{2}$ mm in radius (Table 14); such pyroxenes would be rather inconspicuous when enlarged by further growth from the intercumulus liquid, but a detailed study of zoning might indicate their presence.

The UBG rocks with olivine and pyroxene primocrysts are found near the southern margin and it is tempting to believe that when these rocks were formed, the upward current was adjacent to the southern wall as shown in Fig. 126B.   If this were the case, then the more powerful current would have carried up the large olivines and pyroxenes, as well as the plagioclases, directly to the place where congelation might be able to hold them.   In other places, where the congealing liquid was moving horizontally under the roof, all the primocrysts would be sinking but the heavy ones, such as olivine and augite, would be sinking faster and would have less chance of being caught by the congelation process.

b) *The Banded Division of the Marginal Border Group*

These rocks, like the Upper Border Group, have been explained as congelation cumulates but they differ in having representatives of all the available primocrysts in their make-up.   Apparently, magma carrying both light and heavy primocrysts, of the same general size as the cumulus crystals in the layered series, congealed against the temporary wall, due to loss of heat into the country rocks.   It is interesting to find that F. F. Grout, many years ago, suggested this as a theoretical possibility when discussing some aspects of the Duluth gabbro (1918c, p. 498).   The structures in the banded division of the Skaergaard marginal border group clearly imply that magma was in movement, whereas the structures of the tranquil division, equally clearly, imply an essentially stationary magma. There is no reason to postulate that the total amount of primocrysts in the congealing magma was large in relation to the volume of the liquid. However, it is clear that the primocrysts had had the opportunity of growing to a considerable size, and it is believed that they must have been present in the convection current for some time before being trapped by congelation. Perhaps most of the primocrysts caught in the congealing liquid at high levels in the banded division of the MBG were not from the rapidly descending, intermittent currents, but from the steady current, and had taken part in at least one complete circulation of the magma prior to being caught by congelation.

Owing to the pressure effect, in descending currents there should be either increased crystallization or supersaturation, while in ascending currents, the reverse effect should lead either to some re-solution of crystals suspended in the magma or to undersaturation. As the magma flowed horizontally under the roof, loss of heat would begin to reverse the effects of the ascent of the magma but here, and in the upper part of the descending marginal current, supercooling would be expected to be slight or absent. In the accessible parts of the banded division of the marginal border group, all of which are structurally high in the intrusion (see Fig. 89), there is no evidence of supercooling such as crescumulate structures or, so far as known, adcumulus growth.

Although no direct evidence is available, it has been suggested earlier that the banded division of the marginal rocks is not present at the hidden zone levels. It might have been thought that the tendency for increasing amounts of crystallization at depth, due to increasing hydrostatic pressure, should favour the development of a banded border group. On the other hand, the heat loss into the warmer country rocks would be less than at high levels, which would militate against the development of a marginal border group.

FORMER HYPOTHESIS ON CONVECTION CURRENTS IN MAGMAS

In the original paper on the Skaergaard intrusion (Wager & Deer, 1939, p. 269) it was pointed out that it was not profitable to discuss whether a convective circulation in magma is possible or not, but only whether a study of the rocks themselves provides evidence one way or the other. Convection currents are ephemeral and in plutonic intrusions, evidence for their former existence is necessarily indirect. However, in the case of the Skaergaard intrusion, former currents have left indelible marks and without invoking the hypothesis of convection currents, many features of the Skaergaard differentiation would be inexplicable.

Former currents in magma have often been inferred from structural features of intrusive igneous rocks, but more often these have been considered the result of flow during intrusion, or of external stresses acting soon after intrusion. Becker who, in 1897, was the first to make the suggestion that the differentiation of igneous rocks might be the result of fractional crystallization, also invoked convection currents as part of the mechanism by which fractionation became effective (1897a, b). In approaching the problem, Becker first considered the previously suggested methods of differentiation; in particular, he gave some evidence, for which there is now still more support (Bowen, 1921), that diffusion in magmas is probably too slow a process to allow any considerable concentration of the less fusible constituents by their early crystallization at the cooling surface (the Soret effect). However, if diffusion were inadequate, Becker suggested, in the following words, that convection currents might perform roughly the same function:

> If we suppose a dike in cold rock filled with mobile lava . . . the mass will be subjected to convection currents because the liquid near the walls will be cooler than that near the median plane of the dike. A circulation of lava will take place, the descending flow at the sides being compensated by ascending flow near the central surface . . . If the lava is a homogeneous mixture of two liquids of different fusibility, then the crusts which first form upon the walls will have nearly the same composition as the less fusible partial magma. If one follows mentally a small portion of the liquid in its circulation, it will clearly deposit at each of its early contacts with the growing walls a part of its less fusible component, and at each completed revolution it will have a different composition. This composition will always tend towards that which represents the most fusible mixture of the component compounds.

Becker gave no examples of igneous intrusions in which this process could be seen to have taken place and no certain examples of it in dykes have yet been found. However, if Becker's hypothesis is slightly modified and the early crystallizing minerals are considered to have separated over the floor of the intrusion, due to gravity, as well as attached to the walls, the explanation is essentially the one we have used for the Skaergaard intrusion. The main importance of Becker's two papers is that in them, for the first time, the process which the chemists had called fractional crystallization was applied to the differentiation of igneous rocks, but it is of interest that he visualized convection as the mechanism by which effective crystal fractionation was brought about.

In 1905, the investigation of the Shonkin Sag laccolith of Montana led Pirsson (1905, pp. 181–97) to claim that convection currents had been active in producing the differentiation; later workers have disagreed (Daly, 1914, p. 224; Osborne & Roberts, 1931) but the question is still open. After investigating the Duluth lopolith, Grout also came to believe in the importance of convection currents in magmas (1918c, pp. 495–6) and he successfully used the hypothesis to explain igneous lamination (1918d, pp. 452–7). He did not take the further step of explaining rhythmic layering in terms of intermittent convection currents but thought ' rhythmic effects of cooling, intrusive action, and gas emanation ' were responsible. The hypothesis that convection currents existed in the magma reservoirs responsible for the British Tertiary plutonic centres was used by Holmes (1931) in a paper in which he sought to explain the close association of acid and basic rocks in central igneous complexes. Holmes suggested that convection currents in the basic magmas played a part in melting the crustal rocks to produce acid magma but he did not consider their action during the cooling and crystallization stages. Holmes's hypothesis has been reviewed and extended in a recent paper on the marscoite and related rocks of the Isle of Skye (Wager, Vincent, Brown & Bell, 1965).

After Pirsson's and Grout's attempts to explain certain types of differentiation by convection, and Holmes's paper using convection to aid re-melting, the idea was not taken much further until 1939, when the evidence for the former existence of convection currents in the Skaergaard intrusion was given. Since then, the idea of convection currents has been invoked to explain certain features of the Rhum ultrabasic layered rocks (Brown, 1956) and of the Stillwater layered intrusion (Hess, 1960); and these cases are discussed in later chapters. The hypothesis of convection currents on a global scale, in the solid material of the mantle, was suggested by Holmes and Vening Meinesz, and since then by many others, as an explanation of some of the major features of the Earth. However, convection in the mantle is very different from convection in a fluid magma, emplaced within the solid crustal rocks. The order of magnitude of the velocity of convection currents in the atmosphere is 10^6 m per day; the order of magnitude deduced for the steady currents in the Skaergaard magma is 10^0 m per day, and the estimated order of magnitude of the currents in the solid mantle of the earth is 10^{-4} m per day.

CHAPTER IX

THE SKAERGAARD FRACTIONATION TREND AND POSSIBLE EXPLANATIONS

THE TREND

In the Skaergaard intrusion, the sequence in which the various fractionation products were formed is known from the mapping and petrology. In many igneous complexes, the sequence of the fractionation products can only be presumed from indirect evidence; for example, a fractionation index, based on rock chemical compositions (cf. Poldervaart & Elston, 1954; Simpson, 1954a; Wager, 1956; Thornton & Tuttle, 1960). When the sequence of fractionation products has been established, by one means or other, the nature of the successive products of fractionation can be expressed in terms of the proportions and compositions of the different minerals present, or in terms of the chemical composition of the successive rock fractions. The results of whole-rock chemical analysis are best expressed in a series of variation diagrams which show the percentage of the various elements or oxides plotted against the established time sequence of crystallization, expressed as percentage solidified of the whole: although this integrates much data, the significance of the chemical variation diagram is usually only to be appreciated in terms of the nature and composition of the individual mineral phases composing the rocks.

The changing compositions of the Skaergaard minerals during fractionation have been described and tabulated in Ch. III. The changes have been summarized by plotting certain features of the mineral compositions against height in the layered series (Fig. 14) or against the percentage of the intrusion solidified. The compositions of the cumulus minerals may also be plotted against the changing composition of one of the cumulus minerals showing a regular variation in composition, such as olivine, rhombic pyroxene or plagioclase, and this method is particularly valuable in providing a basis for comparison with other layered intrusions. In the case of the Skaergaard intrusion, the composition of the olivines, the calcium-poor pyroxenes and the augites, in terms of the magnesium/iron ratios, is plotted against the anorthite percentages of the cumulus plagioclases in Fig. 131. The graphs show, conveniently, the compositions of the co-existing mineral phases, and they also show the compositions at which cumulus minerals began or ceased to form as a result of crystal fractionation. The fact that

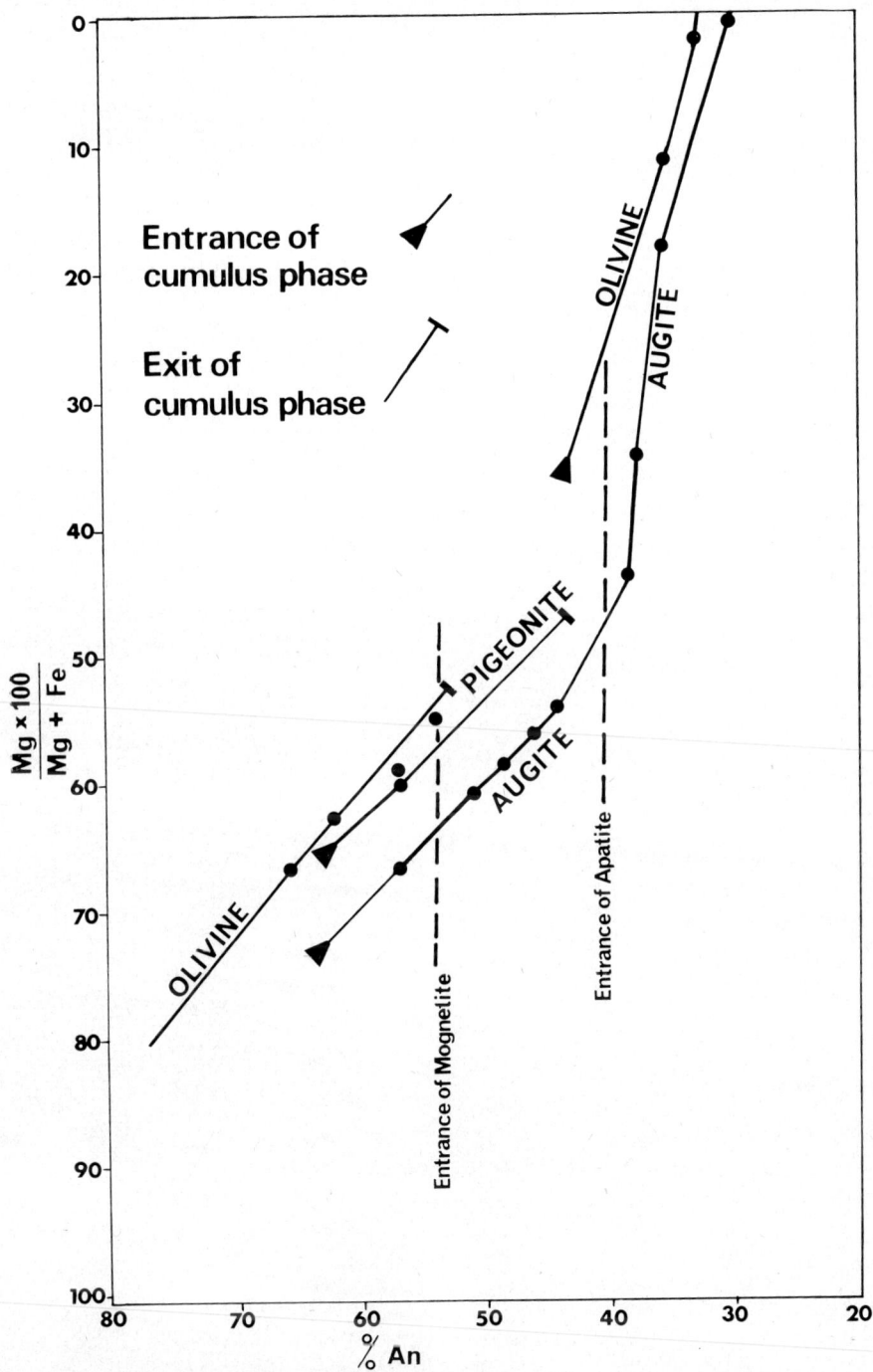

FIG. 131. Compositions and range of cumulus phases of the Skaergaard intrusion, plotted against the compositions of the cumulus plagioclases. The entrances of cumulus magnetite and apatite, also a result of fractionation, are shown.

cumulus olivines have higher magnesium/iron ratios than the co-existing pryoxenes is at once evident from such graphical presentation. The points of entry of other cumulus minerals, such as magnetite and apatite, the compositions of which cannot be expressed in terms of the magnesium/iron ratio, can also be shown conveniently on the graph.

Important aspects of the changing chemical composition of the average rocks, successively formed during the fractionation process, may be summarized on a triangular diagram of the kind shown in Fig. 116 and 132 in which Mg, $Fe^{+++}+Mn$, and $Na+K$

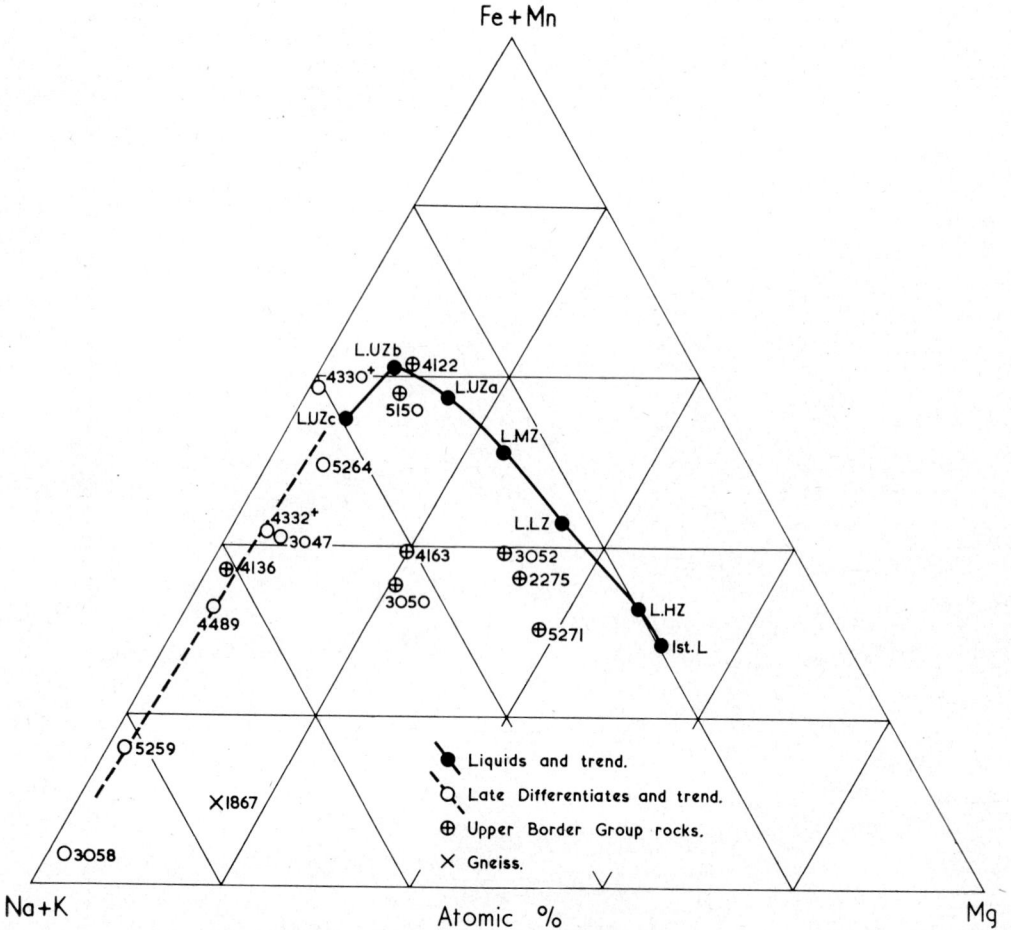

FIG. 132. Successive Skaergaard liquids (calculated compositions) and analyses of Late Differentiate rocks, Upper Border Group rocks, and the acid gneiss of the Metamorphic Complex, plotted on a diagram showing the chief elements of the generally high-, medium-, and low- temperature minerals at the three corners of the triangle. The broken line is the estimated liquid trend, which continues that previously calculated for the layered series (Ch. VII).

are put at the three corners. This particular type of triangular diagram was devised originally for the Skaergaard intrusion (Wager & Deer, 1939) to take account of the marked enrichment in iron which developed in the middle stages of fractionation: it has since been extensively used, either in this form, or with ferric added to the ferrous iron. The early stage, and early-middle stage, average rocks tend to be near the Mg corner because of the relatively high magnesium content of the early ferromagnesian minerals; the late-middle stage average rocks tend to lie near the iron corner because of the enrichment of the later ferromagnesian minerals in iron, and the late-stage granophyric differentiates, which formed only in very small quantities, tend to lie near the sodium and potassium corner because of the rapidly increasing amount of alkali felspars in these rocks. The changing composition of the Skaergaard liquids, obtained by observation in some cases and by calculation in others (see Ch. VII), may also be plotted on the same triangular diagram, and the compositions of the coexisting average rocks may be tied to the liquids from which they were forming (Fig. 116).

The results of physico-chemical experiments on the crystallization of silicate melts are usually expressed on thermal phase diagrams which indicate, among other things, the compositions of a succession of coexisting solid and liquid phases, at gradually lowering temperatures. For the Skaergaard intrusion, the integration of a large number of observations provides similar information but only in relation to one *initial* liquid composition, and the temperatures are only relative, and not absolute. This kind of data for the Skaergaard intrusion corresponds, in fact, to one particular fractionation curve on the usual experimental thermal phase diagram.

In the Skaergaard intrusion, as in any chemical fractionation system, the minerals which separated during cooling have controlled the direction of change in the composition of the liquid. The overall chemical compositions of the successive liquids and of the average contemporaneous crystal cumulates have been expressed already in a series of chemical variation diagrams (Pl. X). It should once again be emphasized that the curves are not smooth, as is often assumed for fractionation series. In the case of the average material separating, there are sharp steps on the curves because of the incoming or outgoing of particular crystal phases; at corresponding points on the curves for the liquids, there are abrupt changes of slope. The curves for the variation in the major elements are what would be expected from the known sequence and composition of the minerals separating; the variation of the trace elements, on the whole, follows that of the major elements, and in the now reasonably well understood way.

The trend in the liquid compositions only diverges to a small extent from that for the average layered rocks, but each rock composition, as would be expected, follows behind that of the contemporary liquid (Fig. 116). The maximum divergences between the liquid and the average rock compositions are shown by Ti and P (see Pl. X) and it is because of the latter that the phosphorus content of any particular rock can be used in estimating the original proportions between trapped liquid and crystal cumulate.

EXTREME FRACTIONATION AS THE SUGGESTED CAUSE OF
THE GRANOPHYRES

The composition of initial Skaergaard magma, if it be accepted that it was the same as the relatively fine-grained, marginal olivine gabbro, is that of a highly reduced, olivine-rich, tholeiitic basalt with a high alumina content (see Ch. VII, p. 151). As a result of extreme fractionation of this magma, iron-rich basic rocks and, ultimately, small amounts of intermediate and acid rocks are developed. The extreme fractionation results from the particularly effective removal of the early crystal phases from positions in which they could react with the bulk of the liquid. For example, the early removal of the magnesian olivines has prevented them from reacting with the bulk of the liquid to give magnesian pyroxene. A calculation of the CIPW norm of the initial Skaergaard magma, by the usual method, gives an approximate indication of the minerals which would crystallize from it, assuming no fractionation. The calculation gives an assemblage including 13·3% olivine, 7·9% hypersthene and, of course, no quartz (Table 16, column I). A norm calculation may also be made in terms of olivine and quartz, rather than hypersthene (Table 16, column II). In this case there is 19·0% olivine and 2·2% quartz, and with strong fractionation there would be a tendency for this association of minerals to form. If the 2·2% of quartz obtained in the second calculation contributed to the formation of granite with about 30% quartz, there would be approximately 7% of granitic rock derived by fractionation from the Skaergaard olivine–tholeiite magma. Early crystallization of abundant orthopyroxene in place of olivine in the hidden layered series may have occurred, however, owing to the pressure effect, and this would have reduced the amount of free quartz for the late-stage rocks (see p. 169). The amount of free quartz will be further reduced because crystallization of the trapped liquid of the cumulates only produces local fractionation; furthermore, crystallization of the liquid from which the greater part of the border group formed also does not produce fractionation changes in the main bulk of the magma. It is clear that any physical conditions influencing the crystallization of orthopyroxene, in place of the equivalent olivine and quartz, would reduce the amount of siliceous residuum, and any reduction in the effectiveness of fractionation will have the same effect. From these arguments it appears that the amount of granitic residuum could vary between 0 and 7%. Earlier it has been assumed that the granitic residuum was only 0·5% of the whole intrusion, and the above discussion shows that this is a possibility, but gives no specific support for this particular figure.

If the amount of acid residuum from fractionation has been seriously underestimated, then the graphs of Fig. 111 and 115 would be changed on the right-hand side of each, where the later stages of solidification are shown. One result would be that the SiO_2 content of the hidden layered series would be lower than previously estimated. It is considered that this alternative is not excluded by the available evidence, but that it is not as likely as the one adopted earlier, that there was about half a per cent of acid residuum. Arguments using norms are not worth refining further at the present time, especially when it is remembered that there is still another unknown factor in the Skaergaard situation; namely, the amount of acid-gneiss inclusions which have been melted down and incorporated in the magma. The main point to be made here is that

TABLE 16

Standard CIPW norm of chilled marginal gabbro EG4507 and calculations of other possible norms, assuming different mineral assemblages and different degrees of oxidation

	I	II	III	IV
	EG4507 Standard CIPW Norm	Norm assuming all available MgO and FeO is present as olivine after allotting FeO to iron oxides	Norm assuming Fe^{++} is converted into Fe^{+++} in sufficient quantity for all iron to be in ilmenite and magnetite*	Assuming Fe^{++} is converted into Fe^{+++} in sufficient quantity for all iron to be in ilmenite and magnetite and assuming all available MgO, after allocation to diopside, is converted into forsterite with no enstatite
Quartz	...	2·16	0·78	5·04
Orthoclase	1·67 ⎫	1·67 ⎫	1·67 ⎫	1·67 ⎫
Albite	19·91 ⎬ 57·14	19·91 ⎬ 57·14	19·91 ⎬ 57·14	19·91 ⎬ 57·14
Anorthite	35·56 ⎭	35·56 ⎭	35·56 ⎭	35·56 ⎭
Diopside				
CaSiO₃	8·35 ⎫	8·35 ⎫	8·35 ⎫	8·35 ⎫
MgSiO₃	5·00 ⎬ 16·25	5·00 ⎬ 16·25	7·20 ⎬ 15·55	7·20 ⎬ 15·55
FeSiO₃	2·90 ⎭	2·90 ⎭	... ⎭	... ⎭
Hypersthene				
MgSiO₃	5·00 ⎫	... ⎫	14·20 ⎫	... ⎫
FeSiO₃	2·90 ⎬ 7·90	... ⎬ ...	... ⎬ 14·20	... ⎬ ...
Olivine				
Mg₂SiO₄	7·98 ⎫	11·48 ⎫	... ⎫	9·94 ⎫
Fe₂SiO₄	5·30 ⎬ 13·28	7·55 ⎬ 19·03	... ⎬ ...	... ⎬ 9·94
Magnetite	1·86	1·86	9·28	9·28
Ilmenite	2·28	2·28	2·28	2·28
Apatite	0·34	0·34	0·34	0·34
Water	1·06	1·06	1·06	1·06

* The figures assumed in this calculation are: 64 mol. prop. of FeO converted to 32 of Fe_2O_3 giving a total of 40 mol. prop. of Fe_2O_3 and leaving 55 mol. prop. of FeO. The amount of oxygen required is 0·5 % by weight.

with strong fractionation, an olivine-rich basaltic initial magma, such as the Skaergaard, can yield a siliceous residue, as Bowen suggested long ago. The amount of siliceous residue produced by such a process, however, appears to be small in bulk, even under the conditions of extreme fractionation as in the Skaergaard intrusion. In intrusions where adcumulates are rare, and in which orthopyroxene is precipitated in abundance,

Q

the amount of free silica in the residual rock fractions may well be even smaller. On the other hand, oxidizing conditions would favour late enrichment in free silica, as discussed below.

The Oxidation State of the Magma in relation to Iron Enrichment, and the Origin of the Ferrodiorite and Calc-alkaline Fractionation Trends

The low state of oxidation of the initial Skaergaard magma is considered to have contributed to the trend towards iron enrichment. In the original memoir (Wager & Deer, 1939, pp. 310–11) it was argued that greater oxidation would not greatly affect the middle or late stages of the fractionation, but this is now believed to be incorrect.

Walker and Poldervaart, in describing the Karroo dolerites (1949, pp. 656–62) showed that fractionation produced a considerable degree of iron enrichment in many dolerites having a high initial Fe^{++}/Fe^{+++} ratio, but that in some quartz dolerites and in basalt lava series with a lower initial Fe^{++}/Fe^{+++} ratio there was less iron enrichment and thus a closer approach to the calc-alkaline trend. In a later paper on the pegmatitic differentiates of basic sheets, Walker (1953, pp. 57–8) reiterated his view and added that where the Fe_2O_3 content is high, 'early elimination of unsilicated iron oxides . . . build up SiO_2 in the rest-magma . . . so that the resulting coarse phase (i.e. the product of fractionation) was distinctly granophyric in character'. The general influence of the degree of oxidation on the trend of fractional crystallization has also been discussed by Goldschmidt (1954, pp. 31–9).

In the Skaergaard intrusion, the highly reduced state of the initial magma must have caused delay in the appearance of magnetite as a cumulus mineral and, in addition, a continued high ferrous iron content which resulted in iron-rich olivines and pyroxenes being abundant. Had the initial Skaergaard magma been more oxidized than the actual chilled marginal gabbro then the same degree of fractionation would, no doubt, have produced less late enrichment in iron, or none at all, because of the early precipitation of magnetite. This can be illustrated if norms are calculated with different assumptions regarding the state of oxidation of the iron. When the FeO in the original magma is assumed to be partly converted into Fe_2O_3 there is, necessarily, an increase in the amount of normative magnetite; a limiting case occurs when the proportion of ferrous to ferric iron is such that all the iron forms magnetite and ilmenite and none is available to enter the silicate minerals. This would happen when 64 molecular proportions of ferrous oxide are converted to 32 molecular proportions of ferric oxide. There would then be a total of 11·6% of magnetite and ilmenite (Table 16, column III), the pyroxene would necessarily be a magnesium-silicate without ferrous iron, there would be no normative olivine because more silica would be available to form the enstatite, and there would be 0·8% of free quartz. On the other hand, if it be assumed that forsterite were formed rather than enstatite (Table 16, column IV), the normative quartz would be

5·0%. To produce the degree of oxidation assumed here it is estimated that the amount of oxygen which would have to be added is 0·5% by weight of the whole intrusion.

The actual minerals which would separate from the postulated, more-oxidized magma (Table 16, columns III and IV) would no doubt approximate more closely to the normative minerals given in column III. Of these, it would be expected that magnetite would crystallize at an early stage and certainly earlier than in the less oxidized magma. There should also be early precipitation, in quantity, of pure diopside and enstatite because these minerals, being free from Fe^{++}, would separate at relatively high temperatures. There would be no possibility of ferrodiorites developing by fractionation because in the case being considered, the ferromagnesian minerals only contain magnesium. (It should be noted that one of us (G.M.B.) disagrees with this particular part of the reasoning.) Silica would be relatively more abundant and would be expected as a separate phase, and finally, in the middle and late stages of the fractionation process, a greater proportion of felspar would be present relative to the magnesian minerals, because the latter would tend to have separated early, at the higher temperatures. In fractionation under strongly oxidizing conditions, the tendency for a large amount of magnesium-rich ferromagnesian minerals to be precipitated early, and of plagioclase to be relatively more abundant in the middle and late stages, may be an important feature in addition to silica enrichment.

The effects of fractionation on a more oxidized basic magma, which was otherwise of Skaergaard composition, have here been considered by the crude method of calculating norms with different assumptions about the degree of oxidation. While such a method does not *prove* anything, it indicates certain possibilities which can be compared with natural occurrences and with experimental data. In general, it would appear that strong fractionation of a reduced magma, like the Skaergaard, leads to ferrodiorites and a small granophyric residuum. Fractionation of a magma similar to the Skaergaard but more oxidized would, presumably, not lead to ferrodiorites, but to rocks of the calc-alkaline series, with *slightly* more abundant acid end-products.

In recent years, the effects of varying partial pressures of oxygen on the fractionation trends of basic magma have been discussed by Osborn (e.g. 1959) and he has used the Skaergaard example as one of the limiting cases. He has shown, from experimental evidence, that fractionation of basaltic magma will produce different results depending on whether there be a constant total amount of oxygen in the system or a constant partial pressure of oxygen, the oxygen being derived from reaction with water. The Skaergaard fractionation corresponds closely to the case in which there is a constant total amount of oxygen and fractionation results in a marked increase in the ferrous iron/magnesium ratio with little change in silica content (Osborn, 1959, pp. 633–6). The contrasting case is the basalt–andesite–rhyolite fractionation series, in which it is believed that the partial pressure of oxygen is maintained, or slightly increased during fractionation, by the dissociation of water and escape of hydrogen. The Skaergaard mineralogy suggests an intrusion of relatively dry magma initially, and the overall chemistry of the intrusion suggests that only a small amount of water was absorbed from its surroundings during cooling. The limited amount of oxidation, leading to the change of Fe^{++} to Fe^{+++} which seems to have taken place during the fractionation, has probably only increased the overall Fe_2O_3 content by about 0·13%; the resulting, slight precipitation of magnetite was clearly not sufficient to prevent the formation of ferrodiorites. If, through the agency of added water and loss of hydrogen, the Skaergaard magma had been oxidized, either initially, or continuously during fractionation, it would, perhaps, have produced small amounts of quartz diorite, granodiorite and granite of the calc–alkaline series, as suggested by Osborn's work (see, for example, 1959, pp. 644–6).

The Skaergaard magma, which in terms of normative olivine was undersaturated, nevertheless produced a small amount of granitic residuum as a result of the strong fractionation. In the original memoir it was argued (Wager & Deer, 1939, p. 313) that the differences between the initial Skaergaard magma and average basic magma were not sufficient to have been responsible for the ferrodiorite trend and that the strong iron enrichment was simply the result of strong crystal fractionation. It was further argued that if the calc–alkaline series of igneous rocks were also produced from basic magma by crystal fractionation, then this was because their initial magma was modified by contamination with sialic material (Wager & Deer, 1939, p. 323). Since ferrodiorites are relatively rare, it was suggested that contamination with acid material, followed by fractionation, was more usual in the Earth's crust than the pure fractionation, leading to ferrodiorites. These arguments were criticized by Burri & Niggli (1945) and are now considered fallacious. The Skaergaard trend, leading first to iron enrichment and then to granophyre, is now accepted as due to strong fractionation of basic magma in a low oxidation state, and it is believed that strong fractionation of a relatively oxidized basic magma of an otherwise similar composition could lead to the calc–alkaline series of rocks, whether there be contamination or not.

The Skaergaard intrusion provides evidence of the production of small amounts of acid rocks as a result of strong fractionation, even from reduced olivine-tholeiite basalt magma. Osborn's work indicates that under more oxidized conditions, slightly greater amounts of acid residuum should result from a similar degree of fractionation. However, we do not regard even the small acid intrusions such as those associated with the Tertiary basic intrusions of Skye, Mull, etc., as the result of fractionation of either reduced or oxidized basalt magma. We believe it is more likely that these acid rocks are mainly the result of re-melting of pre-existing intermediate and acid rocks of the continental crust, as suggested in recent papers (Brown, 1963; Wager et al., 1965; Moorbath & Bell, 1965).

PART II

EXAMPLES OF VARIOUS TYPES
OF LAYERED INTRUSIONS

CHAPTER X

THE RHUM ULTRABASIC COMPLEX, SCOTLAND: FORM OF INTRUSION, UNITS OF LAYERING, AND MINERALS

The igneous rocks of the island of Rhum were mentioned as early as 1819 by Maccullough, but it is to Judd (1874, p. 252; 1885) that we owe the recognition of Tertiary plutonic rocks, which he related to the presence of a central volcano; comparable rocks were related to similar volcanoes in Skye and other neighbouring Hebridean islands (Fig. 133). Subsequent work by Geikie (1897) drew attention to the striking terraced features produced by the almost horizontal layers of ultrabasic rock building some of the central mountains of the island. In 1902 and 1903, Harker mapped Rhum geologically on a scale of 6 inches to 1 mile (together with the neighbouring Small Isles) and a map and Memoir were published in 1908. Harker's work showed that the island consisted, essentially, of 12 square miles of intrusive ultrabasic peridotites and allivalites, together with intimately associated bodies of eucrite and gabbro, and separate complexes of felsite, granophyre and explosion breccia (Fig. 135). The whole of this Tertiary igneous assemblage was intruded into Precambrian (chiefly Torridonian) rocks, occasionally capped by Tertiary basalt flows, and Bailey (1945) later demonstrated the Tertiary age of a ring-shaped fracture within which most of the plutonic rocks were confined. Both Harker and Bailey believed that the layered ultrabasic rocks were formed by sill-like injections of ultrabasic magmas of peridotitic and allivalitic composition.

In 1950 we paid our first visit to the island, and were convinced that the ultrabasic and basic rocks of the Rhum intrusion possessed many features in common with the layered rocks of the Skaergaard, for which an origin of the layering through rhythmic crystal accumulation had been demonstrated. Nevertheless, the Rhum rocks possessed other unique and problematical features; a preliminary note (Wager & Brown, 1951) dealt with one of these, the so-called 'harrisite structure' believed to have been produced through upward growth of crystals from the floor of the intrusion. Investigation of the eastern part of the complex, including the terraced mountains of Hallival and Askival, was carried out between 1950 and 1954 by one of us (Brown, 1956), and the conclusion reached that the eastern ultrabasic complex represented a layered

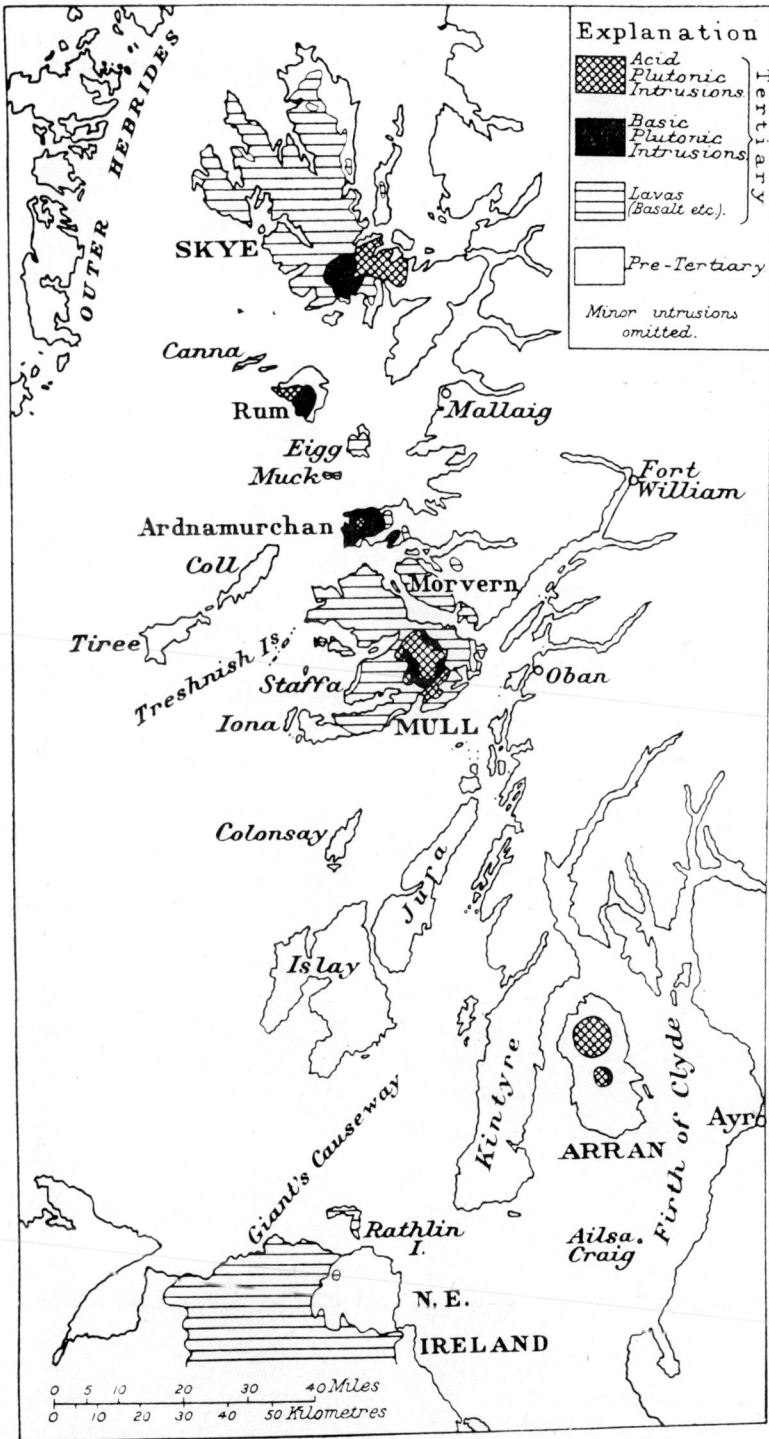

FIG. 133. Sketch-map of the Inner Hebridean islands, Northwest Scotland, showing the position of the Isle of Rhum in relation to the other centres of Tertiary Thulean igneous activity. (From Richey, 1948.)

series formed by rhythmic crystal deposition on the floor of a larger layered intrusion, since faulted and uplifted to its present position along a ring fracture. Later, Wadsworth (1961) studied the western part of the complex and demonstrated a similar history for this sector. Work by Hughes (1960) and Dunham (1962), on the acid rocks, confirmed Bailey's earlier findings that these rocks were emplaced before the basic and layered ultrabasic rocks, and demonstrated that on these and other grounds the acid rocks could not be viewed as the products of fractionation of the layered intrusion, but were probably formed by melting of part of the country rocks.

THE FORM OF THE LAYERED INTRUSION

The outcrops of igneous rock lie within a rough ellipse measuring about 7 miles by 5 miles (Fig. 135). If, however, the large mass of the Western Granophyre is treated as a separate unit, then the rest of the igneous intrusives have a circular outcrop, roughly 5 miles in diameter, which is cut by a north–south fault (the Long Loch fault). The Rhum Tertiary centre is thus comparable in size with others of the British Tertiary Province such as Skye, Mull, and Ardnamurchan (Fig. 133).

The centre is defined quite clearly on its eastern, and the greater part of its northern and southern margins, by the Tertiary ring-fault, and the layered rocks lie within that fault. However, the contacts of the layered intrusion with acid igneous and uplifted Precambrian rocks within the fault are more difficult to define. The layered mass is not enclosed by a marginal border group similar to that found in the Skaergaard intrusion (Ch. V). Instead, the layers often approach close to the country rocks (for this purpose the acid igneous rocks are considered 'country rocks' in that they are believed to have been formed before the layered rocks were emplaced), or are separated from them by a narrow zone of structureless gabbro. To the west, the contact is complicated by back-veining of the layered series or the marginal gabbro, owing to the rheomorphism of adjacent parts of the earlier, Western Granophyre (Fig. 136).

KEY FOR FIG. 134

UIO.	The layered ultrabasic rocks divided into units (U1 ... U15), each consisting chiefly of olivine-rich cumulates (stippled) grading up into felspar-rich cumulates.
+ + +	Fine-grained olivine-gabbro
+	Askival-plateau gabbro
∧ ∨	Barkeval eucrite
∵ ⟨ ⟩	Marginal zone with structureless gabbro
▨	Isolated gabbro
▧	Torridonian sediments
∵ ∴	? Metasomatized Precambrian

⬭	Non-ultrabasic inclusions
⬚	Dykes and cone-sheets
?	Area not mapped in detail
——	Boundary observed, or inferred from topography
- - - -	Boundary tentative
▬▬▬	Main ring-shaped fracture
-·-·-	Other faults
⭨15	Dip of layering
⭨40	Dip of cone-sheets
-··-··-	Contours in feet
······	Footpath
⊗	Analyzed specimens

FIG. 134. Geological map of the Hallival–Askival area, eastern Rhum, showing the disposition of the major rhythmic units of the eastern part of the layered series. (Scale, approx. 4 inches to 1 mile.) (From Brown, 1956.)

The marginal, structureless gabbro appears generally to be in the form of a vertical dyke, parallel to the ring fracture, but locally it spreads into a sheet, inclined at a low angle beneath country rocks which seem to form a local roof to the layered rocks (Fig. 134). These latter are not parts of an original roof, nor is the equivalent of an upper border group present; they are considered, rather, to be blocks carried up on the shoulders of the layered series and brought almost into contact through the expulsion of intervening gabbroic liquid (p. 290). The composition (Table 21) and structure of this gabbro are not reconcilable with a border group envelope. On the contrary, the gabbro is believed to have been a later surge of magma, intruded during the uplift of the layered series along the ring fracture and acting as a lubricant for the uplift of this mass. This episode will be discussed in more detail later, together with more evidence that in the layered series of Rhum we are dealing with a block of layered rocks now divorced from the rest of a larger mass at depth.

CLASSIFICATION OF THE LAYERED SERIES INTO RHYTHMICALLY LAYERED UNITS

Layered rocks are found at sea-level in the western, Harris Bay region, while the highest form the summit of Askival Mountain (2659 ft). The layers dip at about 15° radially inwards towards a centre in Glen Harris (Fig. 138). Although this dip may locally be exceeded, due to disturbances prior to final consolidation, the thickness of the layered series east of the Long Loch fault can be estimated as 2500 ft, and that west of the fault as about 4500 ft. The total exposed thickness is probably more than the sum of these (7000 ft), because no layered units east and west of the fault can be equated.

Within such a thickness of layered rocks, some sort of subdivision is necessary. The method used in the case of the Skaergaard intrusion is based on the cryptic layering, the layered series being divided into zones according to the entry and exit of certain minerals as a cumulus phase. However, cryptic layering is absent throughout the series as a whole in Rhum, except within certain thick units of the western layered series (Table 17). Rhythmic layering, on the other hand, is developed throughout the Rhum ultrabasic rocks; it is a large-scale type, recognized also in the Ultrabasic Zones of the Bushveld, Great Dyke, and Stillwater intrusions, and provides a means of subdivision. On the basis of this large-scale layering, the series can be divided into what have been called ‘major’ rhythmic units or macro-units (in contrast to the ‘zones’ of the Skaergaard intrusion).

The division into units was first carried out for the 2500 ft of layered series in the eastern region (Brown, 1956, fig. 8) and fifteen were recognized, averaging about 170 ft in thickness but ranging, in fact, from 50 ft to 500 ft in individual cases (Table 17; Fig. 134 & 139). A unit consists, essentially, of an olivine-rich rock (ol-cumulate) at the base, grading upwards to a plagioclase-rich rock (plag-cumulate) at the top, although the olivine cumulates generally contain minor amounts of cumulus chrome-spinel, and

FIG. 137

...iew of the layered ultrabasic rocks forming the mountains of Hallival and ...skival, looking south. The layering can also be seen on the lower hillside in ...e background, continuing westward to Barkeval (off the picture). The bare ...lls directly north of the layered intrusion consist largely of felsites and ...plosion breccias (Fig. 135), while the lower exposures just south of the road ...e of slightly metamorphosed Torridonian sediments. The ringfault which ...efines, here, the northern margin of the igneous centre, strikes east–west on ...e soil-covered hillside between Tertiary felsite and Torridonian sediment ...xposures.

View of the Rhum layered intrusion from the west, including (from left to right) the mountains of Barkeval, Hallival, Askival, and Trallval. The layering dips, approximately, radially towards the broad valley in the centre of the picture (Glen Harris). All the rocks in view are ultrabasic, those in the foreground belonging to the Ard Mheall and Dornabac Series (see Fig. 136).

FIG. 138

FIG. 139. View of Hallival mountain from the north, showing the low dip of the layered ultrabasic rocks. Each major rhythmic unit (numbered 7 to 14) consists of olivine-rich cumulates (weathered) overlain by plagioclase-rich cumulates (resistant ribs). Towards the top of the mountain, the character of the units can be seen to change so that the felspathic rocks comprise the greater part of each unit. Sheets of fine-grained olivine gabbro (LG and UG) have intruded the lower units.

the plagioclase cumulates minor amounts of cumulus olivine and clinopyroxene (which may be locally concentrated to give fine-scale rhythmic layering). The distribution of olivine relative to plagioclase in these units is broadly equivalent to grading in gravity-sorted crystal accumulates, but is on such a large scale that its origin is believed to differ from that of the smaller-scale gravity stratification of other layered intrusions. The partly serpentinized olivine rocks of the units weather more easily than the felspathic rocks and this results in the terraced feature of the mountains, in which the tops of the units stand out as pale-coloured ribs against the rest of the layered series (Fig. 139).

Harker (1908) called the olivine-rich and felspar-rich rocks peridotite and allivalite, respectively. Believing them to be produced from injections of magma of the appropriate composition, he stressed the contrast between these two rock types, although he was aware of the existence of intermediate varieties. Here the rocks are viewed as ultrabasic* cumulates and a nomenclature based on this view will be used, although allivalite and peridotite are retained as general descriptive names independent of any theories as to the origin of the rocks.

The division of the western outcrops of the layered series is based on similar criteria, except for the lowest unit. Thus, the series has been sub-divided (Wadsworth, 1961) into four major units, lettered A to D (Table 17 and Fig. 136). Unit A, stratigraphically the lowest exposed in Rhum, is 400 ft thick and consists of olivine–felspar–pyroxene mesocumulates. Unit B consists of 1200 ft of olivine(–spinel) cumulates (unique in the prevalence of olivine crescumulates, see p. 273) overlain by 400 ft of more felspathic cumulates. Units C and D, 1500 ft and 1000 ft respectively, also follow the pattern of

* The term *ultramafic* is inapplicable to most early-stage cumulates, since leucocratic lagioclase is a dominant phase.

unit B in the relative distribution of olivine and felspar, but are free from crescumulates. A problematical Transition Zone occurs between units A and B, while igneous breccias and faults complicate the otherwise regular pattern within each unit. Despite this, the concentration of olivine relative to plagioclase at the base of each unit is a feature common to both the western, and the relatively thinner eastern units.

TABLE 17

The sequence of layered rhythmic units in Rhum

its	Approx. thickness (ft)	Lithological divisions	Principal rock types	Cumulus olivine composition (Fo %)	Cumulus plagioclase composition (An %)
5	180		Each unit consists of olivine cumulates at the base, grading upwards into plagioclase cumulates at the top. Cumulus augite and chrome-spinel present in minor amounts. Lower units chiefly mesocumulates, and upper units adcumulates and heteradcumulates. Total plagioclase/olivine ratio increases above Unit 10 (see Brown, 1956, fig. 8)		Essentially An$_{88-84}$ throughout. Intercumulus zoning extensive in lower seven units (sometimes to An$_{44}$), but rare in the olivine - rich cumulates and absent from the plagioclase-rich cumulates of higher units (see Fig. 147)
4	200				
3	170				
2	50				
1	250				
10	300				
9	120		Essentially Fo$_{86-84}$ throughout, and unzoned	
8	200	Units not sub-divided			
7	50				
6	50				
5	250				
4	30				
3	480				
2	120				
1	>50				
D	1000	U. Ruinsival Series	Chiefly olivine–plagioclase adcumulates, in which olivine crescumulate types are abundant. Unit A, mostly orthocumulates (see Wadsworth, 1961, table 1)	85$\frac{1}{2}$ / 88	84$\frac{1}{2}$ / ...
C	1500	L. Ruinsival Series		86 / 86	84 / ...
B	400 / 1200 / 150	Dornabac Series / Ard Mheall Series / Transition Series		85$\frac{1}{2}$ / 88$\frac{1}{2}$ / 86 / 82	84 / 85 / 84 / 78
A	400	Harris Bay Series		82	78

Thus the division of the Rhum layered series, into units A to D and 1 to 15, is based on the easily recognizable feature of high concentration of olivine, relative to felspar, at the base of each major rhythmic unit. Within each unit, however, many other features of significance are to be found, including the distribution of cumulus pyroxene and chrome-spinel, the development of olivine crescumulate layers, the variable extent of adcumulus growth, fine-scale rhythmic layering, and slump-structures. The mineralogy of the intrusion is described next in this chapter, after which the textures and structures are dealt with, followed by an outline of the whole layered sequence.

THE MINERALS

The layered rocks consist almost entirely of four minerals: olivine, plagioclase, clino-pyroxene, and chrome-spinel. Minor amounts of orthopyroxene and apatite are restricted to certain units and do not have the status of cumulus minerals. Occasional grains of biotite and magnetite are believed to be secondary products, while serpentine, epidote and calcite are clearly of secondary origin.

TABLE 18

Modal analyses of a typical eastern rhythmic unit (Unit 10) in which the proportion of cumulus plagioclase to olivine is the major variant. Minor layering, towards the top, is chiefly due to variation in cumulus plagioclase: cumulus augite ratio

Height above base (ft)	Plagioclase	Olivine	Augite	Chrome-spinel	Specific gravity
295	84·0	15·5	...	0·5	2·83
275	72·5	18·5	9·0	...	2·89
255	61·0	7·5	31·5	...	2·99
220	64·0	16·5	19·0	0·5	2·95
190	44·0	52·0	4·0	...	3·17
170	21·5	66·5	12·0	...	3·26
132	18·5	71·5	8·5	1·5	3·30
92	16·0	76·5	6·5	1·0	3·29
55	10·0	76·5	13·5	...	3·28
10	13·5	77·0	7·5	1·5	3·28

Detailed optical measurements of the olivine, felspar and pyroxene throughout the eastern series showed no regular changes in composition from higher to lower melting-point members of each isomorphous series, but in the thicker, western units, slight but distinct mineral compositional changes were found between and within the units (Table 17). The optical determinations have also been valuable in investigating the amount and extent of the zoning in the felspars. The three chief cumulus minerals, plagioclase, olivine and augite, were separated from an adcumulate of a typical unit (No. 10 of the eastern series) and chemically analyzed, and their carefully determined optical properties then acted as a standard for correlation between compositions and optical measurements for the minerals from other rocks.

a) *Olivine*

The analyzed specimen has the composition $Fo_{86}Fa_{14}$ (Table 20) and is a magnesium-rich member of the Fo–Fa series, i.e. a chrysolite. It is a simple ferromagnesian olivine, with only subordinate amounts of CaO and MnO. The crystals are little serpentinized, but frequently contain exsolved iron ore (magnetite, according to a partial analysis) in the form of the small dendritic plates parallel to (100) which appear to characterize this type of exsolution. A further set of exsolution plates parallel to (110) are brown and translucent, apparently of chromite, which would suggest that whereas

TABLE 19

Modal analyses of rocks from Unit B (Ard Mheall series) of the western layered series, where most of the layering is due to variation in the habit of the olivines

Predominant textural features of adjacent layers within a specimen, or of other, individual thin layers	Olivine	Plagioclase	Augite	Chrome-spinel	Secondary products
Minor layering of three hand specimens					
Small rounded olivines	91·6	4·5	3·3	0·5	...
Large tabular olivines	86·5	6·9	0·6	2·7	3·2
Small rounded olivines	93·8	0·5	0·6	0·7	4·3
Some tabular olivines	81·6	11·1	2·7	0·6	4·0
Large tabular olivines	77·2	17·5	3·0	2·4	...
Small rounded olivines	89·6	5·8	3·8	0·7	...
Some tabular olivines	84·4	9·0	3·3	3·3	...
Other individual minor layers					
Small rounded olivines	99·6	...	tr	0·4	...
Small rounded olivines	94·1	1·5	0·9	0·3	3·1
Some tabular olivines	86·7	11·7	0·4	1·1	...
Layers richer in cumulus plagioclase					
Small rounded olivines	78·4	12·9	7·9	0·8	...
Large tabular olivines	69·5	26·7	0·9	2·9	...
Olivine crescumulates					
Medium grain size	88·6	6·6	1·3	0·6	3·0
Coarse grain size	75·2	11·2	1·9	1·4	10·4

the olivine crystallized in equilibrium with a chrome-magnetite, exsolution of the chromium and ferric iron within the olivine structure took place along different structural planes (Brown, 1956, p. 78).

The full range of olivine composition is from $Fo_{88\frac{1}{2}}$ (unit B) to Fo_{82} (unit A), but the greater part of the layered series rocks (>4000 ft) contain olivine of composition Fo_{86}. Partial analysis of an olivine from the western layered series has been made, to confirm the accuracy of the optical measurements, and this specimen has a composition of Fo_{82} (Table 20). The relationship between composition and the position in the layered sequence is shown in Table 17. No zoning has been detected, even though present in associated felspars, from which it is assumed either that the olivines are more

TABLE 20

Mineral analyses

	1	2	3	4	5
SiO_2	39·87	51·90	47·17	45·22	39·26
Al_2O_3	...	3·40	33·03	35·08	...
Fe_2O_3	0·86	0·53	} 0·82	0·64	...
FeO	13·20	3·70			{ 0·52
MgO	45·38	17·00	0·03	n.d.	16·43
CaO	0·25	21·12	17·05	17·74	41·85
Na_2O	0·04	0·36	1·78	1·35	...
K_2O	0·01	0·03	0·05	0·07	...
TiO_2	0·03	0·46	nil	n.d.	...
P_2O_5	0·01	0·12	n.d.	n.d.	...
MnO	0·22	n.d.	n.d.	n.d.	...
Cr_2O_3	tr.	0·88	n.d.	n.d.	0·27
H_2O^+	0·33	0·13	0·39	0·10	...
H_2O^-	0·10	0·12	0·12	0·07	...
	100·30	99·75	100·48	100·27	98·33
Molecular proportions	$Fo_{86}Fa_{14}$	...	$Or_{0.5}Ab_{15.0}An_{84.5}$	$Or_{0.5}Ab_{11.5}An_{88.1}$	$Fo_{82}Fa_{18}$
Atomic proportions	$Mg_{85.8}(Fe^{++}+Mn)_{14.2}$	$Ca_{43.9}Mg_{49.2}Fe_{6.9}$	...	...	$Mg_{81.7}(Fe^{++}+Mn)$

1. Olivine from plagioclase–clinopyroxene–olivine adcumulate (H5049), Unit 10. Anal: G. M. Brown.
2. Clinopyroxene from H5049. Anal: G. M. Brown.
3. Plagioclase from H5049. Anal: G. M. Brown.
4. Plagioclase from felspathic layer in plagioclase–olivine cumulate, Head of Glen Harris (Unit 11). Anal: I. D. Muir (pers. comm.).
5. Olivine from olivine–plagioclase cumulate (H9789), upper part of Unit D. Partial analysis. Anal: W. J. Wadsworth.

easily made over to a homogeneous composition by reaction with the trapped liquid, or that the equivalent of the narrow compositional range so easily detected in the felspars by extinction angle measurements could not be detected by 2V or birefringence measurements on the olivines.

    The habit of the olivine varies a great deal throughout the intrusion. In general it varies from being well-shaped and equidimensional ($2 \times 1 \cdot 5$ mm) in the olivine cumulates (Fig. 149) to being poikilitic (Fig. 152) in the felspar cumulates, although as it is commonly present as a cumulus mineral, even when in small amounts in the felspar cumulates, the latter habit is rare. A detailed examination of the changing habit of the olivine within a single unit has been illustrated previously (Brown, 1956, fig. 3), from which it was seen that, quite sporadically, tabular olivines (Fig. 140) are precipitated (flattened on (010), with thickness 1/5 to 1/10 of the length), imparting a distinct fissility to the rock over a thickness of a few inches. The reason for this is not properly understood, but it is clearly not related to any post-deposition process, in view of its local, sporadic development. Such olivines frequently show strain lamellae developed parallel to (100), i.e. perpendicular to the tabular plane and therefore to the plane of deposition. It is suggested that the strain developed shortly after crystallization and deposition, at a time when the hot tablets were supported on an irregular surface of underlying, small crystals, so that the weight of the overlying crystal pile was liable to produce differential stresses. Petrofabric analysis of the tabular olivines (Brothers, 1964, p. 263) confirmed that they are orientated with the (010) planes lying within the plane of layering, but failed to show any evidence of preferred orientation within this plane. Other olivine habits are of less widespread occurrence, but prove to be of interest. Thus, certain rocks contain large, rounded olivines which include numerous small and haphazardly arranged felspar crystals, in contrast to the larger and well-orientated felspars outside the olivines (Fig. 141). Such grains are believed to have been deposited as single, composite clusters of plagioclase and olivine, although at first sight their texture might be taken to imply an intercumulus origin for the olivine.

The most spectacular olivine habit on Rhum is, however, that which was originally called the ' harrisite structure ' (Wager & Brown, 1951) as it is well developed in the rocks, called ' harrisites ' by Harker (1908), from the Harris Bay area of western Rhum (chiefly in unit B, but also in unit A). The texture of rocks such as these has now been given the general name ' crescumulate ', so the harrisites of Rhum are called olivine crescumulates. An examination of the numerous clear rock-surfaces within the Ard Mheall series (lower part of unit B) reveals a tendency for the olivines in several layers to be very large, elongated, branching, and orientated with their longer axes roughly perpendicular to the plane of layering. Attention was first drawn to the significance of this habit by us in 1951, the olivines being thought to have grown upward into the overlying liquid in a manner analogous to the inward growth of the felspars in the perpendicular-felspar rock of the Skaergaard intrusion (Ch. V), and they have since been studied in more detail by Wadsworth (1961). Less well-developed layers of olivine crescumulate occur at other horizons in both the western and eastern layered series. The largest branching olivines frequently extend some inches in all three dimensions, and occasionally exceed one foot in length (Fig. 143). The branches tend to be flattened parallel to (010) and often exhibit crystal faces. The (010) plane, and hence the branches (which are picked out on the weathered surface through the resistance of the interstitial felspar) tend to be almost vertical. Within the (010) plane there is a distinct tendency towards elongation parallel to *a*, in which case it is the *a*-axis which is

R

FIG. 140

Olivine–spinel heteradcumulate from Unit B (Ard Mheall series), showing igneous lamination due to the orientation of the large, tabular olivine crystals. Some of the plagioclase crystals may be of cumulus status, particularly in these laminated layers (see Table 19, and the layering with non-laminated olivine cumulates shown in Fig. 158). Cross-nicols. $\times 8$.

FIG. 141

Plagioclase–olivine cumulate, showing a composite cluster of small plagioclases enclosed by a single olivine crystal. The large olivine has settled, rather than grown as an intercumulus crystal. The pile of felspar crystals has been warped by settling of the olivine, while later-deposited crystals have been banked around it. Cross-nicols. $\times 10$.

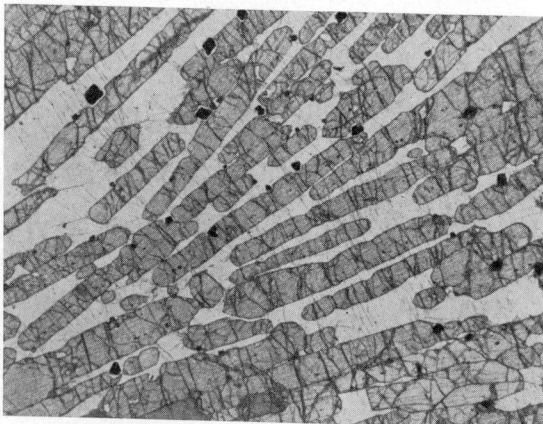

FIG. 142

An unusual type of olivine crescumulate growth. The olivine crystals are, in fact, in the form of very broad tablets (flattened perpendicular to this section) imparting a curved fissility to the rock. Note the embayments at the olivine–chrome-spinel contacts, due to reaction with the liquid. $\times 10$.

closest to vertical. A typical example is illustrated in Fig. 144, while a more skeletal variety is shown in Fig. 146 where an earlier stage in branch development is shown by the tendency of some olivines to 'bud' (Brown, 1956). There are all gradations in size between these markedly elongate olivines, through large but less elongated types, to the small ones of the adjacent normal cumulates, this being related both to mode of growth and to orientation when deposited (Fig. 145).

b) *Plagioclase felspar*

The analyzed specimen has the composition $Or_{0.5}Ab_{15.0}An_{84.5}$ (Table 20), and is a calcic bytownite. Apart from these constituents there is a fairly high content of ferric oxide (0.82%), which is a feature of many calcic plagioclases as well as orthoclases. Another specimen was analyzed independently by I. D. Muir (personal communication) from one of Harker's rocks (probably from unit 12), which has a fairly similar composition, $Or_{0.5}Ab_{11.5}An_{88}$, and also has a high Fe_2O_3 content (0.64%).

The full range of cumulus plagioclase composition in the layered series is broadly from An_{88} to An_{84}, apart from the An_{78} characteristic of the somewhat anomalous unit A (Table 17). As with the olivines, the change in composition over the whole series of units is irregular (Table 17 and Fig. 147) except within some of the thicker western units, where a regular cryptic variation is found. The most likely explanation of the irregularity is discussed later, in connection with the hypothesis (Brown, 1956) that the Rhum magma chamber was periodically replenished by the parental magma. The compositions of the minerals precipitated after the addition of fresh magma would reflect sensitively the composition of the mixture of residual and fresh magma, and its temperature. In the western layered units, Wadsworth (1961) was since able to supply convincing evidence in support of this hypothesis, particularly with reference to the olivines, but also the felspars. Thus, after the formation of unit A (An_{78}) a fresh influx of magma is believed to have given rise to unit B (An_{85}). These units are separated by a 'Transition Zone', in which the felspar compositions (An_{78} upwards to An_{85}) are thought to be indicative of the mixing of the residual and fresh magmas. Further reversals (as shown by the olivine compositions) are found at the base of units C and D, but without the additional evidence of a Transition Zone. We believe that these periodic replenishments with fresh magma would account for the absence of any significant cryptic variation throughout the layered series as a whole, as discussed in a later section.

Zoning in the felspars is of the normal, and never of the reversed or oscillatory varieties. It is absent, or only faintly detectable as an incomplete rim too thin to measure, in the cumulus felspars of the high eastern units (above unit 8). The analysis of a felspar–olivine–augite cumulate (H5049, Table 21), however, suggests a normative felspar composition of An_{81}, in contrast to $An_{84.5}$ for the separated analyzed felspar. This would imply some zoning, were it not that the calculation of normative components assumes a distribution of elements between the minerals which does not, in detail, correspond to that in the actual minerals present. The poikilitic, intercumulus felspars

FIG. 143. Part of a thick layer of very coarse olivine crescumu-late, in which the long olivines have grown less commonly in a vertical direction than found in the higher crescumulate-rich units (Fig. 165, 166). Transition series, southwest Rhum.

FIG. 144. Photomicrograph of olivine cres-cumulate, Unit B. A single olivine crystal is at extinction (dark grey), and is an extreme example of the complex, branching habit. Cross-nicols. $\times 1\frac{1}{5}$.

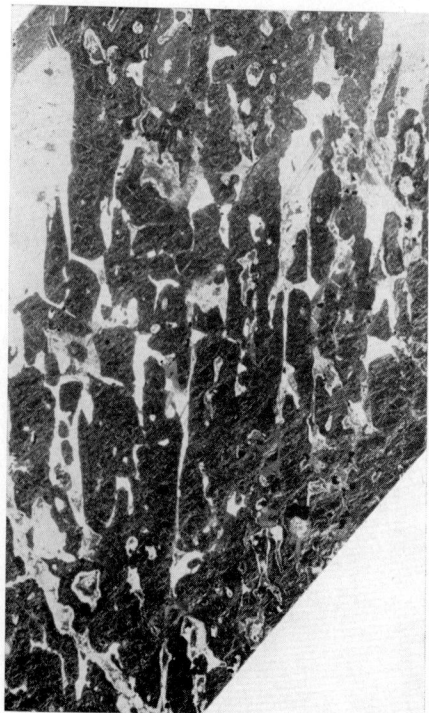

of the olivine cumulates frequently show slight signs of zoning, which is better estimated by chemical analysis of the whole rock (p. 286) than by attempting optical measurements on such large, skeletal crystals.

In the cumulates of the lower eastern layered series (below unit 8), normal zoning is found in which a core of An_{84} is zoned con-tinuously outwards, sometimes to An_{45}. Such felspars indicate appreciable trapping of inter-cumulus liquid, as does the presence of inter-cumulus orthopyroxene and apatite in the cumulates of these units; the rocks are, there-fore, of orthocumulate or mesocumulate type. One such rock, from unit 3, was analyzed

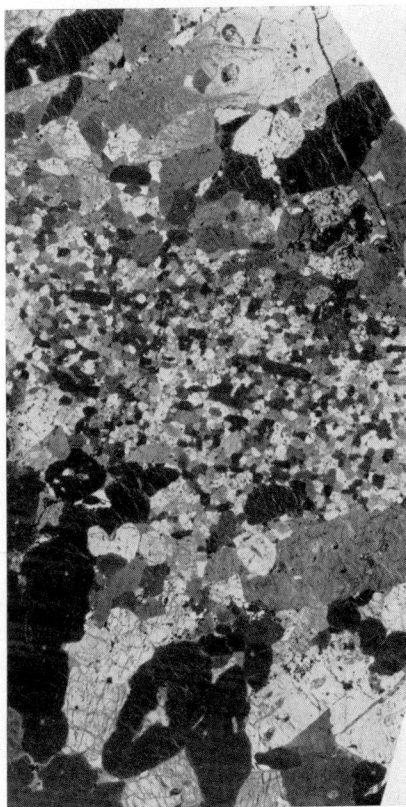

FIG. 145.
Photomicrograph of layered olivine cumulate, Unit B. Layers of large, partly crescumulate olivines alternate with a cumulate of small, discrete olivines. Cross-nicols. $\times 1\frac{1}{3}$.

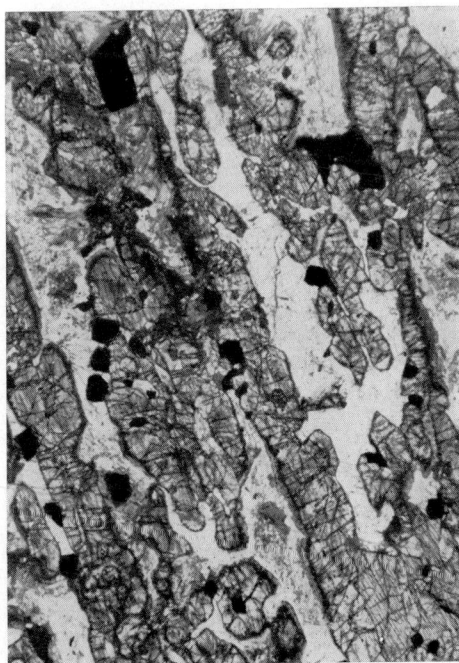

FIG. 146
Photomicrograph of olivine crescumulate, Unit A, Harris Bay. The long, branching olivines have grown almost vertically, perpendicular to the plane of layering. $\times 10$.

(Table 21) in order to attempt an estimate of the composition of the contemporaneous liquid (p. 288). Such orthocumulates have a eucritic composition according to chemical analysis, which led Harker (1908) and Bailey (1945) to map the lower units as separate eucrite intrusions; the subsequent recognition of An_{84} cores to each felspar, however, places these as conformable layers in the ultrabasic layered series. A similar feature is exhibited by the rocks of the western unit A, which were also classed as eucrites but have since been shown to be ultrabasic orthocumulates or mesocumulates (p. 278).

The felspar habit is directly related to its cumulus or intercumulus origin. Where cumulus, the felspar crystals are tabular, flattened parallel to (010) and usually orientated with this plane in the plane of layering, so as to produce igneous lamination. They are small, usually about 1 mm parallel to a and c axes, and the ratio of thickness (parallel to b) to length or breadth averages 1:3, although in rare extreme cases it reaches 1:12. In the plane of lamination no rectangular shape is visible, and therefore there is no obvious lineation, although a petrofabric analysis by Brothers (1964, fig. 3, 4) has revealed a relatively strong preferred orientation of felspar crystallographic c within this plane. This led him to suggest some degree of elongation in the primocrysts when they were being deposited from a flowing liquid. In contrast, the intercumulus felspars in the olivine cumulates are poikilitic crystals which may exceed 10 mm in diameter and have been shown to enclose, in extreme cases, about 10 000 olivine crystals (p. 265).

A single major unit contains large, poikilitic, intercumulus felspars in the lower part, and small, well-shaped felspar tablets in the upper part, according to the levels at which the felspar has been deposited as a cumulus mineral.

c) *Clinopyroxene*

The analyzed specimen is a magnesian augite, having a composition of $Ca_{44}Mg_{49}Fe_7$ (Table 20) and a relatively high chrome content (0·88% Cr_2O_3). When plotted on a Ca–Mg–Fe pyroxene composition diagram (e.g. Fig. 19) it is seen to represent a relatively high-temperature member of the calcium-rich pyroxene series, comparable with those found in the lower rocks of several other layered intrusions, where olivines and felspars of similar compositions coexist. Thus, on the trend-line it plots close to the early Bushveld and Stillwater augites. The earliest pyroxene found in the Skaergaard, in the picrite of the marginal border group, is less magnesian than that of Rhum, and it co-exists with olivine and felspar which are lower-temperature members of their solid solution series than are those in Rhum. It is significant that a cumulus orthopyroxene is not found in Rhum (nor can exsolution be detected in the augites), so it would appear that a subtle variation in the composition of the magma will affect the pyroxene assemblage. The type of magma believed to be the parent of the Rhum layered rocks (Table 21) contains normative nepheline, but this is not a dependable assumption where such small amounts could be due to other reasons, and is not believed to be so significant as to warrant calling it an alkali basalt magma (Yoder & Tilley, 1962, p. 352), because

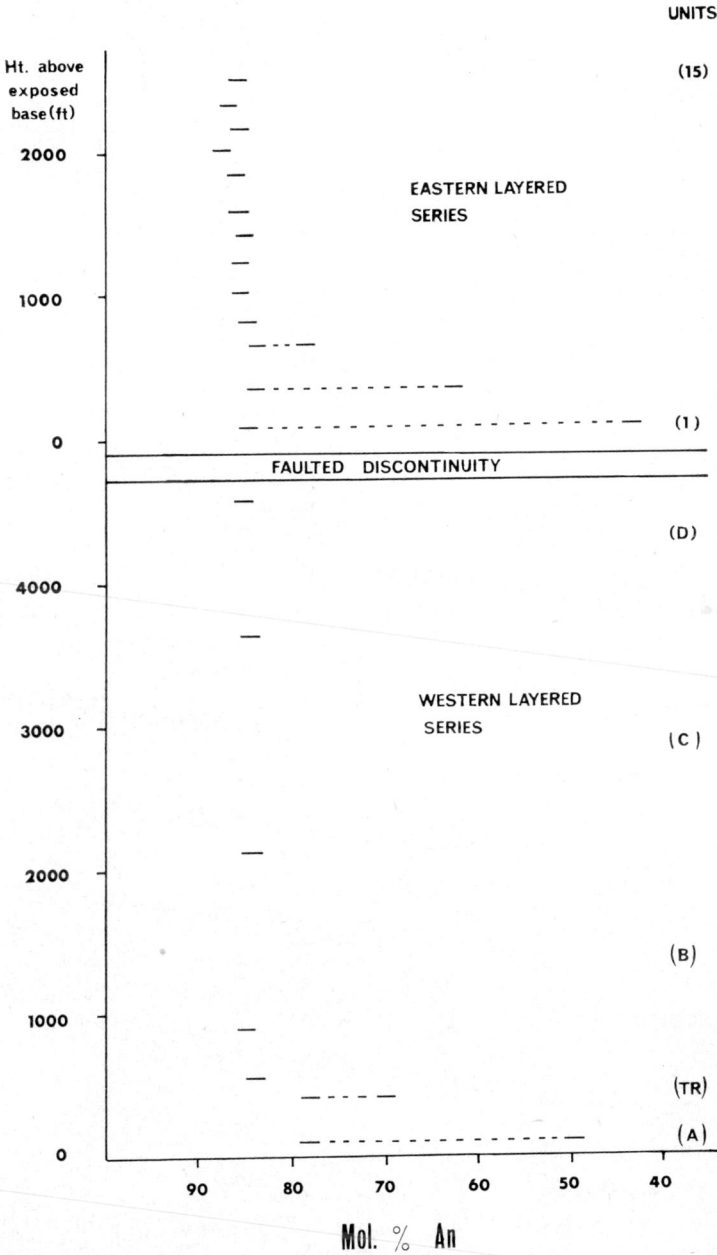

FIG. 147. Variation in plagioclase composition for representative specimens of the Rhum layered series. (The bar for each specimen covers the range of accuracy, $\pm 2\%$ An, in determinations.) The cumulus crystals show slight and irregular, but significant variations in composition with height in the series. The broken lines span the range of zoning in the mesocumulates and orthocumulates, which are confined to the lower exposed eastern and western units. The sequence is broken, through faulting, between Units D and 1.

further fractionation in the trapped liquid of the orthocumulates has resulted in ortho-pyroxene crystallizing with clinopyroxene (p. 271).

The habit of the clinopyroxene varies according to whether or not it is a cumulus phase. Although the Rhum layered series is subdivided on the basis of olivine–felspar content, pyroxene can exist as a cumulus phase in most units and comprises, in certain rare thin layers, 70% of the rock. Such crystals are subhedral and equidimensional, ranging from 0·2 to 1 mm and averaging 0·6 mm across. The cumulus pyroxenes are confined to the upper, felspathic part of each unit, where they participate in the fine-scale rhythmic layering (p. 274; Fig. 159 & Pl. XIIB). This is rather surprising as their density (3·34) is not much different from that of the olivines (3·44), and it is this feature which led Brown (1956) to suggest that in Rhum, the order of crystallization played an important part in producing the layering, rather than differences in mineral densities. In the olivine cumulates, pyroxene is intercumulus and forms large, poikilitic crystals similar in habit to the felspars in the same rock, although rather smaller. Composite grains also occur, with pyroxene playing the rôle shown for olivine in Fig. 141, and there-fore comparable with the Skaergaard example (p. 71). An unusual habit, consisting of upward-growing, tapering pyroxene crystals (Fig. 163), is found at various levels in the eastern layered series, the character being discussed in a later chapter (p. 276).

d) *Chrome-spinel*

This mineral is ubiquitous, being sparsely distributed as small grains in most rocks and, more commonly, as thin concentrations at the base of some of the units (Fig. 156). It has not been chemically analyzed, but brown translucency, only visible at the thin edges of the grains, together with the normative composition of the spinel in the rock analyses, reflectivity measurements, and borax bead tests, suggest that it is a chrome-magnetite. The grain-size is much less than that of the other minerals (0·1–0·2 mm) so that if gravity alone was responsible for their concentration it would be difficult to explain why the chrome-spinels are concentrated at the base of each unit (cf. Table 15). However, if the order of crystallization is the more important factor, then the concentra-tion in this position is understandable.

e) *Orthopyroxene*

This pyroxene occurs only as an intercumulus phase, and in the mesocumulates or orthocumulates from the lower seven units of the eastern layered series. It has a com-position (estimated from optical measurements) of about Fs_{24} and is thus more iron-rich than that likely to be in equilibrium with the cumulus minerals of the Rhum layered rocks.

CHAPTER XI

RHUM TEXTURES, STRUCTURES, AND
LAYERED SEQUENCE

The Cumulate Textures of the Layered Rocks

The textures of the Skaergaard layered rocks, in terms of cumulus and intercumulus material, have been discussed at some length (Ch. IV). Orthocumulates and meso-cumulates are there well-developed, and for these the Skaergaard may be regarded as a type example. The characteristic cumulates of Rhum, on the other hand, are ad-cumulates, heteradcumulates and crescumulates (although a few orthocumulates and mesocumulates are found in the lower seven eastern units, and in unit A of the western units), and it was here that their significance was first appreciated in detail. A typical olivine-rich heteradcumulate and felspar-rich heteradcumulate from the eastern layered series, and an olivine crescumulate from the western layered series, serve to illustrate the commonest textural varieties; further descriptions are given of a pure felspar adcumulate, a chrome-spinel heteradcumulate, a felspar–olivine–augite adcumulate, and a felspar–olivine orthocumulate.

A typical olivine(–spinel)* cumulate occurs near the base of unit 14. Its weathered surface is pitted (Fig. 148), and this feature is an accurate reflection of the rock's distinct-ive texture. Approximate modal measurements indicate, for the mineral assemblage, 79% olivine, 13% plagioclase, and 8% clinopyroxene, with minor (1% or less) amounts of chrome-spinel. The olivine and chrome-spinel, which are well-shaped and in relatively small, discrete crystals, are enclosed by the felspar and pyroxene, both of which are large and poikilitic in habit (Fig. 149, 150). Thus, while the first pair are believed to be cumulus crystals, the others are intercumulus. While the majority of the olivines are 1 mm in diameter (a few are smaller, and may be only 0·2 mm), the poikilitic felspars are each frequently 10–20 mm in diameter. Single felspar crystals usually appear to enclose between 1000 and 10 000 olivines, while each of the smaller pyroxenes encloses about half this number. The whole rock is thus made up of olivine–felspar and olivine–pyroxene areas, and as the latter weather the more easily, the irregular-weathered surface shown in Fig. 148 is produced, each knob being due to a poikilitic felspar and each hollow, to a poikilitic pyroxene. Different parts of a thin section of this rock are illustrated in Fig. 149 and 150.

* The cumulus phase listed in parentheses is in subordinate, often very sporadic amounts.

FIG. 148

Olivine(–spinel) heteradcumulate, showing the character-istic "honeycombed" appearance on the weathered surface (unit 14). The knobs consist of single, large poikilitic crystals of plagioclase (Fig. 149) and the hollows consist of similar poikilitic crystals of augite (Fig. 150), each enclosing several thousand small, cumulus crystals of olivine. $\times \frac{2}{3}$.

FIG. 149

Photomicrograph of part of a single, unzoned, poikilitic crystal of plagioclase enclosing several cumulus olivine and some chrome-spinel crystals. Olivine(–spinel) heteradcumulate (see Fig. 148). $\times 13$.

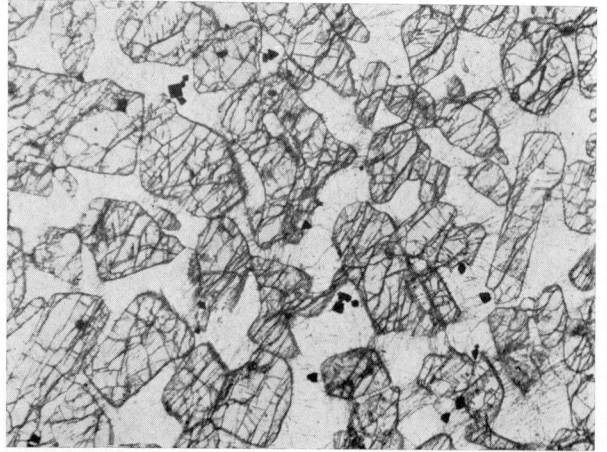

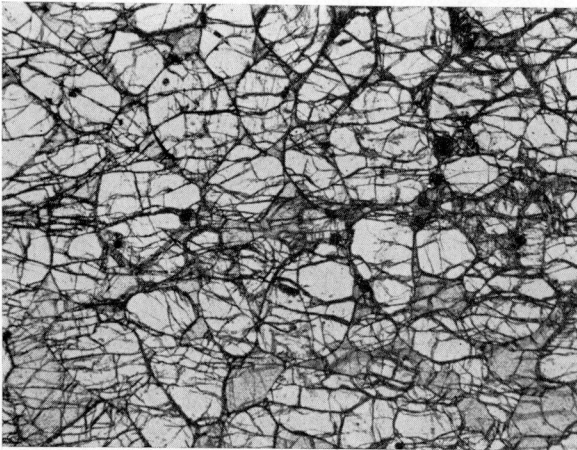

FIG. 150

Photomicrograph of part of a single, unzoned, poikilitic crystal of augite (grey) enclosing several olivine and a few cumulus chrome-spinel crystals. Olivine(—spinel) heteradcumulate. (See Fig. 148.) $\times 13$.

At first glance, from the mode, it would appear that about 80% of this rock is of the cumulus, and 20% of the intercumulus minerals. From the previous discussions (Ch. IV), however, it is clear that if a true picture of the development of this rock is to be obtained, the extent to which the cumulus olivines have themselves been augmented since deposition should be considered. From packing experiments, it is unlikely that the olivines could be packed so closely that the pore liquid occupied less than 40% of the total volume, that is, that there would be more than 60% of cumulus crystals. The absence of zoning, together with the overall magnesian composition of the olivines (Fo_{86}), would imply that continued growth of the olivines took place by the adcumulus process until they formed about 80% of the rock, i.e. by diffusion of appropriate material to the crystals from the overlying liquid at constant temperature. At some stage during this process, felspar and pyroxene apparently nucleated at a few isolated positions in the intercumulus liquid, and diffusion of the appropriate material towards these centres enabled these minerals to grow into large crystals, enwrapping the abundant olivines and the few small chrome-spinels. The surprising feature of these rocks is that even the poikilitic felspars and pyroxenes are virtually unzoned, and closely similar in composition to cumulus felspars and pyroxenes in the same, and adjacent units. Thus there is no evidence for the trapping of appreciable amounts of gabbroic intercumulus liquid which, nevertheless, must once have been relatively abundant in the original, olivine mush. A process resembling adcumulus growth must have taken place not only about the cumulus olivine crystals, but also about the nuclei of pyroxene and felspar formed in the intercumulus liquid immediately after the deposition of the olivines. There may be some slight zoning in the felspars, because extinction is not quite uniform throughout the whole of each large crystal, but whether this is really due to zoning or, instead, to slightly distorted growth, was not determinable. The analysis of the whole rock (Table 21, 22, No. 8) indicates a normative felspar of An_{81}, but this may be slightly more sodic than the true composition (p. 259). Although it is possible that a little intercumulus liquid was trapped before solidification was completed, the poikilitic felspars are remarkably close in composition to the cumulus felspars (An_{85}) of this unit.

These lines of evidence indicate that the olivine(–spinel) cumulates, of the type described, have developed by adcumulus growth of the cumulus olivines and spinels together with an adcumulus type of growth of the poikilitic felspars and pyroxenes, which had nucleated in the positions they now occupy. This added material, having the same composition as cumulus felspar and pyroxene in adjacent layers, must have formed by diffusion between intercumulus and overlying liquid; it is thus similar in origin to the material deposited during adcumulus growth involving only cumulus crystals. For rocks of this kind, the term *heteradcumulate* has been proposed, signifying a heterogeneous type of adcumulus growth, and this type would seem to be the commonest in Rhum. Extreme olivine-rich rocks are rare in Rhum, but olivine adcumulates (dunites) are fairly common in Skye (Ch. XV).

The felspar-rich cumulates are more variable in texture than the olivine cumulates,

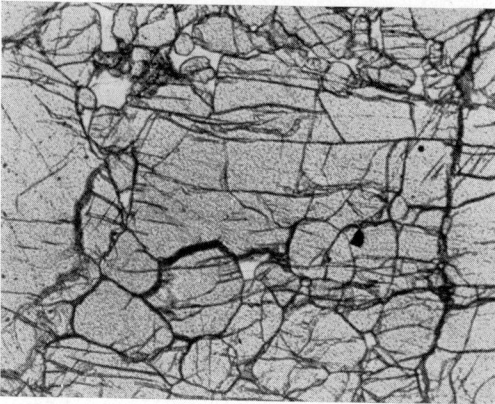

FIG. 151. Olivine heteradcumulate in which olivine growth has produced an almost pure, olivine adcumulate. Such rocks are rare in Rhum, only one or two thin (3–4 inch) layers having been found. ×10.

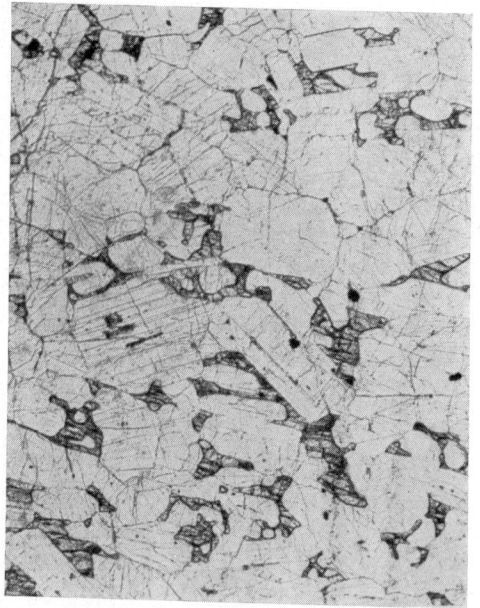

FIG. 152. Plagioclase heteradcumulate (unit 11). The interstitial material is optically continuous and is part of a single, large, poikilitic crystal of olivine, so that this rock is complementary to that shown in Fig. 149. ×13.

some containing only felspar as a cumulus phase while others contain subsidiary amounts of cumulus olivine, augite, or chrome-spinel. The felspar–olivine heteradcumulate of unit 10 (Brown, 1956, fig. 15) is perhaps the commonest type, the amount of cumulus olivine being small compared with felspar, and pyroxene alone being completely inter-cumulus. A type of felspar cumulate from unit 11, which is by no means rare in the eastern rhythmic units, resembles the olivine heteradcumulate just described above except that the rôles of olivine and felspar are reversed. The texture (Fig. 152) is dominated by the small, tabular plagioclase felspars which are crudely orientated to produce an igneous lamination. Olivine, in this case, forms large poikilitic plates up to 10 mm in diameter, as does pyroxene. Owing to the resistance of the felspar to weathering the texture is not apparent from any irregularity on the weathered surface, but on the fresh surface a mottled pattern can be seen. The approximate mode is 85% plagioclase, 10% augite, and 5% olivine. The felspars are not discernibly zoned—nor, apparently, are the poikilitic intercumulus minerals, both of which are of the same magnesian composition as their cumulus equivalents in this unit. The large amount of unzoned felspar, together with the apparently unzoned, intercumulus, magnesian olivine and augite, indicate that this rock is a plagioclase heteradcumulate. Applying the same reasoning as for the olivine heteradcumulate, the rock was probably laid down, initially, as a cumulus of felspar crystals with about 40% pore liquid, and apart from the fact that here the rôles of olivine and felspar are reversed, the subsequent pattern of crystallization would follow that outlined above for the olivine heteradcumulate. The high percentage of felspar, i.e. 85%, although not very different from the 80% olivine in the contrasted rock, may be significant. The olivine percentage is a maximum for all but a few rare olivine adcumulates and heteradcumulates, and the majority contain only about 70–75% olivine, whereas a large proportion of the felspar adcumulates and heteradcumulates contain 85% or more of felspar. Although the difference is slight, and may be due to differing degrees of packing of the initial cumulus assemblage, there may be other reasons connected with the conditions at the time of felspar deposition and adcumulus growth.

Towards the top of certain eastern rhythmic units, notably units 13 and 14, thin monomineralic layers (2–6 in thick) occur (Fig. 153), consisting solely of plagioclase. The texture is illustrated in Fig. 154. The felspars are of the same general size and shape as those in the less extreme felspar cumulates, and are so orientated as to impart an igneous lamination to the rock. Their composition is An_{86-87} and they are virtually unzoned. Clearly this rock does not consist simply of cumulus crystals plus the products of crystallization of the unmodified intercumulus liquid. On the other hand, no satisfactory explanation was to be obtained by suggesting that the intercumulus liquid had been squeezed out by a filter-press type of process, for filter-pressing could not expel all the liquid, down to the final dregs, without deforming the crystals. This problem was facing one of us (G.M.B.) at the time when, in 1951, Professor H. H. Hess paid a visit to Oxford and explained the texture according to the diffusion hypothesis briefly outlined in one of his earlier publications (1939, p. 431). Since then, we have both applied Hess's reasoning to other similar, and often less extreme varieties of rocks in layered intrusions, and described the mechanism as the adcumulus process (Wager et al., 1960).

This extreme type of adcumulate forms the greatest contrast with the orthocumulates, and first led us to consider the mechanisms involved in its formation. Whatever the conditions likely to influence the development of adcumulates, the relatively common occurrence of thin layers of pure felspar adcumulates, and the absence of pure olivine adcumulates, is a significant feature in Rhum. The felspar cumulates occur towards the top of each macro-unit, where fine-scale rhythmic layering is well-developed. The *extreme* felspar adcumulates tend to occur very near, or at the top of each of the higher macro-units of the eastern layered series. Fine-scale layering near the top of each macro-unit suggests current activity at that stage, and this suggests that a factor in the production of the felspar adcumulates is the slow deposition of the felspars from a thin suspension of these crystals in a convecting liquid.

In several cases, the base of a unit is made up of a thin but persistent layer of chrome-spinels, as at the base of unit 12. A specimen from there is illustrated in Fig. 155 and 156, and consists of a felspar heteradcumulate forming the top of unit 11 and a chrome-spinel heteradcumulate, followed upwards by an olivine–spinel heteradcumulate, at the

FIG. 153. One of the thin, pure-felspar layers (plagioclase adcumulate; base of hammer head) in the higher eastern layered units, Hallival. Above it is a characteristically-weathered, olivine-rich cumulate with a sharp base, and a top grading into olivine–plagioclase cumulates showing fine-scale layering. This is part of a minor layered sequence within a much thicker major unit (unit 14) in which felspathic cumulates predominate. The irregularities in the layering are attributed to contemporaneous movement of the crystal mush.

FIG. 154. Photomicrograph of a pure-felspar rock (plagioclase adcumulate), showing igneous lamination. Photographed under cross-nicols to illustrate the size and habit of the crystals, most of which must have grown appreciably since deposition, the original gabbroic, intercumulus liquid having been expelled by adcumulus growth. ×15.

base of unit 12. The base of the spinel layer is more irregular than the top, presumably because the relatively large felspars formed an uneven floor to the settling spinels, whereas the small spinels packed better in forming a floor to the settling olivines. In the rather special case afforded by this thin layer of chrome-spinels, the intercumulus felspar and olivine appear to have grown from the cumulus crystals in the underlying and overlying layers, respectively, so that although the texture at first suggests a heteradcumulate origin, the method of formation may be more closely analogous to that of crescumulus growth (both upward and downward). It could only develop in this way in very thin layers of a particular cumulus mineral, in which intercumulus growth took place soon after the deposition of a crystal cover.

The types of cumulate so far discussed have consisted predominantly of only one main cumulus mineral, and the rock just described consists of three such cumulates recognizable in a layer less than a centimetre thick. Such a contrast is only seen at the junction between major rhythmic units in the eastern layered series. Above the thin, spinel-rich base to each of the eastern major units, the dominant rock type is the olivine (–spinel) cumulate, forming a layer tens of feet in thickness, suggesting that for a long period only olivines, with minor amounts of spinel, were present as primocrysts. In the upper part of most units, however, not only has cumulus felspar become increasingly dominant but cumulus clinopyroxene occurs in relative abundance, while a small amount of cumulus olivine persists. The concentration of one or more of these three cumulus phases into thin layers results in a variety of cumulates, including the felspar adcumulates described above but also other adcumulate and heteradcumulate types.

A felspar–augite–olivine adcumulate, of the type shown in Fig. 157, contains almost equal amounts of each of these three minerals and is, as a result, less obviously a rock formed by crystal accumulation. However, each of the three minerals occurs as discrete, unzoned crystals, and is a high-temperature member of its particular solid solution series. All three minerals are, therefore, cumulus crystals, and the absence of zoning and of any other mineral in poikilitic or other form indicates that the rock is a true adcumulate. The pure felspar rock described earlier shows the effect of adcumulate growth more obviously, but that containing three cumulus phases is, in fact, evidence of the same process, resulting in complete expulsion of pore liquid. Each mineral from this rock was chemically analyzed (Table 20), as was the whole rock (Table 21, Anal. 6).

The cumulates of the lower seven of the eastern units differ from those of higher units in the extent of normal zoning of the felspars (Brown, 1956, pp. 28–30). An analyzed felspar–olivine cumulate from unit 3, for example, contains plagioclase with well-defined cores of An_{84} coated by zones which reach a composition of about An_{65} in several measured cases, and An_{45} in a few. In addition, orthopyroxene (Fs_{24}), an Fe–Ti oxide, and apatite are present in small amounts as pore material, and it is clear that in cases such as this the rock is either an orthocumulate or mesocumulate. Somewhat similar cumulates are found in unit A of the western units. The unit 3 cumulate analysis (Table 21, Anal. 4) was first used by Brown (1956, p. 43) to calculate

272

FIG. 155. A specimen (with polished surface) spanning the junction between two major rhythmic units, the light-coloured, plagioclase adcumulate being the uppermost part of unit 11 (ca. 150 ft thick). The overlying unit 12 has at its base the thin, chrome-spinel heteradcumulate (Fig. 156) visible here, overlain by the olivine(–spinel) heteradcumulate which continues upwards for about 40 ft until it grades into progressively more felspar-rich cumulates and, eventually, to another junction of the type shown here. (Natural size.)

FIG. 156. Photomicrograph of the junction between three types of cumulate, from the rock shown in Fig. 155. The plagioclase(–spinel) and olivine(–spinel) heteradcumulates are similar to those shown in Figs. 152 and 149 respectively. In the chrome-spinel heteradcumulate layer, the poikilitic plagioclases are optically continuous with cumulus plagioclase crystals below, and the few poikilitic olivines are continuous with cumulus olivine crystals above the spinel-cumulate layer. ×12.

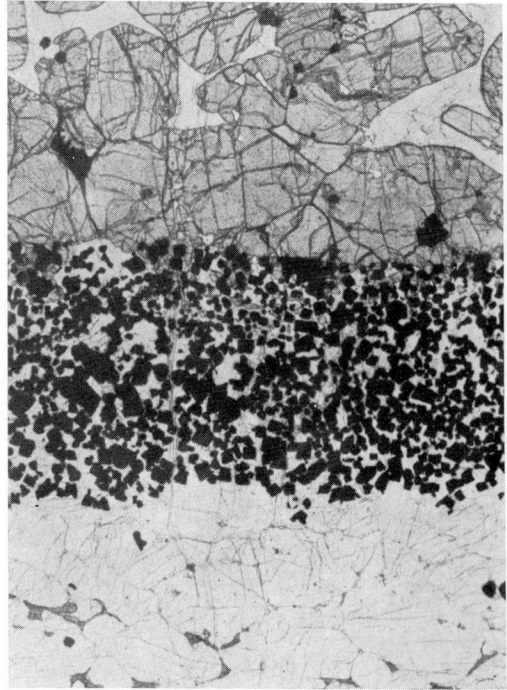

FIG. 157

Photomicrograph of plagioclase–olivine–augite adcumulate from near the top of unit 10. Each cumulus crystal is unzoned and extended by adcumulus growth, this being an example of a polymineralic adcumulate. Each mineral, as well as the total rock, has been analyzed (Tables 20, 21). ×12.

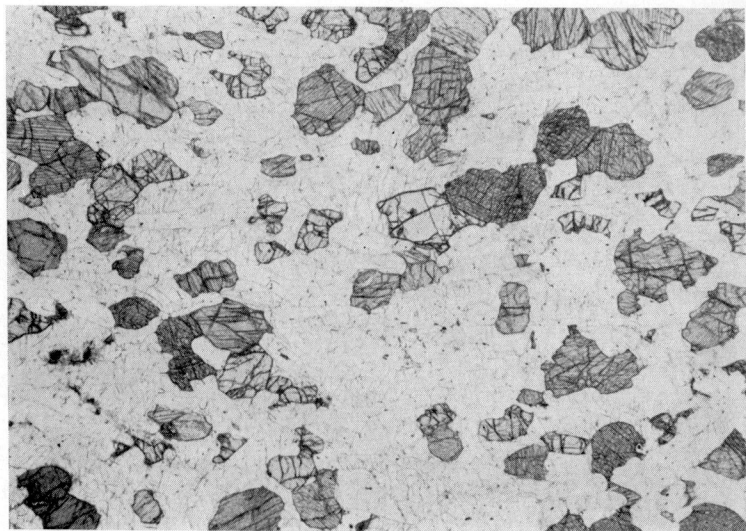

the composition of the pore material and, therefore, to estimate the composition of the contemporary trapped liquid (Table 21, Col. 12) at a relatively early stage in the crystallization history of the intrusion.

The olivine crescumulates are characterized by olivines with the upward-growing habit described earlier, together with subordinate plagioclase, clinopyroxene, and chrome-spinel. Their overall composition is that of a peridotite, i.e. close to the compositions of the olivine heteradcumulates so far described, and a representative chemical analysis is given in Table 21 (Anal. 2). The example for which a mode is given in Table 19 is typical, suggesting that the rocks, on average, contain about 80% olivine, 10% plagioclase, 5% clinopyroxene, and minor amounts of chrome-spinel and secondary alteration products. The olivines in unit B (the unit richest in crescumulates) are Fo_{88} in composition, and are unzoned. The felspars are An_{85} and are little zoned, while the augites are magnesian, virtually unzoned, and of the same composition as those found elsewhere in the layered series. Both the felspars and pyroxenes fill the interstices between the olivines, tending towards a poikilitic texture, and hence these rocks have the mineral compositions of an olivine heteradcumulate. Texturally, however, they are different, due to the acicular habit of the olivines (p. 257), the longer axes of which are perpendicular to the plane of layering (Fig. 144, 146). This texture has led us (1951), and later Wadsworth (1961) to suggest that the olivines have grown upwards into the liquid, from the crystal pile forming the temporary floor of the magma chamber, during periods of non-deposition of cumulus crystals. Optical measurements have demonstrated that such growth took place only on cumulus olivines having a special orientation. Apparently, the olivine primocrysts grew most rapidly along the crystallographic a-axis. From amongst those which were deposited as cumulus crystals on the floor, those with the a-axis approximately perpendicular to the layering grew upwards more rapidly than the others. Crescumulate growth is, in fact, a special type of adcumulus growth, crystals suitably orientated being fed by material of constant composition as they grew upwards into the main body of magma. Any vigorous movement of the magma would tend to break-off such projecting crystals, so we suggest that crescumulate olivines reached their great lengths by growing into essentially stagnant, supercooled magma, while the more common adcumulates developed as a result of the surface of the cumulus layer being bathed by supercooled magma flowing over it. The olivine crescumulates are interlayered with olivine cumulates consisting of small, discrete cumulus crystals (Fig. 145), the latter being viewed as deposited from a suspension of olivine primocrysts in magma moving across the floor. Hence, the layering involving the olivine cumulates and crescumulates in Rhum (Fig. 165–167) is interpreted as evidence for the alternation of movement and stagnation, respectively, in the magma body.

In conclusion, the textures of the Rhum cumulates are interpreted as the result both of deposition of crystals from the magma and of growth of crystals while lying on the floor of the magma chamber. Adcumulates, heteradcumulates, and crescumulates are

S

the predominant types in Rhum, only a few mesocumulates and orthocumulates being found. The Rhum rocks thus have much in common with the ultrabasic layered cumulates in the Stillwater and Bushveld intrusions where, also, the rocks are characteristically adcumulates and heteradcumulates.

The Accumulative Structures of the Layered Rocks

So far, attention has been paid to the division of the layered series into major rhythmic units, the mineralogy of the series as a whole, and the textures of representative rocks from typical units. However, within each unit, structures are developed which provide further evidence for the formation of the series by crystal accumulation.

Fine-scale rhythmic layering is to be found at all levels in the intrusion, from unit B (Fig. 158) to unit 15. It is best developed in the upper, more felspathic parts of each of the eastern units, as in the top 60 ft or so of the 300 ft-thick unit 10 (Fig. 159 and Table 18) and more particularly in the upper parts of units 13 to 15 on the high ground of Hallival and Askival (foreground, Pl. XIIA) where felspathic rocks predominate. A fallen block, probably from unit 14, is shown in Pl. XIIB, alternate layers being dominantly either felspar or pyroxene cumulates. The layers in the block range from two inches to a few millimetres in thickness, while in the rocks near the top of unit 10 from which the analyzed felspar–augite–olivine adcumulate (p. 289) was taken, the layers are each sometimes only one millimetre thick. Yet again, layering such as is shown in Fig. 155 and 156 gives rise to separate thin layers of olivine, chrome-spinel, and felspar cumulates within a thickness of 3 mm at the junction of units 11 and 12, while in other cases layers containing 70% clinopyroxene (e.g. unit 9) or occasionally 100% felspar (e.g. unit 14) are to be seen. The characteristic feature of such fine-scale layering is that the layers are parallel to the predominant major layering and, presumably, to the temporary floor, and that individual layers only a few millimetres in thickness can extend laterally over metres, and often tens or hundreds of metres, without showing any transgressive relationships. In this, the Rhum layered rocks show a feature characteristic of the layered rocks from other, generally larger intrusions.

Irregular layering is to be found in certain areas, in the form of wispy banding and slump-structures, whilst a fairly regular type of undulatory banding is sometimes seen. No structures comparable with the trough banding of the Skaergaard have been recognized. The wispy banding is shown in Fig. 160, and although it comes from close to the easternmost extremity of exposures of the layered series, the margin is faulted and so it cannot be correlated closely with marginal features, as was possible for the Skaergaard. Slump structures occur at several horizons; for example, as low as unit B of the western layered series (Fig. 161) and as high as unit 14 of the eastern layered series (Fig. 162). The thicknesses of each layer of slumped material are usually of the order of 6–10 ft. Hess (1960) has suggested that the slumped layers of the Stillwater intrusion provide evidence for the maximum thickness of unconsolidated material at any one time,

PLATE XII

A. View of the top of Hallival mountain, Isle of Rhum, looking north from Askival. (The neighbouring Cuillins layered intrusion, Isle of Skye, can be seen in the far distance.) The resistant, felspar-rich cumulates towards the summit of Hallival are interlayered with olivine-rich cumulates (Fig. 153), but for convenience all are classed as the top of unit 14. The actual summit is the olivine-rich base of unit 15, which forms a thick unit on the higher Askival mountain and contains abundant felspar-rich cumulates which exhibit impressive fine-scale layering (foreground and B, below).

B. Fine-scale rhythmic layering in plagioclase-augite cumulates from the upper part of unit 15. (Displaced block from near the summit of Askival mountain; see A, above.)

FIG. 158. Fine-scale rhythmic layering in the olivine cumulates of unit B (Ard Mheall series). The layering is due chiefly to variation in the grain-size and habit of the olivines, although the plagioclase content is slightly greater in certain of the coarser layers (see Table 19).

FIG. 159. Fine-scale rhythmic layering involving the sorting of cumulus plagioclase, olivine and augite, near the top of unit 10, Askival plateau. Occasionally, all three cumulus phases may be present in a layer of 'uniform rock' (see Fig. 157).

but this can only be applied to rocks in which there has been little to no adcumulus growth. The development of an extreme adcumulate requires that it becomes solid when lying at the top of the crystal pile, so that under such conditions the unconsolidated material would be of negligible thickness. Slump structures have also been examined in the Bushveld (Fig. 201) and Skye (Carr, 1954), where they occur at only a few horizons, as in the Rhum and Stillwater intrusions. This suggests that the slumping was not connected with the normal slope of the floor of deposition, but was initiated by an especial event such as tectonic movements, either as regional earthquakes or, perhaps, associated with local instability of the magma chamber floor.

A rather unusual structure, found within several units in Rhum, has been termed ' the upward-growing pyroxene structure ' (Fig. 163). Large, poikilitic pyroxene growths protrude into the overlying layers, tapering upwards like the fingers of an outstretched hand. In the thin sections studied it was found that, whereas outside each pyroxene crystal there is the felspar–olivine cumulate which forms the bulk of this local upper layer, only olivines are enclosed within the poikilitic pyroxenes, and the enclosed olivines are in the proportions found in the olivine cumulate below. Furthermore, as seen in Fig. 164, the olivines and felspars outside and adjacent to the pyroxene appear to have been turned out of their usual sub-horizontal alignment so that they now lie almost parallel to the margins of the pyroxene. Several hypotheses have been tried, in order to explain this structure, but none of them provides a convincing explanation for the way in which these remarkable pyroxene growths have developed. Poikilitic growth would have resulted in felspar, as well as olivine, being included in the upward-tapering pyroxenes, which occur in olivine–felspar cumulate layers. Alternatively, crescumulate growth of the pyroxenes, prior to the deposition of the olivine–felspar cumulates around them, would be expected to take place in a stagnant magma free from crystal nuclei, which is contrary to the observation that each pyroxene encloses several olivine crystals.

The most unusual structural feature shown by the layered rocks in Rhum is that involving the olivine crescumulates, and due to the special type of olivine growth, and resultant textures, discussed earlier (pp. 257 & 273). The numerous rock exposures on the slopes of Ard Mheall, near Harris Bay, show rhythmic layering of a type not so far described from other layered intrusions. These rocks, 1200 ft in thickness, have been grouped as the lower part of unit B, and are entirely composed of alternating layers of coarse olivine crescumulates and normal olivine heteradcumulates (Fig. 165–167). The crescumulate layers range from an inch to 8 ft in thickness, but the average is about 2 or 3 ft, while the intervening layers of normal olivine cumulates are of comparable thickness and may consist of two or three sub-layers defined by variable felspar contents. Hence the olivine crescumulates comprise about half of this great thickness of rock, and apart from the several hundred layers found in unit B they are also abundant in unit A, and are even occasionally developed in the higher eastern units (e.g. unit 11).

In the description of the textures, the mechanism of crescumulate olivine growth was discussed. The alternation of crescumulate and normal cumulate layers was also

FIG. 160. Irregular, wispy layering in plagioclase–augite cumulates from near the top of unit 10, Askival plateau.

FIG. 161. Irregular layering, attributed to slumping, in the Dornabac series (unit B). The felspathic layers have tended to break into blocks and lenses, suggesting a greater degree of consolidation in these than in the more mafic cumulates.

FIG. 162. Slump-structures in plagioclase–olivine cumulates from unit 15, Askival mountain. The undisturbed layers in this region dip at low angles (ca. 10–15°) towards the left of the photographed rock face, which is approximately perpendicular to the strike of the layering.

discussed briefly and will be considered, further, in relation to the crystallization history of the intrusion as a whole (p. 292).

THE LAYERED SEQUENCE

The sequence of layered rocks in Rhum cannot be correlated with a regular cryptic variation in the cumulus mineral compositions of the successive layers. For this reason, the rocks and their mineralogy, textures and structures have not been described in an order from bottom to top of the intrusion, but according to the variety of rock types, and the evidence they provide for crystal accumulation. Nevertheless, by precise optical measurements on the minerals, slight cryptic variation within some of the thick western units has been detected (Table 17). Also, a sequence from orthocumulates to ad-cumulates, the latter becoming more felspathic and pyroxenic upwards, has been recognized for the eastern units (Table 17). Faulting and autobrecciation, however, have resulted in disturbed sequences amongst the western units, and between these and the eastern units so that it is desirable to summarize the pattern which has emerged for the Rhum layered sequence in order to interpret the more complicated picture presented by the maps (Fig. 134, 136).

The total exposed thickness is close to 7000 ft, of which the lower 4500 ft has been divided into four major western units (A–D) and the upper 2500 ft into fifteen broadly comparable, major eastern units (1–15). The Long Loch fault, which strikes N–S and separates the western and eastern regions, has apparently resulted in marked differential uplift so that the western, crescumulate-bearing units now lie at the same topographical level as the lower part of the eastern (orthocumulate) units. Hence, the existence of an unexposed thickness of layered rocks, below unit 1 in the east and equivalent to layers formed initially above unit D in the western sequence, cannot be discounted.

A vertical section, together with some mineral compositions, is given in Table 17 and Fig. 147, while mineral and rock analyses for specimens representative of different levels in the layered series are given in Tables 20 and 21. Modal analyses can only be used in a general way for this intrusion, in which fine-scale rhythmic layering has produced rocks containing all proportions of the four minerals having a cumulus status. However, the modes given in Table 19 illustrate the character of the major rock types in the western units, while those in Table 18 indicate the variation found within one of the typical eastern rhythmic units.

Unit A (Harris Bay Series), the lowest exposed member, consists essentially of an olivine–felspar–clinopyroxene orthocumulate in which there is little layering produced by variations in the relative amounts of these cumulus minerals, but a prominent layering produced by the alternation of the normal cumulates with olivine crescumulates. The cumulus and crescumulus olivines are more ferriferous (Fo_{82}) and the cumulus felspars more sodic (An_{78}), than elsewhere in the layered series (Table 17), and although this is taken by Wadsworth (1961) to indicate the presence, at one time, of underlying

FIG. 163

Upward-growing pyroxene structure, unit 8, Askival plateau. The upward-tapering structures are composed of poikilitic augites which have grown at the junction between an olivine-cumulate layer (weathered) and a plagioclase-cumulate layer (resistant). (See Fig. 164.) The largest growth is shown here and is about 6 inches tall.

FIG. 164. Photomicrograph of a section of the upward-growing pyroxene structure shown in Fig. 163. Note the way in which the elongated, cumulus olivine crystals have been drawn towards an almost vertical orientation adjacent to, and within the poikilitic pyroxene growth. × 16.

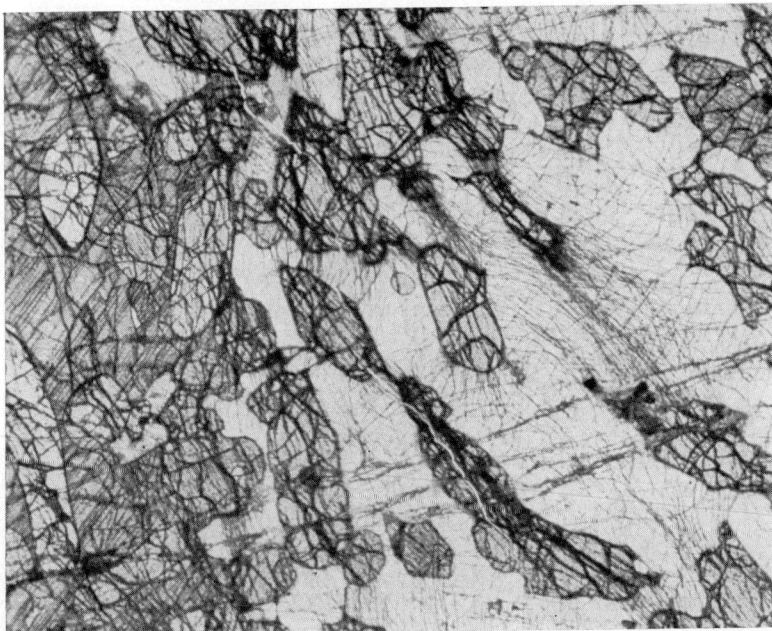

rocks containing more magnesian olivines and more calcic felspars, this would still imply extensive fractionation in this unit. As such relatively ferriferous olivines and sodic felspars are not present in any other unit, unit A remains something of an anomaly.

Unit B consists of 1200 ft of olivine adcumulates (Ard Mheall Series), in which olivine cumulates and crescumulates alternate to produce a special type of layering, overlain by 400 ft of predominantly olivine–felspar adcumulates (Dornabac Series) in which fine-scale layering, sometimes involving minor amounts of cumulus clinopyroxene, is developed. In the lower part of the unit the olivines are more magnesian ($Fo_{88.5}$), and the felspars almost as calcic (An_{85}) as found at any other level in the layered series, and there is a slight but definite change towards lower melting-temperature members of these two mineral series towards the top of the unit (Table 17).

Between units A and B lies a 150 ft thick zone, termed the 'Transition Series' by Wadsworth (1961) in that it consists of olivine crescumulates and olivine–felspar heteradcumulates in which the compositions of the crescumulus and cumulus minerals are intermediate between those of units A and B. Because the upward transition is to more magnesian olivines (from Fo_{82} up to Fo_{86}) and more calcic felspars (from An_{78} up to An_{84}), Wadsworth attributed this feature to the incomplete mixing of a fresh influx of parental magma, capable of precipitating the phases of unit B, with the residual magma which had precipitated the phases of unit A. These features of the Transition zone, as well as the occasional reversals to higher-temperature mineral compositions at the base of some of the major western units, provide evidence which is especially pertinent to the hypothesis (Brown, 1956) that successive magma replenishments account for the major units in Rhum.

From the lowest exposed rocks of unit A to the top of the Ard Mheall series of unit B, the layering is conformable and dips at a low angle of about 15° towards the postulated centre of the layered series, in Glen Harris (Fig. 136). However, the upper, more felspathic part of unit B (Dornabac series) has a much greater dip, eastward at 35° to 40°, while streaky, irregular banding (Fig. 161) or slump-structures (Wadsworth, 1961, fig. 25, 26) also suggest much disturbance at this stage, contemporaneous with deposition and probably due to tectonic events associated with the magma chamber. Further evidence of this is provided by a zone of igneous breccia (Wadsworth, 1961, fig. 28) which separates the Ard Mheall series from the overlying Dornabac series of unit B, and which consists largely of blocks of the Ard Mheall series incorporated in a matrix of minerals characteristic of the Dornabac series.

The olivine–felspar adcumulates of unit B are overlain by a thick sequence of olivine heteradcumulates in which the olivines are only slightly more magnesian than those below (Table 17), and this is taken as the base of unit C. Unit C is 1500 ft thick, and upwards it becomes richer in cumulus felspar. Although it overlies, conformably, the top of unit B in the Dornabac region, the map (Fig. 136) shows that unit C has a faulted contact with units A and B to the south. Wadsworth (1961) interprets this fault as a continuation of the zone of disturbance responsible for the igneous breccia,

Fig. 165

Rhythmic layering of unit B (Ard Mheall series). The layering is due to variation in the olivine habit, layers of fine-grained olivine cumulate alternating with layers of coarse, crescumulate olivine. One crescumulate layer is clearly visible with the hammerhead at about the centre of it.

Fig. 166

A close-up view of the crescumulate layers shown in Fig. 165. The almost-vertical alignment of the long, upward-grown olivine crystals is clearly visible, the shape of the olivines being outlined by the more resistant, interstitial felspars. Note that the top of the layer is more irregular than the base.

Fig. 167

Rhythmic layering of unit B (Ard Mheall series), involving a greater number of thinner olivine-crescumulate layers than shown in Fig. 165.

and suggests that the Dornabac series formed to the east of a prominent fault scarp, being deposited on top of a southerly dipping wedge of plutonic fault scree (the igneous breccia). The southerly dip, superimposed on the previous eastern dip at the time of faulting, results in a tapering out of breccia and Dornabac series to the south as the later cumulates of unit C overlapped them and came to rest against lower members of the layered series (units A and B) to the west of the fault.

The layered rocks of unit C contain cumulus olivine of the same composition throughout, i.e. Fo_{86}, but cumulus felspar becomes relatively abundant towards the top of the unit. A further abrupt change in the cumulus assemblage, to give olivine-rich heteradcumulates once again, is taken as the base of unit D and this, again, is characterized by a return to an olivine of slightly more magnesian composition, Fo_{88} (Table 17). Upwards, unit D (1000 ft thick) also becomes enriched in cumulus felspar, while in addition, the olivine becomes more iron-rich ($Fo_{85\frac{1}{2}}$). Thus units B, C and D, each over 1000 ft thick, are major rhythmic units consisting primarily of olivine-rich cumulates and crescumulates, grading upwards into olivine–felspar cumulates. There is also a slight, but distinct cryptic variation within the individual units.

The marked disturbances which affected units A and B continued during the development of zone C, resulting in a belt of igneous breccia and a NW–SE fault on Ruinsival (Fig. 136). Even more impressive is the Long Loch fault, which runs north–south across the layered intrusion and has brought unit D to the same topographical level as units of a different character east of the fault. This major fault probably represents a zone of weakness developed early in the history of the intrusion, because the layered rocks not only have high dips where close to the fault but slump structures and zones of autobrecciation developed through movement of the crystal accumulates at various horizons. Movement along the Long Loch fault must have continued until after the formation of several more units now exposed to the east of it, and it would appear that the western units were raised relatively high along the fault at the time of final uplift and emplacement of the layered series such that, as previously remarked, it is not possible to correlate between the western and eastern units.

The outcrops of the eastern units encircle the mountains of Hallival and Askival (Fig. 134), the peaks of which are formed by the highest unit of the Rhum layered series, unit 15, while unit 1 (viewed, structurally, as a higher layer than unit D of the western units) lies low on the eastern flanks of the mountains, almost at sea-level.

The rocks of units 1 to 7 are largely orthocumulates or mesocumulates, rhythmic layering of olivine-rich and felspar-rich cumulates occurring occasionally and on a fine scale. The plagioclase felspars are often strongly zoned, and traces of intercumulus orthopyroxene and apatite, absent as cumulus phases in the Rhum layered rocks, bear testimony to the trapping of contemporary liquid to give pore material. The individual units are easily distinguished through having olivine-rich cumulates at the base which grade upwards into felspar–olivine cumulates at the top, but these major rhythmic units (50–150 ft thick) are each much thinner than the crescumulate-bearing units to the west of the Long Loch fault (Table 17).

Units 8 to 12 are similar in thickness to the units below them, but are either adcumulates or heteradcumulates. The part rich in cumulus olivine comprises about 70% of the total thickness of each unit, and the base of some units consist of a thin layer rich in cumulus chrome-spinel. The upper 30% or so of each unit consists of cumulus felspar and subordinate cumulus clinopyroxene as well as cumulus olivine. The overall amount of felspar increases gradually upwards, while olivine decreases, although fine-scale rhythmic layering is always present in the upper part of each unit and results in the formation of thin layers consisting of variable proportions of cumulus felspar, olivine and pyroxene. A typical unit of this group is unit 10, and the modal variation has been shown in Table 18.

Units 13 to 15 are still adcumulates and heteradcumulates, but the olivine-rich cumulates of each unit are here subordinate in thickness to the felspar-rich cumulates and comprise less than 30% of each unit. Fine-scale rhythmic layering reaches the highest degree of development in the upper parts of each of these units of the Rhum layered series, the most spectacular layers being the pure felspar adcumulates.

The sequence of layering exposed to the east of the Long Loch fault, consisting of fifteen broadly similar major rhythmic units which, moreover, differ markedly only in thickness from some of the units to the west of the fault, means that for most purposes. the layered rocks of Rhum can be treated as a single series from unit A to unit 15. The cumulus mineral phases are the same throughout, while similar textures and structures (with the exception of the crescumulates of units A and B, only rarely found in higher units) are found at all levels of the intrusion. Although there is slight cryptic layering within each of the units A, B, C and D, none is detectable within units 1 to 15. Moreover, there is no systematic cryptic variation from the lowest to the highest unit, the cumulus minerals of unit 15 being of virtually the same composition as those in unit B, at least about 7000 ft structurally below. In order to obtain such a constancy in the compositions of the crystallizing phases, the magma composition and temperature must also have been virtually constant throughout the deposition of the exposed layered series, For this reason, it is believed that the Rhum magma chamber was being periodically replenished with fresh magma (pp. 293–97). The individual major units may represent a faithful record of each of these periods, particularly in view of the evidence afforded for some cryptic variation by the mineralogy of units B and D, while the slumped and brecciated structures may be evidence of the periodic disruption caused by the new influxes of magma.

CHAPTER XII

RHUM CHEMISTRY AND GENESIS

THE CHEMICAL COMPOSITIONS OF THE MAIN ROCK TYPES

The rocks of the Rhum intrusion retain an ultrabasic composition throughout the great thickness of the layered series. This fact, together with the marked development of rhythmic layering to produce an assemblage of rocks differing widely in their mineral proportions, diminishes the value of a limited number of chemical rock analyses as an instrument for illustrating overall compositional variation within the intrusion. The analyses of certain rocks, however, are of importance in considering the manner in which the rocks have completed their crystallization by intercumulus growth. In Rhum we are dealing with four cumulus minerals (olivine, plagioclase, clinopyroxene, and chrome-spinel) which differ little in composition throughout 7000 ft of the layered series, but which are sorted into layers consisting, in the main, of varying proportions of the three cumulus silicate minerals. Hence the composition of any rock could, as a first approximation, be obtained by taking varying proportions of these three mineral analyses as given in Table 20 (Anals. 1–3). However, an unconsolidated layer at any one time is unlikely to consist of more than about 50% of cumulus minerals, the rest being made up of the contemporary, intercumulus liquid. According to the extent to which the diffusion mechanism operated to reduce the amount of this liquid, the ultimate rock composition will vary from extremely ultrabasic in the adcumulates towards a less extreme composition in the orthocumulates.

The intercumulus liquid is likely to have been basaltic in character, not only because such a magma would be of the type likely to be in equilibrium with the particular cumulus assemblage found in Rhum (by analogy with certain basalt phenocrysts), but also because of the composition of the trapped pore liquid estimated from the study of an orthocumulate (p. 271). By analyzing one of these orthocumulates from unit 3, and subtracting from it the estimated proportions and probable compositions of the cumulus minerals judged, from the textural evidence, to be present, the composition of the pore material was obtained (Brown, 1956, p. 43 and table 7). Despite the uncertainties inherent in such an estimate, particularly with regard to the total amount of cumulus plus adcumulus material which has to be subtracted from the whole-rock analysis, the calculated composition of the pore material is, broadly, basaltic. Also, the analysis of a

TABLE 21

Rhum rock analyses

	1	2	3	4	5	6	7	8	9	10	11	12
SiO$_2$	43·36	39·20	38·66	45·56	41·06	47·33	43·86	39·96	47·67	48·89	49·12	46·8
Al$_2$O$_3$	9·97	5·11	1·95	21·17	4·82	20·08	32·86	6·60	20·76	19·26	17·76	18·6
Fe$_2$O$_3$	1·99	3·28	3·33	1·10	2·07	0·55	0·96	1·73	1·26	1·68	1·38	·1·4
FeO	10·15	8·24	9·52	5·59	9·46	3·24	0·98	10·25	3·22	6·75	6·10	7·7
MgO	23·58	36·38	41·99	11·48	36·15	12·53	2·71	36·09	8·91	7·46	9·67	10·9
CaO	7·55	2·94	0·64	11·42	4·27	14·47	16·58	3·28	15·19	11·45	12·56	10·6
Na$_2$O	1·32	0·38	0·12	1·99	0·65	1·34	1·43	0·40	1·91	2·52	2·70	2·7
K$_2$O	0·13	0·05	0·05	0·16	0·02	0·07	0·10	0·07	0·07	0·41	0·17	0·3
TiO$_2$	0·54	0·26	0·15	0·40	0·15	0·15	0·19	0·07	0·15	0·72	0·29	0·8
P$_2$O$_5$	0·04	0·03	0·03	0·02	nil	tr.	0·02	0·03	nil	0·05	0·01	0·0(4)
MnO	0·19	0·18	0·20	0·10	0·17	0·08	0·01	0·16	0·07	0·12	0·29	0·1
Cr$_2$O$_3$	0·30	0·45	0·37	0·02	0·51	0·18	nil	0·28	0·23	0·02	0·05	0·0(4)
H$_2$O+	1·49	3·39	2·91	1·21	0·97	0·21	0·31	1·37	1·02	0·58	0·40	...
H$_2$O−	nil	nil	nil	0·07	0·06	0·14	...		0·17	0·13	0·13	...
Total	100·61	99·89	99·92	100·29	100·36	100·37	100·01	100·29	100·63	100·04	100·66	100·00

Key to Specimens

1. Eucritic olivine–plagioclase–clinopyroxene orthocumulate (H9605). Cumulus plagioclase (An$_{78}$) more sodic than in other units. Unit A. Anal. W. J. Wadsworth.
2. Olivine crescumulate (H9659). Unit B. Anal. W. J. Wadsworth.
3. Olivine heteradcumulate (H9656), layer immediately above No. 2. Unit B. Anal. W. J. Wadsworth.
4. Olivine – plagioclase orthocumulate (H5114). Unit 3. Anal. G. M. Brown.
5. Olivine heteradcumulate (average of H5328–34). Unit 10. Anal. E. A. Vincent and R. Hall.
6. Plagioclase – clinopyroxene – olivine adcumulate (H5049). Unit 10. Anal. G. M. Brown.
7. Plagioclase heteradcumulate (H5690). Unit 11. Anal. E. A. Vincent and B. A. Collett.
8. Olivine heteradcumulate (H3223). Unit 14. Anal. E. A. Vincent and B. A. Collett.
9. Contemporaneous ultrabasic vein (H3236) cutting No. 5. Anal. E. A. Vincent and R. Hall.
10. Intrusive eastern marginal gabbro (H5186). Anal. G. M. Brown.
11. Intrusive fine-grained olivine gabbro (H5019). Anal. G. M. Brown.
12. Calculated composition of the pore material of H5114 (Anal. 4), taken as indicative of the approximate composition of the parental magma of the Rhum layered rocks.

fine-grained olivine gabbro, occurring as sheets intruding the layered series and likely to have crystallized from the locally available, undifferentiated parental magma, has a closely comparable composition (Table 21, Columns 11 and 12).

It is sometimes difficult, by a petrographic examination, to tell whether certain large, poikilitic crystals are zoned in their outer parts. So far as they can be employed, the optical measurements indicate that the poikilitic crystals are usually unzoned, and of the same composition as the cumulus crystals in adjacent layers. Since the detection and measurement of the zoning is difficult, chemical rock analysis has been used to estimate the compositions of the poikilitic minerals. The CIPW norms calculated from analyses of typical rocks show that, as the composition of the poikilitic material is close to that of the composition of the corresponding, unzoned cumulus phases in neighbouring layers (see examples below), the poikilitic crystals are probably free from zoning. Such rocks (heteradcumulates) have been described at some length in Chapter XI, and are very common in Rhum.

From this discussion, it will be clear that the chemical analyses of rocks taken from the layered units in an ascending sequence will show no regular pattern of change in the oxide percentages such as is usually associated with a fractionated series of rocks or liquids. Instead, there will be an irregularity which is generally independent of height in the layered series and due not only to the local variations in the character and proportions of the cumulus minerals but also to the degree to which adcumulus and heteradcumulus growth has affected the composition of the initial deposit of cumulus crystals and the amount of trapped intercumulus liquid.

In Table 21, the first eight chemical analyses are listed in an order from bottom to top of the layered series, for convenience of description. Analysis 1 is of an ortho-cumulate from unit A, the lowest exposed unit (p. 278), which contains almost equal proportions of cumulus olivine, plagioclase, and clinopyroxene. From optical measurements, the cumulus parts of the olivines are Fo_{82} and of the felspars An_{78} (p. 278 and Table 17). The norm of the analyzed rock (Table 22) shows that the total olivine has a composition close to Fo_{77} and the felspar close to An_{65}. This contrast with the estimated cumulus composition indicates the presence of pore material and means that the rock is an orthocumulate consisting of an assemblage of eucritic cumulate crystals together with gabbroic pore material crystallized from trapped basaltic pore liquid. It is more difficult to use the oxide percentages, alone, to detect such characteristics, unless they are compared with those of rocks with the same assemblages of cumulus minerals; for example, the amount of an oxide such as Na_2O is influenced to a great extent by the amount of cumulus plagioclase in the rock. As will be shown below, the trapped liquid is most likely to have had a normative felspar composition of about An_{62}. Hence the analyzed rock from unit A must have contained only a small amount of cumulus An_{78} plagioclase and this is unlikely to have been extended by adcumulus growth. However, it is likely that the calculated normative plagioclase composition (An_{65}) is slightly more sodic than the true composition (see p. 259) and

TABLE 22

CIPW norms of analyses in Table 21

	1	2	3	4	5	6	7	8	9	10	11	12
Or	0·78	0·28	0·28	1·11	0·11	0·56	0·56	0·56	0·39	2·22	1·11	1·7
Ab	10·37	3·20	1·00	15·20	2·99	11·32	5·76	2·10	14·04	21·48	20·96	18·9
An	20·88	12·09	3·00	48·37	10·17	48·93	82·84	15·85	47·86	40·03	35·50	37·7
Ne	0·43	...	0·60 (Cor.)	0·85	1·36	...	3·41	0·85	1·14	40·03	1·14	2·3
Wo	6·79	0·96	...	3·48	4·58	9·55	0·23	0·23	11·44	6·96	11·20	6·1
Cl–En	4·84	0·75	...	2·40	3·50	7·26	0·20	0·20	8·50	4·30	7·30	4·0
Fs	1·33	0·11	5·79	0·79	0·59	1·28	...	...	1·80	2·25	3·10	1·7
En	...	3·15	0·81	...	...	...	...	...	...	5·10	...	...
Of	...	0·41	...	...	...	...	...	...	...	2·77	...	...
Fo	37·55	60·44	68·88	18·20	60·32	16·70	4·62	62·51	9·52	6·45	12·04	16·2
Fa	11·49	8·81	10·61	6·22	11·34	3·26	0·51	13·36	2·22	3·90	5·10	7·8
Mt	2·90	4·75	4·85	1·62	3·02	0·79	1·39	2·55	1·85	2·55	2·10	2·1
Ilm	1·03	0·50	0·30	0·76	0·29	...	0·46	0·15	0·29	1·37	0·61	1·5
Chr	0·10	0·07	0·07	...	0·76	...	0·31	0·45	0·33	...	...	0·0(6)
Ap	0·45	0·67	0·54	...	...	...	...	...	...	0·10	...	0·1
Normative mineral compositions												
Plagioclase	An₆₅	An₇₇	An₇₀	An₇₄	An₇₇	An₈₁	An₉₀	An₈₁	An₇₇	An₆₃	An₆₂	An₆₅
Olivine	Fo₇₇	Fo₈₇	Fo₈₇	Fo₇₅	Fo₈₄	Fo₈₄	Fo₉₀*	Fo₈₇	Fo₈₁	Fo₆₂	Fo₇₀	Fo₆₇
Iron ratios $\dfrac{(FeO+Fe_2O_3)\times 100}{FeO+Fe_2O_3+MgO}$	34	24	23	37	24	23	42*	25	33	53	44	45

* Not dependable, due to very low Feo and MgO contents.

that the CIPW norm cannot be used for more than general estimates of the degree of adcumulus growth.

Analyses 2 and 3, of olivine-rich rocks from unit B, show, from the norms, that whereas the olivines have been extended by crescumulate and normal adcumulate growth so as to retain an overall magnesian composition (Fo_{88}), the interstitial felspars have grown in part by heteradcumulate growth and in part from trapped pore liquid, there being more heteradcumulate growth in the normal cumulate (An_{77}) than in the crescumulate (An_{70}).

TABLE 23

Trace element analyses of rocks (major element analyses in Table 21)

	1	2	3	4	5	6
Cr^{+++}	2050	2500	3100	300	4400	1200
Ti^{++++}	4800	2200	2700	2500	1300	960
Ni^{++}	1100	2100	1800	360	1550	300
Co^{++}	79	150	110	45	140	34
Cu^{++}	110	40	120	120	220	32
V^{+++}	135	51	56	96	110	62
Zr^{++++}	53	44	25	18	47	1
Mn^{++}	1500	1400	1250	860	1400	660
Sc^{+++}	53	36	46	35	30	43
Sr^{++}	130	8	31	270	48	250
Ba^{++}	44	33	24	200	72	76
Li^{+}	3	...	1	1	2	1
Rb^{+}	10	10	10	10	10	10
Ni/Co	13·9	14	16·4	8·0	10·7	8·8
Ni/Mg	0·008	0·008	0·008	0·005	0·007	0·004
Ni/Fe^{++}	0·014	0·028	0·028	0·008	0·021	0·012

Numbers of specimens as in Tables 21 and 22.   Spectrographic anal. S. R. Taylor.

Analysis 4 is of an olivine–felspar orthocumulate from one of the lower units (unit 3) in eastern Rhum. The normative compositions, both of olivine (Fo_{75}) and of felspar (An_{74}), indicate the presence in this rock of pore material derived from liquid trapped between the cumulus crystals (Fo_{86} and An_{84}). The calculated composition of the pore material is given in column 12 of Table 21, and is close to that of the later-intruded, fine-grained olivine gabbro (Anal. 11). As discussed on pp. 271 and 284 (and in Brown, 1956), these data are used to obtain an estimate for the composition of the parental magma from which the layered rocks crystallized.

Analyses 5 and 6 are of rocks from near the bottom and top of a higher eastern unit, unit 10. The normative olivines are of identical composition in each, while the felspars are not much different. The contrasted textures indicate clearly, however, that in one rock the felspar is cumulus (Anal. 6) whereas in the other it is intercumulus (Anal. 5), which means that the growth of felspar in the latter, olivine(–spinel) cumulate has been of the heteradcumulate type. The plagioclase from the felspar–augite–olivine adcumulate, which is virtually unzoned, has been separated and analyzed (Table 20) and has a composition of An_{85}, whereas the CIPW normative felspar, estimated from the rock analysis, is An_{81}. For these rocks, this suggests that the calculation of the normative felspar composition from the rock analysis, by the CIPW method, yields a slightly lower An value than the true one. The normative felspar of the olivine(–spinel) heteradcumulate, although broadly similar, is yet more sodic (An_{77}) than the normative value for the felspar–augite–olivine adcumulate, so there must have been a slight amount of trapped liquid left, in pore spaces, after an appreciable period of heteradcumulate growth.

Analysis 8 is of an olivine(–spinel) heteradcumulate (from unit 14) which affords particularly good textural evidence for the growth of large, poikilitic felspars and pyroxenes from the intercumulus liquid (Fig. 148–150). Here again, the poikilitic felspar is calcic, the normative composition (An_{81}) being identical with the normative composition of the felspar, given by Analysis 6, of a rock (see above) in which the actual composition of the separated and analyzed, unzoned cumulus felspar is An_{85}. This is evidence that the rock of Analysis 8 is the ideal type of heteradcumulate, in which the large, poikilitic crystals do not consist, even in part, of material derived from trapped liquid.

The marginal gabbro (Anal. 10), and the fine-grained olivine gabbro (Anal. 11) are later intrusions, both believed to represent magma which was available during, and possibly throughout, the history of this layered intrusion. If it were the parent magma for the layered rocks, it would once have filled the chamber in which the layered series formed and this same magma may, periodically, have continued to enter the chamber, should there have been times when the upper, residual portions of the main body were drawn off to form volcanic eruptions (p. 297). Later, a surge of this magma is thought to have accompanied uplift of the layered rocks to their present position, forming a lubricating, marginal ring of magma which aided their emplacement. The composition of the analyzed marginal gabbro (Anal. 10) is believed, however, to have been partly modified by contamination with the country rocks. The latest episode of magmatic activity in this centre resulted in the high-level intrusion of sills of undifferentiated magma, which cut through the layered rocks to form the fine-grained olivine gabbros (Anal. 11), and these are free from any contamination with country rock. This presumed parental magma is basaltic in composition, and is characterized particularly by its high alumina content (c. 18%). In this respect it is comparable with the Skaergaard original magma, and with the compositions of certain basaltic magmas estimated to be possible parental magmas for some of the other layered basic rocks in the Tertiary Thulean Province.

THE SEQUENCE OF EVENTS IN THE FORMATION OF THE INTRUSION

The layered ultrabasic rocks of Rhum, characterized by certain structures and textures which are indicative of crystal accumulation, lack some of the other features which would help, as is the case in the Skaergaard, in showing this to have been the process responsible

T

for their formation. Thus they are not enveloped by a marginal border group, and in particular by a fine-grained border facies representative of the chilled parental magma; nor is there any of the regular cryptic layering which, in the Skaergaard intrusion, gives so obvious a clue to the fractionation process involved.

a) *The site of the magma chamber*

Rocks with similar compositions, structures, and textures to those in Rhum are to be found in the lowest zones of some of the other layered intrusions, such as the Stillwater and Bushveld. Generally, however, it is this lower zone which is poorly exposed, whereas Rhum provides almost continuous exposure of over 7000 ft of what was almost certainly the lower zone of a layered body. This thick zone cannot be viewed as having been formed in its present position, as part of a thick intrusion from which the roof and later differentiates have been removed by erosion. If this were so, then by analogy with other intrusions some 30 000 ft of gabbroic and later differentiates, as well as the cover, would need to have been eroded during Tertiary times—an untenable hypothesis for the Scottish Hebridean region. In Rhum, the layered rocks extend to within 50 ft or less of the contact with Precambrian country rocks and are not separated from the latter by an original envelope of marginal border rocks, but usually by the coarse gabbro which, although badly exposed, generally has the form of a nearly vertical, annular intrusion. The gabbro appears to have been intruded along a well-defined ring fracture, and where gabbro is locally absent the contrast between the Precambrian country rocks within and outside the fracture zone bears testimony to appreciable upward faulting of the almost circular block, consisting chiefly of the layered rocks.

For these reasons it is believed that the layered rocks have been uplifted to their present position, subsequent to consolidation. Also it is apparent, from the absence of an original margin to the layered series, that the uplifted block represents only a part dissected out of a deeper-seated intrusion. Geophysical evidence, in the form of gravity measurements (McQuillin & Tuson, 1963), indicates the presence in Rhum of a large positive anomaly to the east of the exposed layered rocks, which supports the theory that the block was removed from the western part of the larger intrusion. The convergence of the radial dips of the layers, however, towards a centre close to Glen Harris (Fig. 134–136), is hard to reconcile with the hypothesis that the layered rocks are part of a block dissected asymmetrically from a larger intrusion, unless the present dips were developed during uplift.

The absence of layered rocks outside the ring-fault, and the presence there of country rocks devoid of extensive thermal metamorphic effects, suggest that the rocks outside the fault must have once lain well above the immediate roof of the intrusion. Hence the uplift must have exceeded 3000 ft, the height at which layered rocks now tower above marginal country rocks. Bailey (1945) proposed that the uplift may have been close to 5000 ft in the west (Lewisian inside and Applecross series of the Torridonian outside the fault) and 2000 ft in the east (Diabaig series of the Torridonian inside and Applecross series of the Torridonian outside the fault). The true amount of uplift may lie between these values, because the sedimentary cover inside the fault could have been tilted eastward before uplift, at the time of partial melting of the roof zone. However, as it is believed that the intrusion lay within the Lewisian (to produce acid igneous rocks by partial melting of this material while it formed the roof; see Dunham, 1962), and in view of the absence of marked thermal metamorphism in the Torridonian forming the present marginal country rocks, the uplift was probably closer to 5000 ft. More strongly metamorphosed Torridonian sediments *within* the ring-fault are almost in contact with layered ultrabasic rocks, being separated only by a narrow strip of marginal gabbro. If, at the time of uplift, basic magma still existed above the layered cumulates in the magma chamber, then expulsion of most of this during the beginnings of uplift would bring the layered rocks almost in contact with those lower, well metamorphosed parts of the Torridonian cover. Any acid magma generated by melting of the even lower, Lewisian part of the cover could also have been expelled, probably as the acid intrusions of felsite and granophyre (Hughes, 1960; Dunham, 1962) which preceded, and were then intruded by, the gabbros and uplifted ultrabasic rocks.

Such a relatively small amount of uplift could not be postulated if the layered intrusion, at depth, formed from a closed system of initial basaltic magma (of the type believed to be the parent in Rhum) from which a series of basic, intermediate and acid rocks, complementary to the great thickness of ultrabasic rocks, would be expected to have formed. As discussed above, such an hypothesis is unacceptable because it would require over 6 miles of uplift, and subsequent erosion, to account for the present topographical level of the low-dipping, ultrabasic layered rocks. Thus it has been postulated (Brown, 1956, pp. 44–5) that the Rhum intrusion fed a surface volcano, and that constant surface eruption, attended by replenishment of the magma chamber with basaltic magma from below, resulted in the formation of a thick series of early ultrabasic differentiates deposited at a relatively high and constant temperature. This hypothesis presumed that the replenishments were relatively small in bulk compared with the magma already present, so that no reversals in mineral composition took place in successively accumulated layered units. However, Wadsworth (1961) has since demonstrated reversals of a slight, but significant nature (Table 17) at the base of some of the thick western units (p. 280). This information provides valuable evidence to support the hypothesis advanced, initially, to explain the regular sequence of cumulus minerals within each of the major eastern units (p. 250), and the numerous unit repetitions free from significant changes in mineral compositions.

b) *Crystallization within the magma chamber*

In the postulated, larger magma chamber, of greater lateral extent but probably not of much greater thickness than the present-exposed layered series, the processes of crystallization and crystal deposition operated. The parental magma is believed to have approximated in composition to that of a high-alumina basalt of tholeiitic affinities (see p. 289) and may well have chilled at the margins of the original intrusion, as it seems to have done in the sheets of later-injected, fine-grained olivine gabbro of similar composition which transgress the layered series. Such a magma is capable of precipitating early crystals similar in composition to the plagioclase felspars, olivines, clinopyroxenes, and chrome-spinels of the layered series and it would appear, from the mineralogy of the Rhum rocks, that all four phases were crystallizing at broadly the same time.

Consideration of each major unit in detail, however, shows that chrome-spinel, and then olivine, are likely to have crystallized slightly earlier than the felspar and pyroxene, since mechanical sorting is probably inadequate to explain at least three features of each major unit. Firstly, felspar is not deposited until between 100 and 300 ft of olivine cumulates have formed, in most of the eastern units (and even more in the western units) and to explain this by mechanical sorting would require that the felspars were held in suspension in the magma for a very long time (e.g. almost 1000 years for 300 ft of crystal accumulation; see p. 210). Secondly, the chrome-spinel crystals are so small that, in view of the importance of size relative to density in the rate of crystal sinking, they would not be expected to sink any more quickly than the larger felspars and pyroxenes (Table 14). Thirdly, the pyroxenes have a density and, apparently, primocryst size almost equal to that of the olivines, and yet they are absent as cumulus crystals from the thick olivine cumulates, being concentrated together with the felspathic cumulates towards the top of each unit. These facts suggest an order of crystallization, rather than mechanical sorting, as the cause of the macro-units. Also, the evidence from the thick olivine crescumulates of units A and B suggests crystallization of olivine, by itself, for prolonged periods.

It is unlikely, however, that the order of crystallization within each major unit can be due simply to the changing composition of a static magma near the floor, as proposed by Jackson for the Stillwater ultramafic units (p. 340), because the evidence in Rhum points to the likelihood of periodic movements of supercooled magma down to, and across the floor (p. 273). Any change from the conditions for chrome-spinel and olivine crystallization towards those for felspar and pyroxene crystallization, as well, seems to have been within the magma body as a whole, rather than within a body of stagnant magma lying near the floor and unaffected by limited-depth convection currents. Towards the top of each unit, it seems that felspar, pyroxene, small amounts of olivine and, possibly, traces of chrome-spinel were crystallizing together, and fine-scale rhythmic layering is extensively developed. This is probably an indication of sorting by relatively fast currents of variable velocity. Also, the layers of cumulus olivine which alternate with the olivine crescumulates, in units A and B, suggest that currents periodically disturbed the tranquil conditions necessary for crescumulate growth, at a stage when olivine alone was a cumulus mineral (p. 273). The contrast between the well-layered, and often laminated upper parts of each of the eastern units, and the massive lower parts, is of interest in this connection. A few thin crescumulate layers are to be seen amongst the otherwise massive olivine cumulates, but are relatively rare compared with those in the lower, western units. It is possible that at the later stages of the cooling history, the olivine habit changed to more tabular crystals (Brown, 1956, fig. 3), many of which would tend to lie with their crystallographic a-axes in the plane of layering and, hence, not lead to crescumulate growth during periods of magma stagnation (p. 273).

The repetition of the same cumulus sequence in all the units of the layered series, with only relatively minor modifications, points to there being an underlying set of causes which was repeated many times. Essentially, the sequence consists of a thick olivine cumulate (with minor, cumulus chrome-spinel) which grades upward into a plagioclase–augite cumulate with subsidiary amounts of cumulus olivine and lesser amounts of cumulus chrome-spinel. In the upper type of cumulus assemblage in each unit, consisting of four different mineral phases, there is usually sorting of the cumulus crystals to give thin layers with varying proportions of these phases, and there is often considerable igneous lamination, shown especially by the plagioclases. In all except the lowest five units of the eastern part, extreme adcumulus growth has given rise to adcumulates or heteradcumulates.

The first hypothesis to account for the repeated units (Brown, 1956, pp. 45–9) was that the conditions for the development of a new unit were initiated by injection into the magma chamber of a fresh mass of the parental basaltic magma, from the deeper source levels. This mixed with the residual magma from the previous unit to give a homogeneous liquid, after some of the initial residue had been removed from the chamber as extruded lava at a surface volcano. The mixed magma then began to crystallize spinel and olivine, which sank out of the liquid, and later, as a result of a small fall in temperature, plagioclase and augite crystallized and also sank to the floor. The phase diagram Fo–Di–An (Osborn & Tait, 1952) was used to illustrate the reasonableness of the sequence, assuming that the effect of other components in natural magmas would result in all four phases crystallizing in sequence over a smaller range of temperatures than is shown in the three-component system. This assumption is supported by subsequent experimental work. For example, the effect of adding albite, or water under pressure, to the system Fo–Di–An would be to extend the diopside relative to the plagioclase field and reduce the temperature interval separating the appearance of diopside from that of the other three phases, with cooling in this system. Reference to the system Fo–Di–An–Ab (Yoder & Tilley, 1962, fig. 10) shows that as a simplified system the Rhum magma should plot close to the 'four-phase curve', if we are to account for almost simultaneous crystallization of olivine, plagioclase and clinopyroxene. (This system does not include a chrome-spinel phase which, obviously, has a different stability field from the particular spinel phase in the synthetic system.) The analyzed fine-grained olivine gabbro, thought of as representative of the composition of the Rhum initial magma, has a CIPW normative composition (Column 11, Table 22) which can be expressed almost entirely in terms of the components of this system (Fo+Di+An+Ab

$= 87\%$ of total normative components). The calculated ratios are Fo 14:An 41: Ab24: Di 21, which, despite the low Fo content, would place the liquid composition in the Fo field, and close to the Fo+Pl plane. However, to explain the sequence of a Rhum unit in terms of this system would still require a drop in temperature for the liquid to reach the Fo+Pl+Di (+liquid) four-phase curve. Analogy with more complex phase equilibria systems would still fail to resolve the problem posed by an attempt to explain both the sequence in each unit *and* the frequent repetition of this sequence, because the prime controlling factor would always be a repeated fall and rise in temperature of the liquid, however small in amount. An alternative factor may be slight variations in the amounts of water dissolved in the magma under pressure, and the fresh influxes of magma, together with associated surface eruptions, were believed by Brown (1956) to afford a mechanism for raising the temperature and lowering the water contents of the magma slightly, at the beginning of each successive major unit. Although this process could possibly account for the sequence of cumulus phases, it is difficult to explain in this way the processes of crystallization required for the intercumulus material of the heteradcumulates (see below), and an alternative, less specific mechanism is proposed.

After our visit to the Bushveld intrusion in 1958, the order of crystal nucleation was suggested by Wager (1959) as a mechanism for the development of the major units in the Bushveld and Rhum intrusions. The effect of delay in the appearance of a new phase, resulting in some degree of supersaturation of the magma for this particular phase, has been considered in relation to the mechanism of adcumulus growth in discussing the Skaergaard intrusion (Ch. VIII). The existence of supersaturation was also demonstrated as the cause of the perpendicular felspar rock of the Skaergaard marginal border group (Ch. V & VIII). In Rhum, supercooling of the magma has already been invoked to explain the crescumulates (p. 273), and the concept of delayed nucleation could be used to build a model, as an alternative to that of equilibrium crystallization, to explain the major units of Rhum.

At the beginning of a new major unit, let it be assumed that the conditions of temperature and composition are such that the magma is capable of precipitating chrome-spinel, olivine, plagioclase and clinopyroxene, but that no crystals or nuclei are present in it. With loss of heat to the surroundings, presumably mainly upwards, the magma will become supercooled because crystal nucleation of the four stable crystal phases will not take place immediately. After some degree of supercooling, however, one or other of the crystal phases will nucleate. It is not known which phase of the four which should be present will appear first, but it has been argued (Wager, 1959) that spinel may well be the first, followed shortly by olivine, because these minerals have relatively simple structures. After nucleation they will grow, eliminating the supersaturation in their immediate neighbourhood, and then will continue to grow as heat is lost to the surroundings. At the same time they will sink, being denser than the liquid, and form the first layers of bottom deposit of cumulus crystals. In this way would form the thin layer of chrome-spinel, sometimes found at the base of a unit, and the olivine–spinel cumulate which is the main lower part of the Rhum major units. The further removal of heat will result in clinopyroxene or plagioclase, or both, nucleating, growing, and sinking, thus giving all the cumulus crystals occurring in the upper part of the unit. Except for minor temporary falls, the temperature may remain relatively constant. In this hypothesis, the sequence of cumulus crystals is controlled by the order of nucleation and not, as in Brown's hypothesis, by the order in which the mineral phases crystallize from the liquid under equilibrium conditions with lowering temperature.

The formation of heteradcumulates is, we are agreed, difficult to understand in relation to the hypothesis relating the order of crystallization to cooling under equilibrium conditions. Initially, the olivine–spinel cumulates would be surrounded by liquid from which felspar and pyroxene cannot form until the fall in temperature brings the magma to that appropriate for their precipitation. To account for the composition of the intercumulus plagioclase and pyroxene being the same as their compositions when they appear as cumulus crystals a little higher in each unit, it has been suggested that after having nucleated from the intercumulus liquid, they grew by the adcumulus process in the interstices of the olivine–spinel cumulus pile at, or very near to, the temporary top surfaces. This implies that they formed from the liquid which, at this stage, was precipitating olivine and spinel as cumulus crystals, and this is contrary to the hypothesis that plagioclase and clinopyroxene were not precipitated except with some further cooling. Applying the nucleation hypothesis we do not encounter this difficulty, for the temperature is believed to be such that plagioclase and clinopyroxene, just as much as olivine and spinel, are able to nucleate from the liquid. It is still necessary, however, to have a supplementary hypothesis to account for the nuclei of plagioclase and clinopyroxene forming in the interstices of the olivine and spinel cumulus crystals. Presumably, very slight changes of conditions can cause nucleation and we would suggest that the slight growth of the olivine and spinel crystals may shift the composition of the confined intercumulus liquid so that nuclei of plagioclase and clinopyroxene developed, although supercooling in the main body of magma, of average composition, would not have reached the point where nucleation of these minerals takes place. It is likely that early in the formation of a major unit, convection currents were established which provided the means of heat removal necessary for adcumulus growth and the growth of the intercumulus material of the heteradcumulates. It is probable that the alternating layers of olivine heteradcumulates and olivine crescumulates, found in certain major units, imply single convective overturns of the liquid for cumulus olivine deposition, alternating with periods during which the magma was stationary and the crescumulate olivines grew. In particular, the thin plagioclase adcumulate layers of the eastern part of the intrusion suggest active flow of the magma over the upper surface of cumulus plagioclase crystals, bringing a constant stream of supercooled magma to the growing surfaces.

The extent to which each of the two suggested mechanisms operated in the formation of the major units has not been decided. To solve such a problem would require a detailed knowledge of very subtle variations in the compositions of the various crystal phases, from which it may be hoped to establish the extent of the fluctuations of temperature during the formation of a major unit. The nucleation hypothesis requires a slight fall in temperature, below the equilibrium value, followed by a return to the original temperature during the formation of the unit. For the next unit, it should be noted that if the newly injected magma has moved up rapidly from the depth, as is likely, it may be superheated and this would contribute to the disappearance of all crystal phases in the mixed magma before supercooling started the cycle of crystallization again, to give another major unit. The equilibrium hypothesis requires a steady, slight fall in temperature throughout the formation of a unit, the original temperature conditions being restored for the commencement of the next unit by a new influx of magma. In the Bushveld there is a better opportunity for distinguishing the results of nucleation, because this is a cryptically layered intrusion and the major rhythmic units of the Basal Series (Ch. XIV) must have been precipitated from magma which, through fractionation, has become progressively further removed in composition from any successive fresh influxes of parental magma which might have occurred. Any evidence for a change, within each major unit, towards a lower-temperature mineral assemblage, and a return to a higher-temperature assemblage at the base of the overlying unit, would support the hypothesis of equilibrium crystallization with lowering temperature, and an approach to original temperature conditions through new influx of magma. The overall cooling and fractionation would mean that later influxes, although producing set-backs, would be mixing with progressively lower-temperature residual magmas and so the overall pattern would be of occasional interruptions superimposed on an overall cryptic layering, which seems to be the case (see Ch. XIV). Lack of change in mineral compositions within each major unit would have been more difficult to reconcile with the equilibrium cooling hypothesis, and would have implied that nucleation of the phases in a specific order, at constant temperature, was responsible for the layering within each major unit.

In Rhum, the repetition of the sequence outlined for each major unit took place at least twenty times, and resulted in no regular change in the composition of the mineral phases over a thickness of at least 7000 ft. Such behaviour led to the theory that the Rhum magma chamber was replenished at regular intervals by influxes of parental magma which, it has been suggested (Brown, 1956), could well have coincided with the eruption of magma from the chamber to an overlying surface volcano. The replacement of old by new magma need not have been complete—in fact it is unlikely to have been so, otherwise the layered rocks would have been disrupted—but only sufficient for the relationship between liquid and separating crystal phases (which was delicately poised) to be changed slightly. The likelihood of a more drastic change occurring, occasionally, has been evidenced in the case of the Transition Series (p. 280) and of the more subtle returns to more magnesian cumulus olivine compositions at the base of units B and D (Table 17). These effects, and the slight cryptic layering within units B and D, support the hypothesis of falling temperature during the formation of each unit. Also, despite these replenishments, which resulted in a return to the initial crystallization of olivine and chrome-spinel in each case, there is a progressive increase in the amount of cumulus felspar and clinopyroxene, relative to olivine and chrome-spinel, upwards in the succession of units. Hence, the general trend of crystallization proposed for each unit is only partially upset by each replenishment, and the effect is progressively diminished.

The connection between layered intrusion chambers and surface volcanos was proposed (Brown, 1956) as a general petrogenetic concept, because the changing composition of lavas erupted from central-type volcanos, in contrast to fissure-type eruptions reaching the surface directly, is more than likely due to crystal fractionation and this can only take place in a fairly high-level magma chamber in which slow crystal separation is taking place. It would appear, therefore, that whereas certain layered intrusions, such as the Skaergaard, were formed by a single episode of magma emplacement and provide an example of a closed system, others, such as Rhum, formed in magma chambers lying along the path of magma on its way to the surface. Erupted magma would come from this magma chamber and fresh supplies would flow into it from below. Periods of volcanic eruption would result in periodic changes both in physical and chemical conditions in the magma chamber and, in turn, leave their imprint on the character of the crystallizing layered rocks. Such a volcano as that postulated on Rhum would have continued to erupt basalt during the whole period represented by the exposed layered series, because the minerals found in these ultrabasic rocks are the characteristic phenocrysts of many basalts, the groundmass of which is still capable of precipitating a wide range of calcic plagioclase, magnesian olivine and magnesian pyroxene.

CHAPTER XIII

THE STILLWATER INTRUSION, MONTANA, U.S.A.

INTRODUCTION

The Stillwater is a Precambrian intrusion of layered basic and ultrabasic rocks out-cropping in a belt 30 miles long and, at a maximum, 5 miles wide, on the northeast margin of the Beartooth Mountains in Montana (Fig. 168). It is bordered to the south by Precambrian metasediments which locally have been thermally metamorphosed to hornfels by the Stillwater intrusion and then invaded by a granite of Precambrian age (1530–1580 million years). To the north, Middle Cambrian sediments lie unconform-ably on the eroded surface of the steeply-dipping upper layered rocks of the basic intrusion, and although drilling beneath this Palaeozoic cover would be likely to reveal more of the layered rocks, present available data are restricted to the exposed part of the intrusion. The disposition of the exposed rocks has led to the belief that the intrusion crystallized in the form of a horizontal sheet and that the present high dips of the rhythmically layered rocks are the product of later tectonic movements, which resulted in a complex pattern of folding and faulting (Jones *et al.*, 1960) but in the preservation of the floor and most of the original lateral extent of the intrusion.

Work was begun on the intrusion in 1930, by J. W. Peoples and A. L. Howland (under the direction of E. Sampson) and in 1935, H. H. Hess initiated a detailed study of the mineralogy of the complex. The history of the investigation has been given by Jones, Peoples and Howland (1960, p. 283). For the purpose of writing this chapter we have drawn freely from the recent excellent accounts given by Hess (1960), Jones *et al.* (1960), and Jackson (1961), with only brief reference to other papers. The intrusion was visited by one of us (G.M.B.) in 1955, photographs being taken and rocks collected from the Stillwater Valley (Fig. 170) and a traverse of the East Boulder Plateau (Fig. 169).

In view of the nature and extent of the previous work it is not incumbent on us to argue afresh that the Stillwater intrusion is a body of layered igneous rocks, crystal deposition on the floor of the magma chamber having been the prime factor responsible for the well-described layered structures, textures, and cryptic layering, for all those who have worked on the intrusion are agreed on that point. Rather, we have drawn together

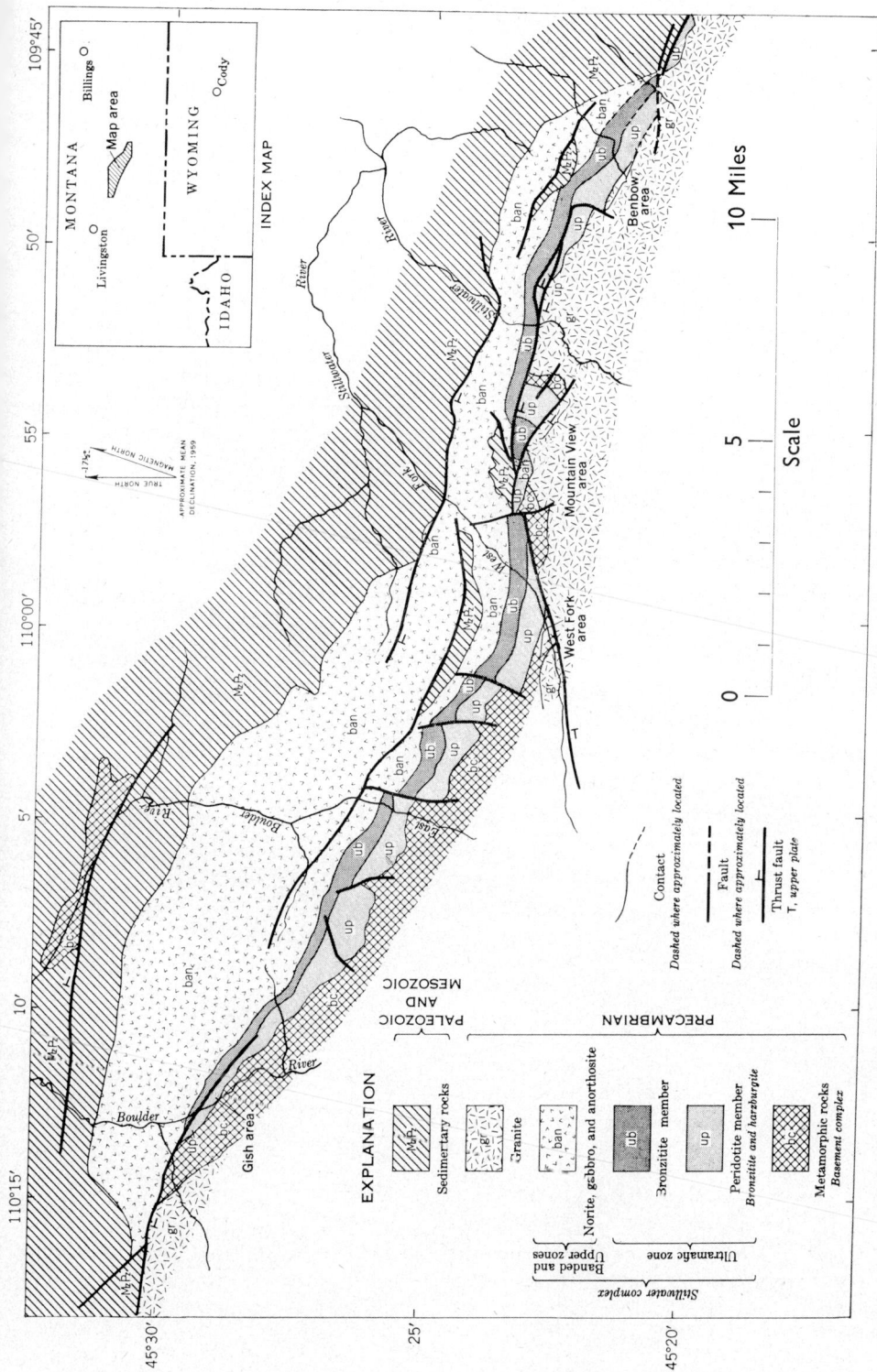

FIG. 168. Generalized geological map of the Stillwater intrusion, with inset showing its geographical locality. (From Jackson, 1961; with the original scale, in error, corrected). A more detailed map is given by Jones et al. (1960, pl. 24).

EXPLANATION

Sedimentary rocks — PALEOZOIC AND MESOZOIC

Granite

Norite, gabbro, and anorthosite — ban — Banded and Upper zones

Bronzitite member — ub — Ultramafic zone

Peridotite member
Bronzitite and harzburgite — up

Metamorphic rocks
Basement complex — Stillwater complex

PRECAMBRIAN

Contact
Dashed where approximately located

Fault
Dashed where approximately located

Thrust fault
T, upper plate

Scale

0   5   10 Miles

INDEX MAP

MONTANA
Livingston
Billings
Map area

WYOMING
Cody

IDAHO

the published information and presented it in a way designed to facilitate direct comparison with the descriptions of the other layered intrusions. In particular, we have drawn freely from Hess's Memoir and Jackson's paper, and also have benefited greatly from private discussions with Professor Hess and Dr Jackson.

THE FORM AND SUBDIVISION OF THE INTRUSION

The later tectonic events have so modified the original shape of the intrusion (Fig. 168; see also Jones *et al.*, pl. 25) that it is not possible to find, in any single traverse across the complex, a fully representative rock sequence; this is further complicated by lateral variations in the thicknesses of certain layered zones. However, detailed mapping of much of the complex by the U.S. Geological Survey has permitted reconstruction of the stratigraphic section in most areas (E. D. Jackson, personal communication, 1963). Both Hess (1960) and Jackson (1961) are agreed that the layered features originated as semi-horizontal structures and that later tilting to the north has produced the high dips, and occasional overturned layers, now seen. These high dips provide exposures of layered rocks amounting to a maximum of about 20 000 ft according to Jackson and Jones *et al.*, and 16 000 ft according to Hess.* The greater value is based on a consideration of the complex as a whole, whereas Hess's estimate relates to his detailed study of the East Boulder Plateau where at least one thrust fault has shortened the section.

Subdivision of the complex has been made independently by the authors of the three recent papers. It has been one of our more troublesome tasks to attempt to reconcile the three views, all presented within a year and apparently stemming from independent writing. Jones *et al.* (1960, pp. 287–8) accept the following zones, as originally defined by Peoples: Basal (50–2000 ft); Ultramafic (4000–6200 ft); Banded (7400 ft); Upper (6700 ft). Jackson (1961, pp. 2–3) uses this nomenclature, and gives the same thicknesses for the Banded and Upper Zones, but states that the Basal Zone reaches a maximum thickness of 700 ft and is locally absent, while the Ultramafic Zone averages

---

* As shown in Fig. 171, a thickness of over 17,000 ft is a reasonable estimate in view of Jackson's revised estimates for the Ultramafic Zone.

FIG. 169 ↗

View of the western flanks of the East Boulder Plateau (skyline) taken from the Clydehurst Ranch, Boulder River Valley. The northward dip of the layering results in the contrasted exposure of the Norite and Anorthosite Zones (light), and the Ultramafic Zone and basement metamorphic rocks (dark).

FIG. 170 →

View of the Beartooth granite, structurally below the Stillwater intrusion. The Ultramafic zone at the base of the intrusion lies in the area occupied by the chrome-mining settlement, the strike of the floor being approximately parallel to the low granite-ridge behind the settlement. The view is looking south, up the valley of the Stillwater River.

3500 ft. This revision of thicknesses is due to Jackson's re-definition of what constitutes each zone. Hess (1960, p. 50), on the basis of a detailed mineralogical study, proposed the following zones (values rounded off to the nearest 50 ft): Border (400 ft); Ultramafic (2500 ft); Norite (2750 ft); Lower Gabbro (2200 ft); Anorthosite (6200 ft); Upper Gabbro (2150 ft).

The choice of nomenclature and zone thicknesses accepted in this account (Fig. 171) has been influenced by the necessity for considering both the large amount of quantitative data presented by Hess and the detailed study of the Ultramafic Zone by Jackson. For the Ultramafic Zone, an approximate average of Jackson's values for the thicknesses, 3500 ft is taken, and Hess's values for the thicknesses of the higher zones: an average of 550 ft is taken for the Border Zone, as a compromise between Jackson's and Hess's values. Hess's nomenclature for the upper four zones is used; for the lower two zones, the term Ultramafic is undisputed while the term ' Border ' is chosen in preference to ' Basal ' because of its wider possible implications.

The lowest rocks exposed, in contact with the hornfelsed floor sediments, form the Border Zone and possess many of the features to be expected in what may be part of the border envelope of this layered intrusion. Two specimens examined in detail have been thought to represent the chilled border facies, but one is now regarded as a marginal hornfels. The composition of the other specimen (see Table 24), however, is taken by Hess to be representative of the chilled initial Stillwater magma. Above the fine-grained rocks there is a coarser-grained facies of the Border Zone which has so far received less attention than it may deserve. Felspathic bronzitites, with fairly well-developed layering approximately parallel to the floor of the intrusion, contain streaks and segregations of coarse norite–pegmatite, parallel to the layering. Upwards, these rocks grade into a bronzitite and thence, across a sharp junction, into the olivine-rich layers of the Ultramafic Zone. The recorded structures of this coarse inner part of the Border Zone, and the fact that the minerals are of a composition intermediate between those of the chilled rock and the Ultramafic Zone rocks, prompt us to suggest an origin through some marginal supercooling, similar to the Marginal Border Group of the Skaergaard intrusion (Ch. V).

FIG. 171   →

The sub-division into zones, and the compositional variation of cumulus minerals, in the Stillwater intrusion. The broken lines indicate horizons at which the minerals occur only as intercumulus phases.

BZ = Border Zone; UmZ = Ultramafic Zone (P = Peridotite Member and B = Bronzitite Member); NZ = Norite Zone; LgZ = Lower Gabbro Zone; AZ = Anorthosite Zone (A1, G1 . . . = Anorthosite and Gabbro Sub-Zones); UgZ = Upper Gabbro Zone.

The compositions of chemically analyzed plagioclases and pyroxenes are shown (see Table 25) as representative examples, a few optically determined specimens being shown in parentheses. The analytical data for the olivines are not yet published (see footnote, p. 311). The data given for the Border Zone minerals are for the chilled marginal gabbro (see p. 332 & Table 25).

STRUCTURAL
HEIGHT
(ft)

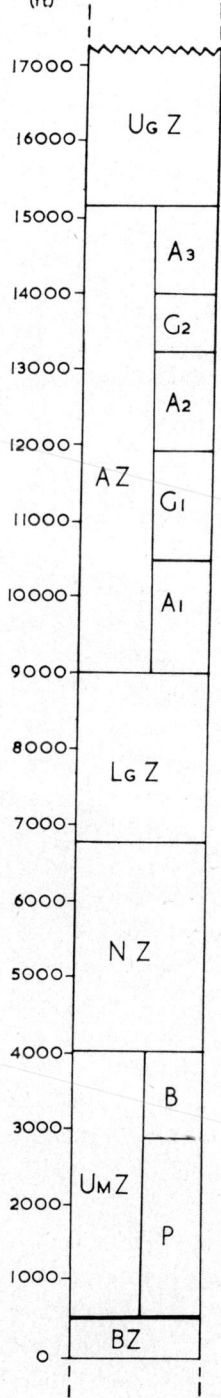

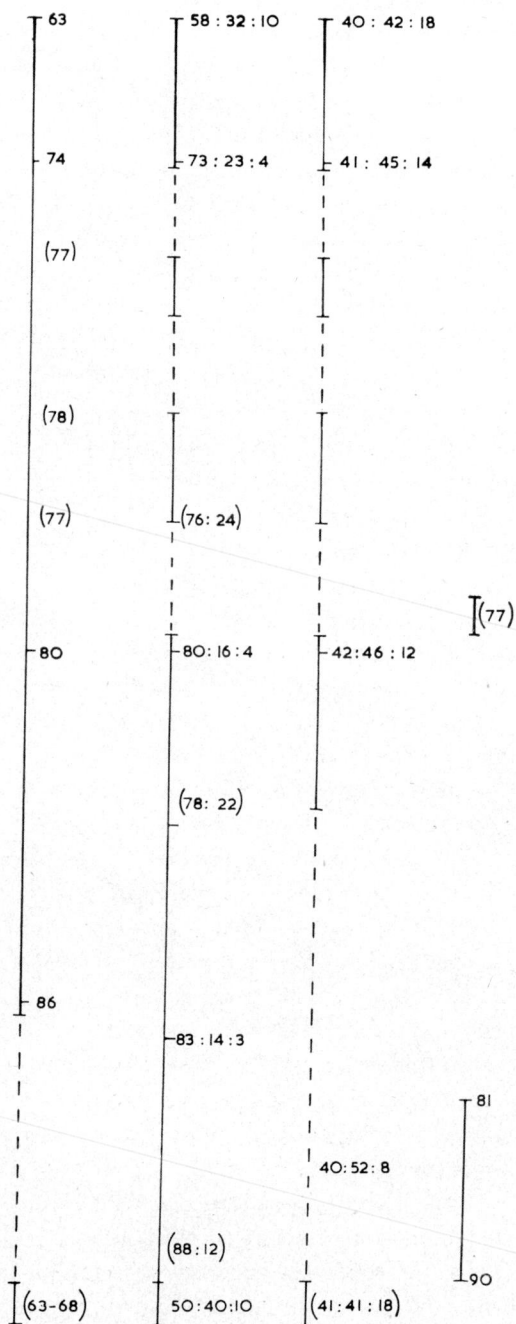

PLAGIOCLASE
(An%)

Ca-POOR
PYROXENE
(Mg:Fe:Ca)

Ca-RICH
PYROXENE
(Ca:Mg:Fe)

OLIVINE
(Fo%)

17000
16000
15000
14000
13000
12000
11000
10000
9000
8000
7000
6000
5000
4000
3000
2000
1000
0

U_G Z
A_3
G_2
A_2
A Z
G_1
A_1
L_G Z
N Z
B
U_M Z
P
BZ

63
74
(77)
(78)
(77)
80
86
(63-68)

58 : 32 : 10
73 : 23 : 4
(76 : 24)
80 : 16 : 4
(78 : 22)
83 : 14 : 3
(88 : 12)
50 : 40 : 10

40 : 42 : 18
41 : 45 : 14
42 : 46 : 12
40 : 52 : 8
(41 : 41 : 18)

(77)
81
90

The beginning of the layered series is defined by the horizon at which olivine cumulates first appear. Up to the level at which cumulus plagioclase first appears, the layers are classed as belonging to the Ultramafic Zone. Both the base and top of this zone are sharply defined, according to Jackson (1961, p. 4), although the base may rest either on the fine-grained dolerite, on the coarser bronzite-rich rocks of the upper part of the Border Zone or, where this zone is absent, directly on the country rock. More detailed work on the lower contact is necessary before the nature of the changeover from inward congelation of magma to crystal precipitation on the floor can be established. The average thickness of the Ultramafic Zone is 3500 ft, of which the lower two-thirds (the Peridotite Member) is composed, broadly, of harzburgite, chromitite, bronzitite, and dunite layers, and the upper third (the Bronzitite Member) of, virtually, a single thick layer of bronzitite. The thickness reaches a maximum in the Mountain View area where the Peridotite Member, about 4000 ft thick (the Bronzitite Member being 3000 ft), has been subdivided by Jackson into fifteen cyclic units. Each unit begins with an olivine cumulate and ends with a bronzite cumulate, and may reach a maximum of 600 ft in thickness; thus they are comparable, in scale, with the major rhythmic units of Rhum (Ch. X). The details of the Ultramafic units will be discussed later (p. 339), but at this stage it is worth noting that a distinctive pattern of rhythmic layering (a term preferred to ' cyclic ' layering, see p. 545) is well developed in the lowest rocks of this intrusion. Chromitite layers are common and are concentrated at specific levels in each unit, the thickest, economically-important chromitite horizons (G and H) occurring near the top of the Peridotite Member. Irregular bodies of secondary dunite and serpentinite are fairly abundant in the lower part of the Ultramafic Zone in the East Boulder Plateau and Boulder River regions; they will not be discussed in this chapter as they are probably a late-stage and subordinate feature of the complex (see Hess, 1960, pp. 62–9; Jackson, 1961, p. 18).

The incoming of cumulus plagioclase marks the sharply defined base of the overlying Norite Zone, about 2700 ft in thickness. The only cumulus phases are plagioclase felspar and orthopyroxene; olivine and chromite are absent, while clinopyroxene and quartz are present only as intercumulus phases. Small amounts of rounded sulphide grains (chiefly chalcopyrite and pyrrhotite) are found throughout the zone, and are thought to have formed as immiscible liquid droplets. Rhythmic layering is best developed in this zone; fine-scale layers range from a few inches to a few feet in thickness, while layers of anorthosite or norite of fairly uniform composition, 50 to 100 ft in thickness, are also found. According to Hess, about two-thirds of the fine-scale layers show gravity stratification.

The Lower Gabbro Zone is distinguished from the underlying Norite Zone through the appearance of augite as a cumulus phase, and this mineral is in excess of orthopyroxene in the estimated average-rock composition of the zone. Plagioclase, as in the Norite Zone, is the commonest mineral, and together with the pyroxenes comprises the whole of the cumulus material. Quartz, ilmenite, magnetite, and sulphides are present as intercumulus and accessory minerals in small amounts, while olivine continues to be absent. Rhythmic layering is present, but except for a special type called 'inch-scale layering' (Fig. 187) it has not been described in detail because of poor exposures. A separate traverse has been made by Hess of the upper part of the Norite Zone and lower part of the Lower Gabbro Zone, comprising 42 specimens collected within 1220 ft, in order to investigate the rapidity with which compositional and textural changes took place (Hess, 1960—'The Dead Tree Section').

The Anorthosite Zone, 6300 ft in thickness, is characterized by the presence of three thick layers (each 1300–1500 ft) in which plagioclase felspar is the only cumulus phase. One such layer occurs near the bottom, and one near the top of the zone, and the three are separated by two sub-zones of gabbro in which clinopyroxene and, to a lesser extent orthopyroxene, also occur as cumulus minerals. Apart from this large-scale layering, no fine-scale rhythmic layering has been described from this zone, each of the five thick units being relatively homogeneous. Olivine returns, in small amounts, as a cumulus phase, while traces of quartz and ilmenite–magnetite are found as intercumulus phases. Three thin layers unusually rich in sulphides (1–2% of the rock) occur near the top of the zone. The highly felspathic nature of this zone (92% in the anorthositic layers, 63% in the gabbros, and 84% in the zone as a whole) presents a special problem, and is discussed later. An unusual olivine-rock, termed the 'pillow troctolite', occurs at the base, while olivines also occur at several levels within the gabbro sub-zones.

The Upper Gabbro Zone consists of 2130 ft of layered rocks, the base being marked by the return of cumulus pyroxenes and by a sharp contact with the underlying, thick anorthosite layer. The reason it is not classed as a gabbroic sub-zone of the Anorthosite Zone would seem to be due to the relative thickness of rocks within which few anorthosite layers are found. The cumulus phases are plagioclase, augite, and orthopyroxene, the latter being an inverted pigeonite in the upper 800 ft of the zone. However, in the rocks selected for description by Hess, plagioclase is the only mineral having cumulus status; this is proof that rhythmic layering is present, even though Hess (1960, p. 88) found it to be generally faint or absent in the field. Igneous lamination is more pronounced in this zone than lower in the layered series, apparently as a result of the more tabular habit assumed by the felspar at this level (although the platy clinopyroxenes impart some igneous lamination to the rocks of the Lower Gabbro Zone). Quartz was found in only one specimen from near the base of the zone, which is an indication of paucity in pore material.

The upper 800 ft of the Upper Gabbro Zone are layered hypersthene (inverted from pigeonite) gabbros, the rest of the intrusion being covered by Palaeozoic sediments. It is likely, from the structural evidence, that more layered rocks underlie this cover and Hess has called them the 'Hidden Zone', the estimated average composition being that of a quartz ferronorite (p. 337). Some terminological confusion naturally arises from a comparison between the Skaergaard and Stillwater intrusions, because the term 'Hidden Zone' has a different meaning in each case; in the Skaergaard, the postulated hidden

U

zone consists of the earliest differentiates and in the Stillwater, of the latest differentiates. For this reason, the Stillwater upper hidden rocks will be referred to as the ' Upper Hidden Zone '.

THE MINERALS

The chief constituent minerals of the Stillwater intrusion are plagioclase felspar, orthopyroxene, and clinopyroxene, while olivine and chromite are of local significance. These minerals may be of either cumulus or intercumulus status, whereas quartz, iron oxides, biotite and, rarely, garnet are always intercumulus, and scarce. Sulphide minerals are not uncommon and generally seem to be polymineralic units, presumably crystallized from immiscible droplets of sulphide liquid. Secondary minerals, such as serpentine and epidote, are developed locally but are not discussed here. The overall pattern of rhythmic layering is the result either of mechanical sorting or of variation in the supply of the cumulus phases, while varying types of intercumulus growth, also, are responsible for the distribution and textures of the minerals. Cryptic layering, involving both continuous changes in the compositions of the cumulus minerals and abrupt appearances and disappearances of particular cumulus phases, is also a characteristic feature of the intrusion. The variability produced by rhythmic and cryptic layering can only be appreciated when the structures, textures, and mineral compositions are viewed as a whole, but in order to discuss the textural features a brief outline of the mineralogy is given here.

a) *Plagioclase felspar*

This is the commonest mineral of the intrusion and is present as a cumulus phase in all but the Ultramafic Zone, where it occurs as an intercumulus mineral. The normative felspar of the chilled Border Zone rock (Table 24) is An_{68},* while the full range of composition in the layered series is from An_{86} at the base of the Norite Zone to An_{63} at the top of the Upper Gabbro Zone (Fig. 171). Eleven plagioclases have been chemically analyzed (Hess, 1960, table 10), covering the range from the most calcic to the most sodic, four representatives of the range being given in Table 25.

One of the most significant differences between the Skaergaard and Rhum intrusions on the one hand, and the Stillwater (and possibly Bushveld) on the other, is in the degree of zoning of the felspars. Zoning is absent, or hardly discernible, throughout most of the Stillwater layered series, and only present as a narrow fringing zone to the felspars in the rocks above the Norite Zone. The absence of zoned rims to the cumulus felspars should, according to the previous discussion (Ch. IV) be indicative of extreme adcumulus growth, and the recorded overall change from An_{86} to An_{63} taken as indicative of cryptic layering of the cumulus felspars. Furthermore, little zoning has been detected in the intercumulus felspars of melanocratic layers, so that heteradcumulate growth, also, is

---

* Jackson (personal communication, June 1963) has obtained a value of $An_{62.5}$ from the average of two new, unpublished analyses of chilled Border Zone rocks.

a common phenomenon. However, the change in cumulus felspar compositions from bottom to top of the layered series is by no means regular (see Hess, 1960, pl. 9 & 10). Numerous careful measurements by Hess and Smith (Hess, 1960; Appendix by J. R. Smith) leave little doubt that the fluctuations in the compositions of virtually unzoned crystals are real, and are also exhibited by the pyroxenes. To explain this phenomenon, Hess correlates the fluctuations in *whole-crystal composition* with varied rates of crystal deposition (and sometimes, as appropriate, with whether he is dealing with either cumulus or intercumulus growth). Because the whole crystals, rather than only the rims, appear to fluctuate in composition, he proposes that whereas during slow accumulation the diffusion mechanism operated to produce unzoned felspars with the same composition as the original cumulus crystals (as discussed in Ch. IV), during rapid accumulation there was trapping of contemporary liquid which *reacted* with the cumulus crystals to convert them into an unzoned, lower temperature assemblage.

If the reaction hypothesis is, in fact, tenable (see p. 314) then some fluctuation in felspar composition would be due to differences in the amount of trapped liquid. If there is no trapped liquid, i.e. the rock is an adcumulate or heteradcumulate, then the felspar would have the maximum possible anorthite content and, with increasing amounts of trapped liquid, a progressively lower anorthite content.

The intercumulus felspar of the Ultramafic Zone is more common, according to Jackson's observations (1961, table 1) than Hess's descriptions of rock types from this zone (1960, pp. 56–62) would lead one to suspect. It is supposed to vary in composition from An_{65} to An_{85} (Jackson) or An_{60} to An_{80} (Hess, p. 60), but in view of the data shown on Fig. 171, An_{86} would be the minimum cumulus An value likely in this zone. There is no detailed information on the distribution of the somewhat anomalous range of values given by Jackson and Hess, but if it is accepted, from the evidence of the chill, that the intercumulus liquid has a normative composition of An_{63},[*] then only the compositions much more calcic than An_{63} could be an early precipitate, the rest being parts of zoned crystals. The supposed range, An_{85-60}, may well be due to the amount of zoning and, therefore, to the previous presence of trapped liquid: indeed, Jackson (1961, fig. 76 & p. 71) records the presence of strongly zoned felspar in some of these rocks. The felspars of the coarse rocks from the Border Zone are An_{77-70}, and it is suggested here that some supercooling would produce felspars more calcic than those of the more quickly-cooled chill zone, but less calcic than those in equilibrium with the liquid in the inner, layered part of the intrusion.

b) *Orthopyroxene*

This mineral is found as a cumulus phase in every zone of the layered series, and its predominance over olivine provides the most obvious distinction between the exposed rocks of the Stillwater (and Bushveld) intrusions on the one hand, and the Skaergaard and Rhum intrusions on the other. It is generally a bronzite, either of cumulus origin

[*] See footnote on p. 306.

('peri-euhedral' crystals, as defined by Hess), intercumulus primary origin, or inter-cumulus secondary origin (through replacement of olivine). Towards the top of the layered series, generally in the upper 800 ft, a hypersthene formed through sub-solidus inversion of pigeonite is found, the only other occurrence of inverted pigeonite being in the chilled rock of the Border Zone.

The composition ranges from about $Mg_{88}Fe_{12}$ in the Ultramafic Zone to about $Mg_{64}Fe_{36}$ in the Upper Gabbro Zone. The values are quoted as calcium-free in order to facilitate comparison with optically derived values although, in fact, the more magnesian variety contains about 3% Ca, while the most ferriferous variety is an analyzed inverted pigeonite with the composition $Ca_{10}Mg_{58}Fe_{32}$ (Fig. 171). Three bronzites within the range Mg_{83} to Mg_{73} have been analyzed, together with two inverted pigeonites (see Table 25). The chemical analysis of the inverted pigeonite from a chilled Border Zone rock (which, however, may be contaminated) gives the value $Ca_{10}Mg_{50}Fe_{40}$, while optical determination of a similar mineral from the rock accepted as an uncontaminated chill (Table 24) indicates a composition close to $Mg_{58}Fe_{42}$ (i.e. $Ca_{10}Mg_{52}Fe_{38}$, by inference). The relatively ferriferous composition of these two specimens suggests the effect of chilling, being comparable with the Skaergaard case (Fig. 23).

The compositional range of the orthopyroxenes indicates appreciable iron enrichment with fractionation, and is broadly comparable with the pattern of cryptic variation shown by the felspars. However, accurate optical measurements of numerous specimens throughout the layered series reveal the same complex pattern of fluctuating composi-sitional changes observed for the felspars. Thus Hess (e.g. table 15; pl. 9 & 10; text) records variations, irregular with height, from Mg_{88} to Mg_{77} (Ultramafic Zone), Mg_{83} to Mg_{72} (Norite Zone), Mg_{81} to Mg_{73} (Lower Gabbro Zone), Mg_{79} to Mg_{60} (Anorthosite Zone), and Mg_{75} to Mg_{64} (Upper Gabbro Zone). In general, there is a trend within each zone towards a more iron-rich member, but it is not a simple trend with a reversal at the base of each zone; rather, the reversals are gradual and may occur within the zones (e.g. Hess, 1960, pl. 10). In view of the absence of more specific information on the dis-tribution of the compositional fluctuations in relation to modes and textures, we shall not attempt to comment further on their particular significance. However, Hess has made it clear that certain extreme types (e.g. Mg_{72} in the Dead Tree Section of the Norite and Lower Gabbro Zones) are the result of a considerable amount of trapped liquid, the cumulus phases in adjacent rocks being more magnesian (Mg_{78}). In other cases, as for the felspars in the Anorthosite Zone, he relates violent fluctuations in the compositions of the cumulus orthopyroxenes to major convective overturns of the magma. The rest of the complexity is attributed to the varying degrees to which the cumulus phases have been subjected to reaction with, and 'making over' by, the pore liquid, i.e. to differing rates of crystal accumulation and, a further factor, to the amount of a particular cumulus mineral available, through crystal sorting, for reaction with a given amount of trapped pore liquid. When one considers these complications, to which may be added the development of a second generation of orthopyroxene through reaction of olivine with the liquid, it is not surprising that the orthopyroxenes, in which the relative ease of Mg–Fe substitution would tend to favour the development of homogeneous crystals, show rather erratic fluctuations in composition. In the Skaergaard intrusion, the distinction between the cumulus and adcumulus material on the one hand, and the pore material on the other, can be drawn with fair confidence; compositional fluctuations, due to the varying amounts of pore material, are reflected only in the total rock analyses. In the Stillwater, however, even the analyses of separated, apparently unmodified cumulus phases may need to be viewed in relation to subsequent modification according to the conditions of deposition of each layer.

The textural relations between the orthopyroxenes and the olivines have been studied in detail by Jackson (1961) and will be discussed in a later section (p. 324). Both Hess (1960) and Jackson state that according to its distribution, olivine was the first mineral to crystallize in the Ultramafic Zone. Hess, however, implies that olivine deposition was followed simply by orthopyroxene, whereas Jackson, although he divides the zone into a lower Peridotite and upper Bronzitite Member, finds settled bronzite crystals throughout the Peridotite Member. However, he recognizes 15 units within that member, each unit beginning with a poikilitic harzburgite, in which olivine is cumulus and bronzite intercumulus, passing through olivine–bronzite cumulates, and ending with a bronzite adcumulate, so that viewing *each unit* as a crystallization sequence it is true that bronzite follows olivine. In more general terms, it is clear that even in the lowest part of the Ultramafic Zone bronzite occurs as a cumulus phase, either together with olivine or alone. The coexistence of cumulus magnesian olivine and ortho-pyroxene in the granular harzburgites is a fact of particular interest, and is proof that the reaction relationship in terms of the system $Fo–SiO_2$, although borne out by some reaction textures, is an over-simplification when applied to rocks crystallizing at depth.

The details of orthopyroxene mineralogy have been given primarily by Hess in his important papers on pyroxenes (Hess & Phillips, 1940; Hess, 1941, 1952; Poldervaart & Hess, 1951), and in his Memoir (1960, pp. 23–34). In regard to their nomenclature, however, the classification of their relationships to pigeonite, by Hess and others, makes it desirable to dispense with the terms ' Bushveld ' and ' Stillwater ' types of ortho-pyroxene. Hess (1960) defines the Bushveld type as orthopyroxene with a low Ca content (c. 3%) and fine-scale exsolution lamellae of augite parallel to (100), and the Stillwater type as orthopyroxene with a higher Ca content (c. 10%) and, mostly, broad exsolution lamellae of augite following the relict (001) plane of an original monoclinic pigeonite. It is now widely accepted (e.g. Hess, 1941; Poldervaart & Hess, 1951; Brown, 1957) that the former is probably an orthopyroxene directly crystallized from the magma and that the latter is formed through pigeonite crystallizing directly, exsolving augite whilst monoclinic and, later, inverting to orthopyroxene at sub-solidus temperatures. For most mafic magmas, the changeover takes place at about $Mg_{70}Fe_{30}$, and this change of phase is found in both the Stillwater and Bushveld intrusions, so that both types of pyroxene are found in each intrusion. More than that the Bushveld contains an appreciable thickness of later differentiates containing the so-called 'Stillwater type' pyroxene (Ch. XIV), as does the Skaergaard, whereas the Stillwater contains hardly any, having been eroded to just above the level at which this pyroxene phase (i.e. inverted pigeonite) first appeared. Hence, we feel there is a strong case for dispensing with these misleading names and referring to the two pyroxenes as either orthopyroxene or inverted pigeonite.

c) *Clinopyroxene*

The base of the Lower Gabbro Zone is defined as that level at which augite first appears as a cumulus mineral, and thereafter it is found throughout the rest of the layered series. It occurs only as an intercumulus phase in the Ultramafic and Norite Zones, although a local exception is recorded through its appearance in 'peri-euhedral' form in a rock 1750 ft below the top of the Norite Zone (Hess, 1960, p. 73).

The composition of the intercumulus augite in the Ultramafic Zone is nearly constant at $Ca_{40}Mg_{52}Fe_8$, being an emerald-green, chrome-diopside containing about 1% Cr_2O_3. It is possible that cumulus augites, had they been present in this zone, might have been more magnesian than this; on the other hand, the intercumulus augites may be of heteradcumulate origin in which case, as in Rhum (Ch. XI), they would have the same composition as potential cumulus augites. In subsequent zones (Table 25), the calcium proportion remains constant at Ca_{40-42} while the replacement of Mg by Fe^{++} shows a progressive increase towards the top of the layered series, the most ferriferous member analyzed having the composition $Ca_{40}Mg_{42}Fe_{18}$. These values only approximate to those quoted by Hess (1960, e.g. p. 60, table 14, and fig. 25) because, as with certain other recorded data on the minerals (e.g. felspars, table 14 and p. 60, and orthopyroxenes, table 14 and p. 27), there are conflicting values cited in different sections of the Memoir. In the case of the clinopyroxenes, the discrepancy between the calcium contents of analyzed and optically measured specimens may be due, in part, to the exsolution of a calcium-poor phase, the host then having a more calcic composition than the whole crystal. The chilled Border Zone augite has the composition $Ca_{41}Mg_{41}Fe_{18}$, and that of the coarse mafic norite of the border facies, $Ca_{41}Mg_{46}Fe_{13}$, both suggestive of super-cooling. It is interesting that the pyroxenes and felspars of the chill zone are very close in composition to those of the uppermost rocks of the exposed layered series, indicating that a great proportion of the intrusion is unaccounted for in terms of a crystallized, closed system.

Despite the overall iron enrichment from Fe_8 to Fe_{18}, the clinopyroxene compositions show irregular fluctuations with height in the sequence, in the same way as the felspars and orthopyroxenes. In fact, fluctuations between Fe_{10} and Fe_{18} are found both within the Dead Tree Section of the Norite and Lower Gabbro Zones, and within the Anorthosite and Upper Gabbro Zones (Hess, pl. 9 & 10).

The clinopyroxene exsolution textures have been described in detail by Hess (e.g.

1960, pp. 35–9), and consist of lamellae of orthopyroxene parallel to (100) in the more magnesian range and of pigeonite parallel to (001) in the more ferriferous range. The trend of crystallization followed by the clinopyroxenes, the character of the coexisting, calcium-poor pyroxenes, and the sub-solidus exsolution textures in all three pyroxene phases, are characteristic of pyroxenes crystallizing from tholeiitic-type basaltic magmas and provide the most obvious link between the mineralogy of the Stillwater and of the Skaergaard or Bushveld intrusions. However, in view of non-exposure of the later differentiates, the Stillwater trends are very limited in extent, especially for the Ca-rich pyroxenes.

d) *Olivine*

Olivine is restricted, in occurrence, to the Peridotite Member of the Ultramafic Zone (where it is the chief mineral) and to the Anorthosite Zone. It may occur as a cumulus phase at both of these widely separated horizons, but is not present either as a cumulus or an intercumulus phase in the other zones. Its absence, both modally and normatively, from the chilled Border Zone rocks provides further evidence that the Stillwater magma was not undersaturated, in contrast to the Skaergaard chilled olivine gabbro (Table 7).

Olivine in the Stillwater intrusion has received but scant attention, and no chemical analyses have been published. From optical measurements, Hess (1960, table 15) records a range in composition, irregular with height, of Fo_{88-80} for the Ultramafic Zone. A more detailed study by Jackson (1960), using the X-ray powder method, placed the range at Fo_{95-80},[*] although the various compositions are not related to height within the zone. Presumably, the olivines show irregular fluctuations in composition with height, perhaps through reaction with the liquid, in the same way as do the other silicate minerals. Within the Anorthosite zone, olivine returns as a cumulus phase and may constitute up to 25% of a few of the rocks (Hess, 1960, pl. 10), but the crystals are usually too much altered for measurement and Hess (1960, table 23) quotes only a tentative composition, i.e. Fo_{80-75}. At one horizon at the base of the Anorthosite Zone, termed the ' Pillow troctolite ' (Hess, 1960, p. 83), there are rocks with a special texture, containing about 30% olivine of composition Fo_{77}. The texture (Hess, 1960, pl. 5, fig. 3) is suggestive of a cumulus origin for the olivine, being reminiscent of the composite clusters from Rhum (Fig. 141).

From this information, it is apparent that the olivines of the Stillwater intrusion have not only a limited range of occurrence but a narrow range of composition, although the broad limits quoted, Fo_{90-75}, indicate iron enrichment with fractionation. The olivines in the lower layered rocks of the Bushveld intrusion (Ch. XIV) show a broadly similar type of distribution, but in the higher Bushveld rocks there is a reappearance of more fayalitic olivine, suggesting that the Upper Hidden Zone rocks of the Stillwater

---

[*] Jackson (personal communication, June 1963) has obtained a range from $Fo_{89.7}$ to $Fo_{80.5}$ by chemical analysis of these olivines.

may well show a return to olivines more ferriferous than those in the exposed part of the layered series.

e) *Chrome-spinel*

According to Jackson (1961, 1963), chromite-rich layers occur in thirteen of the rhythmic units of the Peridotite Member of the Ultramafic Zone, always being in the lower, predominantly olivine-cumulate part of each unit. Most of the thinner, 1–2-in layers can be traced laterally for over 10 miles, while the thickest layer (G), varying from $1\frac{1}{2}$ to 12 ft in thickness, can be traced for nearly 30 miles. Jackson (1963) has analyzed 95 chromite samples in order to determine both vertical and lateral compositional variation. The full analytical data are not published, attention being focussed on variations in total iron $(Fe^{++} + Fe^{+++})$ and oxidation ratios $(Fe^{+++}/Fe^{+++} + Fe^{++})$. However, as Al^{+++}, $(Mg^{++} + Fe^{++})$, and $(Cr^{+++} + Al^{+++} + Fe^{+++})$ are each nearly constant, variations in Fe^{++} or Fe^{+++} affect the amounts of Mg^{++} and Cr^{+++}, respectively.

The results of Jackson's study show two significant variables. Firstly, in the vertical sequence of layers the chromites show no change in oxidation state but a variation in total iron such that the amount decreases, and then increases, upwards. Secondly, lateral variation for a particular layer shows little change in total iron, but a distinct increase in oxidation state towards the margins of the intrusion. Both these features are difficult to explain in terms of what is known otherwise about the intrusion's crystallization history, but a similar pattern is found in the Bushveld and Great Dyke intrusions (Jackson, 1963). The lateral variation in oxidation state is attributed by Jackson to a possible increase in oxidation towards the margins at the time of chromite crystallization, and not to any secondary effects. The initial, upward decrease in total iron is somewhat surprising, as this also imples an increase in Cr and/or Mg, and this clearly requires further detailed study in relation to the textures and ferromagnesian silicate compositions at each chromite horizon.

f) *Other minerals*

Quartz is found as an intercumulus mineral in certain rocks from every zone—a feature observed in the Bushveld and, to a less marked degree, in the Skaergaard intrusion. The amount varies from 0–5% in rocks of the Border Zone; in the Ultramafic Zone it is never found in the olivine-bearing rocks, but only in the bronzitites (Jackson, 1961, pp. 5–8). Although information is not available as to its precise distribution within the other zones, the presence of quartz and scarcity of olivine above the Ultramafic Zone suggests that, as would be expected, the quartz and forsteritic olivine show antipathetic relations within the intrusion. The norm of the uncontaminated chilled rock contains no quartz, so it is to be assumed that its presence in certain layered rocks is due to late crystallization of trapped liquid, and that this liquid reacts with magnesian olivine, where present, to produce orthopyroxene, and precipitates quartz only in the absence of cumulus olivine.

Other minerals to be expected as products of a relatively late stage of fractionation, by analogy with the Skaergaard or Bushveld intrusions, are iron–titanium oxides and apatite. With trapping of pore liquid they would be expected as interstitial grains along with quartz, but in fact they are present only in negligible amounts and have not been correlated, by Hess or Jackson, with the presence of quartz. Biotite, on the other hand, is associated with quartz and is believed to have crystallized from trapped pore liquid (Jackson, 1961, p. 71). Hess (1960) believes that the exposed layered rocks represent only 60% of the total solidified, and that at this stage the conditions of phosphorus saturation and, oxidation, necessary for the precipitation of apatite and iron oxides, respectively, had not been reached. In regard to the trapped pore liquids, however, which should crystallize a wide range of minerals on slow cooling, it may be suggested that at these early stages, contemporary liquids were still so basic that only a minute amount of apatite and iron ore would form in the small amount of pore material, and may be overlooked in the presence of sulphides and secondary alteration products. The most unusual mineral to be found, as a pore-material mineral, is garnet. According to Jackson (1961, pp. 5 & 71, fig. 78–9) minute amounts of ' grossularite–pyrope ' garnet occur together with quartz and biotite in the bronzitites of the Ultramafic Zone (see Fig. 179) and appear, in part, to be replacing bronzite.

Sulphides (chalcopyrite, pyrrhotite, and pyrite) occur in traces throughout the intrusion, and are occasionally concentrated in thin, rusty-weathering layers. The content of sulphides in the rocks is usually less than one per cent, and according to Hess their habit suggests crystallization from immiscible sulphide droplets. Appreciable concentrations of sulphides also occur near the floor of the complex, which would suggest the early separation of a sulphide-rich liquid. This needs to be reconciled with Hess's observation (1960, p. 45) that the sulphides higher in the layered series seem to have formed from trapped residual pore liquid, it being unlikely that an observable amount of sulphides would crystallize from the pore liquids, which represent only a small part of the contemporary magmas. From the two contrasted occurrences, it would appear that the initial Stillwater magma was relatively rich in sulphur, so that an immiscible sulphide liquid formed early in its cooling history, and that the remaining sulphur was precipitated late, as settled droplets of immiscible sulphide liquid when, perhaps, certain chalcophile elements reached a saturation level. A study of the composition of the sulphide phases and the sulphur content of the chilled Border Zone rock would help to elucidate this problem.

THE TEXTURES OF THE LAYERED ROCKS

a) *Nomenclature*

The textural features of the layered rocks are similar in most respects to those already described for the Skaergaard and Rhum rocks, and both Hess (1960) and Jackson (1961) have considered them to be the result of crystal accumulation of primary precipitate minerals followed by crystallization of the interprecipitate liquid. These authors have used the terms primary precipitate or settled crystals, and interprecipitate or interstitial growth, to describe what are broadly two generations of crystalline material, but the meanings attached to the terms are not always consistent, or clear to us (see below). They have also retained the conventional petrographic names to describe the various rock-types. Thus the main types are dunites, harzburgites, chromitites, anorthosites, pyroxenites, norites, troctolites and gabbros, together with rocks termed olivine chromitite, felspathic harzburgite, etc., and such names cover, broadly, the range produced by sorting of the primary precipitate crystals. The interprecipitate rôle of a mineral is not shown by this nomenclature, although Jackson's division of the harzburgites into granular and poikilitic varieties illustrates the need for extra definition, because the bronzite is primary precipitate in the first, and interprecipitate in the second variety.

The rocks are cumulates according to our nomenclature (Ch. IV), and can generally be described in relation to the cumulus and intercumulus phases. In addition, the ways in which the rocks completed crystallization can generally be considered as processes leading towards the formation of orthocumulates, mesocumulates, adcumulates, or heteradcumulates. The complication in the case of the Stillwater rocks, mentioned briefly already, is due to the reaction which appears to have taken place between cumulus crystals and intercumulus liquid. If the intercumulus liquid has reacted with the felspar, pyroxene, and olivine to make-over each of them to lower-temperature solid solutions, on the olivine to produce new orthopyroxene and clinopyroxene, and on the orthopyroxene to produce new clinopyroxene (Hess, 1960; Jackson, 1961), then those discrete crystals which, according to habit and freedom from zoning would normally be classed as entirely cumulus in origin may, in fact, be modified cumulus crystals. A pure felspar rock consisting of unzoned, relatively calcic plagioclase crystals is generally to be viewed as an adcumulate, but if there has been reaction between initially more calcic cumulus plagioclases and intercumulus liquid, the resultant plagioclase will be relatively sodic and, therefore, not of adcumulus origin. The fluctuations in the compositions of the unzoned plagioclases in the anorthosite layers of the Stillwater intrusion are attributed, by Hess (1960), largely to fluctuations in the amount of such reaction. In the absence of other minerals as late-stage pore material in these anorthosites, however, none of them can be viewed as orthocumulates or mesocumulates. In fact, the suggestion that such rocks are in part due to a reaction process requires more detailed consideration of the possible mechanism involved, to make it convincing. We find it impossible to reconcile the adcumulus growth of the felspars, to produce a pure felspar rock, with reaction to produce a more sodic felspar. The latter would have required an appreciable lowering in temperature and change in composition of the liquid involved in the reaction, and yet if this is postulated to have taken place only by the agency of trapped liquid, several minerals other than felspar should ultimately have crystallized.

Similarly, a ' poikilitic harzburgite ' consisting of subhedral olivines enclosed by anhedral, unzoned bronzites would be classed as a heteradcumulate according to our nomenclature. However, Jackson (1961) has proposed that in many such cases some bronzite has grown at the expense of partially resorbed olivine, this implying that the initial amount of cumulus olivine was greater than that to be inferred from the textures. Again, however, such reaction requires cooling of trapped pore liquid, and yet the rocks may contain only olivine and bronzite, and be free from the other minerals which ought to have crystallized from trapped and cooled pore liquid. Although Jackson may have envisaged reaction followed by adcumulus or heteradcumulus growth, such a process is hard to envisage. Appreciable cooling of trapped liquid would occur only after the deposition of much overlying cumulus material, and by this time any opportunity for adcumulus growth, in the buried layer, would have vanished.

Thus we are reluctant to accept the hypothesis of reaction between cumulus crystals and intercumulus liquid, to account for the compositions and textures of some of the Stillwater rocks, until an explanation for the absence of other products of the trapped liquid can be provided. However, several interesting textures described by Jackson (1961) are mentioned here, and Jackson's explanation is given since, at present, no other is available. Apart from these special cases, most of the Stillwater rocks are either adcumulates or heteradcumulates.

b) *Cumulus material*

In the earlier chapters, we have generally described the textural features of the cumulus minerals in relation to the rock textures as a whole, including such properties as shape, size, and orientation. However, Jackson (1961) has made a special, important study of certain textural aspects of the Ultramafic Zone rocks which, amongst work on layered igneous rocks, is new in method and interpretation. Only a brief summary can be given here, as the study concerns only the Ultramafic part of the intrusion; other aspects of his work, on intercumulus textures, will be discussed later in this section.

For chromite, olivine, and bronzite, every gradation in shape exists between euhedral crystals, completely bounded by crystal faces, and anhedral grains: the influencing factors are the amount of adcumulus growth, and reaction with intercumulus liquid, of each cumulus crystal (see Jackson, 1961, fig. 57–62). Grain sizes are remarkably consistent within the layering plane, whilst layers characterized by particular grain sizes are parallel to the compositional layering (e.g. Fig. 172). A detailed statistical study of size distribution has shown that in most of the layered rocks containing only one settled mineral, the sorting is remarkably good. Thus, grain-size distribution curves approach lognormal for these rocks and are very similar to those of well-sorted sandstones. For rocks containing two settled minerals, the sorting appears to be poor, i.e. bimodal, providing that no distinction is made between the mineral species (the common practice with detrital sediments); however, treatment of such species separately shows good sorting for each. The presence in rocks such as granular harzburgites (olivine–bronzite cumulates) or olivine chromitites (olivine–chromite cumulates) of two settled mineral species which are each well sorted, and yet different in size, might lead to the assumption that density differences have been responsible for bringing together two size groups. However, Jackson has shown that this is not so, the particles not being 'hydraulically equivalent' except in certain current-bedded rocks, which approach hydraulic equivalence more closely than do structureless rocks containing the same cumulus minerals.

A petrofabric analysis of either cumulus bronzite or cumulus olivine showed igneous lamination in five out of six measured specimens, of which one showed a weak unimodal lineation and another a weak bimodal lineation of the long axes within the plane of lamination. From this it can be concluded that providing the habit of the cumulus crystal is not isotropic, igneous lamination occurs in the lower Stillwater layered rocks, but that lineation within the depositional plane is rare.

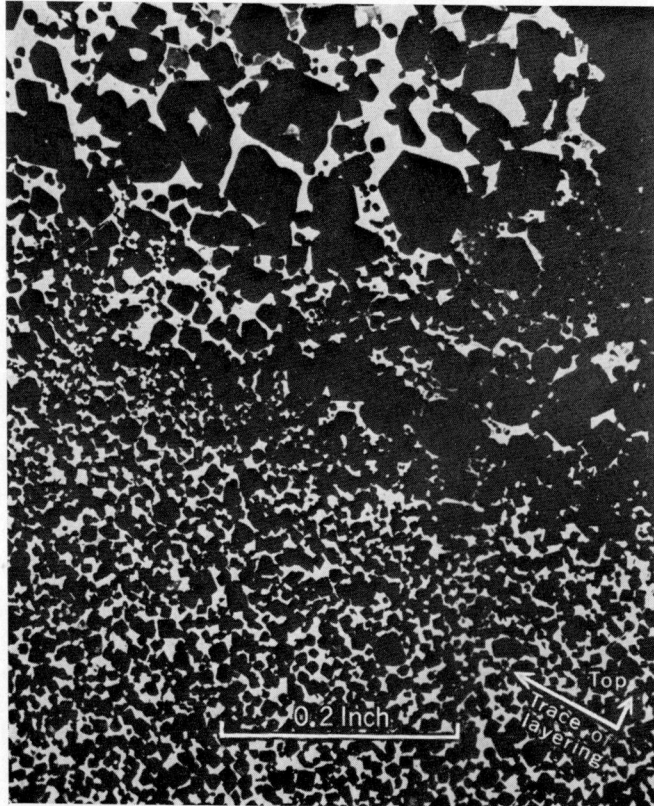

Fig. 172. Photomicrograph of a chromite heteradcumulate layer (with intercumulus plagioclase), Ultramafic Zone, showing the abrupt change to a larger grain-size towards the top of the layer. (From Jackson, 1961.)

FIG. 173. Photomicrograph of an olivine(–chromite) adcumulate, Ultramafic Zone (cross-nicols). A few of the olivines show translation lamellae. (From Jackson, 1961.)

FIG. 174. Photomicrograph of a bronzite adcumulate, Ultramafic Zone (cross-nicols). Fine-scale exsolution lamellae of augite, ∥ (100), can be seen in suitably orientated crystals. (From Jackson, 1961.)

c) *Intercumulus material*

The intercumulus material of the Stillwater layered rocks can be divided according to the relative amounts of adcumulus and pore material. Thus there is evidence for homogeneous extension of settled crystals by adcumulus growth (termed 'secondary enlargement' by Jackson, 1961), for homogeneous extension of poikilitic crystals by heteradcumulate growth, or for the presence of pore material in the form of outer zoned fringes and new mineral phases such as quartz. In addition, however, there may have been some reaction between trapped pore liquid and certain cumulus crystals to produce a different mineral phase or, perhaps, to modify the composition of certain cumulus crystals (i.e. discontinuous or continuous reaction of Bowen, 1928).

Estimates of the original porosity of any particular rock are difficult to make if reaction has occurred, and impossible to make if adcumulus growth has taken place. Jackson (1961) has made a systematic attempt at such estimates by avoiding areas in a rock where oikocrysts,* presumably formed by discontinuous reaction, are present, and

---

* A term used by Jackson (1961) for poikilitic crystals enclosing the relict crystals of other minerals, from which they are derived by reaction between the latter and the magma.

FIG. 175. Poikilitic bronzite crystals, characterized by light reflection from the cleavage surface of each large crystal, each enclosing numerous small, cumulus olivine crystals. Olivine heteradcumulate (i.e. poikilitic harzburgite), Ultramafic Zone. Near Chrome Mt., East Boulder Plateau.

by assuming that the preservation of an euhedral outline is indicative of the absence of adcumulus growth. Presumably, continuous reaction would not affect estimates of porosity if the euhedral outlines of the cumulus crystals are still preserved, whereas if they are destroyed then it is immaterial whether it is adcumulus or reaction growth that is responsible for the subhedral to anhedral overgrowths, for in either case it would be impracticable to make porosity estimates. Although, therefore, the Stillwater intrusion contains a greater variety of rocks in which porosity estimates are impracticable than do the Skaergaard or Rhum intrusions, judicious selection helps to give a general estimate. For the Ultramafic Zone, Jackson (1961) estimates that the porosity (i.e. percentage of intercumulus liquid) varied between 20 and 50%, and averaged 35%. A decrease in the porosity, he suggests, would be produced by three factors: an increase in the grain size of the cumulus crystals, an increasing tendency towards a tabular habit, and a poor degree of size-sorting. The estimates are in general accordance with our values for the Skaergaard intrusion, except that we believe the higher value (c. 50%) to be closer to the likely figure. Hess (1960, p. 109) examined a rock with cumulus bronzite (71%) and intercumulus plagioclase and augite (29%) and concluded, initially, that the porosity was close to 30%. In a footnote, he accepted our suggestion that intercumulus bronzite added onto the cumulus crystals should not be neglected, and that the initial porosity might have been closer to 50%.

Adcumulates are probably fairly common in the Stillwater intrusion, for rocks closely approaching a monomineralic character are frequently found, such as dunites, bronzitites, and chromitites in the Ultramafic Zone (Fig. 173 & 174) and anorthosites in the higher zones. Also, there are certain granular harzburgites, norites, and other polymineralic cumulates which, apparently, are free from zones to the cumulus crystals and from other pore material. Certain poikilitic harzburgites (Fig. 175), peridotites (Fig. 180 & 181), and gabbros (Fig. 183) are clearly of heteradcumulate origin. Both Hess and Jackson relate this to a slow rate of accumulation, and Jackson's observation (1961, p. 62) that some adcumulus growth probably took place at, or near the surface of the crystal mush is in agreement with our hypothesis (Ch. IV). Jackson (1961, pp. 60 & 62) has been able to detect a significant lateral variation in the rate of crystal accumulation, within a single horizon, by observing the character of the intercumulus material. That is, the Peridotite Member shows a lateral variation in thickness, and where it is thickest the amount of pore material is greatest, reflecting a higher rate of deposition in a local, basin.

Orthocumulates are generally recognized by zoning of both cumulus crystals and poikilitic crystals, and by the presence of low-temperature phases which are absent as cumulus minerals at that horizon, while mesocumulates differ from these only in the evidence provided for some adcumulus growth prior to the crystallization of pore material. In the Stillwater layered series, such cumulates appear to be rare if the zoning criterion is applied. This is not so if account is taken of the relative abundance of interstitial quartz, or of evidence which is taken to suggest that the cumulus crystals

320

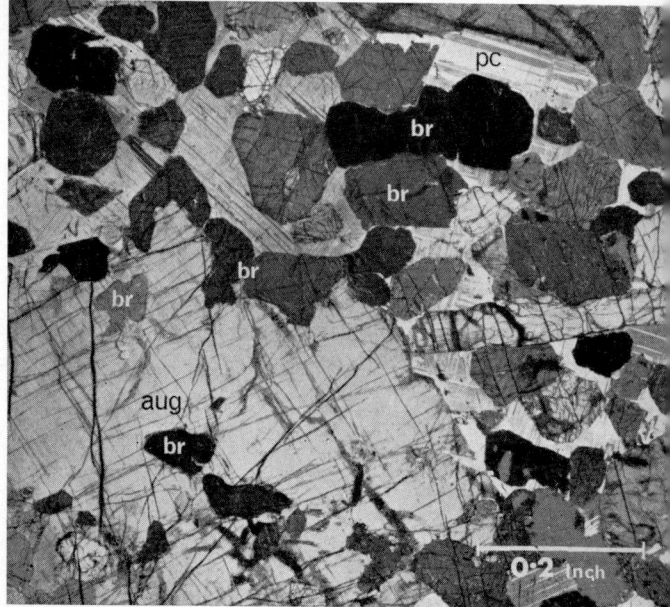

Fig. 176. Photomicrograph showing crystals of cumulus bronzite (br) which are euhedral where poikilitically enclosed by plagioclase (pc) and rounded, embayed, and isolated where poikilitically enclosed by augite (aug). Ultramafic Zone. Cross-nicols. (After Jackson, 1961.)

Fig. 177. Photomicrograph showing cumulus bronzite crystals (br) with embayments where enclosed by augite (aug), the same crystals showing original crystal faces where in contact with plagioclase (pc). Ultramafic Zone. Cross-nicols. (After Jackson, 1961.)

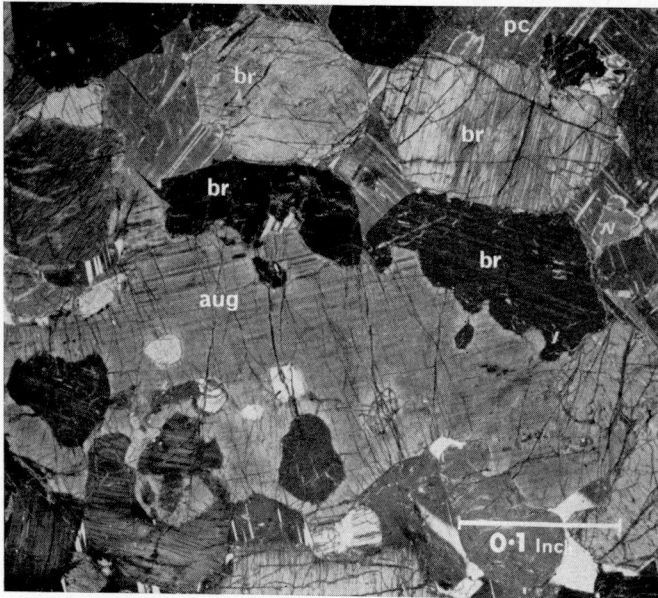

have been affected by reaction with trapped pore liquid. There seems to be little likelihood, however, that extreme orthocumulates are present, those rocks providing evidence for the previous presence of some trapped liquid being probably mesocumulates (Fig. 179). Within the Ultramafic Zone, such mesocumulates have been studied in detail by Jackson (1961). A useful indication of their likely abundance is to be found in his table 5 and fig. 69, where an estimate of the average initial porosity of each important rock type is given, and then the interprecipitate material is sub-divided into the average

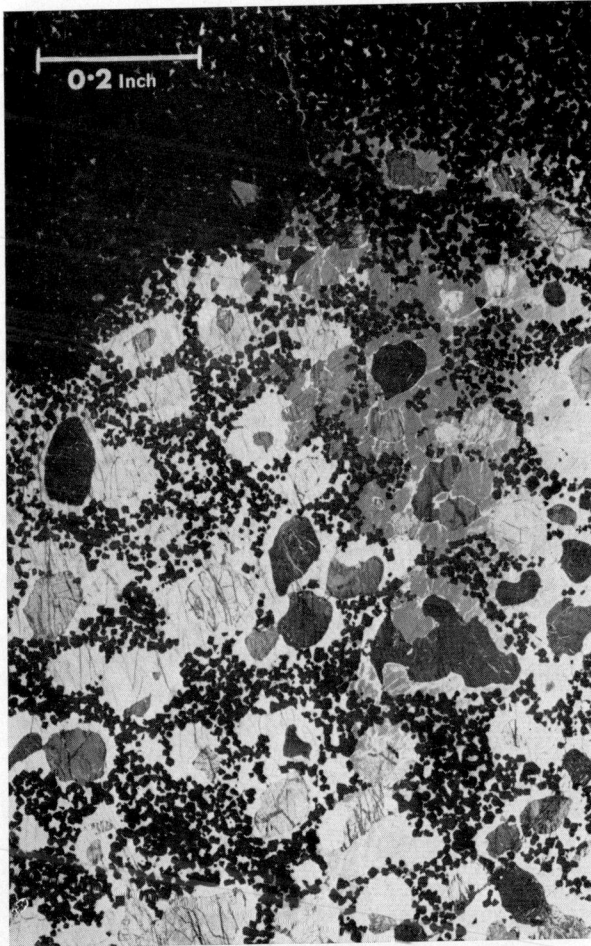

FIG. 178. Photomicrograph of a layered olivine–chromite cumulate, Ultramafic Zone. The way in which the chromite crystals are distributed serves to illustrate the probable size and shape of the original cumulus olivine crystals. Reaction between the latter and the intercumulus liquid has since produced irregular-shaped relics of the original olivines, of greatly varying sizes, surrounded or completely replaced by bronzite. Cross-nicols. (After Jackson, 1961.)

X

amounts of ' secondary enlargement ' and ' interstitial ' material. In general, he believes that the average porosity was about 35%, and that 20% of this space is now filled by secondary enlargement and 15% by interstitial material (Fig. 180 & 181). This evidence in itself would indicate that all the rocks studied in this way are mesocumulates, because 35% of secondary enlargement would be necessary for the rock to be an adcumulate and 35% of interstitial material for it to be an orthocumulate. However, Jackson has omitted from this table the almost monomineralic rocks described elsewhere in his account (e.g. fig. 59 & 62; see our Fig. 173 & 174), perhaps because he believes the true adcumulates to be of subordinate occurrence.

The ' interstitial material ' of the mesocumulates would be equivalent to what we have called ' pore material ' were it not for the presence, amongst it, of material believed by Jackson to have been produced by the replacement of some of the cumulus crystals by reaction with the pore liquid. Such reaction is of three types; olivine→bronzite,

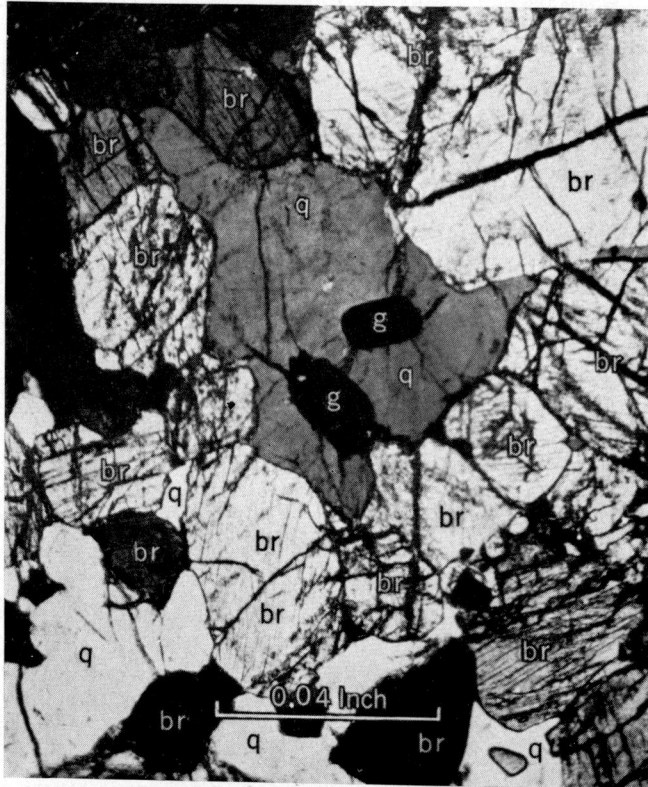

FIG. 179. Photomicrograph of a bronzite (br) mesocumulate, Ultramafic Zone. Both quartz (q) and garnet (g) are believed to be part of the crystallized pore material, although the garnet is an unusual product for a reasonably high-level layered intrusion. Cross-nicols. (From Jackson, 1961.)

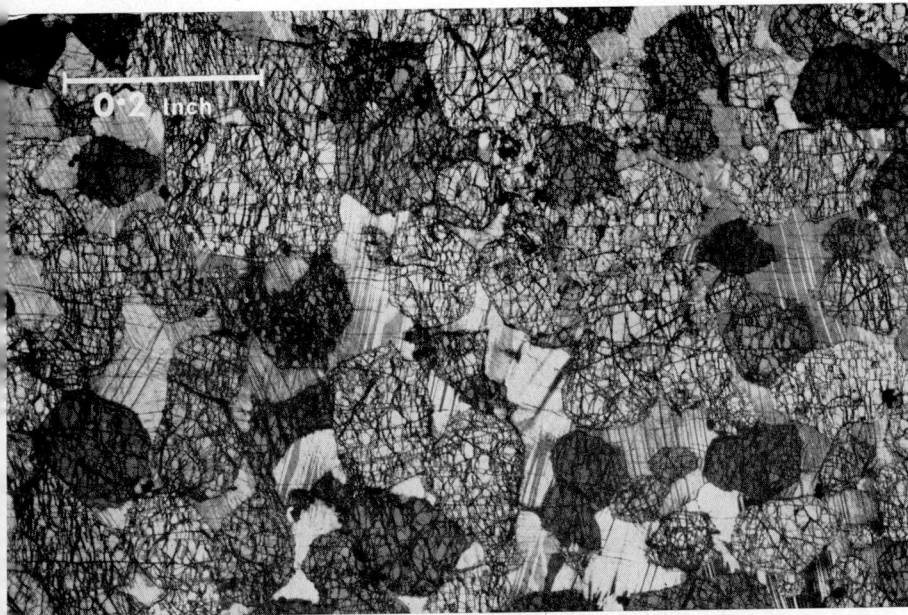

FIG. 180. Photomicrograph of an olivine heteradcumulate, Ultramafic Zone. Olivine is the only cumulus mineral, and the poikilitic plagioclase is virtually unzoned (cf., Fig. 149). Cross-nicols. (From Jackson, 1961.)

FIG. 181. Photomicrograph of an olivine heteradcumulate, Ultramafic Zone. Similar to that shown in Fig. 180 except that here the cumulus olivines have been enlarged, by adcumulus growth, to a greater extent. Cross-nicols. (From Jackson, 1961.)

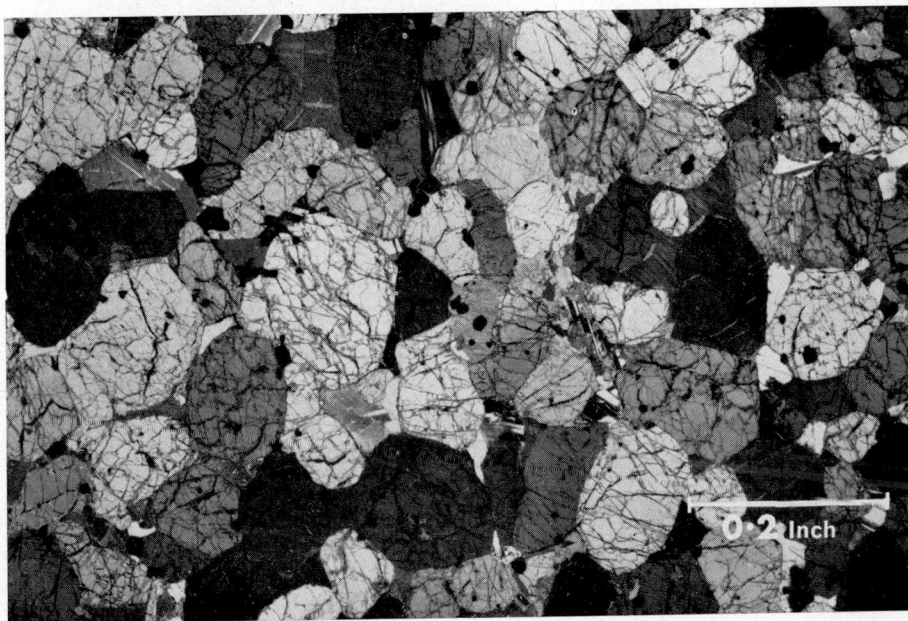

olivine→chromian augite, and bronzite→chromian augite. Evidence in support of these reactions is provided by Jackson (1961, fig. 50–55), the criteria for textural recognition being: (1) the rounding and embayment of crystals totally enclosed by the reaction product ('oikocryst') and otherwise euhedral in the rock (Fig. 176); (2) the embayment only of that edge of a crystal which is partly enclosed by the reaction product (Fig. 177); and (3) a reduction in the volume of the settled crystals where enclosed by the reaction product, despite the number of centres being the same (Fig. 178). From measurements made on the relative volumes of crystals inside and outside these oikocrysts, Jackson estimates that on the average, where surrounded, 35% of olivine is replaced by bronzite, 25% of olivine by augite, and not more than 40% of bronzite by augite.

Apart from this special type of material, the more conventional type of pore material

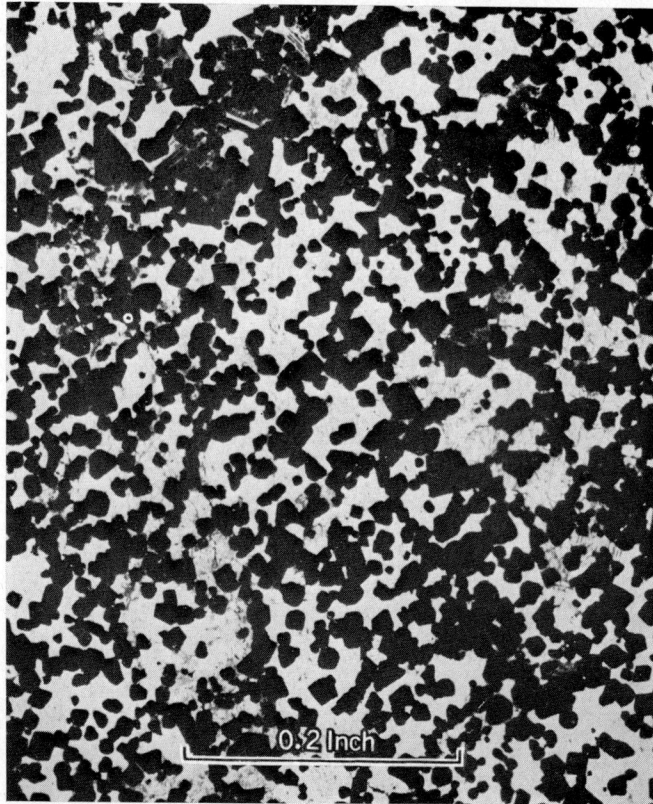

FIG. 182. Photomicrograph of a chromite heteradcumulate, Ultramafic Zone, showing the 'chain structure' characteristic of most chromite cumulates and suggesting mutual attraction between adjacent, settling crystals. Plagioclase, bronzite and augite, which cannot be distinguished in this photograph, together constitute the intercumulus material. Cross-nicols. (From Jackson, 1961.)

is common, consisting of minerals absent as a cumulus phase and yet normatively present in the pore liquid.   They may either be lower-temperature equivalents of the minerals otherwise present as cumulus minerals in adjacent layers, minerals locally absent as a cumulus phase through cryptic layering, or (rare in the Stillwater) zoned outer fringes to the cumulus minerals.   Neither Hess nor Jackson has attempted to give this material a name, it being grouped together with the reaction products, by Jackson, as interstitial material.   In the Ultramafic Zone, plagioclase, quartz, biotite and rare garnet (Fig. 179) are the minerals otherwise absent as cumulus phases, whilst in the higher zones, quartz is the only described pore material in this category. From Jackson's account, zoning would appear to be more frequent in the Ultramafic Zone pore material than, according to Hess, in the higher zones.   The compositional range of the zoning is not given, but extreme zoning in plagioclase and thin, iron-rich outer zones to olivines and bronzites are recorded (Jackson, 1961, fig. 76, and pp. 54 & 74).   Finally, the presence in the pore material of minerals otherwise available as cumulus minerals in adjacent layers is described in detail for the Ultramafic Zone, for which an order of intercumulus crystallization, broadly analogous to that believed to have operated for the cumulus phases, is proposed (Jackson, 1961, fig. 80).

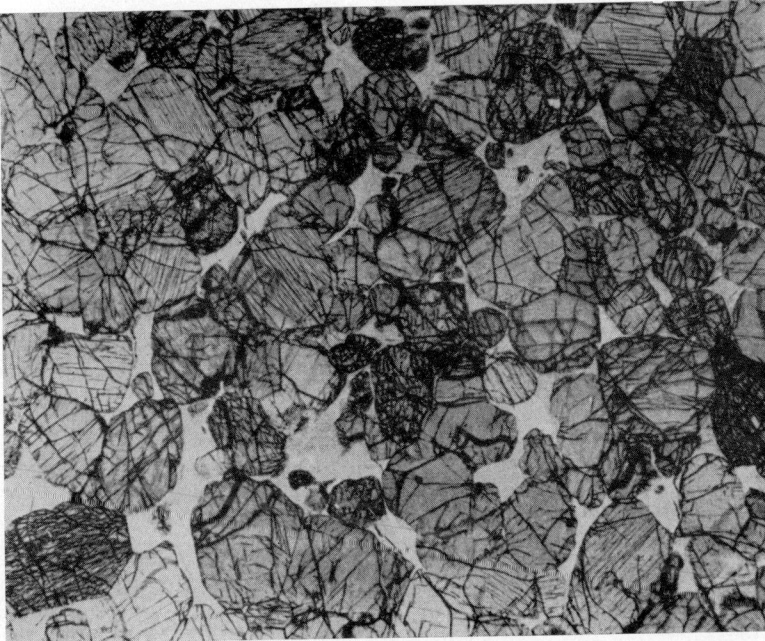

FIG. 183. Photomicrograph of a bronzite–augite heteradcumulate, Lower Gabbro Zone.   Both pyroxenes are of cumulus status, and the intercumulus plagioclase is unzoned.   (From Hess, 1960.) $c. \times 5$.

The Structures of the Layered Rocks

Apart from the thin Border Zone, the Stillwater intrusion consists of a series of layered rocks characterized, in particular, by the property most easily recognized in the field, i.e. rhythmic layering. Other structures characteristic of igneous rocks formed through crystal accumulation are also found, including gravity stratification, igneous lamination, and slump structures.

The rhythmic layering is present on both a coarse and a fine scale (Fig. 184–186 & 188), the rhythmic units ranging from several hundred feet to a few inches in thickness. The problem of deciding what to include in a unit of rhythmic layering is sometimes difficult, but in general is taken to comprise two adjacent layers differing appreciably in their cumulus assemblage together with, when present, a layer transitional between the two extremes. However, the term ' unit ' of rhythmic layering, first used by one of us for Rhum (Brown, 1956), can conveniently be applied either to a thickness of rather uniform layering or, as in the macro-units of Rhum or the gravity-stratified, thinner units of Skaergaard, to a group of such thicknesses. As Hess (1960, p. 125) points out, a thick layer of anorthosite underlain by a pyroxenite may conveniently be taken as one unit, but if a thin pyroxene layer, only one or two crystals thick, is present in the thick anorthosite then this sequence ought strictly to be divided into two units. The same problem occurs in Rhum, but in that case the term ' major rhythmic unit ' or ' macro-rhythmic unit ' was applied to the thick, major rhythms, and it was recognized that several tens of minor rhythms may occur within the fine-scale layered part of each macro-unit. For the Skaergaard, fine-scale rhythmic layering is so common that no attempt was made to count or classify the minor rhythmic units, and the division of the layered series into zones, rather than units, was based on cryptic layering as it has been for the Stillwater layered series, by Hess.

Within two of the Stillwater zones, several thick rhythmic units have been recognized, each of which is similar in thickness to those in Rhum (Ch. X). The Ultramafic Zone has been subdivided by Jackson (1961, fig. 19) into 15 such units, the general character of each (maximum thickness 600 ft) being such that olivine cumulates pass upwards, through a transitional zone of olivine–bronzite cumulates, into bronzite cumulates. In addition, cumulus chromite is confined to the lower part of each unit and is particularly common in the lower to middle part of the olivine cumulates. However, fine-scale layering is common within each major unit, and is particularly well shown by the olivine–chromite cumulates (Jackson, 1961, figs. 9, 10 & 30; see Fig.184). The Anorthosite Zone has been subdivided by Hess (1960, fig. 19) into three thick anorthosite layers (1300–1500 ft each) and two gabbro layers (800 and 1400 ft), but it is probably undesirable to apply to these very thick, problematical layers the term ' rhythmic unit '.

Fine-scale rhythmic layering is best developed in the Norite Zone (e.g. Fig. 186), and Hess (1960, pp. 127–9) has made a detailed study of this occurrence. For example, he found that a 217-ft section contained 75 layers, the average thickness being, therefore,

FIG. 184. Layering in olivine–chromite cumulates, Ultramafic Zone. (From Jackson, 1961)

FIG. 185. Chromite layers (dark) which give the appearance of branching, probably due to deposition of chromite and olivine crystals on an irregular floor. (From Jones *et al.*, 1960.)

about 3 ft. Such sections frequently contain thicker layers of ' uniform rock ', and six such layers were present in the case examined. If they are subtracted from the total thickness, then there are 69 fine-scale layers, of average thickness 16 inches, although we would prefer to see *units* defined where possible, so as to include both the six uniform units and the units consisting of sets of layers. A series of measurements on another two sections, each 250 ft thick, showed that the uniform units constituted about 70–80% of the total thickness and that in one clear case the thickness ratio of a uniform layered unit to the group of thinly layered units above, ranged from 5:1 to 11:1.

The great lateral extent of an individual sheet, compared with its thickness, is an impressive feature of these structures in the Stillwater, and Hess has observed that the length or breadth of the Stillwater sheets are of an order of several hundred to perhaps a thousand times their thicknesses. However, ultimately they pinch-out laterally and some may do this over a short distance (Fig. 185), so that Hess prefers to call them lenses.

Most layers appear to be sharply bounded and regular, and no cross-cutting structures are recorded, apart from the ' transgressive ', structure which is due to the pinching-out of thin layers and is probably formed where deposition took place on the irregular surface of the preceding deposit. Layers with a sharp base and top appear to be the commonest, as in other layered intrusions, but gravity-stratified units, in which a melanocratic rock with a sharp base grades upwards into a leucocratic rock with a sharp top, are developed in about one-third of the Stillwater layered series, according to Hess (1960, p. 128). Some ' reversed gravity stratification ' also is observed, however (Fig. 186), while Jackson's analysis of size, density and habit of the minerals in the Ultramafic Zone (1961) has led him to conclude that the sorting there is not according to hydraulic equivalence and that any apparent graded-bedding cannot, therefore, be attributed solely to gravitational effects. The gravity stratification of the Stillwater intrusion is, however, one of the more convincing cases amongst layered igneous rocks (e.g. Jones et al., 1960, pl. 28) and indeed, the term was first applied to this phenomenon by Buddington (1936) and by Peoples (1936) with reference to the classic example featured from the west side of the Stillwater valley. An example from this area is shown in Fig. 186.

Igneous lamination is conspicuously developed at several horizons in the Stillwater (Hess, 1960, pl. 6; Jones et al., 1960, pl. 26A; Jackson, 1961, fig. 6) and is due chiefly to the local development of a tabular habit for felspar, pyroxene or olivine. Both Hess and Jackson believe that such a structure could be produced through tranquil settling of tabular crystals, and that only linear orientation within the plane of lamination would be evidence for current movement. We believe, however, that some movement of currents along the floor is necessary to produce strong igneous lamination, and if this is so then evidence for such currents is abundantly present in the Stillwater intrusion.

A special type of fine-scale layering, found near the top of the Lower Gabbro Zone, has been termed ' inch-scale layering ' by Hess (1960, pp. 51, 77, 133, and pl. 8) and ' grating structure ' by Jones et al. (1960, pl. 29B); an example is further illustrated in Fig. 187 and 188. The structure consists of alternating layers, uniform in thickness, of pyroxene and felspar, the distance from the centre of one pyroxene layer to the next being about one inch. The wavy contacts between adjacent layers, and the lack of euhedral crystal forms in each layer, prompted Hess to suggest that this type of layering formed by a process different from that producing normal rhythmic layering. He comments only briefly on its possible origin (1960, pp. 77 & 133), but suggests growth of crystals on the floor during periods of non-deposition of settled crystals. If this is so, then the inch-scale layering may be of crescumulate origin and comparable in origin to the olivine crescumulates of Rhum (Ch. XI), although the textural relationships between the pyroxene and felspar layers of the Stillwater need further study before this view can be substantiated. The inch-scale layered structures do not present a picture of individual layers of upward-growing crystals alternating with layers of accumulated crystals, as in Rhum. However, the ragged textures both of pyroxene and felspar layers might be due to mutual growth interference through rhythmic nucleation and crystallization at the floor, occasionally interrupted by crystal accumulation (Fig. 145).

Slump structures can be seen at several horizons, but most commonly in the Norite

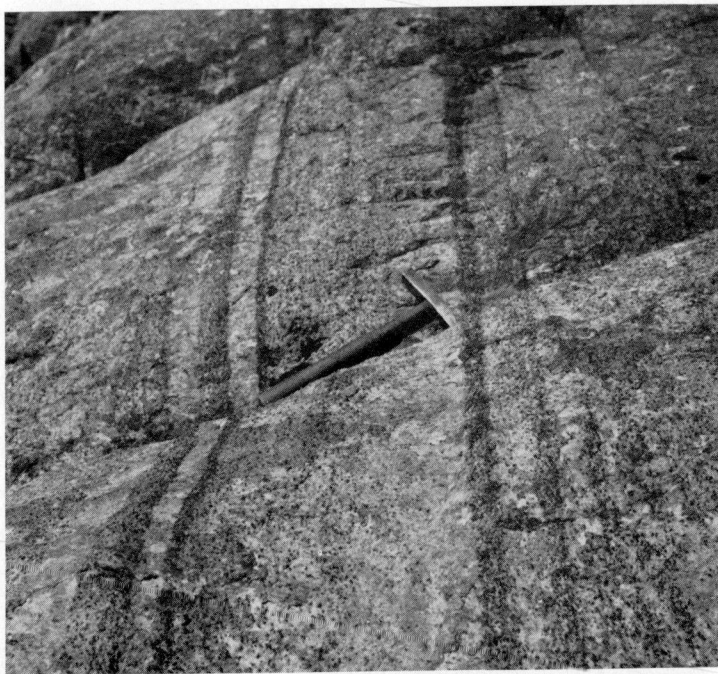

FIG. 186. Rhythmic layering in the Norite Zone, west side of the Stillwater River Valley. The relatively steep dips are to the north (towards the right of the picture). Cumulus bronzite and plagioclase are involved, one unit (on which the hammer rests) showing gravity stratification while others show grading in the reverse direction. (Photo. by E. D. Jackson.)

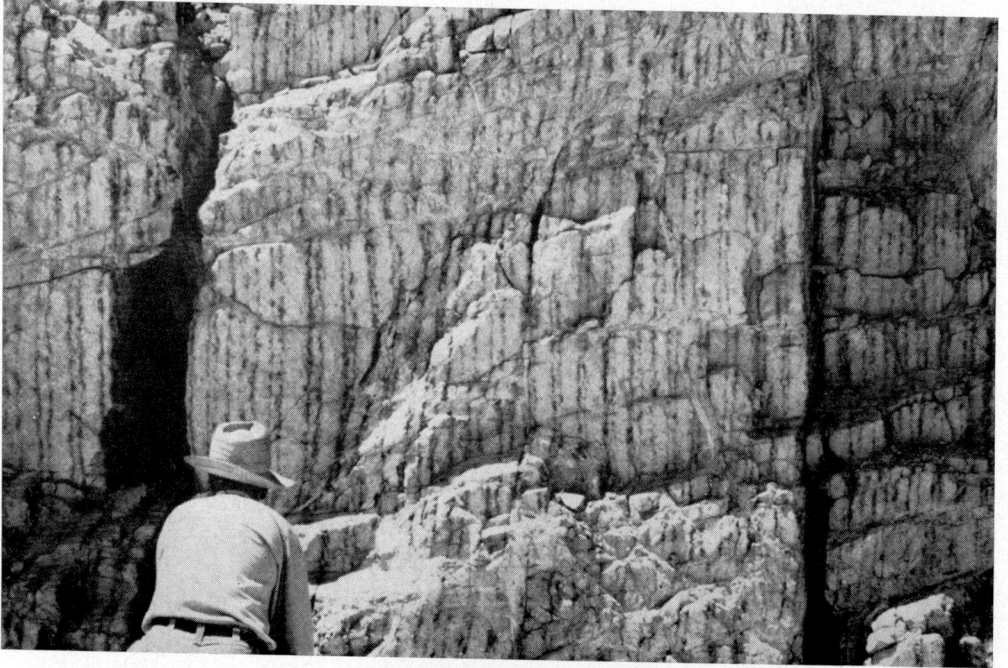

↑

FIG. 187

'Inch-scale' layering, involving
pyroxene and felspar, in the
Lower Gabbro Zone, west of the
Stillwater River Valley. Post-
consolidation flexuring has produced the present, almost vertical
dips.

FIG. 188   →

Normal-scale layering alternating
with 'inch-scale' layering. Near
locality of Fig. 187. (Photo. by
E. D. Jackson.)

Zone. This is probably because rhythmic layering is commonest in that zone; slumping is detected through disturbance of regular layering and operates over such a limited scale of thickness that it cannot, of course, be detected if it has affected a layer of uniform material. The slump structures described and illustrated by Hess (1960, pp. 129–31; pl. 7; fig. 27–8) are very similar to those found in Rhum (Fig. 162) and the Bushveld (Fig. 201). The thicknesses of the Stillwater slumped zones (3–10 ft and averaging 5 ft) are comparable with those in Rhum (4–5 ft) and suggest that the rate of consolidation was similar in the two intrusions, although quicker consolidation must have taken place at horizons where adcumulates were being formed. Hess (fig. 27) shows how certain better-consolidated layers have responded to the movement by the fracturing and thrust-faulting of brittle slabs rather than by the overturning of folds, and a similar feature was noted in the Skye gabbros by Carr (1954) and is seen in certain disturbed layers in Rhum (Fig. 161).

The Chemistry of the Intrusion

Chemical data on the Stillwater intrusion are provided by Hess (1960), the majority of the analyses being of separated minerals (Table 25). Hess was clearly aware of the restricted value of whole-rock anayses in demonstrating cryptic variation within a rhythmically layered series of igneous cumulates, and has used mineral analyses for this purpose. In general, the simplest way of studying cryptic layering is to analyze the minerals of a series of adcumulates or heteradcumulates, or to estimate the compositions of the cores of zoned crystals (by optical methods) and make a whole-rock analysis, for mesocumulates or orthocumulates. In this way, the changing compositions both of cumulus phases and, in the latter case, of the contemporary trapped liquids can be measured. However, in the Stillwater intrusion the possibility of reaction between cumulus crystals and intercumulus liquid is a complicating factor, and only in a few cases can the analyses of separated minerals be taken as representative of the compositions of the initial cumulus phases; it is equally rare to find a layered series rock which, when analyzed, could be used to obtain an estimate of the composition of the contemporary magma.

Twelve rock analyses are presented in Hess's memoir, of which four are of secondary dunites and serpentinites and will not be considered in this account, and one of a monomineralic bronzitite which can be treated as a mineral analysis. The other seven analyses consist of two fine-grained Border Zone rocks, three gabbros from the Upper Gabbro Zone, one composite sample from the Ultramafic Zone, and one composite sample from the remaining zones of the exposed layered series (Table 24).

The hypersthene dolerite collected by J. W. Peoples from the Border Zone (Table 24) is believed, by Hess, to be from the chilled margin of the intrusion and to represent the composition of the original Stillwater magma. Another analyzed ' dolerite ' (Hess, 1960, table 12) was thought to be either contaminated hypersthene dolerite or marginal hornfelsed sediment, and rejected in favour of Peoples's specimen. The chilled marginal rock has the composition of a saturated basaltic magma with a high alumina content and tholeiitic affinities. It is similar to an analysis of the Bushveld marginal facies (Table 26), but differs from the Skaergaard chilled marginal gabbro (Table 7) in having more

TABLE 24

Chemical analyses of rocks from the Stillwater intrusion

	1	2	3	4	5	6	7
SiO_2	50·68	47·73	49·70	50·12	51·11	51·12	52·0
Al_2O_3	17·64	4·82	22·04	20·01	15·36	16·34	14·8
Fe_2O_3	0·26	2·94	0·66	0·80	1·22	1·09	5·0
FeO	9·88	6·54	4·02	4·29	6·63	5·84	12·0
MgO	7·71	28·98	7·03	7·91	10·04	8·49	3·4
CaO	10·47	2·44	13·59	13·97	12·63	13·63	8·6
Na_2O	1·87	0·19	1·79	1·74	1·82	2·18	2·4
K_2O	0·24	0·02	0·07	0·05	0·07	0·06	0·5
H_2O^+	0·42	4·91	0·82	0·69	0·90	0·49	...
H_2O^-	0·06	0·49	0·09	0·04	0·04	0·07	...
Cr_2O_3	0·04	0·48	0·03	...	...	...	...
TiO_2	0·45	0·12	0·16	0·19	0·22	0·19	0·9
P_2O_5	0·09	0·01	0·02	tr	tr	0·03	0·2
MnO	0·15	0·17	0·09	0·09	0·08	0·15	0·2
Total	99·96	99·84	100·11	99·90	100·12	99·68	100·0
Normative (mol. %)							
An	68	85	76	75	67	64	53
En	59	89	77	73	70	68	40
Fo	...	87	74	...	...	...	...

Key to analyses

1. Chilled Border Zone, hypersthene dolerite. Anal.: R. B. Ellestad. (Hess, 1960, table 12, no. 1.)
2. Composite sample of 26 specimens from the Ultramafic Zone. Anal.: R. B. Ellestad. (Hess, 1960, table 31).
3. Composite sample of 40 specimens from the Norite, Anorthosite, and Gabbro Zones. Anal.: R. B. Ellestad. (Hess, 1960, table 32.)
4. Plagioclase – augite – hypersthene cumulate, EB43, 400 ft above base of Upper Gabbro Zone, East Boulder Plateau. Anal.: R. B. Ellestad. (Hess, 1960, table 27.)
5. Plagioclase–augite cumulate, EB42, 1200 ft above base of Upper Gabbro Zone, East Boulder Plateau. Anal.: A. H. Phillips. (Hess, 1960, table 28.)
6. Plagioclase–augite–pigeonite cumulate, EB40, 2150 ft above base of Upper Gabbro Zone, East Boulder Plateau. This is the uppermost exposed level of the layered series (cf. EB41, Table 25). Anal.: A. H. Phillips. (Hess, 1960, table 29.)
7. Estimated composition of the Upper Hidden Zone, assuming that this represents 40 % of the total volume of the intrusion, and calculated by using analyses 1, 2 and 3 given here (Hess, 1960, table 36). The calculated normative composition (approx.) is Qz, 8·7 %; Plag, 51·3 %; Di, 10·8 %; Hy, 19·7 %; Mt, 7·4 %; Ilm, 1·7 %; Ap, 0·4 %.

SiO_2, less TiO_2 and Na_2O, and a lower Fe_2O_3/FeO ratio. Both quartz and olivine are absent from the CIPW norm, so that the presence of both these minerals in the layered rocks is indicative of marked crystal fractionation. The normative plagioclase is An_{68}, and the hypersthene (actually an inverted pigeonite) is En_{59} (Table 24).

The fractionation of the Stillwater magma led to enrichment of the successive residual liquids in alkalis and iron, relative to calcium and magnesium, as shown by the changing compositions of the precipitated crystals in equilibrium with these liquids, i.e. the plagioclases, pyroxenes, and olivines. There is also evidence for an acceleration in compositional changes towards the top of the exposed layered series, as in the Skaergaard and Bushveld intrusions. Thus the analyses of the constituent minerals from gabbros of the Upper Gabbro Zone show a change in the composition of felspar from An_{74} to An_{63}, and of hypersthene from En_{73} to En_{64} (expressed Ca-free), over a thickness of only 1750 ft (see Fig. 171 and Tables 24 & 25).

It would have been of the greatest interest to have found an intrusion in which the changing compositions of the successive residual liquids could be compared with those of the Skaergaard, and in some respects the Stillwater seemed promising because of the presence of a chilled marginal rock and a thick succession of layered rocks equivalent, in many respects, to what we expect to exist in the lower, Hidden Zone of the Skaergaard. Unfortunately, the rocks are not amenable to such a study (p. 331), and the absence of exposed later differentiates further precludes any direct comparison. However, Hess has attempted to estimate the likely composition of the Upper Hidden Zone of the Stillwater and it is this calculation, using the analyses of the chilled rock and the two composite samples, which forms the greater part of the bulk chemical study of the intrusion.

If a layered intrusion can be assumed to have crystallized as a closed system, then from a knowledge of the composition of the original magma, the volume of the magma chamber, and the volume of the exposed rocks, the composition of the unexposed part of the intrusion can be calculated. Unfortunately, the shape and size of the original Stillwater magma chamber cannot be estimated, so that calculations involving relative volumes are not practicable. Instead, Hess has made an estimate of the bulk composition of the exposed layered series, and by comparing this with the initial magma composition he arrived at two limiting values for the proportionns of exposed to un-exposed material, and a likely intermediate value. Thus he was able to propose, with certain reservations, a probable composition for the Upper Hidden Zone.

The analyzed composite sample from the Ultramafic Zone consists of chips of twenty-six specimens collected by Peoples from the eastern part of the intrusion, while a second composite sample was made up of forty specimens from higher zones of the East Boulder Plateau traverse, weighted with respect to the thickness of the section represented by each specimen (Table 24). Assuming that the zone thicknesses are proportional to their volumes, these two analyses were then combined to provide an estimate of the composition of the exposed layered series (Hess, 1960, tables 31–33).

TABLE 25

Chemical analyses of representative minerals from the Stillwater intrusion

	Pl 1	Pl 2	Pl 3	Pl 4	Cp 1	Cp 2	Cp 3	Cp 4	Op 1	Op 2	Op 3	Op 4	Op 5
SiO_2	46·34	47·67	49·19	52·95	52·55	52·61	51·83	51·86	55·36	55·20	53·67	52·81	50·24
Al_2O_3	33·36	33·46	31·82	29·48	2·98	2·72	3·07	2·33	1·72	1·50	1·65	1·65	4·52
Fe_2O_3	0·54	...	0·55	0·56	1·17	1·36	1·38	1·60	0·54	0·84	1·14	1·13	0·41
FeO	0·38	0·35	0·41	0·10	3·98	5·85	7·21	9·45	9·91	11·86	15·30	18·13	21·81
MgO	...	...	...	...	18·21	15·97	16·00	14·50	29·79	28·14	25·37	19·81	15·81
CaO	17·31	16·23	14·95	12·70	19·23	20·50	19·21	18·92	1·63	1·93	1·81	5·13*	4·39
Na_2O	1·55	2·19	2·78	4·12	0·32	0·28	0·27	0·23	...	...	...	0·05	0·18
K_2O	0·05	0·07	0·09	0·13	...	0·02	0·02	...	...	...	...	0·00	0·04
H_2O^+	0·23	...	0·28	0·10	0·14	0·19	0·47	0·37	0·18	0·30	...	0·54	1·42
H_2O^-	...	...	...	...	0·08	0·04	0·11	0·09	0·10	0·06	0·09	0·16	0·04
TiO_2	...	...	...	0·05	0·24	0·44	0·49	0·55	0·17	0·22	0·29	0·37	0·69
Cr_2O_3	...	...	...	...	1·18	...	...	0·01	0·47	0·07	0·06	...	...
MnO	tr	...	...	...	0·12	0·18	0·17	0·24	0·22	0·28	0·33	0·38	0·43
NiO	...	...	...	...	0·05	...	...	0·02	0·07	0·07	0·05	...	...
Total	99·76	99·97	100·07	100·19	100·25	100·16	100·23	100·17	100·28	100·47	99·76	100·16	99·98
Mol %													
An	86·0	80·2	74·4	62·9									
Ab	13·7	19·4	25·0	36·3									
Or	0·3	0·4	0·6	0·8									
Atom %													
Ca					40·0	42·3	41·0	39·5	3·2	3·7	3·6	10†	10
Mg					52·0	46·0	45·0	42·5	83·3	79·7	73·1	58	50
Fe					8·0	11·7	14·0	18·0	13·5	16·6	23·3	32	40

* CaO content too high, due to slight augite impurity in separate.

† Recalculated after subtraction of estimated amount of augite impurity.

If the Upper Hidden portion amounted to less than about 28% of the whole then, as indicated by a graphical solution, no MgO would be left for this part and a minimum value is obtained.   If, on the other hand, the Upper Hidden portion comprised 50% of the whole, then the normative plagioclase of the liquid present after crystallization of the exposed layered series would be about An_{60} (because the chill normative felspar is An_{68} and that of the average exposed layered series about An_{76}).   This is too close to the composition of the cumulus plagioclase of the highest exposed layered rocks (c. An_{63}) to be acceptable, and any value higher than 50% for the proportion of hidden material would mean that the Upper Hidden Zone liquid would be richer in normative anorthite than the plagioclase crystals first precipitated from it.* Hess concludes that the Upper Hidden Zone probably constitutes about 40% of the total intrusion, implying a total thickness for the intrusion of 26 800 ft, and uses this percentage to make an estimate of the bulk composition of the Upper Hidden Zone (Hess, 1960, table 36).   The greatest

* If, however, the chill has normative An_{63} (see footnote, p. 306), the UHZ liquid would have normative An_{50}, which would alter the basis of this neat argument.

Key to mineral analyses (Table 25)

Pl 1. Plagioclase from lower part of Norite Zone (earliest appearance as a cumulus phase), East Boulder Plateau. EB18.

Pl 2. Plagioclase from upper part of Lower Gabbro Zone, East Boulder Plateau. EB38.

Pl 3. Plagioclase from lower part of Upper Gabbro Zone, East Boulder Plateau. EB43.

Pl 4. Plagioclase from upper part of Upper Gabbro Zone (uppermost exposure of layered series), East Boulder Plateau. EB41.

　　　(Analyses 1–3 by A. H. Phillips; Analysis 4 by E. H. Oslund. Hess, 1960, table 10).

Cp 1. Augite (intercumulus phase) from Ultramafic Zone (locality not known). I52.

Cp 2. Augite from upper part of Lower Gabbro Zone (appears as cumulus phase at base of this zone), East Boulder Plateau. EB175.

Cp 3. Augite from lower part of Upper Gabbro Zone, East Boulder Plateau. EB43.

Cp 4. Augite from upper part of Upper Gabbro Zone, East Boulder Plateau (uppermost exposure). EB41.

　　　(Analyses by R. B. Ellestad. Hess, 1960, table 9).

Op 1. Bronzite from the upper, bronzitite layer of the Ultramafic Zone, East Boulder Plateau. EB13.

Op 2. Bronzite from upper part of Lower Gabbro Zone, East Boulder Plateau. EB38. (Slightly higher level, structurally, than Cp2.)

Op 3. Bronzite from lower part of Upper Gabbro Zone, East Boulder Plateau. EB130. (Equivalent level, structurally, to Cp3.)

Op 4. Inverted pigeonite from upper part of Upper Gabbro Zone, East Boulder Plateau (uppermost exposure). EB41.

Op 5. Inverted pigeonite from dolerite, chilled facies (? contaminated), Border Zone, East Boulder Plateau. EB89.

　　　(Analyses 1–3 by L. C. Peck; Analyses 4–5 by R. B. Ellestad. Hess, 1960, tables 3–4.)

Note—Coexisting phases are Pl4–Cp4–Op4 and, approximately, Pl3–Cp3 (−Op3) and Pl2–Op2 (−Cp2), the minerals in parentheses being from adjacent, but not identical rocks to those in which the other two minerals occur (see Hess, 1960, pl. 2).

source of error in such a calculation lies in the assumption that the thickness of each zone is proportional to its volume, and Hess calculates an alternative composition assuming that the floor is saucer-shaped and that, therefore, the Ultramafic Zone thickness should be halved in order to reflect its smaller lateral extent. This, and other calculations based on 35% and 45% of total volume for the Upper Hidden Zone (Hess, 1960, tables 37–38), show that the alternative values have little effect on the estimated composition of the unexposed material.

The estimated bulk composition of the Upper Hidden Zone (Table 24) corresponds, according to Hess, to that of a quartz-bearing ferronorite. Thus the norm (see Table 24,

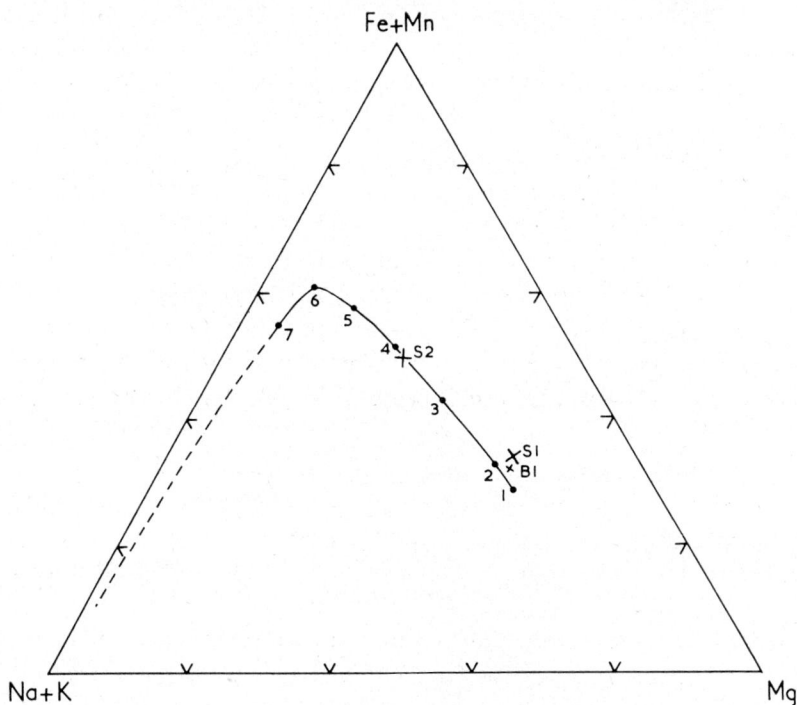

Fig. 189. Mg–(Fe+Mn)–(Na+K) diagram for two likely successive liquids of the Stillwater intrusion. S₁ = First liquid (chilled Border Zone dolerite; Anal. 1, Table 24). S₂ = Composition estimated by Hess for the Upper Hidden Zone liquid (Column 7, Table 24), presumed to have been present at the stage when the highest exposed layered rocks (top of the Upper Gabbro Zone) were crystallizing.

For comparison, the successive liquids of the Skaergaard intrusion (1 to 7) are shown (see Fig. 132), and the first liquid (B1) of the Bushveld intrusion (see Fig. 217). The first (initial parental) liquids of all three intrusions have similar compositions according to the components plotted, although the Stillwater and Bushveld magmas are slightly more advanced, and compare more closely with the second (Lower Hidden Zone) liquid of the Skaergaard intrusion.

The probable course taken by the Stillwater second liquid, and the rocks likely to have formed in the Upper Hidden Zone, are discussed in the text.

note 7) contains about 9% quartz, 51% plagioclase (An_{53}), and 20% hypersthene (En_{40}), as well as clinopyroxene, Fe–Ti oxides and apatite. Hess (1960, p. 103) has not gone beyond estimating the *bulk* composition of the Upper Hidden Zone rocks, but we feel it important to emphasize that the individual, hidden late-differentiates of the Stillwater intrusion are not to be visualized as quartz-ferronorites. For example, the term 'ferronorite' is misleading because the bulk material under consideration contains a high Fe^{++}/Mg ratio such that a hypersthene would be unstable in the bulk of these late differentiates and instead a ferropigeonite, succeeded by a fayalitic olivine plus quartz, would be the more likely equilibrium assemblage (together with a ferroaugite). Also, as seen from the norm (Table 24, note 7), it is likely that Fe–Ti oxides and apatite would appear in relative abundance in the later differentiates. By analogy with the Skaergaard and, more particularly, the Bushveld (in view of the similar initial magma compositions and early cumulates), it is likely that the Upper Hidden Zone rocks of the Stillwater would have included ferrodiorites, and also a small volume of fayalite–hedenbergite granitic rocks.

The likely initial magma of the Stillwater plots very close to that of the Bushveld, and fairly close to that of the Skaergaard, on an Mg—(Fe+Mn)—(Na+K) diagram (Fig. 189). Of particular interest, also, is the fact that Hess's estimated Upper Hidden Zone liquid (see Table 24, column 7) plots on the Skaergaard trend, and close to the Middle Zone liquid of that trend. The analogy between the two hypothetical liquids cannot be drawn in detail, except that pigeonite is apparently in equilibrium with both; they differ in that the Stillwater liquid was not in equilibrium with a ferriferous olivine at the top of the exposed layered series, and that it was in equilibrium with a more calcic plagioclase (An_{63}). However, factors which may have influenced these differences, such as SiO_2, CaO, Al_2O_3, and Fe_2O_3:FeO ratio, are not included in the parameters of Fig. 189. The main conclusion to be drawn from Fig. 189 (and, in relation to the Bushveld, from Fig. 217), is that the estimated Upper Hidden Zone liquid ('second liquid') of the Stillwater is of such a composition, and relative bulk, that a continuation of the fractionation process so well exhibited by the exposed part of the intrusion can be expected to have yielded a series of later differentiates comparable with those found in the Skaergaard and Bushveld intrusions. These would include both iron-rich diorites and, ultimately, iron-rich granitic rocks. Hess (1960, p. 176) argues that the latter product is unlikely because of the low K_2O content of the Stillwater initial magma but it is, in fact, as rich in K_2O as both the Skaergaard and Bushveld initial magmas (p. 355), both of which show enrichment in K_2O (and Na_2O) in the late-stage granophyres. Also, the Stillwater initial magma is less undersaturated than the Skaergaard, and the early differentiates show no obvious indication of the oxidation state being lower than in either the Skaergaard or the Bushveld intrusions. Hence, the early separation of magnesian olivine and a non-silicate (chromite), and the abundance of adcumulates, together emphasize the likelihood that silica enrichment would have led to the formation of quartzose late-differentiates.

Y

The Sequence of Crystallization

In order to account for the layered features characteristic of the Stillwater intrusion, both Hess (1960 and earlier papers) and Jackson (1961) have furnished accounts which not only describe these in detail, but contain several interesting and ingenious theories to account for their origin. No single theory is capable of explaining, for all layered intrusions, the origin of a feature such as rhythmic layering because, despite overall similarities, the details differ from one intrusion to another. For this reason we present in this section the views expressed by Hess and Jackson for the Stillwater intrusion, only modified either where conformity in terminology is desirable or, in a few instances, where alternative explanations can be offered.

The original form of the intrusion is believed to have been sheet-like, with a diameter in excess of 30 miles and a thickness of about 5 miles. A single act of injection filled the magma chamber, and thereafter the intrusion crystallized as a closed system. The thickness of original cover is not known, although Hess estimates that it was at least 1 km. The initial magma was a silica-saturated, tholeiitic basalt, presumed to have been at a temperature of about 1225° C, which chilled against the wall rocks. From the estimated rate of heat loss, Hess (1960, pp. 144–6) has postulated that it took about 50 000 years for the exposed 60% of the intrusion to crystallize, by which time the temperature had fallen only to about 1100° C. The length of time taken for the deposition of the exposed 16 000 ft or so of layered cumulates has led Hess to postulate, therefore, an average rate for crystal accumulation of about 1 ft in three years (i.e. 10 cm per year). This is a valuable contribution to the study of crystal accumulation in layered intrusions, and at present Hess's figure is accepted as the best available estimate, in making other calculations (see Ch. VIII). Hess assumed that at 1225° C the initial magma was superheated and that, on the basis of pyroxene inversion relationships, crystallization began at about 1125° C and had only reached 1100° C after 60% of the intrusion had crystallized. The scant available thermal data on pyroxenes hardly justifies their use as a geological thermometer, however, and from the data given by Tilley et al. (1963, fig. 12) for the Stillwater initial magma, and by analogy with the temperature range given by them for crystallization of the Skaergaard intrusion, it is more likely that the exposed 60% of the Stillwater intrusion crystallized over about 120° C, and that crystallization began at about 1245° C. However, as Hess's calculations took into account the heat loss from 1225° C to 1125° C, which in any case is small compared with the total, and as the latent heat of crystallization is independent of the temperature of initiation of crystallization, the calculated rate of crystal accumulation is probably of the right order.

The chilled gabbro of the Border Zone is found only at the lower contact, although it is thought to have once formed an encircling envelope to the layered series. It is a fine-grained, ophitic rock (Jackson, 1961, fig. 87; Hess, 1960, pl. 4) in which the textures, mineral assemblages, and compositions suggest chilling against cool country rock. This facies of the Border Zone is estimated to be about 100 ft thick, the upper part sometimes consisting of a slightly coarser-grained hypersthene dolerite, with fairly well-developed layering parallel to the floor of the intrusion, which contains streaks and segregations of coarse norite–pegmatite parallel to the layering. This is overlain locally by bronzite cumulates, 0–700 ft thick, which are also grouped, by Jackson, with the Border Zone. These bronzite cumulates, although coarser-grained than the chilled dolerite, are considerably finer-grained than the olivine cumulates of the Ultramafic Zone which immediately overly them (Jackson, *op cit.*, p. 21), the more magnesian varieties occurring upward in the section. These features suggest that the degree of supercooling probably decreased towards the inside of the Border Zone. Jackson, in his model (1961, pp. 96–9), postulates that the supercooled magma was warmed by latent heat of crystallization to such a temperature that for the bronzite cumulates of the upper part of the Border Zone, crystal–liquid equilibrium was eventually attained.

In considering the formation of the layered series, it is necessary to deal with the Ultramafic Zone and the higher zones separately, and in that order. The reason is that whereas Hess's work on the higher zones deals with the effect of current activity in crystal sorting and the formation of fine-scale layering, Jackson believes that the thick layers of the Ultramafic Zone were formed by a different process and that the higher zones, only, were produced at the level of relatively continuous convection.

The Ultramafic Zone consists of a lower Peridotite Member (thicker than 2500 ft) and an upper Bronzitite Member (thicker than 1000 ft). The lower member is divisible into 15 units which we would term macro-*rhythmic* units (stressing the inter-relationship, rather than the cyclic repetition found *within* each unit), as they are similar in magnitude to the macro-rhythmic, ultrabasic units of Rhum (Ch. X). The cumulus crystals of a typical unit, from bottom to top, are olivine, chromite, chromite–olivine, olivine–chromite, olivine, olivine–bronzite, bronzite–olivine and bronzite. Jackson believes that in view of the absence of current structures, lineation, size–density sorting, and hydraulic equivalence between coexisting settled crystals, convection currents were not operative in that part of the magma chamber in which the Ultramafic units were being deposited. For this reason he presents a new hypothesis (Jackson, 1961, pp. 93–9) to explain the repetition of units and the changes within each unit, according to which the units were formed by the bottom accumulation of crystals separating from successive, relatively thin layers of stagnant magma adjacent to the floor. Loss of heat from the floor would permit crystals to form, and sink over the relatively short distance envisaged, until such times as the latent heat of crystallization had raised the temperature, locally, sufficient to prevent further crystallization. After crystal sinking had freed this now hotter magma of suspended crystals it would have a lower density than the overlying, main body of magma and a convective overturn would occur, the process of cooling and crystallization then being repeated to form the next ultramafic unit. The reader should refer to Jackson's account of the details of his hypothesis, where he also considers that crystallization from a stagnant liquid could best account for the upward transition, with silica enrichment, from olivine-rich to bronzite-rich cumulates within each unit. Bearing the above-mentioned points in mind, the main part of Jackson's hypothesis is explained in the following statement (1961, p. 98):

> In the favored hypothesis, the cyclic units are pictured as depositional products of periodically refreshed stagnant magma which became stabilized by bottom crystallization. Because the bottom magma during each crystallization cycle was stagnant, it became enriched in silica as crystallization proceeded, and the early chromite and olivine were followed by bronzite. Each overturn brought a fresh supply of magma to the bottom and the cycle was repeated; but after such selective crystallization of a number of overturns, the composition of the entire magma was gradually but appreciably changed, so that higher cycles in the Peridotite member systematically contained smaller proportions of olivine. In this manner olivine and chromite finally ceased altogether to be crystallization products, and the magma crystallized bronzite alone in the Bronzitite member, and bronzite plus plagioclase at the base of the norite section above.

There are difficulties in applying Jackson's hypothesis, particularly in isolating the stagnant magma from the overlying magma so that silica enrichment could proceed, and bronzites crystallize, despite the proposition that the temperature of the stagnant magma was gradually increasing. We feel that gradual chemical and temperature changes would not easily be maintained in an isolated part of the magma, and that diffusion and convection, affecting the whole body of magma, would be continuous processes tending to eliminate steep compositional and temperature gradients. Jackson accepts that convective overturn of the main body of magma took place once for each macro-unit, and also that after the deposition of the Ultramafic layered macro-units a convective system operated throughout the magma chamber and accounted for the layering in the rest of the Stillwater layered rocks. We do not feel that the evidence cited for the absence of convection currents during the formation of each Ultramafic unit entirely justifies Jackson's conclusion. Current structures and lineation, for example, are rare in layered rocks even where fine-scale rhythmic layering is present. The scarcity of size and density sorting of crystals, and hydraulic equivalence, is also commonplace, even according to the cruder observational methods. A rigorous petrographic analysis, as used by Jackson, would reveal more information on other layered rocks than is at present available, but first one should consider what interpretation is to be placed on the results. Particularly, we have presented a case for convection currents operating in the Skaergaard intrusion during the formation of layers of *uniform* rock (p. 213) and if this is true, then crystal settling through such thick layers of slowly-moving magma would not produce hydraulic equivalence in the layers of deposited crystals because crystals of variable density and size would be forming in, and settling from, various levels within the liquid layer.

For these reasons, we feel that the origin of the layering in the Ultramafic Zone of the Stillwater intrusion is still an open question, but that continuous convection, due to the difficulty of heat loss through the floor, is more likely to have taken place than long periods of local magma stagnation.

After the deposition of the Ultramafic Zone, the magma composition had by now reached the stage at which plagioclase crystallized as a cumulus phase, while olivine and chromite had ceased to crystallize. Here, in the Norite Zone, bronzite and plagioclase crystals are well-sorted to produce the best fine-scale rhythmic layering of the intrusion. To explain the fine-scale rhythmic layering, Hess has postulated the presence of subsidiary currents in the magma immediately above the floor, or perhaps within a few hundred feet of it. On the other hand the larger ' megacycles ' characterized by the thick sub-zones of the Anorthosite Zone, and perhaps the occasional violent fluctuations in cumulus mineral compositions within that zone, are attributed by Hess to intermittent, complete convective overturns of the magma. In considering layering in the Stillwater, account must also be taken of the layers of ' uniform rock ' which commonly occur intercalated with the thin differentiated layers (p. 328). We have found a satisfactory explanation for this feature in the Skaergaard intrusion (Ch. VIII), and this could be applied to the Stillwater, also, in view of the broad similarities. If this is so, then it could be that the convection currents were of variable velocity and character, and this seems preferable to a distinction between local and complete convective overturns.

Hess explains the features of the fine-scale rhythmic layers, including cases of gravity stratification, ' reversed gravity stratification ', and layers with sharp contacts, by calculating the effect which upward and downward moving currents would have on the settling rates of pyroxene and plagioclase crystals (1960, fig. 29). This assumes hydraulic equivalence in the gravity-stratified bands, however, whereas Jackson has found no evidence for this (p. 316).

The well-layered Norite Zone is succeeded by the Lower Gabbro Zone, in which clinopyroxenes began to crystallize as a cumulus phase. Inch-scale layering is found at certain levels within this zone and may be indicative of rhythmic crystallization from supercooled magma near the contemporary floor, during periods of non-deposition. The Lower Gabbro Zone is succeeded by the thick Anorthosite Zone, the base of which is defined as the level at which olivine returns as a cumulus phase. The troctolites at the base of the zone contain about 30% olivine (Fo_{77}) and 70% plagioclase, and have a unique texture in that the large olivine crystals (5–10 mm) include many plagioclase crystals near their rims. These large crystals, which Hess calls ' pillows ', are set in a matrix of small plagioclase crystals; we are tempted to call them ' composite clusters ', by analogy with those found in Rhum (Fig. 141), and to suggest that the large olivines are not poikilitic crystals (cf. Hess, 1960, pl. 5) but are parts of settled, composite grains of olivine and felspar.

The Anorthosite Zone is characterized by the presence of three thick layers, each about 1500 ft in thickness, of almost pure plagioclase cumulates. Hess finds these rocks difficult to explain, but suggests that convection currents may have been moving at such a rate that they would have kept felspars in suspension and carried them into hotter portions of the magma where they would be resorbed. A continuation of this process would produce a magma so rich in the felspar component that it would then proceed to crystallize only this mineral over a long period. An alternative hypothesis, we suggest, is that slightly undercooled magma moved downwards during convective overturn and on reaching deeper levels, crystallized both felspar and mafic minerals at these relatively higher pressures. The currents moving across the floor and turning upwards may have had such a velocity that few felspars had time to settle out (see Fig. 127) and the felspar crystals would be carried up to levels of lower pressure, where they may have melted. Whether melting took place or not, the magma would become progressively enriched in the felspar component until only felspars crystallized or settled from it for a long period. The latter hypothesis differs from Hess's in that we do not expect there to be a particularly hot zone of magma, at this late stage in the history of the intrusion, at the temperature necessary for melting a large amount of plagioclase felspar.

The uppermost exposed zone of the layered series is the thick Upper Gabbro Zone, in which plagioclase, clinopyroxene, and orthopyroxene are abundant as cumulus phases, the orthopyroxene giving place to an inverted pigeonite towards the top of the zone. Cryptic layering is more noticeable than in the lower zones, while the rocks have a composition and mineralogy similar to those found in the Lower Zone of the exposed part of the Skaergaard intrusion, except that Fe–Ti oxides have not become a cumulus phase and olivine is absent.

At this stage, 60% of the Stillwater intrusion is thought to have crystallized according to Hess, and the remaining 40% of the Upper Hidden Zone has the calculated bulk composition of what he has termed a quartz-ferronorite. We have suggested (p. 337) that continued fractionation would probably have yielded rocks such as ferrodiorites and fayalite–hedenbergite granophyres in the unexposed part of the intrusion. Such rocks may be preserved beneath the present cover of Palaeozoic sediments, together with an Upper Border Zone, or may have been removed by erosion during the tectonic disturbances which have resulted in the present, complex form of the Stillwater intrusion.

CHAPTER XIV

THE BUSHVELD INTRUSION, SOUTH AFRICA

INTRODUCTION

The Bushveld Igneous Complex, which is Precambrian in age,* comprises a vast and varied assemblage of intrusive and extrusive rocks. First grouped together by Molengraaff (1901), they lie essentially within a basin-shaped structure of the Transvaal System, north of Pretoria in South Africa. The main Complex forms an elongate, roughly oval area with its major axis striking approximately east–west. In general, the major axis measures about 300 miles and the minor axis about 150 miles although these are close to maximum values and the total area generally accepted for the Complex is about 25 000 square miles. The major components of the Complex, as described by Hall (1932, p. 143), include lavas and pyroclastic rocks, sills, and major and minor plutonic phases. The large mass of basic, coarse-grained rocks, predominantly of norite, was termed the 'Norite Zone' of the Complex by Hall, although geologists have some-times since tended to refer to this body alone as the Bushveld Complex. It is this body of rock which will be described in the present chapter, and will be referred to as the Bushveld Layered Intrusion or the Bushveld Intrusion. Hall (1932, p. 147) has estimated that this intrusion accounts for about 42% of the visible part of the Complex.

Our belief that this huge body of layered igneous rocks formed by bottom-accumulation of settled crystals is based largely on a critical examination of the recorded observations of a large number of investigators. In addition, in 1958 we spent two months examining and collecting from the more important sections of the Eastern and Western Bushveld.

The history of investigation of the Bushveld basic intrusion is a long and complex one for by virtue of its size alone, few geologists have been able to do more than pay attention to specific problems related to limited areas. Hall (1932, pp. 21–139) has given a useful summary of the contributions made in 184 papers up to that date, the greater proportion dealing with the basic intrusion and the exploitation of the chromite- and platinum-bearing rocks within it. Although Hall's classic memoir and map (1932)

---

* Isotopic age of the Red Granite is 1950±50 million years, according to Nicolaysen et al. (1958, p. 137) and Schreiner (1958, p. 112).

constitute the most important single contribution to the literature on the Bushveld, an impressive amount of work has continued since that time, particularly with regard to the character and origin of the layered rocks.

The greater part of the intrusion is well layered, the terms ' pseudo-stratification ' and ' rifting ' being used by Hall (1932, p. 264) to describe this '. . . the most remarkable and most conspicuous feature in the entire lopolith.' In addition, the broad composition of the layered rocks changes upwards from ultrabasic, through basic, to intermediate and acid varieties. Hence the main features of a layered intrusion, that is rhythmic and cryptic layering, are remarkably well displayed in the Bushveld intrusion. Although this has been explained according to the hypothesis of crystal sorting and accumulation by groups from both the earlier and recent workers on the intrusion, the origin of the Bushveld rocks has attracted more varieties of hypothesis than for that of any other basic intrusion. This is not because of any abnormalities in the compositions and mineralogy of the rocks but is due, rather, to the size of the intrusion and to some major and minor structural complexities. These have precluded lateral correlation between layers throughout the whole intrusion and particularly, between the eastern, western, and northeastern lobes. Moreover, a large proportion of the geologists working on the Bushveld have been concerned with mapping isolated, specific regions, often for their economic interest, and hypotheses suiting local evidence have been applied in preference to one aimed at explaining the intrusion as a whole. For example, regions where the layers are irregular, wedge out laterally, and may include blocks from adjacent layers, are more easily mapped according to the simpler hypothesis of separate injections of contrasted magmas than by trying to explain the complexities according to crystal sedimentation. Those geologists, however, who have been concerned with the intrusion as a whole have been more impressed with the overall regularity of the layering than with local irregularities, while they have also found it necessary to view the whole system according to what is known of the likely physico-chemical relationships that existed within this huge magma body.

It is important, therefore, to consider previous work not in historical sequence but according to the proposed hypotheses, each of which has had supporters at various periods. Broadly, it is significant that geologists with extensive experience on a large part of the intrusion have favoured the hypothesis that the layered rocks originated by crystallization differentiation, the crystals being sorted and deposited in successive layers during cooling of the intrusion as a whole. Of the earlier workers, Hall (1932) and Wagner (e.g. 1929) published lucid accounts based on this concept, Hall (op. cit., p. 267) appealing to '. . . a rhythmical " rain " of specific minerals leading to local piles of accumulation . . .' Since then, Lombaard (1934) and his students at Pretoria (e.g. Berning, 1941; Boshoff, 1942; Van der Westhuizen, 1945) developed this hypothesis further, particularly in relation to the regular changes in mineral compositions throughout the thickness of layered rocks. More recently, detailed studies by Cameron and others on the rocks of the Critical Zone (e.g. Cameron & Emerson, 1959; Cameron, 1963) have emphasized the extent to which lateral correlation between layers is possible, while they and Ferguson and Botha (1964) have explained certain local irregularities in the layering according to the general hypothesis of crystal accumulation. Hess (1960) has given a summary of his own work and views on the Bushveld, in which he favours an origin through crystal accumulation from a basic magma. Finally, Willemse (1959, 1964), advantageously working from Pretoria University as did Lombaard, has presented two summary accounts in which he advocates the application of this hypothesis to explain most features of the Bushveld basic intrusion.

The most popular alternative hypothesis for explaining local irregularities is that of multiple injections of contrasted magmas of similar compositions to the individual layers (e.g. Reuning, 1927; Truter, 1955; Schwellnus, 1956; Coertze, 1958). The obstacles to applying this hypothesis to the intrusion as a whole appear to us to be overwhelming, particularly when account is taken of the enormous distances over which individual layers remain parallel, of the regular upward change in the mineral compositions of successive layers, and of the abundant experimental evidence which informs us of the excessively high temperatures, and temperature differences, that would be required for relatively anhydrous chromitite, pyroxenite, or anorthosite magmas. The hypothesis that the Bushveld basic rocks are transformed Transvaal System sediments (Sandberg, 1926; van Biljon, 1949) is even more difficult to relate to the mineralogy and chemistry of these rocks.

There seems little doubt, from all the evidence now at our disposal, that the Bushveld basic rocks are the products of crystal accumulation from a basaltic magma. Much more work is still needed, however, before an overall picture can be obtained of the mechanisms which operated in this vast intrusion. It has recently been proposed (Truter, 1955; Willemse, 1964) that the intrusion should be viewed as having at least four separate centres, and more geophysical and other evidence along these lines may well provide an answer to the problems which exist in attempting to describe the intrusion as a single body, in the sense of lateral correlation between widely separated areas. For example, the sequences in the western (Zeerust) and in the northeastern (Potgietersrust) areas are less complete than in the eastern (Sekukuniland) area, while over shorter distances the sequences also differ (Cameron, 1963, p. 100). It is likely that despite these irregularities, all the layered rocks formed during one phase within this broad basin, and that irregularities of the floor within the basin led to the development of isolated pockets within which the specific crystallization products differed slightly. Further detailed mineralogical studies are needed before a correlation between events within each of these areas can be made, while specific attention to textural relationships are needed in order to establish the critical changes in *cumulus* mineral compositions.

Layered intrusions have not all behaved as did the Skaergaard, as a type of closed system, the Rhum intrusion serving as a type of open system in which periodic replenishment of initial, parental magma occurred. Lombaard (1935) favoured the latter hypothesis in general, but not in detail* for the Bushveld, although Cameron (1963) has since disagreed with the necessity for such an hypothesis. In the case of Rhum (Ch. XII) and Kap Edvard Holm (Ch. XV) the evidence is strongly in favour of parental magma replenishments, but more work on the cumulus minerals of the Bushveld is needed before the number of compositional fluctuations, and their meaning, is established.

In writing this chapter we have been confronted on the one hand with a vast amount of published information and, on the other hand, with a lack of information on certain critical features. Our own field observations have been supported by an examination of the rock textures and the mineralogy of those rocks collected from traverses in representative areas. These were made chiefly in the Jagdlust, Steelpoort, Lulu Mountains, Dwars River, Magnet Heights, Blood River, Rustenburg, and Northam regions (Fig. 190). On the basis of this and the published information a synthesis has been attempted, the layered rocks being described and explained according to an origin by crystal accumulation. In doing this we have needed to make some generalizations with regard to the layered sequence, but where anomalies appear to exist we feel that further work can result in these being explained according to the crystal accumulation hypothesis, in the same way that we have given explanations for certain once-problematical features of a structural and textural nature.

The Form of the Intrusion

The Bushveld intrusion has generally been viewed as a type-example of a lopolith, and both Hall (1932, p. 201) and Daly (1928) drew a comparison with the form of the Duluth complex. The well-known diagrammatic sections of Molengraaff (1905) and Daly (1926) depicted the intrusion as a basin-like structure formed by down-sagging of the floor, the layered rocks being visualized as being almost parallel to the synclinal floor rocks of the Transvaal System. Later sections by du Toit (1926) and Hall (1932, fig. 22 & 23) showed the floor structure to be more complex through folding and faulting, and from du Toit's east–west section the possibility emerges that the intrusion may have formed in at least two subsidiary, but connected basins. Since then, much attention has been paid to the structural relationships at the floor and margins, the picture being too complex (e.g. Willemse, 1959, fig. 4) for a synthesis by us, although Willemse (1964) has recently given a generalized cross-section depicting the form of intrusion (shown here on Fig. 190). The general impression to be gained from studies such as those of Schwellnus (1956) and Hiemstra and van Biljon (1959) in the eastern Transvaal is that layered rocks of the intrusion transgress across marginally folded and fractured country rocks, and that in certain cases the layered rocks have since been slightly folded. The

---

* Lombaard postulated replenishments from a differentiating body at depth, which is more difficult to accept than the simpler mechanism of influxes of parental magma generated in the same way as the initial surge.

relationship between layered and floor rocks is certainly far from conformable on a small scale, and on a larger scale it has been suggested by Schwellnus (1956, fig. 23) on the basis of gravity data, and Wilson (1956) from mine data, that the marginal contact is steeply dipping. Willemse (1959, 1964) concludes that the basic intrusion is funnel-shaped and that the mechanism of emplacement is essentially that of cone-sheet formation. The suggestion that this type of intrusion be termed a funnel-intrusion, rather than a lopolith, was also made by us (1957) after reading the observations on marginal relationships made by Wilson (1956). In view of the present state of knowledge, it would be unwise for us to attempt to consider a series of sections across the intrusion, apart from giving the generalized relationships in Fig. 190. Not only are the marginal relationships complex (Fig. 191), but the shape of the intrusion in either an east–west or a north–south direction is uncertain.

For the purposes of the present account, the intrusion can be viewed essentially as a series of layered rocks with dips ' directed into the interior of the Bushveld ' (Hall, 1932, p. 264). The predominant dips are around 10–15° although these increase towards the margins, and in certain areas are complicated by local floor irregularities (e.g. Coertze, 1958) and subsequent folding. The predominantly low dips, together with evidence of higher dips at the marginal contacts, invites comparison with the analogy of stacked-saucers in an inverted cone, as applied to the Skaergaard intrusion (Ch. II), although it has been proposed (Gough & van Niekerk, 1959) that the layering was originally horizontal. The map (Fig. 190) shows how difficult it is to visualize the original form of the Bushveld magma chamber. In the western portion, west of the Pilanesberg Complex, floor rocks are exposed abundantly and there is the impression here of a relatively shallow

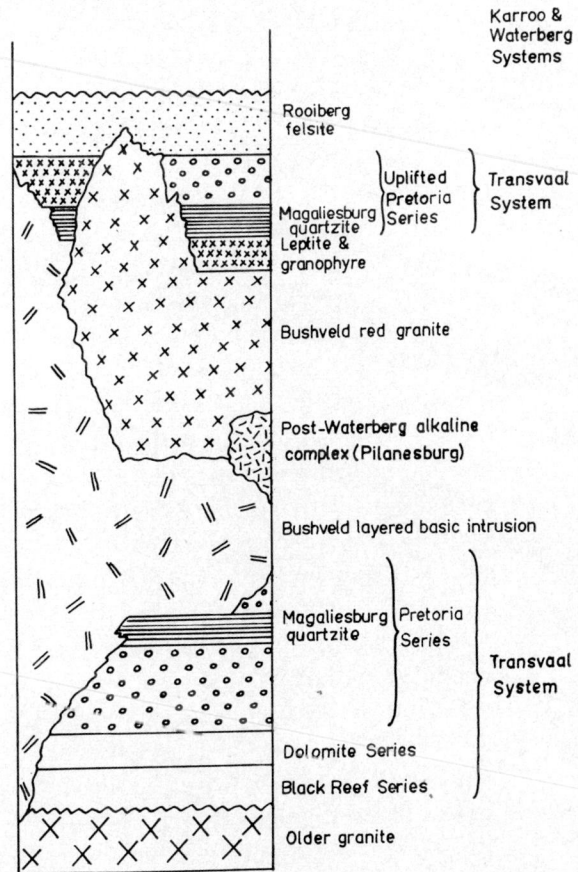

FIG 191.

Schematic diagram of the relationships between the rocks of the Bushveld Complex (modified from Willemse, 1964, pl. 1).

magma-chamber. In the central parts, north and northwest of Pretoria, the great extent of cover, Red Granite, and the problematical Crocodile River fragment together preclude drawing a simple correlation between the eastern and western portions of the intrusion. Here it is possible that a barrier existed between the eastern and western magma chambers which may occasionally have been bridged by magma flowing over it. That this barrier was not always operative is suggested by the presence of extensive, easily recognizable layers such as the chromite layers, magnetite layers, and Merensky Reef in both portions. To the northeast, in the Potgietersrust area, and to the southeast, near Belfast, two irregular extensions of the intrusion are to be seen (Fig. 190). For these, Hall (1932, pp. 207–8) evoked the mental image of chipped rims to the saucer, the magma spilling from the main chamber into these subsidiary basins. This seems quite likely, for the fact that in each of these areas the lower zones of the layered series are missing, would imply that the magma migrated here at a relatively late stage of its fractionation within the main chamber. This hypothesis of an inter-connected group of magma chambers has advantages over attempting to view the Bushveld intrusion as a single body, although in considering the layered rocks they are to be thought of as essentially the product of the same magma. The extent of each magma chamber cannot be estimated, in view of the extensive blanket of later rocks; there is the possibility, for example, that the large blanket between Pretoria and Belfast is underlain by basic rocks; Hall (1932, p. 204) suggests the presence of 'windows' of Bushveld rocks here, while Truter (1955) has postulated an extension in this direction. Even within the relatively uniform eastern Bushveld region, local differences in crystallization conditions apparently operated (Cameron, 1963, pp. 100–1; see also p. 360).

The roof of the intrusion is even more difficult to define than the floor because the central, and stratigraphically highest rocks of the layered series are generally in contact with the complex mass of granites, felsites and granophyres. In certain areas, however, such as the extensive Signal Hill belt in the eastern Bushveld, semi-horizontal sediments and felsites of the Rooiberg Series form a conformable capping to the upper layered rocks. Here it is difficult to distinguish between acid rocks of the roof zone (including possible melted sediments) and late differentiates of the Bushveld intrusion. In other areas, particularly in the western Bushveld to the north of Rustenburg, the upper layered rocks are in direct contact with the huge mass of Red Granite and some of the associated granophyres, which represent later transgressive intrusions also roofed, as for example at Rooiberg, by the Rooiberg Series. The Pretoria Series forms most of the floor of the intrusion and therefore it is likely that only the Rooiberg succession of sediments and lavas can have formed the roof, because the succeeding Waterberg sediments are clearly post-Bushveld in age. The maximum exposed thickness for the Rooiberg Series is about 10 000 ft in the Middelburg District, which does not seem to represent a cover of sufficient thickness to have permitted the slow crystallization of such a huge body of basic magma as the Bushveld intrusion. Daly (1928), in fact, allots only about 1000 ft of Rooiberg sediments to the original roof, believing that where this thin roof foundered '. . . the compound red granite–granophyre–main norite magma, roofed by the air and its own solidified outflows (i.e. felsites), thus appears as a huge lava flow . . .' Daly (1928, p. 762) was of the impression that the Red Granite preceded the basic rocks and provided a hot roof to them, whereas present-day evidence supports the reverse age-relationship (Willemse, 1964, p. 112). As a roof other than granite magma must once have overlain the basic intrusion, it is suggested that a greater thickness of Magaliesburg and Rooiberg sediments may once have formed this roof. Uplift, rather than sagging would characterize a funnel-intrusion, and although a large number of sedimentary blocks have foundered back into, and are preserved in, the intrusion, a large amount of material would have been pushed beyond the periphery and since eroded. The heat supplied to such an agglomeratic roof of siliceous sediments and felsites may well have resulted in partial melting, and the production of a relatively low-temperature acid magma capable of remaining liquid until after most of the basic rocks had crystallized.

THE SUBDIVISION OF THE INTRUSION

The rocks of the Bushveld layered intrusion, broadly termed the ' Norite Belt ' by Hall (1932), comprise a great differentiated suite ranging from ultrabasic rocks such as peridotites and pyroxenites, through basic gabbros, norites, anorthosites, etc., to intermediate diorites and, ultimately, to basic granophyres. The greater part of the intrusion is well layered (e.g. Fig. 197, 204, 207 & 209) and the presence of easily recognizable, laterally extensive, specialized bands within the layered sequence led Wagner (1929) to use the term ' Horizon ' for the levels at which chromite-rich, magnetite-rich, and platinum-rich (Merensky Reef) layers occurred.

Hall (1926) divided the intrusion into the following five units from below upwards: the Chill Phase, the Transition Zone, the Critical Belt (Differentiated Series of Wagner), the Main Belt, and the Upper Belt. Later, however, in the Memoir (1932, p. 271), Hall introduced the term ' Zone ' for these five units, an alternative name for the Chill being the Basal Zone. Lombaard (1934, pp. 22–3) has since replaced the terms Chill and Transition Zone by Norite–diabase and Basal Zones, respectively, while Schwellnus (1956, p. 85) has proposed a modification of Lombaard's terminology for the zones in the Eastern Transvaal, introducing the term Marginal Zone to include the Norite–diabase and part of the Basal Zone, and selecting different horizons for the delineation of the Basal and Critical Zones.

The criteria according to which these various zones have been distinguished and limited have varied. In general, the emphasis has been on easily recognizable marker horizons for drawing zone boundaries, such as the chromite-layers, the Merensky Reef, and the magnetite-layers, although it is never clear whether these layers themselves are placed at the top of a zone or at the base of the succeeding zone. Hall's zone boundaries (1932, pp. 273–7 and fig. 25), reproduced by Cameron (1963, fig. 3) are, in particular, vaguely defined, the marker horizons being unrelated to the zonal subdivisions drawn on the cross-section, and the base of the Main Zone being defined only as the eastern foot of the Lulu Mountains escarpment.

Clearly it is desirable that some agreement be reached on a zonal subdivision of the Bushveld, even though slight disagreements may still exist over the precise levels and distinctions. In order to describe the layered sequence, we have devised a zonal nomenclature which has been found useful for our purposes and may find general approval, although subsequent work is needed to establish the overall validity of the proposed levels at which cumulus phases enter and leave the crystal assemblages, with fractionation. It is obviously impracticable to devise a general scheme of zonal sub-division applicable to all the major layered basic intrusions, because the co-existing mineral phases vary. The zonal names for the Bushveld are now well-established, and in general we have kept these names (Fig. 192). The intrusion consists mainly of a layered series of rocks, there being no exposures of an extensive marginal envelope consisting of marginal and upper border group rocks, of the kind preserved in the Skaergaard intrusion. The only marginal rocks found in the Bushveld are the fine-grained hypersthene gabbros at the eastern margins, which appear partly to underlie the layered ultrabasic rocks. Some of these probably represent a chilled margin to the intrusion and are distinguished from the zones of layered rocks by calling them the Marginal Group.

The rocks of the layered series are generally viewed as being divisible into four zones, the Basal, Critical, Main, and Upper Zones, and we have retained this nomenclature with only a few modifications. The choice of an appropriate horizon for drawing zone boundaries has always been difficult in view of the gradations found between the rocks of this, as in any other layered series. For the Bushveld, geologists found it convenient to use the easily mappable horizons marked by three distinctive and laterally persistent layers: the Main Chromitite layer (the lowest, thick Steelpoort Seam), the Merensky

FIG. 192 →

Stratigraphic sequence of cryptic layering in the Bushveld intrusion, based on traverses chiefly in the eastern part. The cumulus compositions are from optical determinations except for the pyroxenes, the values for which are given from analytical data (see Fig. 214 & Table 28) or (in parentheses) optical data. Broken lines indicate the intercumulus status of a mineral (chromite in the Basal Series may be largely intercumulus, but this is difficult to decide) while the olivine value in parenthesis refers to its occurrence at this horizon only in the western Bushveld. The first appearance of a cumulus phase is indicated (Pig+ for pigeonite appearance, etc.) except for the Integration Stage, where local fluctuations may occur and the vertical lines are used to indicate the overall distribution of each cumulus phase.

MgZ, Marginal Zone (Group); MCR, Main (lowest Steelpoort) Chromitite layer; MR, Merensky Reef; MaZ, Main Zone; UZ, Upper Zone.

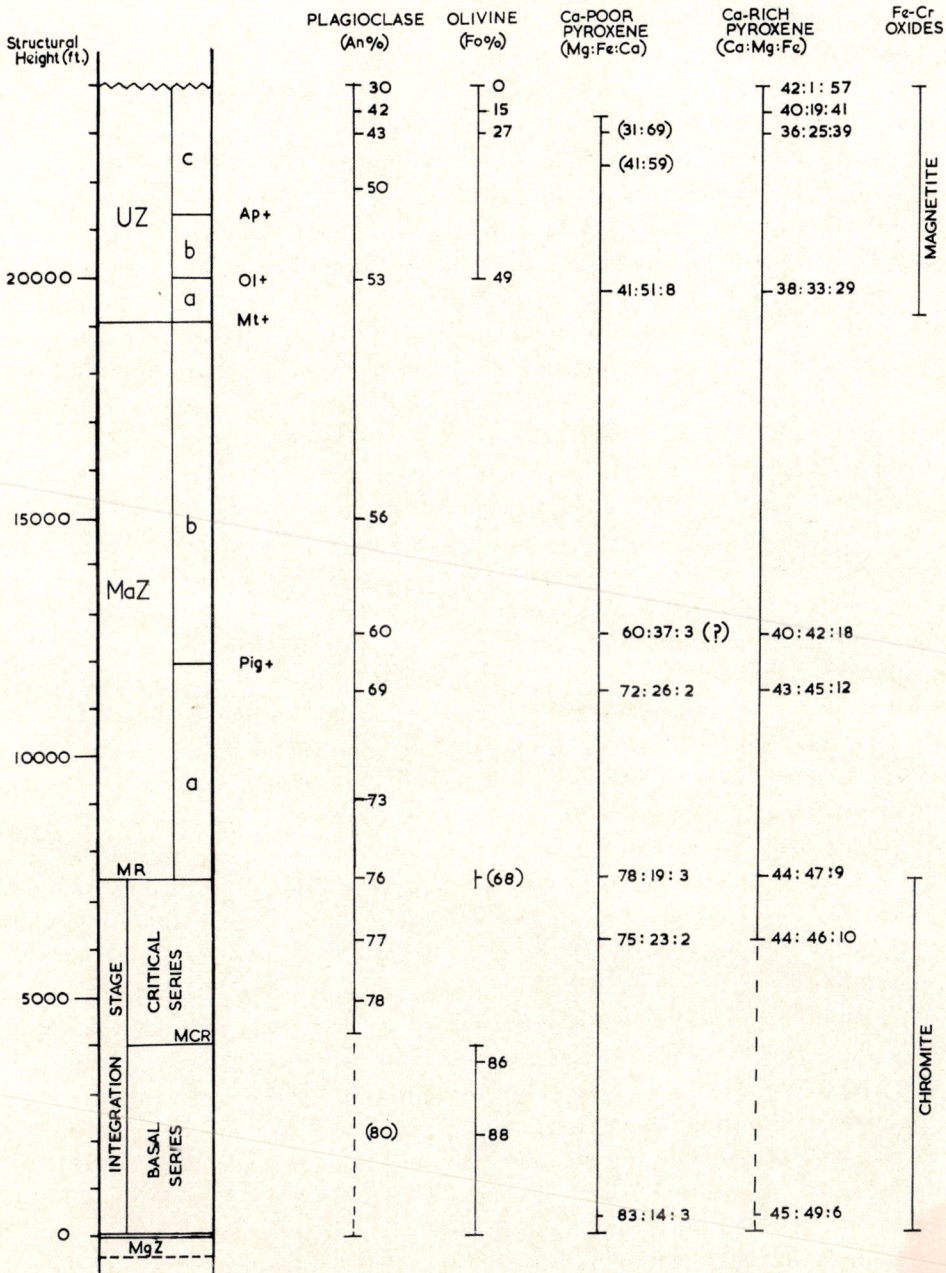

FIG. 192.

Reef layer, and the lowest, thick Magnetite layer. Generally, but not always, these have been taken to mark the base of the Critical, Main, and Upper Zones respectively, rather than the top of the Basal, Critical, and Main Zones. An alternative method of delineating zones and sub-zones of a layered series, used first for the Skaergaard and, later, for some other layered intrusions described in this book, is according to the entry or exit of a particular cumulus mineral phase. The two methods do not give the same results in detail, but only in general for the Bushveld layered series, and a choice needs to be made between the two. The Main Chromitite layer marks the horizon at which cumulus chromite becomes an abundant cumulus phase, but thin, chromite-rich layers occur at lower levels (Cameron & Emerson, 1959, fig. 6). Similarly, the Merensky Reef is a distinctive layer in which sulphide and platinum-rich minerals first enter in abundance, but similar layers occur a short distance below the true Merensky Reef layer (Fig. 202). The lowest Magnetite layer is not open to the same criticisms although rare, cumulus magnetite crystals occur sporadically in rocks below this horizon.

Despite these difficulties it does not seem practicable to advocate a major revision of the criteria according to which the Bushveld layered series can usefully be subdivided into zones. Few geologists working on this huge intrusion would find it practicable to map according to the presence or absence of cumulus crystals of a mineral which, though critical from a genetic viewpoint, would be difficult to detect, and to recognize as a cumulus phase, in small amounts. This applies not only to the oxide and sulphide minerals used at present, but also to minerals such as olivine, pyroxene and plagioclase, which could also be used, according to the same argument, for the Bushveld. A second, particularly serious objection to the rigorous use of cumulus phases for subdividing the Bushveld layered series is that the cumulus assemblages are slightly but significantly different in the eastern and western parts of the intrusion. For example, cumulus olivine is not found above the Main Chromitite layer in the eastern Bushveld, but it occurs in rocks up to and including the Merensky Reef in the western Bushveld. No system of zonal nomenclature could take into account all these problems, but providing that the local exceptions are borne in mind then the existing system is broad enough to cause less difficulties, in its usage, than a more rigorous system of subdivision.

We have proposed three modifications to the manner in which the layered series is subdivided. The first change is designed to take into account the differences between the eastern and western layered series, and some irregularities in the cumulus mineral compositions, with height, in the Basal and Critical zones. We believe that at the time of formation of these early-stage cumulates there were periodic fluctuations in the character and compositions of the separating crystals. In Rhum, and in the ultrabasic parts of other layered intrusions, this characteristic feature is attributed to periodic influxes of parental magma into the crystallization chamber (Brown, 1956) and the same hypothesis could be applied to the lower part of the Bushveld layered series. The differences between the eastern and western series are, presumably, due to their formation in separate, adjacent magma chambers in which the periodicity and character of the magma influxes differed slightly. We have called this the *integration stage* of the Bushveld's crystallization history (Fig. 192), the magma present in the chamber and the magma added, periodically, being integrated by mixing. Cooling and fractionation would continue, despite these occasional reversals, and the layered series formed at this stage show both an overall regularity and slight irregularities in the succession of cumulus minerals deposited. We have retained the use of the Main Chromitite layer to sub-divide the integration stage of the lower layered series into a *Basal Series* and a *Critical Series* although, as noted above, the distinction between the two is not so rigid as a division into zones, and will vary in different parts of the intrusion. Within the integration stage, the entry of cumulus plagioclase and augite, and of chromite in abundance, as well as the exit of cumulus olivine and chromite, are features of appreciable significance but occur at levels that vary throughout the lateral extent of the layered series.

The second change in the subdivision of the layered series concerns the Upper Zone. This zone is much thicker and more variable (see Boshoff, 1942) than has generally been appreciated, and represents the latest exposed differentiates produced by a process of crystal fractionation that was, apparently, relatively regular after the integration stage and gave rise to the Main Zone and Upper Zone cumulates. The base of the Upper Zone is still defined as the level at which cumulus magnetite becomes abundant, to form the lowest of the distinctive magnetite-rich layers. Above this level, a detailed study by Atkins (1965) has shown that the Upper Zone can usefully be sub-divided into UZa, b, and c, according to the levels at which an iron-rich olivine and apatite enter as cumulus phases (Fig. 192). Within the Upper Zone, other phase changes include the exit of cumulus Ca-poor pyroxene near the top of UZb and the presence of small amounts of quartz, hornblende, biotite and alkali felspar, viewed as pore material and varying according to the extent of adcumulus growth.

The third change in the subdivision concerns the problematical Merensky Reef horizon, which is generally shown as occurring either at the top of the Critical Zone (Series) or at the base of the Main Zone. Here again, the level varies slightly throughout the intrusion and, for the reasons discussed above, we feel that the Merensky Reef layers should be viewed as having formed during the integration stage; that is, at the closing stages of formation of the Critical Series (Fig. 202) and before the conditions had settled down to a regular, uninterrupted process of crystal fractionation. The sub-division of the Merensky Reef horizon and a discussion of its formation are given later (pp. 365–72).

THE MARGINAL GROUP

This group of rocks has been termed the Basal or Chill Zone by Hall (1932) and the Norite–diabase Zone and part of the Basal Zone by Lombaard (1934). The term Marginal Group is here proposed because these rocks are not part of the layered series, sub-divisible into zones, and because the term 'basal' implies that the rocks are

z

restricted in their initial distribution to the floor of the intrusion, in the sense of being conformable with the dip of the layering. The term Basal Series is used, instead, for the lowest exposed rocks belonging to the layered series. The terms ' chill ' and ' norite–diabase ' are restrictive in describing what may ultimately turn out to be a complex assemblage of variable rock-types.

The study of these rocks has been hindered owing to the difficulties of distinguishing between those belonging to the main intrusion and those of the preceding, sill-phase of the Bushveld Complex. No clear example of steeply-inclined, marginal-envelope rocks has been described from any part of the main intrusion, the so-called chilled facies having been collected chiefly from a fine-grained norite dipping conformably with, and overlying, the Magaliesburg quartzite of the Pretoria Series (Fig. 193) at the eastern margin of the intrusion, in the Lydenburg district (Hall, 1932, pp. 20 & 34). Hall estimated this group to be about 400 ft thick, although much of it may be a sheet of norite (the Hendriksplaats norite) which could be an intrusive sheet in the locally uplifted floor of the intrusion (Willemse, 1964, p. 103 and frontispiece). In the western Bushveld, a ' diabase quartz norite ' of this group is particularly extensive in the Marico district (Hall, 1932, p. 308) and well-developed southwest of the Pilanesberg (Wagner, 1924). However, as pointed out by Hall (1932, pp. 308–9) these fine-grained diabases, as in the eastern Bushveld, are almost indistinguishable from the extensive basic sills intrusive into marginal or floor sediments.

In view of the difficulty in detecting a true chill to the Bushveld intrusion, or in defining a marginal facies, the likely nature of the parental magma cannot easily be estimated. However, the possibility remains that some of the early sills injected into the Pretoria Series represent earlier intrusions of the Bushveld parental magma or, in a few cases offshoots from the margin of the main intrusion (see Hall & du Toit, 1924, p. 91). In that case, the well-sampled noritic diabases from the eastern margin, whether a true chilled facies or, instead, marginal offshoots or sills, would serve a similar purpose. In slight support of this contention the specimens so far analyzed have compositions close to that of high-alumina tholeiitic basalts, closely comparable with the analyzed chilled facies of the Stillwater intrusion (Table 26) and appropriate to an intrusion in which a great thickness of norites have crystallized. The most-quoted analysis is of a fine-grained hypersthene gabbro from near Pretoria (Daly, 1928), given in Table 26, the normative composition being slightly oversaturated (2·5% quartz). A similar rock, texturally, is described from the eastern contact at Groothoek (Hall, 1932, p. 24).

In a recent traverse of the Lydenburg section of the intrusion, Dr F. B. Atkins paid special attention to the eastern zone and we have made a new analysis of a fine-grained hypersthene gabbro (Fig. 212), apparently free from modal olivine or quartz, which he believed was as representative of a chilled facies as was possible to obtain in that region (Table 26). The rock is of particular interest in that the CIPW norm shows about 5% normative olivine, diminishing the likelihood of contamination with siliceous sediments. An analyzed diabase from a sill intruding the country rocks in this region (Daly, 1928, table 16, anal. 1) contains minor amounts of modal olivine, so it is still possible that Atkins may have sampled from one of the sills which occur near the contact zone. Nevertheless, these rocks are more likely to represent the uncontaminated, parental basic magma of the Bushveld Complex than the varieties containing biotite,

TABLE 26

Chemical analyses of fine-grained gabbros from the Marginal Group of the Bushveld intrusion, with comparisons

	1	2	A	B
SiO_2	50·55	51·45	50·68	48·08
TiO_2	0·66	0·34	0·45	1·17
Al_2O_3	15·23	18·67	17·64	17·22
Fe_2O_3	1·04	0·28	0·26	1·32
FeO	10·07	9·04	9·88	8·44
MnO	0·23	0·47	0·15	0·16
MgO	8·30	6·84	7·71	8·62
CaO	11·30	10·95	10·47	11·38
Na_2O	2·24	1·58	1·87	2·37
K_2O	0·19	0·14	0·24	0·25
Cr_2O_3	0·01	...	0·04	...
P_2O_5	0·12	0·09	0·09	0·08
H_2O^+	0·24	0·34	0·42	1·01
	100·18	100·19	99·90	100·17

1. Fine-grained hypersthene gabbro (SA1087), Marginal Group. 4½ miles east of Klip River, along road between Lydenburg and Roos Senekal. Anal.: Miss H. Hegedus, Univ. of Oxford.
2. Fine-grained hypersthene gabbro, Marginal Group. 1 mile south of Sjambok Railway Station, northwest of Pretoria. Anal.: E. G. Radley (Daly, 1928, No. 129, p. 727).
A. Hypersthene dolerite, Chilled Border Zone, Stillwater intrusion. (Hess, 1960, table 12.)
B. Chilled olivine gabbro (EG4507), Marginal Border Group, Skaergaard intrusion. (Wager, 1960, table 2.)

quartz, apatite, hornblende and apatite (Liebenberg, 1942; Wagner, 1924, p. 32; Hall & du Toit, 1924, p. 93). The new analysis is plotted on an Mg–Fe–(Na+K) diagram (Fig. 217) and later discussed in relation to the chemistry of the intrusion (pp. 400–03). Expressed in terms of the components of this diagram, the analyzed Bushveld rock is very similar in composition to the rocks taken to represent the original magma of the Stillwater and Skaergaard intrusions. Although contamination cannot be ruled out for the Bushveld rock, the Al_2O_3 and Fe_2O_3 contents are not unusually high, and the SiO_2 content is such that olivine appears in the norm. Hence, despite the lack of convincing field relations the recently analyzed, fine-grained hypersthene gabbro serves as an example of the type of locally available, undifferentiated basic magma which may have been the parental magma of the Bushveld layered intrusion.

TABLE 27

CIPW norms of analyses from Table 26

	1	2	A	B
Qz	...	2·94	...	...
Or	1·12	0·83	1·39	1·67
Ab	18·95 } 51·03	13·37 } 57·64	15·72 } 56·17	19·91 } 57·14
An	30·96	43·44	39·06	35·56
Di { Wo	10·16	4·30	5·04	8·35
Di { En	5·41 } 20·00	2·13 } 8·51	2·60 } 9·95	5·00 } 16·25
Di { Fs	4·43	2·08	2·31	2·90
Hy { En	11·30 } 20·57	14·91 } 29·50	16·70 } 31·82	5·00 } 7·90
Hy { Fs	9·27	14·59	15·12	2·90
Ol { Fo	2·78 } 5·29	...	...	7·98 } 13·28
Ol { Fa	2·51	...	...	5·30
Mt	1·51	0·41	0·46	1·86 } 4·14
Ilm	1·25	0·65	0·91	2·28
Ap	0·28	0·21	0·20	0·34
Chr	0·02	...	0·07	...
Plagioclase	An_{60}	An_{75}	An_{68}	An_{62}
Hypersthene	En_{55}	En_{51}	En_{53}	En_{63}
Olivine	Fo_{62}	...	...	Fo_{60}

THE BASAL SERIES

The lowest-exposed layered rocks of the basic intrusion are well seen in the eastern Bushveld, just to the south of the Olifants River and to the east of Jagdlust (Fig. 190). Here the layering dips to the southwest, from about 35° near the margins to about 20° further in. The total exposed thickness of this series is about 4000 ft, the layers consisting of ultramafic bronzitites, harzburgites and dunites, with thin layers of chromitite, particularly towards the top. Exposures are good along the south bank of the Olifants River, whilst along the road to the south the bronzitite layers, dipping at about 30°, have weathered to produce the striking row of dip and scarp features generally known as 'the Pyramids' (Fig. 194). The layers above the Pyramids are poorly or non-exposed, as shown in an aerial photograph of this area (Willemse, 1964, frontispiece). Further exposures of the Basal Series in other regions are mentioned by Willemse (1964, p. 104), the exposed, minimum thickness varying from about 2000 ft in the Rustenburg region to about 5000 ft at Northam.

The Basal Series of the integration stage has not previously been described to the

FIG. 193. Westward-dipping sediments of the Transvaal System, eastern floor of the Bushveld. They dip at about the same angle as the layered igneous rocks of the Bushveld intrusion.

FIG. 194. Bronzite cumulates of the Basal Series, Jagdlust region, dipping southwestward. The preservation of the well-known ' Pyramids ' is due to the easier weathering of the interlayered olivine cumulates. (The mountains in the background, behind the Olifants River valley, are of marginal country rocks.)

same extent as the overlying Critical Series, although an unpublished thesis by Schwellnus (1956) contains some relevant observations. The layering is on a large scale and the series consists of a few thin layers of olivine cumulates (each *c.* 20 ft thick) alternating with thick layers of bronzite cumulates. Although Schwellnus associated the olivine-rich layers with the local intrusive plugs of peridotite, postulating multiple injection for the alternating layers, our own observations lead to the conclusion that the layers are crystal accumulates. In addition to the more obvious olivine- and bronzite adcumulates, there are bronzite–olivine adcumulates (harzburgites) and olivine heteradcumulates (with poikilitic bronzite and, occasionally, poikilitic plagioclase and augite). The scale of the layering is similar to that found in the Ultramafic Zone of the Stillwater intrusion (Ch. XIII) in which olivine, bronzite and chromite are also found as the only cumulus phases. The ultrabasic macro-units in Rhum (Ch. X) are also similar in scale, but differ in the presence of cumulus plagioclase and augite. Hypotheses relating to the development of macro-rhythmic units, which are particularly characteristic of ultramafic layering, have been discussed in Ch. XII and XIII.

The cumulus olivines and bronzites show little variation in composition with height in the layered sequence, the olivine changing upwards from Fo_{88} to Fo_{86} according to optical measurements (Fig. 192). The bronzite varies from En_{87} to En_{83}, but irregularly with height. Although a few of the rocks are monomineralic adcumulates, the majority are heteradcumulates. Thus most of the so-called bronzitites consist of cumulus bronzite enclosed by large, poikilitic plates of diopsidic augite and calcic plagioclase. Both of the latter are the highest-temperature members of their solid solution series to be found in the Bushveld intrusion. The plagioclase (from optical measurements) is An_{80}, while the analyzed augite (Hess, 1949) is $Ca_{45}Mg_{50}Fe_5$.

Towards the top of the Basal Series, numerous thin layers of chromite-cumulates occur (Cameron & Emerson, 1959, fig. 6; Cameron, 1963, fig. 4). Their grouping with the Basal Series is purely for convenience in using the overlying, thick Steelpoort Seam (Main Chromitite layer) as the base of the Critical Series. Certainly cumulus chromite is present throughout the Basal Series, not only as sporadic crystals in the olivine- and bronzite cumulates but even as a thin layer, 3500 ft below the well-developed chromitite layers (Willemse, 1964, p. 104).

THE CRITICAL SERIES

These rocks are characterized by the most impressive development of fine-scale layering found in the Bushveld intrusion. From the base of the Steelpoort (Main) Chromitite layer to the top of the Merensky Reef layers, the total thickness reaches a maximum, in the eastern Bushveld, of about 3500 ft. Aspects of the layering have been described recently by Cameron and Emerson (1959), Cameron (1963), and Ferguson and Botha (1963).

Amongst the cumulus minerals liable to be present in this series are, firstly, chromite, concentrated to form the Steelpoort and Leader seams but persisting as a cumulus phase, and forming other seams, throughout the series. Cumulus bronzite may occur at any level of the Critical Series, as in the underlying Basal Series. Although cumulus olivine disappears just below the base of the Critical Series in the Eastern Bushveld, it is found as a cumulus mineral in rocks as high as the Merensky Reef, at the top of the Critical Series, in the Western Bushveld. At about 100 to 200 ft above the base of the series, cumulus plagioclase first appears, while at 1000 ft above this level, cumulus augite first appears. Thus, throughout most of the series, five cumulus minerals, some of greatly differing densities, are present (chromite, orthopyroxene, plagioclase and, less commonly, augite and olivine). The strongly contrasted, often adjacent layers of chromite- and plagioclase cumulates give rise to the most impressive layering, a similar contrast not being encountered in the Main Zone but next found with the incoming of cumulus magnetite at the base of the Upper Zone.

The best impression of the degree of layering in this zone, even down to a scale of a few millimetres, is to be found in the descriptions of core samples and rock surfaces by Cameron and Emerson (1959) and Cameron (1963), from studies in the Jagdlust region of the eastern Bushveld. We have visited this region, examining, also, the large number of available core-sections, and feel that the rocks can best be described as cumulates in order to avoid the allocation of specific rock names, such as anorthosite, mafic norite, felspathic pyroxenite, anorthositic norite and so on, to thin layers which often grade one into the other. The rocks consist of varying proportions of cumulus minerals which can be listed in order of their relative abundance, while this nomen-clature also allows consideration of the processes of intercumulus crystallization (see Ch. IV). In regard to the latter point, the Bushveld nomenclature becomes increasingly confusing with the use of extra field names such as 'spotted anorthosite' and 'mottled anorthosite'. For example, in terms of the cumulus phases, the pure bronzitites are bronzite adcumulates; the anorthositic norites are bronzite–plagioclase adcumulates; the spotted anorthosites are plagioclase–bronzite adcumulates; and the pure anorthosites are plagioclase adcumulates. Thus an ideal, graded unit involving only cumulus bronzite and cumulus plagioclase can be more simply and logically expressed according to the varying, inter-related cumulus assemblage than from the use of the four distinct rock names. In such a graded unit, intercumulus minerals may also be present, as poikilitic plates. The 'mottled anorthosite' is a good example of a heteradcumulate, with large, poikilitic bronzites enclosing cumulus plagioclase crystals, and it has often been recorded from horizons where such a rock would be expected according to the concept of crystal accumulation, that is, between plagioclase–bronzite adcumulates ('spotted anorthosites') and overlying, plagioclase adcumulates ('anorthosites'). This type of sequence is discussed, later, in considering the macro-units of the Merensky Reef horizon as an example (Fig. 202).

The lateral extent of the Critical Series is well known because of the attention paid to the economically important chromitite layers. In the eastern Bushveld, in particular, the rocks are well exposed in a central sector from the Olifants River south to Steelpoort

(Cameron, 1963, fig. 2), while to the south of Steelpoort are exposed the classic, layered, chromite- and plagioclase cumulates of Dwars River. In the region north of Steelpoort, individual layers of chromite cumulates (e.g. a 3-ft layer of the Steelpoort seams) can be traced for about 40 miles along strike, while Cameron (1963, fig. 4) has shown that despite slight lateral variation in thickness, the sequence of layers involving, also, plagioclase- and pyroxene-rich cumulates can be traced over similar distances. South of the Steelpoort River, a slightly different sequence can be traced for about 15 miles and Cameron has suggested (1963, p. 101) that this southern sector may have been down-faulted. Certain features, well shown in the Dwars River and Driekop regions, are discussed below and considered in relation to the possibility, also mentioned by Cameron, that the layers of the southern sector may have formed under special conditions operating nearer the edge of the magma chamber.

The rocks of the Critical Series provide evidence for the general uniformity of the Bushveld intrusion. Although the likelihood of crystallization in separate, interconnected magma chambers exists, especially in view of the distribution of cumulus olivine, broadly similar layered sequences are to be found in areas hundreds of miles apart. Chromitite layers of the Critical Series, associated with similar layers of plagioclase–pyroxene cumulates, can be traced westward into the Brits and Rustenburg regions, and from there around the western edge of the Pilanesberg complex into the distant, northwestern region of Northam (Fig. 190). An even greater impression of regularity is obtained when the full sequence of the Critical Series, up to and including the uppermost layers of the Merensky Reef horizon, are considered. In the eastern Bushveld, the chromitite-bearing Merensky Reef can be traced laterally for about 50 miles along the approximate north–south strike, while it persists, with the underlying assemblage of plagioclase–pyroxene cumulates and chromite cumulates, for about 200 miles when traced into the exposures of the western Bushveld (Willemse. 1964, pl. 1). However, the sequence is missing, entirely or in part, in the lateral extremities of the intrusion in the northeastern (Potgietersrust), northwestern (Northam), and southeastern (Belfast) regions. This could be due to magma overflowing into separate but inter-connected magma chambers at varying stages of fractionation, or to each magma chamber having had a different history, in relation to fresh influxes of parental magma, during this integration stage. For example, near Potgietersrust the Critical Series is missing and blocks of the Basal Series occur amongst the Main Zone cumulates (Willemse, 1964, p. 116). This could have occurred through the influx of magma, at the Main Zone fractionation stage, into a magma chamber which, after the precipitation of Basal Series cumulates, had been depleted of magma (probably by lateral migration) and in which, therefore, Critical Series cumulates had not been deposited. Similar non-sequences are found at other horizons (e.g. in the Northam region, described by Coertze, 1958) and are discussed later.

Layering is rare, or poorly exposed, in the bronzite cumulates of the upper part of the Basal Series, although fine-scale layering involving cumulus bronzite and olivine occurs in a thin zone near the top (Cameron, 1963, fig. 8). With the incoming of cumulus chromite in abundance to form the Steelpoort (Main) Seam at the base of the Critical Series (Fig. 195), fine-scale layering becomes a predominant and spectacular feature of the Bushveld rocks. This involves, chiefly, the concentration in thin layers of varying proportions of cumulus chromite, bronzite, and plagioclase (e.g. Cameron, 1963, fig. 6, 7 & 8). Adjacent layers, according to published descriptions, rarely show a gradation upwards from chromite-, through bronzite-, to plagioclase cumulates, either on a small or a large scale. Small-scale gradations are rare in most intrusions and where present, may alternate with examples showing the reverse relationships (Ch. XII, Fig. 186). However, as Cameron points out (1963, p. 105), the layers in the Critical Series of the Bushveld are characterized by having gradational *contacts*, irrespective of the cumulus assemblage below and above, and this in itself is significant as a property of igneous cumulates. Regarding the layering as viewed on a larger scale, the vertical sections given by Cameron (1963, fig. 4, 5 & 10) show a broad

FIG. 195. Steelpoort chromite-cumulate layer (dark seam, ca. 4 ft thick, with centre at about head-level) overlain and underlain by chromite–bronzite cumulates grading outwards into bronzite cumulates (grey). Another dark, chromite cumulate (the Leader seam, ca. 1 ft thick) can be seen at about 3 ft above the top of the dark Steelpoort layer. Base of the Critical Series. (The mine entrance is that of Jagdlust.)

FIG. 196. General view of the Dwars River bed, showing the extent of exposures of well-layered, chromite-rich and plagioclase-rich cumulates of the Critical Series.

FIG. 197. The regular layering of chromite-rich and plagioclase-rich cumulates, Dwars River. Poikilitic bronzites can be seen as darker areas in the felspathic layers. The small felspathic inclusion in a chromitite layer and the pinching-out of certain layers are attributed to penecontemporaneous slumping, and deposition on an irregular floor, respectively (see text).

alternation between chromite-, bronzite-, bronzite–plagioclase-, and plagioclase cumulates, but a more detailed study of the variations in cumulus mineral proportions and sizes, and of intercumulus mineral assemblages, is needed in order to establish the nature and character of any regular repitition of rhythmic units which may be present.

The most spectacular type of layering in igneous rocks is found where alternating layers consist of almost monomineralic adcumulates, particularly those involving leucocratic and melanocratic minerals. Extreme cases of such layering are to be found in the Bushveld, both in the Dwars River region of the Critical Series (chromite- and plagioclase cumulates) and in the Magnet Heights region of the Upper Zone (magnetite- and plagioclase cumulates). The strongly contrasted nature of the adjacent layers in these areas has tended to perplex many geologists working on the Bushveld and to lead to extreme hypotheses as to their origin (including separate injections of contrasted magmas) which have been extended to include the less extreme types of layering. As we have pointed out in other parts of this book, contrasted cumulates owe their origin to overall processes similar to those producing gabbroic cumulates of a uniform and unremarkable appearance (e.g. Ch. VIII), particular processes of extreme crystal sorting, followed by adcumulus growth, leading to the formation of extreme variants. In this respect the Dwars River cumulates are no more a phenomenon than the layered bronzitites and dunites of the Basal Series, or the bronzitites and chromitites from other levels of the Critical Series.

The layering at Dwars River (Fig. 196–199) consists chiefly of alternating layers of chromite cumulates and plagioclase cumulates. The layers are characterized mainly by their remarkable parallelism, which can be traced laterally without any signs of sharp discordance in the shape of high-angle, cross-cutting relationships. However, as seen in Fig. 197 and 198, the layers tend to pinch-out laterally, this feature having led to suggestions both that chromitite liquids have intruded anorthosite layers and that anorthosite liquids have intruded chromitite layers. Neither hypothesis accounts for the overall parallelism of the layers or takes into account the problems associated with the mechanics of emplacement or the physico-chemical properties of the liquids involved. We are convinced that the layering is the result of crystal accumulation of cumulus chromite and cumulus plagioclase on the contemporary floor of the magma chamber. It is unlikely that this floor, at any one time, would be so flat that the base of an individual cumulate layer would be perfectly horizontal and retain this character for more than a few yards: the fact that many layers in the Bushveld show this property is all the more remarkable. The uppermost, thick chromitite layer shown in Fig. 197 possesses this latter quality, the base being flat and the top, slightly undulating. This suggests that the underlying plagioclases were packed to form an even floor on to which the chromites settled, whereas the later batch of plagioclase crystals was deposited on a slightly uneven floors of chromite crystals. Other chromitite layers (Fig. 197 & 198) give the impression that one thick layer has bifurcated into two thinner layers (illustrated, also, by Cameron, 1963; Ferguson & Botha, 1963). Here, we would suggest that the

FIG. 198. Local irregularities, more extensive than those shown in FIG. 197, in the layering at Dwars River (see text).

FIG. 199. Displaced layers of chromite cumulates at Dwars River, attributed to penecontemporaneous movement and fracture of solified adcumulates, inter-layered with only partly solidified cumulates which responded by slumping (see text).

chromites and plagioclases were deposited on an undulating floor, the apparent bifurcation of the chromite layer, or pinching-out of the felspar layer, being due to the upper deposit of chromite being deposited on top of the lower deposit of chromite on the crests, and on top of plagioclases where the latter had covered the lower deposit of chromite in the hollows.

Apart from this local type of irregularity in the layering at Dwars River, there is evidence for contemporaneous slumping and faulting of the crystal accumulates. Small blocks of partly consolidated felspar cumulates have moved and been deposited with cumulus chromite crystals (Fig. 198) while in other cases, individual layers have been displaced and partly invaded by a more mobile mush of cumulus felspars (Fig. 199). Slumping, involving plagioclase–pyroxene cumulates, has been observed at Dwars River (Fig. 201) and is very similar to the type seen in Rhum (Ch. XI, Fig. 162). Further evidence of contemporary movements is also to be found in the textures of these rocks, the plagioclase frequently showing bent and fractured twin lamellae.

Despite these minor irregularities observed in the layering at Dwars River, the overall regularity is a remarkable feature. Quite apart from the evidence for a regular alternation of cumulus crystals, there are fine examples of the growth of large, poikilitic pyroxenes from the intercumulus liquid (Fig. 200). These extensive, feathery growths appear at first sight to be restricted to individual plagioclase cumulate layers but, in fact, they pass into the adjacent chromite cumulate layers and are in optical continuity across the layer boundaries (as observed, also, by Ferguson & Botha, 1963, p. 4). This is particularly compelling evidence for the layers being an alternating series of heteradcumulates rather than having independent origins. A special characteristic of the Dwars River region is the evidence for appreciable contemporaneous slumping and faulting, and in the lenticular nature of the deposited piles of crystals. These latter features are reminiscent of those found near the margins of the Skaergaard layered series (Ch. V), and support the suggestion made by Cameron (1963, p. 101) that the Dwars River rocks crystallized near the margins of the magma chamber in the Eastern Bushveld.

THE MERENSKY REEF HORIZON

Few parts of the Bushveld intrusion equal this platiniferous horizon for its world-wide repute and economic importance. Since the first prospecting by Hans Merensky in the 1920s, it has been perhaps the most important of the world's sources of platinum. The horizon consists of a series of igneous cumulates, of which the main Pt-bearing layer (the ' Reef ') varies only between 3 and 15 ft in thickness and yet can be traced laterally over the greater part of the intrusion. From the surface outcrops and mine data, it extends for about 80 miles along strike in the eastern Bushveld, and about 120 miles in the western Bushveld (Fig. 190). The platinoid and associated sulphide minerals (pp. 399–400) always occur in the same coarse ' felspathic pyroxenite ' (a bronzite-olivine– chromite cumulate) and must, therefore, have been deposited with the cumulus

FIG. 200. Close-up of the junction (see Fig. 197) between layers of chromite-rich cumulates (dark) and plagioclase-rich cumulates (light) at Dwars River. The poikilitic bronzites (grey) in the felspathic layer appear to be confined to that layer, but in thin-section are seen to extend into the adjacent chromite-rich layers. This means that the layers are not independent in origin but, rather, that intercumulus growth extended across the junctions between the contrasted layers of cumulus crystals.

FIG. 201. Slumping in bronzite–plagioclase cumulates at Dwars River.

crystals at this particular horizon (Fig. 202 & 203). Similar cumulates at other, adjacent horizons may contain small amounts of platinum or be barren, so that neither the type of cumulate nor the precipitation of platinum can be viewed as restricted to the Merensky ' Reef ' horizon itself.   Nevertheless, taking the thicker horizon as a whole (Fig. 202), the physical and chemical conditions in the Bushveld magma chambers during this relatively brief episode in the crystallization sequence must have been unique.

Detailed studies of the Merensky Reef horizon at the Rustenberg Platinum Mine (Schmidt, 1952) and the Union Platinum Mine (Feringa, 1959) have revealed such complexity that it is necessary to consider the relatively thin, Pt-bearing layer in relation to a thickness of about 500 ft of layered rocks at this horizon (Fig. 202). Apart from the local structural complexities, the nomenclature for the rock types has become so unwieldy as to create difficulties in describing and interpreting the relatively regular layered features.   Feringa (1959, fig. 2) has presented the first systematic subdivision of the horizon into layered groups and sub-zones, but has retained a host of rock names such as ' felspathic pyroxenite ', ' spotted anorthosite ' and ' pyroxenitic diallage norite ', as well as the mining terms for layers such as ' Reef ', ' Merensky ', ' Pseudo-Reef ', and ' Bastard Reef '.

The non-platiniferous layers consist, essentially, of varying proportions of cumulus chromite, bronzite and plagioclase, with sporadic crystals of cumulus augite and olivine in rare, thin layers.   As mentioned earlier and outlined below, the cumulate nomenclature describes the assemblages satisfactorily and indicates the character of the repeated units more clearly than do the isolated rock-names.   Also, the mineral nature and crystal size of the intercumulus phases provide a consistent picture which is otherwise obscured by the use of prefixes such as ' spotted ' and ' mottled '.   We have taken Feringa's data and applied the cumulate nomenclature to each rock type (Fig 202), from which it can be seen that a regular series of five major rhythmic units occurs at this horizon, termed ' macro-rhythmic units (MRU) ',   The units differ slightly from Feringa's sub-zones, the basis for our definition being according to a regularly repeated cumulus assemblage.

Each unit consists of a mafic lower part with cumulus chromite, bronzite and, less commonly, olivine or augite, and a felsic part rich in cumulus plagioclase. The upward gradation in most layers is well shown, cumulus plagioclase increasing at the expense of cumulus bronzite until bronzite occurs only as an intercumulus phase. In terms of these two minerals alone, the sequence within each unit (listed from top to base) can be generalized as follows, the earlier rock-names being given in parentheses:

Cumulus plagioclase; intercumulus bronzite as large, poikilitic crystals (mottled anorthosite).

Cumulus plagioclase in excess of cumulus bronzite (spotted anorthosite).

Cumulus bronzite in excess of cumulus plagioclase (anorthositic norite).

Cumulus bronzite; intercumulus plagioclase as large, poikilitic crystals (felspathic pyroxenite).

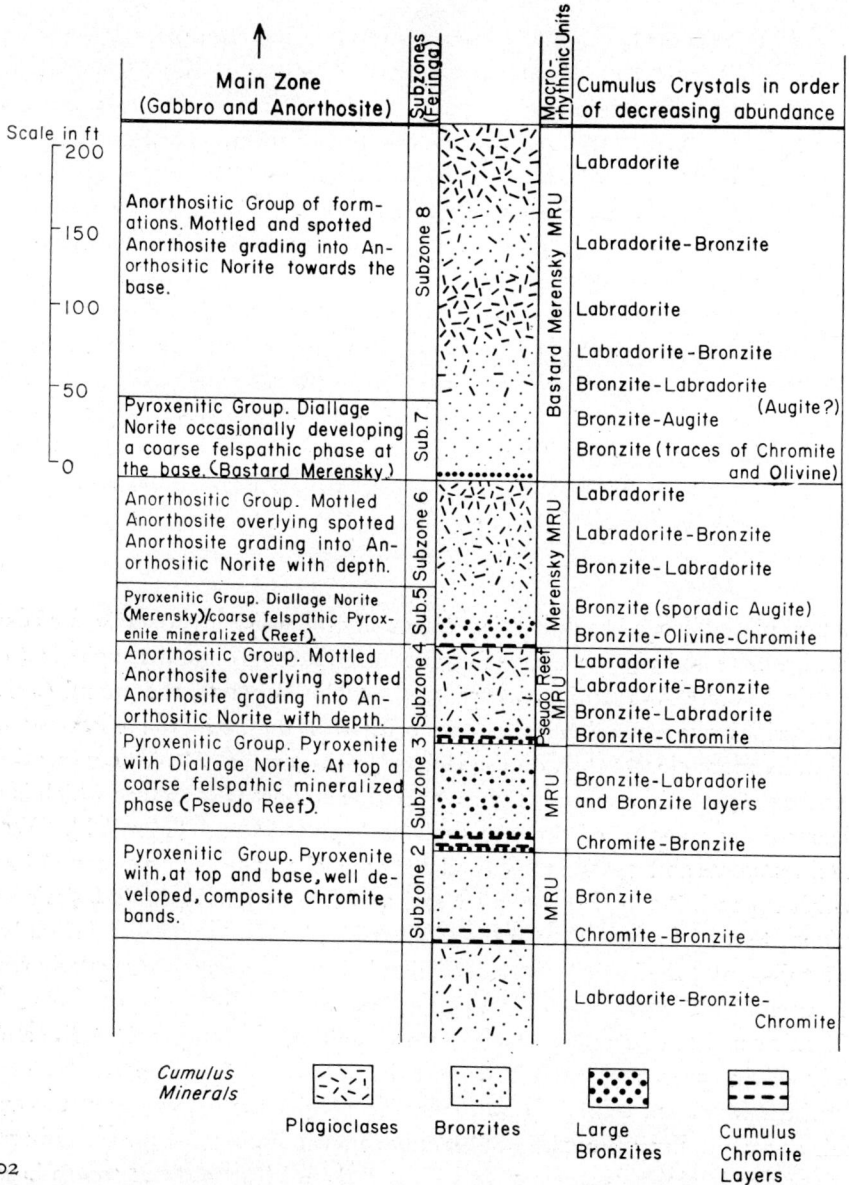

Scale in ft

Main Zone (Gabbro and Anorthosite)	Subzones (Feringa)	Macro-rhythmic Units	Cumulus Crystals in order of decreasing abundance
200 —			Labradorite
Anorthositic Group of formations. Mottled and spotted Anorthosite grading into Anorthositic Norite towards the base.	Subzone 8	Bastard Merensky MRU	Labradorite-Bronzite
150 —			
100 —			Labradorite
			Labradorite-Bronzite
50 —			Bronzite-Labradorite
			Bronzite-Augite (Augite?)
Pyroxenitic Group. Diallage Norite occasionally developing a coarse felspathic phase at the base. (Bastard Merensky.)	Sub. 7		Bronzite (traces of Chromite and Olivine)
0 —			
Anorthositic Group. Mottled Anorthosite overlying spotted Anorthosite grading into Anorthositic Norite with depth.	Subzone 6	Merensky MRU	Labradorite
			Labradorite-Bronzite
			Bronzite-Labradorite
Pyroxenitic Group. Diallage Norite (Merensky)/coarse felspathic Pyroxenite mineralized (Reef).	Sub.5		Bronzite (sporadic Augite)
			Bronzite-Olivine-Chromite
Anorthositic Group. Mottled Anorthosite overlying spotted Anorthosite grading into Anorthositic Norite with depth.	Subzone 4	Pseudo Reef MRU	Labradorite
			Labradorite-Bronzite
			Bronzite-Labradorite
Pyroxenitic Group. Pyroxenite with Diallage Norite. At top a coarse felspathic mineralized phase (Pseudo Reef).	Subzone 3	MRU	Bronzite-Chromite
			Bronzite-Labradorite and Bronzite layers
			Chromite-Bronzite
Pyroxenitic Group. Pyroxenite with, at top and base, well developed, composite Chromite bands.	Subzone 2	MRU	Bronzite
			Chromite-Bronzite
			Labradorite-Bronzite-Chromite

Cumulus Minerals

Plagioclases   Bronzites   Large Bronzites   Cumulus Chromite Layers

FIG. 202

Diagrammatic section showing the macro-rhythmic units (MRU) of the Merensky Reef horizon (Union Mine) at the top of the Critical Series. The left-hand part shows the subzones and their description as given by Feringa (1959), to which we have added a diagram re-interpreting the data in view of the distribution of the cumulus phases. From the distribution a pattern of rhythmic units can be recognized, each having a melanocratic base grading upwards into a leucocratic top. From the pattern defined here, the two platinoid-bearing layers occur at the *bases* of the Merensky and Pseudo Reef macro-rhythmic units, associated with rock textures and cumulus silicates similar to those at the base of the Bastard Merensky and the other macro-rhythmic units.

Elsewhere in the intrusion, but not at this particular horizon, extreme bronzite adcumulates (pyroxenites) and plagioclase adcumulates (anorthosites) are found and, with the above sequence idealized, they would occur at the base and top of the sequence, respectively. With such a nomenclature available, the other cumulus phases may also be listed where present (Fig. 202) although a further distinction between adcumulates and heteradcumulates (and mesocumulates and orthocumulates where present) would need to be made in describing particular rocks, in order to take account of the phases crystallized from the intercumulus liquid.

The important Pt-rich layer is generally referred to broadly as the 'Merensky Reef' (Fig. 203) but, in fact, is more specifically called the 'Reef' (Feringa, 1959, p. 4) by the mining geologists, the overlying, finer-grained layer (with less important platinum) being called the 'Merensky'. The Reef layer varies laterally from about 3 to 15 ft in thickness, the platinum being generally concentrated in the top 30 inches although sometimes in the bottom 24 inches as well. The layer consists chiefly of large cumulus bronzites poikilitically enclosed by plagioclase, but with a thin layer of

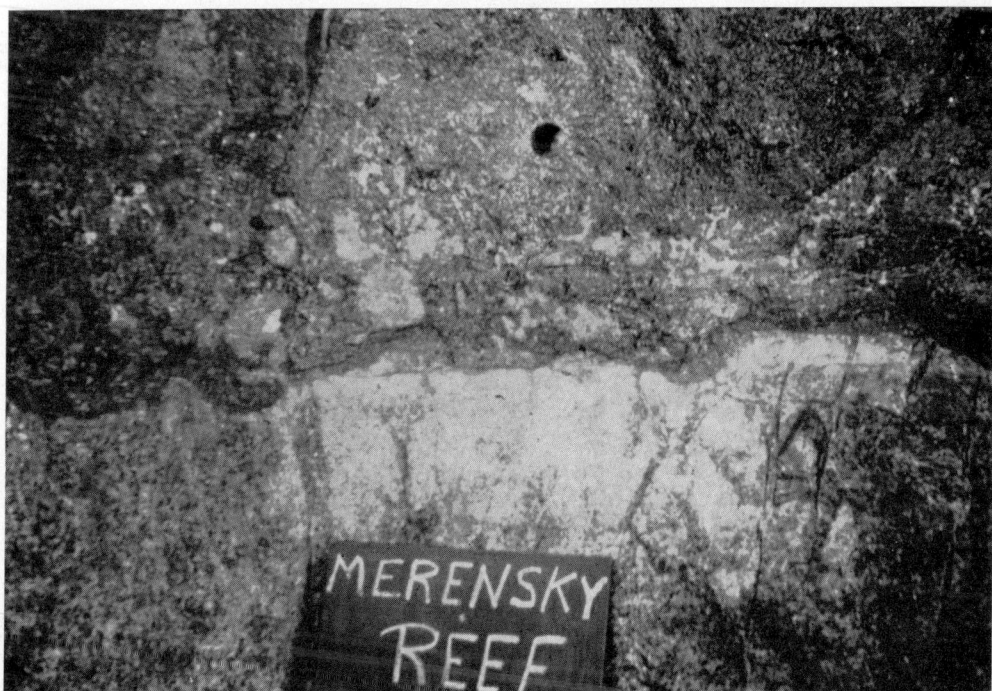

FIG. 203. The platinoid-bearing 'Reef' (a coarse, bronzite-rich cumulate) at the base of the Merensky Reef macro-rhythmic unit and overlying the plagioclase-rich cumulate at the top of the Pseudo Reef macro-rhythmic unit (see Fig. 202). (For scale, the poster-board is about 2½ ft long. Photograph taken underground at Rustenburg, by courtesy of the Rustenburg Platinum Mines Ltd.)

2 A

cumulus chromites at the base (Fig. 202) and with occasional crystals of cumulus olivine. The silicate compositions are shown on Fig. 192, and the Pt-minerals and associated sulphides described on pp. 399–400. The overlying Merensky layer is similar, except that the cumulus bronzites are smaller and the rock is characterized by large, sporadically distributed crystals of augite ('diallage'), probably representing cumulus 'composite clusters'.

Although it is expedient for the mining geologists to define the Pt-rich Reef layer with precision, we have grouped it with the overlying Merensky layer (Fig. 202) because both layers are bronzite cumulates. Above, the Merensky macro-rhythmic unit becomes progressively more felspathic, as discussed in connection with the general sequence for each unit. The macro-unit underlying the Reef is very similar, the bronzite cumulates at the base having been called the 'Pseudo Reef' by the mining geologists because although texturally similar to the Merensky Reef layers, they are non-platiniferous. Similarly, the 'Bastard Merensky' layers (Fig. 202) are similar in texture to the underlying Merensky Reef layers, but are only mildly platiniferous. Hence, the repetitive nature of the macro-units at this general horizon is emphasized once again, and there is clearly no case for suggesting that the thin, Pt-rich layer is otherwise unique in relation to the layers and the macro-units above and below it. As shown in Fig. 202, the seven sub-zones of Feringa (1959) have been reviewed and in their place, a series of five macro-rhythmic units has been defined.

When examined in this sort of detail, the general practice of using the thin 'Merensky Reef' as a horizon to mark the base of the Main Zone, or the top of the Critical Zone (Series), becomes difficult to apply. It has been suggested (e.g. Hess, 1960) that this Pt-rich and sulphide-rich layer may represent the first deposit from a fresh surge of basic magma, and although this is a valuable concept for explaining the whole series of units shown in Fig. 202 it cannot be applied so rigorously as to divorce the Pseudo-Reef from the Bastard Reef. Clearly all three 'Reefs', with the associated layers of each macro-unit, have formed under similar conditions and the magmatic event visualized by Hess must be taken to involve conditions in which precipitation took place under unstable equilibrium for an appreciable period. Hence, the whole sequence shown in Fig. 202 is viewed as the upper part of the Critical Series.

It has been suggested earlier (p. 353) that the whole of the Basal and Critical Series are best viewed as having developed during an 'integration stage' when the Bushveld magma chamber(s) were unstable and periodically replenished by fresh supplies of parental magma. The earlier workers introduced the term 'critical' and this appropriate term has even more significance when related to the conditions operative towards the closing stages of formation of the Critical Series, particularly as the specific, critical conditions probably varied slightly from one subsidiary chamber to the other. The overlying Main Zone differs in that layering is less well-developed, while augite and plagioclase become the significant cumulus phases rather than bronzite and chromite.

According to this grouping the Pt and sulphide minerals appear to have separated along with the silicate minerals, by bottom accumulation in distinct layers. The vast and remarkable extent of these thin inter-related layers, throughout the greater part of the Bushveld intrusion, obviously precludes any suggestion of local mineralization. It is now clear that any attempts to explain them should be based on the fact that the ore-bearing layers are part of a rhythmically repeated series of layered units, and that where the ore minerals are found they occur at specific, comparable levels within each unit.

Despite the overall regularity, there are certain peculiar structural features at localized Merensky Reef Horizons. Schmidt (1952) has described 'dome structures' and 'circular depressions', and Feringa (1959) has described broad fold-structures and unconformities. From these accounts it appears that there is occasionally a marked discontinuity between the Pt-rich 'Reef' layer and the overlying, Pt-poor 'Merensky' although the higher, Pt-poor 'Bastard Merensky' layer is also locally unconformable with the underlying 'Merensky' layer (Feringa, 1959, figs. 3-9). None of these features suggests that the Pt-rich layer can be separated from the other layers according to some special magmatic event. However, the structures do suggest that during the formation of the whole sequence shown in Fig. 202, the conditions in the magma chamber were relatively unstable and that warping of the floor occurred during crystal accumulation. The 'dome structures' and 'anticlinal structures' described by Schmidt and by Feringa are apparently due to warping of almost-solid layers, subsequent cumulates being either banked around them or not deposited on the crests. The 'circular depressions' (Schmidt, 1952, fig. 3), on the other hand, appear to have been scoured from unconsolidated cumulates. Schmidt suggested that the scouring could be attributed to eddy currents in the magma, and this would seem to be a likely result of the mixing of residual and fresh, hotter supplies of magma.

Hess (1960) has proposed that the Merensky Reef horizon marks the stage at which a fresh surge of parental magma entered the chamber, precipitating the ore-minerals early, presumably from an immiscible, sulphide-rich liquid. This, applied to the five units as a whole, would also account for some of the structural discontinuities likely to develop during a turbulent and unstable episode but it is difficult to account for other features. For example, if the fresh magma was relatively small in bulk compared with the residual magma, the large amount of deposited Pt would be difficult to explain, as would the appreciable reversals in the compositions of the felspars and pyroxenes. If, on the other hand, the relative magma bulks were in the reverse order, olivine and chromite should be precipitated in abundance. Instead, olivine is rare and chromite (abundant in the Critical Series just below) is not found above the Merensky Reef Units. Despite these difficulties, Hess's explanation is the best available, further work being needed to test it in the light of the features discussed above. In particular, a study of the Pt-distribution in minor or trace amounts, at lower and higher horizons and in the marginal rocks, would be of interest. Also, the bulk of deposited Pt could be estimated roughly in relation to the bulk of cumulates, either in the intrusion as a whole or above the Merensky Reef Units, and this value related to experimental studies on the solubility of Pt in basaltic magmas. If only a moderate amount of new magma accounted for the precipitation of the Pt deposits, as seems most likely, then this magma must have been relatively depleted in Cr, and probably enriched in Pt, relative to the earlier magmas of the Bushveld. Undoubtedly, the Bushveld region as a whole shows evidence that the igneous activity stemmed from a source in which Pt enrichment was proceeding, as shown by the fact that the later, hortonolite-dunite pipes are particularly rich in Pt minerals (Stumfl, 1961).

The Main Zone

This is by far the thickest zone of the intrusion, varying from about 17 000 ft in the eastern Transvaal to about 10 000 ft in the western, Rustenburg area. The base and top of the zone have been defined differently according to various authors, as described in an earlier section. We have placed the Merensky Reef Units at the top of the Critical Series rather than at the base of the Main Zone, as proposed also by Feringa (1959, fig. 2) and discussed above. Hence, as shown on Fig. 202, the base of the Main Zone has no obvious characteristic other than that the macro-rhythmic units of the Merensky Reef horizon give place to a sequence of cumulates in which layering is poorly developed or absent. Also, we have placed the lowest of the magnetite layers at the base of the Upper Zone rather than the top of the Main Zone, for reasons discussed earlier. This means that the Main Zone, as shown on Fig. 192, is slightly thinner than other accounts show it to be, although the thickness which we have given is also a generalized version of the sequences observed, chiefly, in the eastern Transvaal.

The Main Zone rocks, according to this classification, show few layered features of an interest comparable with those in the Critical Series and Upper Zone. In particular, this may be attributable to the absence of a cumulus ore mineral (Fig. 192) and of cumulus olivine, the rocks being broadly gabbroic with plagioclase, clinopyroxene and, to a lesser extent, orthopyroxene or inverted pigeonite as the only cumulus phases. The zone is generally referred to as being massive and unlayered, but this is only in relation to the other zones. The majestic exposures such as those forming the Lulu mountain-range in the eastern Transvaal display the most impressive single array of scarp and dip features seen in a layered intrusion (Fig. 204), and this in itself bears testimony to the presence of conformable, low-dipping layers of variable mineralogy and, therefore, variable resistance to weathering. Despite this broad feature, however, individual scarp slopes of the Lulu mountains present a monotonous appearance, as do western exposures of the Main Zone. The best fine-scale layering of the Main Zone is found in the Bon Accord quarry near Pretoria, and in the Olifants River near the main road between Lydenburg and Pietersburg (Ferguson & Botha, 1963, pp. 8–9, and pl. 6, fig. 1). The layering in these areas consists of alternating pyroxene-rich

FIGURE 204

Panoramic view of the distinctive scarp-and-dip topography of the Main Zone, eastern Bushveld. The feature (once termed 'rifting,' or 'pseudo-stratification') is due to the layering of the Bushveld cumulates, and to the relative resistance to weathering of pyroxene-rich and plagioclase-rich layers.

and felspar-rich cumulates (Fig. 205), as would be expected from the cumulus assemblage present throughout the Main Zone.

The most striking feature of the Main Zone rocks is the igneous lamination, to be seen at all horizons, which is better developed here than in any other zone. This is a reflection of the habit of the cumulus phases, particularly the plagioclases which must have crystallized as markedly tabular primocrysts. Van den Berg's detailed petrofabric analysis of the Bon Accord rocks (1946) emphasized the nature of the igneous lamination, the cumulus plagioclases and orthopyroxenes showing, also, a slight preferred orientation within the plane of layering. These features were contrasted with the absence of any orientation of the augites, as would be expected in layers where the augite crystallized as an intercumulus phase. The mineral compositions are discussed in a later section, but from Fig. 192 it can be seen that the most important phase-change taking place within the Main Zone is the disappearance of cumulus orthopyroxene, its place being taken by a cumulus pigeonite (since inverted to hypersthene).

The field relationships between the underlying Critical Series and the overlying Upper Zone are not elsewhere as conformable as in the Lulu mountains region of the eastern Transvaal. Evidence exists for the lateral transgression of magma, both towards the beginning and the end of the Main Zone of crystal accumulation, as summarized

FIG. 205. Small-scale layering in a plagioclase–bronzite–augite cumulate near the base of the Main Zone. (Hand specimen, $5'' \times 4''$.)

by Willemse (1964, pp. 116–7). In the Potgietersrust district, the exposures of the intrusion are almost entirely of Main Zone and Upper Zone rocks (Berning, 1941), the few underlying exposures being apparently of Basal Series rocks. Further to the south, in the region of Zebediela, xenolithic bodies of Critical Series rocks occur in the lower part of the Main Zone. Both these observations indicate that the magma from which the Main Zone cumulates were being precipitated transgressed into a subsidiary chamber in this northeastern part of the complex. In doing so, blocks of the already-deposited Critical Series cumulates were probably torn from the floor and incorporated as xenoliths in the Zebediela region, while towards the northeastern extremity, in the Potgietersrust district, the magma may have entered a subsidiary chamber in which Critical Series cumulates had not been deposited. Towards the end of the period of Main Zone formation, a further lateral transgression has been described from the western Bushveld, in the Northam region (Fig. 206). According to our zonal classification, this transgression occurred within the period of Main Zone deposition while Coertze (1958) has proposed that this local episode (including cumulates well below the lowest magnetite layer) should mark the base of the Upper Zone. The transgressive relations are well seen in this region, the strike of the later layered rocks being almost perpendicular to the strike of the Critical Series layers with which they

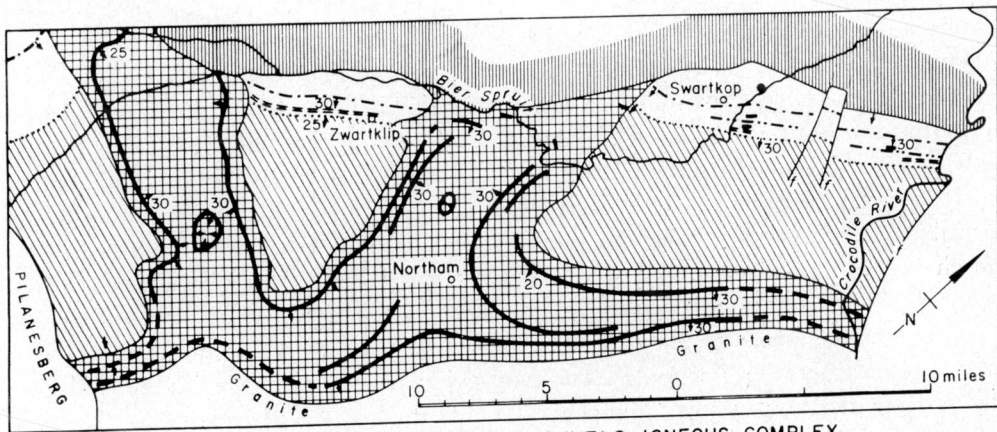

THE MAFIC PORTION OF THE BUSHVELD IGNEOUS COMPLEX
NORTH OF PILANESBERG

A ▥ Superficial deposits   3 ▦ Ferrogabbro (Upper zone)   a ⟋ Magnetite   c ⤳ Upper Chromitite

2 ▧ Gabbro (Main zone)   h ⋯ Merensky Reef   d ⤳ Middle Chromitite

1 ▭ Pyroxenite, anorthosite and norite (Critical and basal series)   e ⤳ Main Chromitite

FIG. 206. Sketch-map showing the locally transgressive relationships in the Northwestern Bushveld (after Coertze, 1958). The local transgression of magma at the time when Upper Zone rocks began to accumulate indicates instability of the magma chamber in this region, as indicated by the local pattern of dips.

are locally in contact. The transgressive, contemporary magma must have cut across earlier Main Zone cumulates to reach this low level of Critical Series cumulates, after which further Main Zone cumulates and magnetite- and apatite-bearing layers of the Upper Zone were locally deposited. The complexity in this western region, in contrast to that in the eastern, Magnet Heights region, is further testimony to the fact that the Bushveld probably crystallized in more than one subsidiary magma chamber.

THE UPPER ZONE

The base of this zone has been defined as that level at which cumulus 'magnetite' (see p. 397) first occurs in significant amounts (Fig. 192) to form the lowest of the spectacular magnetite layers. An horizon such as this provides an ideal basis for mapping the zonal boundaries over great distances, as does the Main Chromitite layer used to define the base of the Critical Series. However, as with chromite in the Basal Series, it is likely that sporadic crystals of cumulus magnetite were deposited in some of the upper layers of the Main Zone, prior to their appearance in large quantities.

The thickness assigned to the Upper Zone varies according to the origin attributed to the acid rocks comprising the uppermost exposures of the intrusion. A maximum thickness of 6300 ft has been quoted by Boshoff (1942) for the Eastern Bushveld, including certain fine-grained granites of uncertain origin (see below), while values of about 3000 ft for the Western Bushveld are at a minimum owing to the marked transgression of the later, Bushveld Red Granite in this region. The thickness shown in Fig. 192, of about 5000 ft, is based largely on traverses of the well-developed sequence in the Tauteshoogte and Blood River regions of the Eastern Bushveld, and does not include certain acid rocks, probably representing late differentiates, which are transgressive and of unknown bulk. In the Western Bushveld, the Upper Zone rocks may reach a maximum thickness of about 6000 ft in the Northam region although local transgressive relationships between Upper Zone, and underlying Main Zone and Critical Series rocks complicate the picture (Fig. 206). As with the Main Chromitite and Merensky Reef horizons, the Magnetite horizons provide striking evidence for the enormous lateral extent of the layering, being found across most of the intrusion (Fig. 190).

In general, the well-layered Upper Zone is characterized by the incoming of magnetite, iron-rich olivine and apatite as important cumulus phases, and of quartz, alkali felspar, hornblende and biotite in minor amounts. The extreme iron-enrichment, closely comparable with that found in the Upper Zone of the Skaergaard intrusion, is not only evidenced by the magnetite layers, unique to the Bushveld, but also by the high Fe^{++}/Mg ratios in the olivines and pyroxenes. The plagioclases are chiefly of andesine composition, so the rocks as a whole should be termed 'ferrodiorites'. Certain acid rocks, free from layering, occur as transgressive sheets within the roof-zone country rocks of the Eastern Bushveld and contain fayalite, ferrohedenbergite and andesine–oligoclase in minor, but significant amounts.

207. Dark layer of titanomaghemite-rich cumulates overlying a pale-
red layer of plagioclase-rich cumulates. The dark layer (c. 6 ft thick) is
y referred to as the Main Magnetite and is here taken as the base of the
r Zone. Magnet Heights, Eastern Bushveld. (N.B. Despite a half-way
parallel to the plane of layering, the ore-rich layer extends from the
e, almost to the top of the photograph.)

FIG. 208. View of the Blood River valley, looking north to the Tauteshoogte
plateau. This small area in the Eastern Bushveld is particularly important in
that traverses up the slopes on the north and south sides of the valley show
continuous exposures of the ferrodiorites and later, acid differentiates of the
Upper Zone. The large boulders are of olivine-bearing ferrodiorite, the
underlying, magnetite-rich layers being worked by the roadside. The top of
the plateau is formed by the resistant, pre-Bushveld, Rooiberg felsites forming
the roof zone.

The best exposure of the lowest layers of the Upper Zone is at Magnet Heights in the Eastern Transvaal. Layers rich in cumulus magnetite alternate with layers rich in cumulus plagioclase (Fig. 207) and dip at a low angle towards the centre of the intrusion. The lowest melanocratic layer, called the Main Magnetite, is about 6 ft in thickness, about five similar but thinner layers occurring at this horizon. According to Hall (1932, fig. 31 & 34) the magnetite layers can be sub-divided into a Lower Set of three layers (72, 15 and 12 in thick) and an Upper Set, 60 ft higher, of three layers (70, 15 and 12 in thick).

The lowest, Main Magnetite layer has a slightly irregular floor, suggesting deposition of magnetite crystals on an irregular, perhaps scoured surface to the underlying anorthositic layer. Although the magnetite layers present a massive appearance in the field (Fig. 207) they consist of discrete crystals and are to be viewed as magnetite adcumulates. Hall (1932, fig. 35) has shown that the distribution of cumulus magnetite and plagioclase in the layers is of the graded type, the amount of cumulus plagioclase increasing gradually towards the top of each of the melanocratic layers. In the more leucocratic layers the magnetite forms poikilitic crystals, so the succession at this horizon can be viewed as comprising six macro-rhythmic units, each of which has a melanocratic (magnetite-rich) base and a leucocratic (plagioclase-rich) top. However, as in the case of the chromitite–anorthosite layers at Dwars River, it is surprising that cumulus pyroxene is not found amongst these remarkable, extreme types of layers.

The Upper Zone succession above the Magnetite Horizon is best seen in the Eastern Bushveld, about 100 miles northeast of Pretoria. There the valley of the Blood River (Fig. 208) runs approximately east–west and magnetite layers outcrop at the roadside along the floor of the valley. The slopes forming the sides of the valley, i.e. the Tauteshoogte Heights to the north (Fig. 208) and the Bothasberg Heights to the south, are formed of Upper Zone ferrodiorites, the higher ground and plateau surfaces consisting of a complex of acid rocks belonging chiefly to the roof-zone. Ferrodiorites are also well exposed in the Western Bushveld near Northam (Coertze, 1958) but there, as in the Potgietersrust region of the Eastern Bushveld (Berning, 1941), later acid intrusions have obliterated part of the succession. The Tauteshoogte region has been described in detail by Boshoff (1942) and the Bothasberg region only briefly by Van der Westhuizen (1945), but as both these reports are unpublished theses (supervised by B. V. Lombaard, University of Pretoria) these significant, iron-rich differentiates of the Bushveld intrusion have received little general appreciation outside South Africa.

Olivine enters as a cumulus phase at about 1000 ft above the base of the Upper Zone and is an iron-rich member with the composition Fo_{49} (Fig. 192). Thus although olivine is absent as a cumulus phase from about 12 000 ft of Bushveld cumulates, its exit and re-entry are generally according to the pattern found in the Skaergaard intrusion and explicable in terms of fractionation in the system $MgO–FeO–SiO_2$. Olivine ferrodiorites outcrop in the lower ground of the Blood River valley (Fig. 208) but are best seen in a stream-section to the south, terminating at a steep cliff of the Bothasberg

FIG. 209. Layered ferrodiorites of the Upper Zone, on the hillside (Bothasberg) at the south side of the Blood River valley.

FIG. 210. Layered ferrodiorites upstream from the locality shown in Fig. 209, containing large clusters of plagioclase crystals.

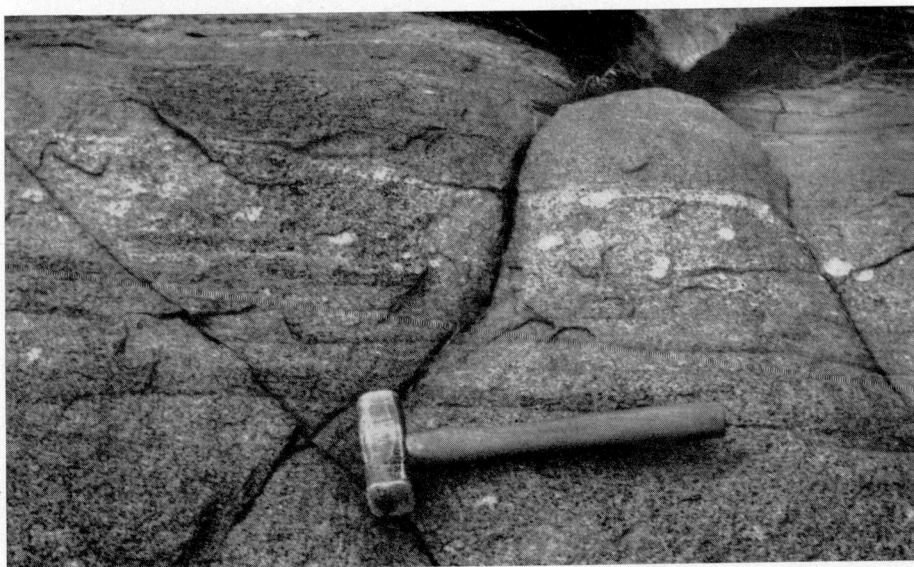

Heights. Layering is well developed on a fine scale (Fig. 209), the chief cumulus phases being andesine, ferroaugite, hortonolite, and magnetite. Pigeonite (inverted to hypersthene) is present in small amounts, disappearing at about 400 ft below the top of the zone, while apatite enters as a significant cumulus phase (max about 5%) at about 2500 ft above the base of the zone. Quartz (7–10%) and microperthite (3–5%) occur particularly in the ferrodiorites of the uppermost 500 ft or so of the Upper Zone, and these rocks have been called either syenodiorites, granodiorites, or perthite-bearing diorites by Boshoff (1942) and Van der Westhuizen (1945). Apart from these discontinuous-cryptic phase changes, continuous-cryptic variation is remarkably well developed in the Upper Zone (Fig. 192), the olivine eventually becoming pure fayalite and the clinopyroxene a ferrohedenbergite almost free from magnesium (Fig. 214).

The mineralogical evidence (discussed at greater length on pp. 406–7) suggests that the uppermost ferrodiorites represent the latest layered fractions of the Bushveld intrusion, both the olivines and the clinopyroxenes being the most ferriferous members of their solid-solution series. Similar rocks occur in the UZ of the Skaergaard intrusion (Sandwich Horizon), above which occur the Upper Border Group rocks which have crystallized downwards. Certain rocks of the Skaergaard intrusion (the melanogranophyres) contain pyroxenes with compositions suggestive of Upper Border Group (UBGγ) affinities (Brown & Vincent, 1963, fig. 2) although on other evidence (bulk chemistry and field relations) they may alternatively represent filter-pressed differentiates later in the fractionation sequence than the Sandwich Horizon fayalite-ferrodiorites. If this is so, then the Skaergaard trend is such that extreme iron enrichment is succeeded by silica and alkalis enrichment (Fig. 132), although we have no explanation for the pyroxenes becoming slightly less ferriferous in the process. These features are important in helping us in an attempt to understand some of the more complex relationships at the upper part of the Bushveld intrusion.

There is no evidence for an Upper Border Group in the Bushveld intrusion. The ferrodiorites, in the stratigraphically higher exposures (Fig. 210), contain an abundance of large, evenly distributed clusters of plagioclase which may represent crystals that floated towards the top of the intrusion and then sank only a short distance subsequent to aggregation. Such a stage would approach the termination of the usual conditions of bottom accumulation, and the compositions of the rocks at this and higher horizons probably approach closely the compositions of the residual Bushveld magmas (Fig. 217). If this is so, then some of the thick felsites and granophyres that overlie the extreme ferrodiorites at Bothasberg and Tauteshoogte are the only rocks which could represent the original roof of the intrusion. The contact is often very sharp (Fig. 211) and in such cases, represents an intrusive contact between rocks of greatly different compositions. To assume that this represents the upper contact of the intrusion would be contrary to the concepts regarding the crystallization history of layered intrusions. At the time when the initial magma was emplaced it would have been broadly of basaltic composition, would probably chill against the roof rocks, and would lose heat from the roof zone to such an extent that crystallization of basic material would continue while some melting of less refractory roof-material took place. Even if some of the Upper Border Group rocks had been removed as blocks fallen into the magma chamber, it would be fortuitous in the extreme for the roof region to be so denuded that the extreme, ferrodioritic residual liquid crystallized directly in contact with pre-existing, roof-zone felsites.

The acid rocks of Tauteshoogte and Bothasberg consist of a variable complex of felsites, granophyres, microgranites and leptites, as well as the massive Bushveld Red Granite. The complexity is due to the fact that we are dealing with Pre-Bushveld quartzose sediments and Rooiberg felsites (which in part have probably been remelted by the basic intrusion), possible late acid differentiates of the Bushveld basic intrusion, and Post-Bushveld intrusions (the Red Granite complex). The Red Granite is now accepted as being later and unrelated to the basic intrusion, while the Pre-Bushveld

FIG. 211. Contact between ferro-syenodiorite and pre-existing rock forming the local roof zone at Tauteshoogte. The roof-zone rock is generally termed a felsite, though it now has a hornfels texture due to thermal metamorphism.

acid rocks are characteristically free from iron-rich olivines and pyroxenes and contain alkali felspar rather than an andesine–oligoclase felspar. Boshoff (1932) made this clear distinction and then considered the problematical melanogranophyres containing ferrohedenbergite and, sometimes, fayalite, which he called the 'Fine-grained acid' group. From the available data it is clear that apart from an area described by Lombaard (1950) in which they grade downward into ferrodiorites, these rocks occur as irregular lenses. At Tauteshoogte they generally occur between the perthite-bearing ferrodiorites and the roof felsites (Boshoff, 1942, p. 33) but are locally absent from this level (Fig. 211). Also, consideration of the chemical analyses of a series of specimens collected in a vertical traverse of Tauteshoogte Heights (Liebenberg, 1960, table 21) shows that melanogranophyres and ferrodiorites may occur both below and above extreme acid (felsite) horizons. It is proposed, therefore, that the latest differentiates of the Bushveld intrusion are certain iron-rich acid rocks (e.g. the melanogranophyres) which crystallized from residual liquids that were filter-pressed from the main intrusion and have invaded the roof-zone acid rocks in an irregular, transgressive fashion. If this is the case, then the Upper Zone of the Bushveld intrusion may be taken to include certain acid rocks containing over 75% quartz+microperthite, minor plagioclase (zoned from An_{30} to An_{20}) and sporadic, minor amounts of fayalite and ferrohedenbergite. As their thickness is irrelevant to the layered sequence and their bulk unknown, they have not been shown on Fig. 192. Their chemical compositions are represented broadly on Fig. 217 but further chemical and mineralogical work is needed in order to establish their compositional range and limitations, while a study of the strontium and oxygen isotopic ratios may lead to a clearer distinction between Upper Zone differentiates and remelted country rocks.

Textural, Mineralogical, and Chemical Relationships

The chemical compositions of the Bushveld rocks and of their constituent mineral phases need to be considered in relation to the textures of the rocks concerned, in view of their cumulus origin. In this way an attempt can be made to distinguish between cumulus and intercumulus phases, and to consider the chemistry of the adcumulates and heteradcumulates in a different way from that of the orthocumulates and mesocumulates.

a) *Cumulate types*

The cumulus minerals found in the layered rocks of the Bushveld intrusion are plagioclase felspar, olivine, Ca-rich pyroxene, Ca-poor pyroxene, chromite, magnetite, and apatite. As shown in Fig. 192, none of these minerals has a cumulus status in all the zones, although the Ca-poor pyroxene (bronzite or pigeonite) is found in all but the uppermost 500 ft of the Upper Zone. The chief cumulus minerals found at various horizons are: Basal Series, bronzite and olivine; Critical Series, bronzite, plagioclase, augite, and chromite; Main Zone, pigeonite, plagioclase, ferroaugite, olivine, magnetite, and apatite. Rhythmic layering results in the concentration of some cumulus phases at the expense of others. While this may be due to crystal sorting in the case of the fine-scale layering, the thick layers of cumulus bronzite or olivine (Basal Series), plagioclase (Critical Series, Main Zone, and Upper Zone), chromite (Critical Series) and magnetite (Upper Zone) are more likely to be due to the temporary crystallization of only one or two phases at these horizons.

Most rocks of the Basal and Critical Series contain either one or two cumulus phases. The Main Zone, in which layering is rare, is characterized by rocks which present an 'average gabbroic' appearance in that they generally contain all three of the cumulus phases present in this zone, i.e. the two pyroxenes and plagioclase. The Upper Zone rocks, in which six cumulus phases may occur, generally contain at least three, and usually more cumulus phases, despite some sorting into relatively melanocratic and leucocratic layers. Hence, the Bushveld layered succession displays an important feature of the crystal fractionation of basaltic magmas, i.e. a gradual increase in the number of crystalline phases in equilibrium with successive liquids. Ultimately, in the uppermost rocks of the Upper Zone, quartz and alkali felspar have crystallized, presumably in equilibrium with the material forming the outer rims to many of the other cumulus phases. This general increase in the number of mineral phases in equilibrium with liquid is only apparent from the broad pattern in this great thickness of layered rocks, on which is superimposed local variations due to magma replenishments, crystal sorting, or non-equilibrium nucleation. The sulphide and Pt-bearing phases probably crystallized in equilibrium with a separate liquid, immiscible with the main silicate-rich liquid.

Adcumulates are amongst the commonest cumulate types in the Bushveld, of which the most obvious are those in which only one cumulus phase is present (e.g. the dunite, pyroxenite, anorthosite, chromitite, and magnetite layers). Rocks such as the norites, felspathic pyroxenites, and pyroxenitic anorthosites are usually adcumulates containing two cumulus phases, i.e. bronzite and plagioclase, while the gabbros of the Main Zone are usually adcumulates with three cumulus phases, i.e. plagioclase, augite, and hypersthene (inverted pigeonite). Another common type of cumulate is the heteradcumulate, in which minerals locally present as a cumulus phase, but absent as such from particular layers, occur in these layers as large, poikilitic, unzoned crystals. The felspar cumulates of the Critical Series, for example, frequently contain large poikilitic bronzites of the same composition as cumulus bronzites in adjacent layers, giving rise to a rock originally called a 'mottled anorthosite'. At lower horizons, certain minerals occur exclusively with this poikilitic habit, prior to their appearance as a cumulus phase in the cryptically layered sequence. For example, the most calcic plagioclases and magnesian augites found in the layered series occur in the olivine heteradcumulates, bronzite heteradcumulates, and chromite heteradcumulates of the Basal Series and lower part of the Critical Series (Fig. 192).

Orthocumulates and mesocumulates are rare in the Bushveld layered series, and in this important respect the textures are similar to those in the Stillwater intrusion. A full, systematic study needs to be made before the distribution of cumulate types is fully understood for the Bushveld intrusion. In view of its enormous lateral extent and the likelihood that crystallization occurred in separate, inter-connected magma chambers, it is likely that the conditions for adcumulus growth at any one horizon varied according to the distance from the margins. Occasional, small patches of quartz are found in a few olivine-free cumulates from the lower zones, suggesting that mesocumulates, with a small amount of pore material, occur inter-layered with adcumulates and heter-adcumulates. In general, however, rocks of the Basal and Critical Series, and of most of the Main Zone, are either adcumulates or heteradcumulates, whereas the Upper Zone ferrodiorites, above the magnetite adcumulates, are mesocumulates frequently containing small amounts of quartz and alkali felspar, and strongly zoned plagioclase felspars.

Some zoning of plagioclase felspars in the Critical Series and the Main Zones was originally attributed to the addition, to cumulus crystals, of material from the trapped pore liquid. However, from optical measurements, since confirmed by electron micro-probe analysis, this zoning has been shown to be reversed and is discussed below. Crescumulates are virtually absent from the Bushveld, as from the Stillwater and Skaergaard layered series, and appear to be a feature almost unique to the Rhum cumulates. The conditions necessary for crescumulate growth appear to have been

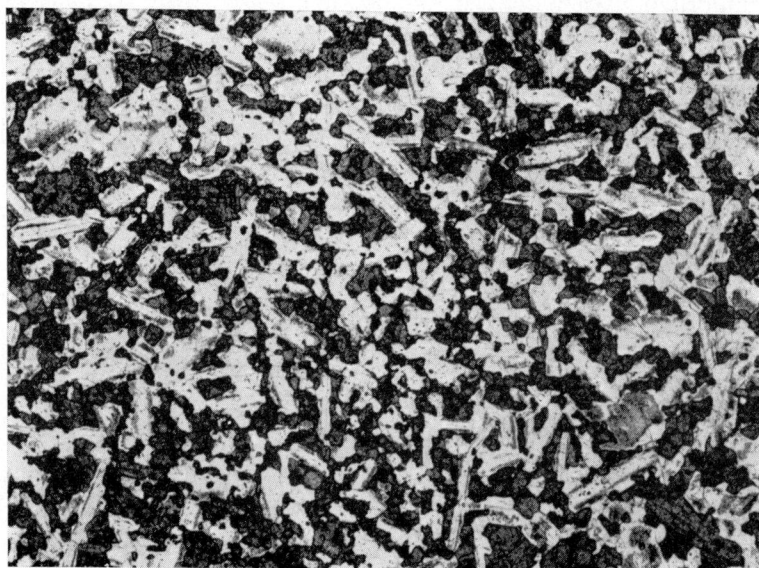

FIG. 212. Photomicrograph of the analyzed, fine-grained hypersthene gabbro (No. 1, Table 26). Marginal Group, Eastern Bushveld. (×20.)

realized only at one horizon in the Bushveld, i.e. amongst the Merensky Reef Units. There, the cumulates are particularly coarse-grained (cf. the coarse harrisitic layers in Rhum), and some of the olivine and bronzite crystals have grown in an elongate, crescumulate fashion.

b) *Plagioclase felspar*

The olivine–bronzite (–chromite) cumulates of the Basal Series are chiefly adcumulates, and only rarely are other mineral phases present. Where plagioclase is present it occurs as large poikilitic plates, usually seen in thin-section as optically continuous, small, interstitial patches. Optical determinations are difficult, particularly as the twin-lamellae are often bent and tend to pinch-out, and refractive index measurements in separated grains do not provide information on the nature and extent of any zoning that may be present. However, the values obtained from optical methods give fairly consistent results, indicating a general composition for the Basal Series plagioclases of An_{78-80}.

Cumulus plagioclase first appears at a horizon corresponding approximately to the base of the Critical Series, as defined by the incoming of cumulus chromite in abundance to form the Main Chromitite Layer (Fig. 192 & 195). There are undoubtedly local exceptions to this, but in general terms it is clear that plagioclase is abundant and occurs as discrete, euhedral crystals, frequently concentrated in felspathic layers, above the base of the Critical Series and not below it. Cryptic variation is only slight throughout the Critical Series, the range being from An_{78} near the base to An_{76} near the top. Minor compositional reversals appear to occur with height, although a more detailed study would be needed in order to establish the pattern of variation. The bronzites show this pattern more obviously, and for this reason we do not consider that the Basal and Critical Series should be viewed as two zones showing regular cryptic variation. Instead, they are viewed as having formed during an integration stage of the Bushveld crystallization history, fluctuations in cumulus mineral compositions being due to repeated influxes of parental magma.

Cumulus plagioclase is found throughout the Main Zone and Upper Zone cumulates, there being a regular cryptic variation from An_{76} to An_{30} (Fig. 192). A few apparent minor exceptions (Atkins, 1965, fig. 9) are probably due to the problems of correlating our traverses accurately in regions where faulting has occurred, such as in the Lulu mountains region.

No systematic study of the plagioclase felspars, in terms of chemical analyses, zoning, and structural state, has yet been undertaken. Preliminary work on the zoning has shown, however, that reversed zoning (first referred to by Liebenberg, 1942) is relatively common in the Bushveld rocks, and to the best of our knowledge this phenomenon has not been recorded from other basic layered intrusions. We were reluctant to accept, without other confirmation, the results obtained entirely from the indirect, extinction-angle measurements, in case the twinning orientations had been mis-identified.

2 B

However, electron microprobe measurements (by G.M.B.) have since confirmed the optical determinations, both qualitatively and quantitatively.

In the rocks of the Critical Series, the unzoned plagioclase cores (An_{76-78}) have narrow rims of An_{82-83}. In the Main Zone, the plagioclase zoning is more complex because a rim of normal-continuous zoning around the unzoned, cumulus cores is succeeded outwards by a thin, reversed zone. In general terms we have:

(1) Large unzoned cores representing the cumulus plagioclase crystals, varying regularly, with height in the layered series, from An_{76} to An_{56} (Fig. 192).

(2) A mantle of continuously zoned plagioclase, becoming up to 10% richer in albite than the cores and attributable to the cooling of trapped pore liquid.

(3) A thin (up to 0·5 mm) outer rim of more calcic plagioclase which, from present estimates, appears to lie within the range An_{79} to An_{74} in composition, becoming less calcic with height in the Main Zone.

Thus plagioclase crystals low in the Main Zone can be depicted, broadly, as consisting of three zones with An compositions from the core outwards of 76: 76→66: 79; while those high in the Main Zone are closer to 56: 56→46: 74. The intermediate zones are not always present, the majority being similar to the adcumulates of the Critical Series in having a large core and an outer thin, reversed rim. In the rocks of the Upper Zone the pattern of zoning is similar to that in the Main Zone, although the increasingly more sodic compositions of the cores, associated with rims which remain as calcic as those in the Main Zone, makes the contrast even more marked. For example, one Upper Zone cumulate contains plagioclase with an An-content zonal pattern of 55: 55→43: 83. The rocks of UZc are exceptional in that no reversed zoning has been detected in the plagioclases, the crystals being either unzoned (in the adcumulates) or normally zoned (in the mesocumulates).

We have given some thought to this perplexing problem but a satisfactory explanation has yet to be obtained. The general application of the experimental data on the system anorthite–albite would lead us to expect that with cooling, any plagioclase added to the cumulus cores would be progressively more sodic, as in the usual orthocumulates. An hypothesis involving re-heating above the temperature at which the cumulus crystals and, in many cases, some intercumulus material formed is clearly the most difficult to accept and is not borne out by any other textural evidence. Two alternative hypotheses are advanced at present, each of which deserves further consideration:

(1) The pressures exerted in the locally-confined pore spaces may result in the pore liquid crystallizing a relatively calcic plagioclase. The snag is that if this were so then the phenomenon would be expected to be found in other layered intrusions where evidence points equally to a relatively deep level of crystallization, likely to be augmented by the local pressure effect. In connection with the effects of pressure it should be noted that an alternative hypothesis, that the cumulus crystals were carried down to a level where a more calcic plagioclase may be stable, would not

account for the fact that many cumulus crystals show normal zoning, developed at the level of deposition.

(2) As discussed by Bowen and Tuttle (1950, p. 508) the crystallization of melt with moderate amounts of dissolved water, under sufficient load pressure, would result in an increase in water and P_{H_2O} in the remaining liquid. The effect of increased P_{H_2O} on the anorthite–albite system is to lower the liquidus–solidus temperatures of the system (Yoder, Stewart & Smith, 1957). Hence the outer parts of crystals which have, say, an An_{70} composition would attain equilibrium with this liquid by being dissolved or made over. The solid in equilibrium with a cooling liquid of $(An_{70}Ab_{30}).H_2O$ composition, locally, would be more calcic than An_{70}, the extent depending on the local P_{H_2O}. Such a recrystallization process would result in the release of SiO_2, and it is significant that the calcic rims are characteristically myrmekitic (Fig. 213). The process would also be accompanied by a slight build-up of Na_2O and H_2O in the interstitial liquid but apart from a few observed specks of scapolite these constituents may be occult in serpentinization and uralitization.

Although the latter hypothesis is the best at present available, it requires that conditions in the Bushveld were probably unique in that the load pressure and water content of the magma were appropriate to the wide development of calcic, myrmekitic rims to the plagioclase crystals. As we have pointed out, the phenomenon requires much further consideration in the light of the available compositional evidence, and a more detailed study of the compositional and textural features of the Bushveld plagioclase felspars.

FIG. 213. Photomicrograph of plagioclase from a Main Zone cumulate. The dark corner of the crystal is more calcic than the core, and is distinctly myrmekitic. (Cross-nicols, ×60.)

c) *Olivine*

Cumulus olivine, together with cumulus bronzite, occurs in the lowest exposed layered rocks of the Basal Series, the relative amounts of the two minerals resulting in the macro-rhythmic layering displayed in the Olifants River region of the Eastern Bushveld. Pure olivine adcumulates (dunites) comprise some layers, while other adcumulate, olivine-bearing layers consist mainly of olivine and bronzite in variable proportions (harzburgites). Towards the top of the Basal Series in the Eastern Bushveld, fine-scale layering of olivine-rich and bronzite-rich cumulates is found (Cameron, 1963, fig. 8) above which, neither cumulus nor intercumulus olivine is found in the rocks of the Critical Series and Main Zone (i.e. over a vertical thickness of about 16 000 ft). In the Western Bushveld, where the Basal Series is not exposed, cumulus olivine occurs sporadically in the Critical Series, up to and including the Merensky Reef Units, but is absent from the Main Zone (i.e. over a vertical thickness of about 12 000 ft). In both Eastern and Western Bushveld, an iron-rich olivine appears as a cumulus phase at about 1000 ft above the base of the Upper Zone, and continues to be present up to the top of the exposed layered series.

The olivines of the Basal Series show an overall change in composition, upwards, from Fo_{88} to Fo_{86}, although a more detailed study may reveal fluctuations, as in the bronzites. In the Critical Series of the Western Bushveld, the most iron-rich variety recorded is Fo_{68} (Lombaard, 1956). The olivines of UZb become gradually more iron-rich with height, from Fo_{49} to Fo_{35}, as do those of UZc, from Fo_{35} to Fo_0. The compositional variation thus shows two significant features: a variation between the extremes of Fo_{88} and Fo_0, and a compositional break between Fo_{86} and Fo_{49} for the Eastern Bushveld, and between Fo_{68} and Fo_{49} for the Western Bushveld.

The marked enrichment in Fe^{++} exhibited by the olivines, resulting in the ultimate crystallization of pure fayalite, is a feature shared by the Bushveld and Skaergaard layered rocks and, together with the extreme enrichment in Fe^{++} shown by the pyroxenes, is now generally attributed to the low state of oxidation of the magmas. Nevertheless, the Upper Zone rocks of the Bushveld include several thick magnetite layers, indicating differences between the physico-chemical conditions of crystallization of the two intrusions. Another difference lies in the fractionation stage at which the exit and re-appearance of olivine takes place, using the An values of the coexisting plagioclases as a basis for comparison (Fig. 131 & 216). It has been pointed out (Wager, 1967) that this varies from one intrusion to another, and in the Bushveld it seems to have varied, also, from one subsidiary magma chamber to the next, as shown by the difference between the Eastern and Western Bushveld. The reason for the temporary cessation of olivine crystallization in tholeiitic basalt differentiates is explicable by analogy with the synthetic system $MgO–FeO–SiO_2$ (Bowen & Schairer, 1935). Slight differences in olivine behaviour is probably due to differences in magma compositions, particularly initial SiO_2 contents, but the extent of fractionation in terms of adcumulus growth, and the state of oxidation affecting Fe^{++}/Mg ratios in the liquids, would also be important factors. The depth and pressure at which crystallization took place may have been higher in the Bushveld than in the Skaergaard, which would extend the field of orthopyroxene stability relative to that of olivine (Davis & England, 1963) and could account for the early exit of magnesian olivine in the Bushveld. However, this cannot be an important factor in explaining the differences between the Eastern and Western Bushveld, where olivine is likely to have crystallized under similar depths of cover.

d) *Pyroxenes*

In the Skaergaard intrusion the pyroxenes have provided an interesting story (Ch. III) and certain aspects of the Bushveld pyroxenes, mentioned by Hess (1941, 1949, 1960) and Boshoff (1942), led one of us (G.M.B.) to plan a detailed study of these and the co-existing minerals. A chemical study by F. B. Atkins (1965) has supplemented the six previously-published pyroxene analyses (by Hess) with eighteen new analyses. This work, including a detailed study of exsolution and inversion textures and of unit-cell dimensions, is in the process of being published and only a summary account of the distribution and chemistry is given here.

A Ca-poor pyroxene occurs as a cumulus mineral throughout most of the exposed layered series, from low in the Basal Series up to within about 500 ft of the top of the Upper Zone. At about the 11 000 ft level (i.e. half-way up the exposed section) orthopyroxene gives place, according to textural evidence, to inverted pigeonite (Fig. 192) which gradually decreases in amount until, prior to its disappearance, it is present only as rare, sporadic crystals. Augite, the Ca-rich pyroxene phase, is present as rare, poikilitic crystals in the Basal Series and most of the Critical Series, but does not appear as a cumulus mineral until a relatively late stage, at about 1000 ft below the top of the Critical Series. The fact that cumulus augite appears appreciably later than cumulus olivine, orthopyroxene, and plagioclase is also a feature of the Stillwater, Great Dyke and, probably, the Skaergaard layered series. Thereafter, augite is present as a cumulus mineral up to the top of the exposed Bushveld layered series.

The main character of the Bushveld pyroxene assemblage is the marked iron enrichment shown by the three pyroxene phases, represented on the conventional Di–Hed–En–Fs diagram as Fig. 214. Only a few representative examples are plotted (their analyses being given in Table 28), the trend lines being based on 24 analyses to be published soon by Atkins. (For convenience in relating future papers, the numbers given in Fig. 214 correspond to those used by Atkins, also indicating those of the series not plotted here.) In the augite series, No. 1 is an intercumulus, diopsidic augite from a bronzite cumulate of the Basal Series (Hess, 1949, p. 645). The earliest appearance of cumulus augite is towards the top of the Critical Series (Fig. 192), represented by No. 3. The Main Zone augites are represented by a series including No. 9 (Hess, 1949, p. 650), and the Upper Zone ferroaugites by a series including Nos. 11, 14, and 15.

TABLE 28

Chemical analyses of some Bushveld pyroxenes (plotted on Fig. 214)

	1	3	9	11	14	15	1a	3a	9a	10a
SiO$_2$	54·07	52·90	51·39	49·39	47·92	46·76	55·94	54·32	51·4	48·9
Al$_2$O$_3$	2·08	2·41	2·45	1·59	1·72	1·14	1·61	1·83	1·8	1·5
Fe$_2$O$_3$	0·56	1·03	1·26	2·71	2·31	2·34	0·97	1·28	1·4	2·1
FeO	2·53	5·10	11·63	19·03	21·92	29·21	7·15	13·44	20·7	28·7
MnO	0·09	0·16	0·32	0·52	0·52	0·80	0·19	0·29	0·5	0·6
MgO	17·39	16·18	14·21	11·13	6·35	0·15	32·12	27·56	19·4	13·8
CaO	22·12	21·46	18·12	15·52	18·99	18·82	1·48	1·18	4·1	3·6
Na$_2$O	0·41	0·34	0·27	0·21	0·22	0·23	0·01	0·05	0·1	0·2
K$_2$O	0·01	0·02	0·02	0·04	0·01	0·02	0·01	0·02	n.d.	0·0$_6$
H$_2$O$^+$	0·04	n.d.	0·03	n.d.	n.d.	n.d.	0·09	n.d.	0·4	n.d.
H$_2$O$^-$	0·06	0·02	0·07	n.d.	n.d.	n.d.	0·01	0·01	...	...
TiO$_2$	0·21	0·37	0·41	0·35	0·65	0·69	0·11	0·25	0·3	0·5
Cr$_2$O$_3$	0·98	0·26	n.d.	n.d.	n.d.	n.d.	0·45	0·12	n.d.	n.d.
NiO	0·04	n.d.	n.d.	n.d.	n.d.	n.d.	0·07	n.d.	n.d.	n.d.
Total	100·58	100·25	100·18	100·49	100·61	100·16	100·18	100·35	100·1	99·96
Atomic %										
Ca	45·4	44·0	37·7	32·3	40·4	42·7	2·8	2·3	8·4	7·6
Mg	49·6	46·1	41·0	32·1	18·8	0·5	85·0	75·0	55·4	40·7
ΣFe	5·0	9·9	21·3	35·6	40·8	56·8	12·2	22·7	36·2	51·7
% Al in Z	2·9	3·4	3·7	3·6	4·1	2·6	2·7	3·1	3·2	3·5
% Ti in Z	...	...	...	0·2	0·3	...	...	...	...	...

Note: ΣFe = Fe^{+++} + Fe^{+++} + Mn.

Analysts: F. B. Atkins except for nos. 1, 9, 9a (R. B. Ellestad) and 1a (L. C. Peck).

Key to pyroxene analyses

1. Diopsidic augite of intercumulus status in bronzitite (7666) of UZa Basal Series, Malips Drift. (Hess, 1949, p. 645.)

1a. Cumulus bronzite from same rock as No. 1. (Hess, 1960, p. 25.)

3. Cumulus diopsidic augite from gabbro of the Critical Series (S.A.660). Structural height 600 ft. Top of ridge above Jagdlust Chrome Mine, Eastern Bushveld.

3a. Cumulus bronzite from same rock as No. 3.

9. Cumulus augite from gabbro (7493) of Main Zone, Pretoria district. (Hess, 1949, p. 650.)

9a. Cumulus inverted pigeonite from same rock as No. 9. (Hess, 1949, p. 579.)

10a. Cumulus inverted ferropigeonite from ferrogabbro of UZa (S.A.139). Structural height 19 900 ft. Near Magnet Heights, Eastern Bushveld.

11. Cumulus ferroaugite from ferrodiorite of UZc (S.A.1139). Structural height 23 400 ft. Farm Duikerkrans 104 (250 yds north of Steelpoort River), Eastern Bushveld.

14. Cumulus ferroaugite from ferrodiorite of UZc (S.A.1143). Structural height 23 700 ft. Farm Duikerkrans 104 (2 miles north of Tauteshoogte), Eastern Bushveld.

15. Ferrohedenbergite from ferro-syenodiorite of UZc (S.A.1149). Structural height 24 000 ft. Near top of scarp from which No. 14 was collected.

(Specimen numbers pre-fixed by S.A. refer to the Bushveld collection, Department of Geology and Mineralogy, Oxford.)

The Ca-poor pyroxene series includes, firstly, a series of orthopyroxenes showing iron enrichment from No. 1a, the lowest exposed cumulus bronzite of the Basal Series. From petrographic evidence the intercumulus augite of this rock is probably of heteradcumulus origin, so Nos. 1 and 1a can be viewed as a coexisting pair. Nos. 3 and 3a are both cumulus, and therefore a coexisting pair of pyroxenes. No. 9a (coexisting with No. 9) is an inverted pigeonite, and from Atkins' more extensive data it is clear that the phase-change from bronzite to pigeonite occurred, in the Bushveld, at a composition between $Mg_{70}Fe_{30}$ and $Mg_{65}Fe_{35}$, and within the Main Zone rocks. Inverted pigeonite is found throughout the UZa and UZb rocks as a cumulus phase, and as an intercumulus phase to within 600 ft of the top of UZc (Fig. 192). The most ferriferous cumulus pigeonite to be analyzed is No. 10a, although optical measurements of intercumulus pigeonites indicate iron enrichment to compositions up to, and slightly beyond, No. 11a. (Augite No. 10 is not plotted here because of an anomalous composition yet to be resolved).

The most striking feature of the trends shown in Fig. 214 is the close similarity with trends established by Brown (1957) and Brown & Vincent (1963) for the Skaergaard pyroxenes (Fig. 19). These are the only two layered intrusions of tholeiitic affinities for which we now have analytical data on a wide range of pyroxenes in the exposed series of layered rocks, and the Bushveld data serve to emphasize that in neither case is the picture unique or unusual, but that taken together the pattern must have a fundamental significance. Characteristic features of the pattern shown in Fig. 19 and

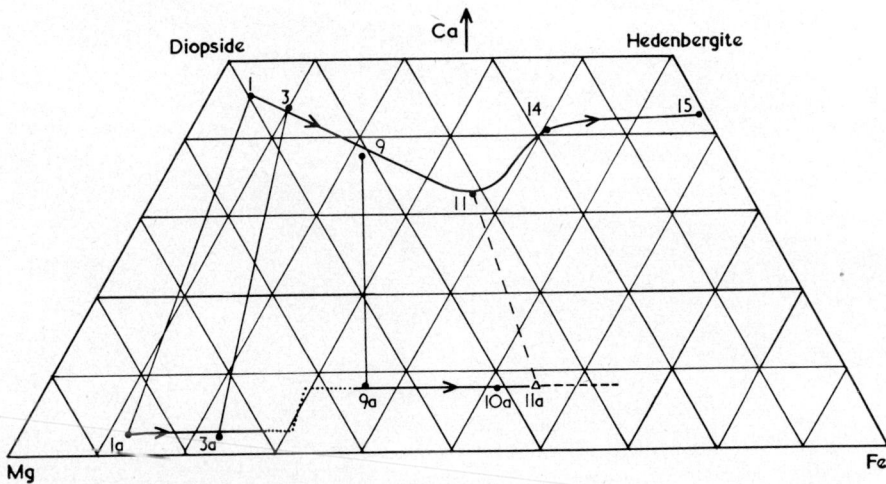

FIG. 214. Trend of the Bushveld Ca-rich and Ca-poor pyroxenes, obtained from new chemical analyses. (See Table 28 for the analyses and localities except for No. 11a, estimated from optical measurements.) It should be noted that the trends have been defined from a large number of analyses (Atkins, 1965), only a few representative examples, numbered from the full sequence, being plotted prior to the publication of Atkins' results. (Cf. Fig. 19.)

214 include the beginning, shape and ending of the Ca-rich pyroxene trend, and the beginning, break, shape and ending of the Ca-poor pyroxene trend. In detail, the differences between the Skaergaard and Bushveld are as interesting as the similarities, and will ultimately be of value in understanding the reasons for the shape of the augite–ferroaugite trend, the phase-change from bronzite to pigeonite, and the stability of ferropigeonite+liquid in relation to fayalite+silica+liquid. Other differences, not shown on Fig. 214, include the apparent absence of uninverted pigeonite and of inverted ferriferous β–wollastonites in (with inverted tridymite) the Bushveld layered series.

The size, certainly, and the depth of crystallization and time for cooling and crystallization, probably, are greater in the Bushveld than in the Skaergaard. These factors are likely to be of greater importance than the slight differences in initial magma composition in accounting for the pyroxene crystallization history, although initial differences in, and progressive variability of P_{O_2} would have an important contributory

FIG. 215

$\downarrow \rightarrow$

Photomicrographs of a representative variety of Bushveld pyroxenes, illustrating sub-solidus exsolution and inversion textures (from Atkins, 1965). The nature of the exsolved material is at present being investigated by electron-probe microanalysis.

A. Bronzite with fine-scale exsolution lamellae of augite ‖ (100) and blebs of exsolved augite. Critical Series. (Cross-nicols, ×50.)

B. Inverted pigeonite twinned on (100), with augite exsolved as broad lamellae ‖ (001) before inversion and as narrow lamellae ‖ (100) after inversion to hypersthene. Upper Zone *a*. (Cross-nicols, ×50.)

C. Inverted ferropigeonite with well developed, closely-spaced augite lamellae and blebs exsolved ‖ (001). The crystal is interstitial to plagioclase and augite, enclosing two crystals of the latter which coalesce with the augite lamellae. Upper Zone *c*. (Cross-nicols, ×70.)

D. Augite twinned on (100), with pigeonite (? inverted) exsolved as broad lamellae ‖ (001) and hypersthene exsolved as narrow lamellae ‖ (100). Main Zone. (Cross-nicols, ×70.)

E. Ferroaugite with well developed, closely-spaced lamellae of pigeonite (probably uninverted) exsolved ‖ (001). Upper Zone *c*. (Cross-nicols, ×80.)

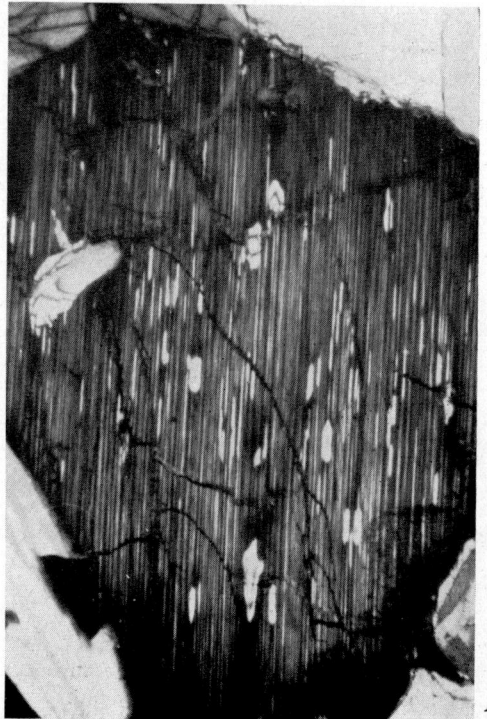

A

B

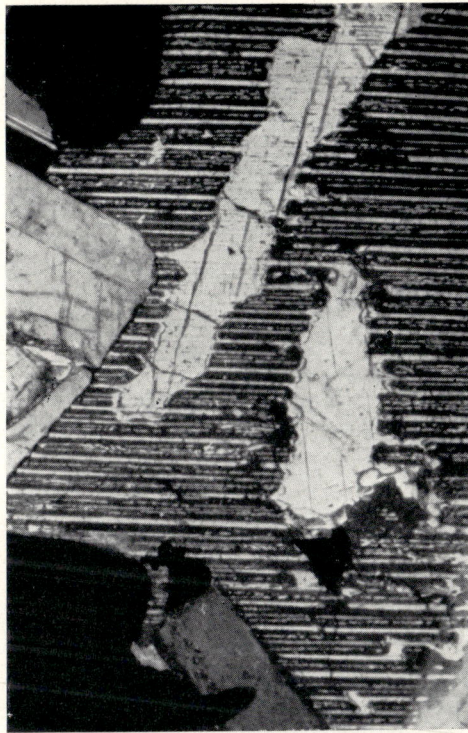

C

D

E

effect. Insofar as the rates of cooling and crystallization are important, the exsolution textures of the Bushveld pyroxenes, although broadly similar to those in the Skaergaard (Brown, 1957; see Ch. III), are generally on a coarser scale. Some examples are shown in Fig. 215A–E.

e) *Ore minerals*

The opaque minerals of the Bushveld fall into three categories, concentrated at three specific levels of the layered series. The chrome-spinels occur at the lower horizons, chiefly in the Critical Series, while the Fe–Ti oxides are found at the upper horizons, chiefly near the base of the Upper Zone. Between these levels of oxide concentrations, at the top of the Critical Series, the Merensky Reef Horizon carries a wide variety of complex Pt and Pd minerals together with other metallic sulphides, native metals, and chrome-spinels. The Bushveld is unique amongst layered intrusions in the striking display of opaque minerals which is not only of paramount economic importance but also of great genetic significance. There is no doubt that the ore minerals owe their origin and distribution to the same geological process as that which gave rise to the silicate minerals, and that in this respect the Bushveld is the world's classic example for observing the relationship between silicates, oxides, and sulphides, each deposited in a regular sequence during the slow cooling and crystallization of a large body of basic magma. The opaque minerals are found in distinct layers (e.g. Fig. 197, 202 & 207) and the most reasonable conclusion is that they are igneous cumulates. However, despite the attention paid to these economic deposits of the Bushveld, experimental data on relevant synthetic systems are not available and the petrologist does not have the advantage of being able to interpret this aspect of the crystallization of the intrusion in the way made possible by studies on synthetic silicate systems. Instead, it can be accepted that the pattern of events outlined for the formation of the silicate layers needs to encompass the formation of the other layers, within the same bounds of physico-chemical conditions. This means that the mining geologist cannot assume a licence to propose hypotheses on the genesis of the ore bodies, such as metasomatic replacement or crystallization from injections of strongly contrasted magmas, that do not at the same time take into account the crystallization history of the contemporary olivines, plagioclases and pyroxenes. The authors would like to see an extensive study of the array of Bushveld opaque minerals, to be viewed as a vast layered sequence of crystal fractions produced in equilibrium with a cooling basic magma, taking into account the evidence provided by the silicate minerals for cryptic variations and local fluctuations amongst the cumulus minerals, the depositional processes, and the varied types of intercumulus crystallization. A broad pattern of cryptic variation is already evident for the spinels, the chromites becoming enriched in iron relative to chromium upwards in the layered sequence; then there is a gap where no cumulus spinel crystallized, and this is succeeded by the crystallization of abundant titaniferous magnetites. This is a significant and striking pattern, but the compositional changes have not yet been studied in sufficient detail. The sulphides of the Merensky Reef probably owe their formation to the immiscibility of a sulphide-rich liquid but we are convinced that if so, the sulphide-rich droplets accumulated rhythmically in layers, together with silicate and oxide crystals (Fig. 202). Such an origin requires explanation in terms of the concentrations and temperatures at which the sulphide-rich liquid was in equilibrium with both the basic magma and the silicate phases at the Merensky Reef horizon, and from which sulphides and associated minerals crystallized before the overlying crystal pile was thick enough to squeeze the sulphide-rich liquid as veins into adjacent layers.

The chromites have been studied more than the other opaque minerals, recently by Cameron and Emerson (1959) and earlier, for example, by Sampson (1932), van der Walt (1942), and de Wet (1952). Other information of especial significance is given by Lombaard (1956) and Jackson (1963). Each writer has tended to concentrate on a particular compositional aspect, from which the origin of the chromite deposits can be argued, including Cr:Fe, Mg:Fe, and Fe_2O_3:FeO ratios in the chromites, and Mg:Fe ratios in coexisting chromites and orthopyroxenes. The attention varies from considerations of broad vertical or lateral variations throughout the complex, through local variations in western or eastern Bushveld sections, to variations within a single seam. The resulting information, however, is difficult to use, chiefly because the significance to be attached to subtle variations in compositions depends upon the accuracy of the earlier analyses. Also, the opaque minerals are not amenable to the type of optical study which, for the silicates, enables one to detect vertical and lateral fluctuations in cumulus compositions, as distinct from the variable extent of zoning resulting from differing amounts of adcumulus growth. This problem cannot be resolved until the opaque crystals have been studied with the electron-probe micro-analyzer, but at present it is as well to realize that three analyzed chromite samples from one general level may represent unzoned (adcumulus) crystals, zoned cumulus crystals, or intercumulus material, each representing different ranges of crystallization temperature and, therefore, of composition.

The main information to emerge from the analytical data (Table 29) is that there is a broad tendency towards a decrease in Cr:Fe and Mg:Fe upwards in the layered sequence (Lombaard, 1956, p. 71). Cameron and Emerson (1959, p. 1195) suggest that this trend may be confined to each seam, there being a reversal to higher Cr:Fe and Mg:Fe ratios at the base of each of the two seams studied by them. Jackson (1963, p. 53) proposes that there may be a variation in the Bushveld similar to that in the Stillwater chromites (see p. 312), with a decrease followed by an increase in total iron upwards, and a lateral increase in oxidation state towards the margins. The second general point (e.g. Cameron & Emerson, 1959) is that a comparison between chromitite and adjacent pyroxenite layers shows both chromite and orthopyroxene to have higher Mg:Fe ratios in the chromitite layers.

If, as seems likely, the fractionation trend of the Cr–Fe spinel solid-solutions shows iron enrichment, the crystals will be richer in Cr (and probably in Mg) than the magma from which they crystallized. Hence the cumulus crystals will be richer in Cr and Mg than those formed entirely from trapped pore liquid, while adcumulates and heter-adcumulates will contain unzoned chromites richer in Cr and Mg than the zoned crystals in possibly-adjacent orthocumulates and mesocumulates. Such factors would explain local variations in the bulk composition of analyzed chromites, and especially the relative iron enrichment in the partly intercumulus chromites of pyroxenite layers. In making this generalization we may be overlooking some of the fine points raised by

TABLE 29

Representative compositions of chromites (freed from silicate impurities) from the Critical Series, Bushveld intrusion

	LG	MG	UG	MR
Total iron as FeO	26·33	27·16	29·79	37·72
MgO	9·7	8·4	8·8	7·27
Cr_2O_3	45·94	44·74	42·31	38·22
Cr: Fe	1·54	1·45	1·26	0·89
$\dfrac{\text{FeO} \times 100}{\text{FeO} + \text{MgO}}$	73	76	77	84

LG. Average (11 analyses) from two seams of the Lower Group of chromitite layers, Critical Series.
MG. Average (19 analyses) from eight seams of the Middle Group of chromitite layers, Critical Series.
UG. Average (12 analyses) from five seams of the Upper Group of chromitite layers, Critical Series.
MR. Average (4 analyses) from the chromitite layer in the platiniferous Merensky Reef, top of the Critical Series.
(From Lombaard, 1956, table 3, summarized chiefly from data of van der Walt, 1942.)

Cameron and Emerson (1959) but we are in general agreement with all their conclusions (*op. cit.*, p. 1211) except the statement that '. . . partial resolution of settled crystals seems required as a supplementary process'. They believe that the chromites of the chromitite layers were partially melted in the hotter zone of deposition at the floor of the intrusion, the residue being therefore enriched in Cr relative to Fe. We find this difficult to accept, not only because a hot liquid near the floor would be unstable and tend to rise through convection but, also, because we believe that for the Bushveld rocks the textural evidence is strongly in favour of adcumulus growth having taken place at the site of cumulus deposition, frequently to produce monomineralic rocks. As discussed in Chapter VIII, such adcumulus growth requires extension of the cumulus crystals by crystallization at the floor, with loss of heat to the overlying, convecting liquid. Jackson (see Ch. XIII) also believes that in the Stillwater intrusion the conditions require that the liquid near the floor is temporarily cooler than the overlying liquid, the latent heat of crystallization resulting in occasional convective overturns. We would suggest that the chromitite layers of the Bushveld are adcumulates and that adcumulus growth was responsible for the chromite crystals, in bulk, having a higher Cr:Fe ratio than those which crystallized partly or entirely from the contemporary pore liquid. According to the same argument, the pyroxenes of the chromitite layers would be of adcumulate or heteradcumulate origin, and therefore richer in Mg relative to Fe than those that are zoned or made over by reaction with trapped pore liquid. We have come to the conclusion, however, that a factor such as the increase in P_{H_2O} of the pore liquid may have resulted in partial dissolution and re-precipitation of plagioclase in certain Bushveld rocks (p. 387) and it may be that the two processes of crystallization and dissolution occurred intermittently near the floor, for reasons not yet clear. We make this point because despite our arguments regarding adcumulate growth, outlined above, we are aware of cases where reaction and recrystallization appear to have occurred, and of the textural difficulties in merely postulating much trapped liquid to account for the relatively iron-rich character of the pyroxenes in the pyroxenite layers. Finally, the chromitite veins mentioned by Cameron and Emerson (1959) are not easily attributed to the lateral migration of a crystal mush and may be due to the generation of a low-temperature, chromite-rich liquid where local volatile concentrations have resulted in a lowering of liquidus temperatures.

The iron-titanium oxides are confined chiefly to the thick ' magnetite ' layers occurring at the base of the Upper Zone of the layered series, but two local exceptions to the apparently wide gap between chromite and titaniferous magnetite crystallization are found as a layer in the Critical Series (Lydenburg district) and one near the base of the Main Zone (Rustenburg district). With the exception of two papers (Frankel & Grainger, 1941; Schwellnus & Willemse, 1944) the thick deposits of Bushveld iron–titanium oxides have received little attention by workers in this field of mineralogy, apart from brief mention by Ramdohr (1953, 1956) and Buddington et al. (1955, table 12).

Ilmenite appears to be rare in the Bushveld, occurring as cumulus crystals only in the rather exceptional occurrence low in the layered sequence, i.e. the Critical Series (Schwellnus & Willemse, 1944, pl. 3, fig. 1). This is probably a significant feature, ilmenite occurring only as intergranular patches and exsolution lamellae in the ' magnetite '-rich layers of the Upper Zone. The Upper Zone ore-rich layers are up to 6 ft in thickness (Fig. 207) and in their lower parts are, apparently, adcumulates consisting almost entirely of cumulus oxide minerals extended by adcumulus growth, to the exclusion of all but a trace of silicate minerals. According to the chemical analyses (Table 30) the dominant oxide of these layers is particularly rich in Fe_2O_3, with lesser amounts of FeO, TiO_2, and V_2O_3. In reflected light, the dominant ore mineral is isotropic and whitish-grey to whitish-blue and this, together with its magnetic properties and staining characteristics, led Frankel and Grainger (1941, p. 103) to accept the name ' maghemite ' given to it earlier by Wagner. Schwellnus and Willemse (1944, pl. 5, fig. 1) have shown that the maghemite has developed at the expense of titaniferous magnetite, the latter being present as ' moth-eaten ' cores to the large areas of maghemite. Such a process suggests secondary oxidation of an original titaniferous magnetite and it is likely, also, that the ilmenite lamellae may have developed in a similar way from exsolved ulvöspinel, shown to be present by Ramdohr (1956, pl. 6, fig. 12). Some of the analyses given by Schwellnus and Willemse (1944, table 1) are very similar to an analyzed, Ti-bearing, γ–Fe_2O_3 from the Atumi dolerite (Kushiro, 1960, p. 150) and broadly similar to other examples given by Akimoto and Kushiro (1960), in which

TABLE 30

Representative compositions of titanomaghemites from the Upper Zone Bushveld intrusion

	LG	MG	UG
Fe_2O_3	69·15	64·09	59·86
FeO	8·56	10·82	10·09
TiO_2	12·63	16·56	18·19
V_2O_3	1·2	0·5	0·2
SiO_2	1·76	1·59	3·59
Al_2O_3	3·35	3·36	2·84
MgO	1·04	0·75	0·70
MnO	0·21	0·25	0·37

LG. Average (5 analyses) from the Lower Group of ' magnetite ' layers, Upper Zone.
MG. Average (3 analyses) from the Middle Group of ' magnetite ' layers, Upper Zone.
UG. Average (5 analyses) from the Upper Group of ' magnetite ' layers, Upper Zone.

(From Buddington *et al.*, 1955, table 12, summarized from data of Schwellnus & Willemse, 1944, table 1.)

case the Bushveld mineral would better be termed a titanomaghemite. To the best of our knowledge, no X-ray structural measurements have been made. As shown by Kushiro (1960), titanomaghemite is stable over greater ranges of pressure and temperature than maghemite and, rather than titanhematite, it is likely to be formed early in the cooling history of a fairly deep-seated intrusion, and at the expense of titaniferous magnetite.

Apart from the oxidation state of the iron, the iron-rich oxides of the Upper Zone of the Bushveld show an interesting feature in the distribution of TiO_2 and V_2O_3 (Schwellnus & Willemse, 1944, table 1). If the layers are grouped as a Lower, Middle, and Upper set, then between the averages of these three there is a distinct increase in TiO_2 and decrease in V_2O_3 upwards in the sequence (Table 30) which, as pointed out by Buddington *et al.* (1955, table 12) is similar to the trend found in the Skaergaard titaniferous magnetites. In regard to the TiO_2 in the Bushveld magnetic oxides, the overall amount (averaging about 12–20%) is similar to that attributed generally by Buddington *et al.* (1955, tables 6 & 10) to titaniferous magnetite segregations in stratiform basic intrusions, so it would seem appropriate to view the development of the titanomaghemite as an oxidation process operating in a closed system and thereby not affecting the amounts of TiO_2 and V_2O_3 contained in the original titaniferous magnetite. In considering the Bushveld mineralogy it is worth noting that although the iron-rich oxides show an increase in TiO_2 with fractionation, this cannot be explained according to the Buddington *et al.* hypothesis (1955, fig. 2) because according to the available data the minerals do not, apparently, coexist with a separate Ti-rich phase such as ilmenite.

The sulphide and platinoid minerals occur in the layers of the Merensky Reef horizon, although minor amounts of sulphides (chiefly of copper) are recorded from higher levels of the layered series, particularly in the oxide-rich layers of the Upper Zone. As shown on Fig. 202, the opaque minerals occur at the base of each rhythmically layered macro-unit of the unique Merensky Reef horizon at the top of the Critical Series, at least two of which units contain ore minerals other than chromite. There is little available information on the textural relationships between the sulphide or platinoid minerals and the silicate or oxide minerals, but it is likely that the former crystallized from a liquid phase that was immiscible with the main body of magma from which the latter crystallized, and which tended to concentrate along with chromite and bronzite, rather than with the less dense plagioclase, at the base of each unit.

The variety of minerals found in these thin, ore-rich layers is quite remarkable,* a brief summary being given in two articles in the *Platinum Metals Review* (Cousins, 1957; Beath *et al.*, 1961). Together with the platinoid minerals, sulphides of iron, nickel and copper are common, a list of accessories including chalcopyrite, pyrrhotite, nickeliferous pyrite, pentlandite, cubanite, graphite, millerite, and violarite. The platinum is partly in the form of native metal, invariably alloyed with iron (ferroplatinum) and partly as the sulphide, arsenide, and sulph-arsenides. Platinum forms the major component of the platinum-group metals, plus gold, with palladium, ruthenium, rhodium, irridium and osmium in descending order of importance. In mining the material, the gravity concentrates consist of native platinum, native gold, sperrylite (platinum arsenide), cooperite (platinum sulphide), braggite (platinum–palladium–nickel sulphide), stibiopalladinite (palladium antimonide) and laurite (ruthenium sulphide), mixed with chalcopyrite, pyrrhotite, pentlandite, other sulphides, chromite, and graphite.

Platinoid minerals also occur in relative abundance in the dunitic pipes which cut the Bushveld intrusion near its eastern margins, and there is probably a close genetic relationship between the source of these bodies and the fresh pulse of magma that is believed to have replenished the main Bushveld magma chamber immediately prior to the deposition of the Merensky Reef layers (p. 371). A detailed study of the platinoid mineral assemblage in the Driekop pipe has been made by Stumpfl (1961), using an

* The assemblage recorded by Hawley (1962) from the Sudbury intrusion, although not present in a distinct layer, is similar to that in the Bushveld.

electron-probe microanalyzer. While this assemblage may differ slightly from that in the Bushveld layered rocks, most of the mineral assemblage mentioned by Stumpfl is identical, with the addition of at least ten new minerals, so far without names, which are chiefly compounds of Pt, Pd, Ir, and Os with As, Sb, and Bi.

f) *Chemical compositions of rock types*

   More rocks have been chemically analyzed from the Bushveld than, probably, from any other igneous intrusion (e.g. Hall, 1932; Liebenberg, 1960). Despite the wealth of data, however, there has been no systematic chemical study aimed at an under-standing of the fractionation processes likely to have operated in this vast intrusion. For this purpose, information would be needed on the composition of the initial magma, on the compositions of successive, coexisting cumulus mineral phases, and on the calculated successive *liquid* compositions. It is unlikely that we shall ever obtain a sufficiently clear picture of the shape and size of the magma chamber, of the volume of successive crystal fractions, and of the degree to which the Bushveld behaved as a closed system, to be able to estimate, as for the Skaergaard, the course of successive liquid fractions. Without this information, the most useful chemical data for the layered series are on separated and analyzed cumulus phases, the limited information so far available having been given in the previous part of this section.

   The composition of the initial Bushveld magma, termed ' the Bushveld 1st liquid ', has been discussed earlier (pp. 354–6) and despite the problems associated with the recognition of a true chilled marginal rock, the new analysis (Table 26, anal. 1) of an olivine-normative rock is used here. The large number of available analyses of layered rocks from the Basal and Critical Series, and from the Main Zone, are generally of extreme cumulates such as chromitites, peridotites, pyroxenites and anorthosites, and where of gabbroic cumulates the textural evidence shows them, also, to be adcumulates or heteradcumulates. For the reasons discussed earlier in this book, such analyses cannot be plotted in order to define any sort of fractionation trend because the rhythmic layering and adcumulus growth cause local variations which completely mask any regular variation that may be present. On a broader scale, the averages of available analyses from each zone could be plotted on an Mg–(Fe+Mn)–(Na+K) diagram, and wide fields indicated (Fig. 217). These suggest iron enrichment relative to mag-nesium, and felspar enrichment relative to mafic minerals, upwards in the layered series. However, this type of diagram is generally unsatisfactory for cumulates because of the different Fe:Mg ratios of the various mafic minerals (which are sorted into different, sampled layers) and because the absence of Ca as a component means neglect of the changing composition of the plagioclase felspars. At the higher levels of the layered series, where diorites occur in the Upper Zone, there is a better case for plotting the whole-rock analyses. The later fractionation stages of basic intrusions differ from the earlier stages in the increase in the rate of compositional change for each cumulus phase with height in the layered series (e.g. Ch. III). Also, the differences between the compositions of liquids and solids in equilibrium have decreased, and are approaching the stage when the two must ultimately be identical. This is not to say that the Upper Zone diorites are not crystal accumulates (see Fig. 209) but only that some general guide as to the trend of the liquid line of descent can, in the absence of better data, be obtained by plotting the analyses of relatively uniform rocks. For similar reasons, the Upper Zone acid rocks are plotted and compared with the early

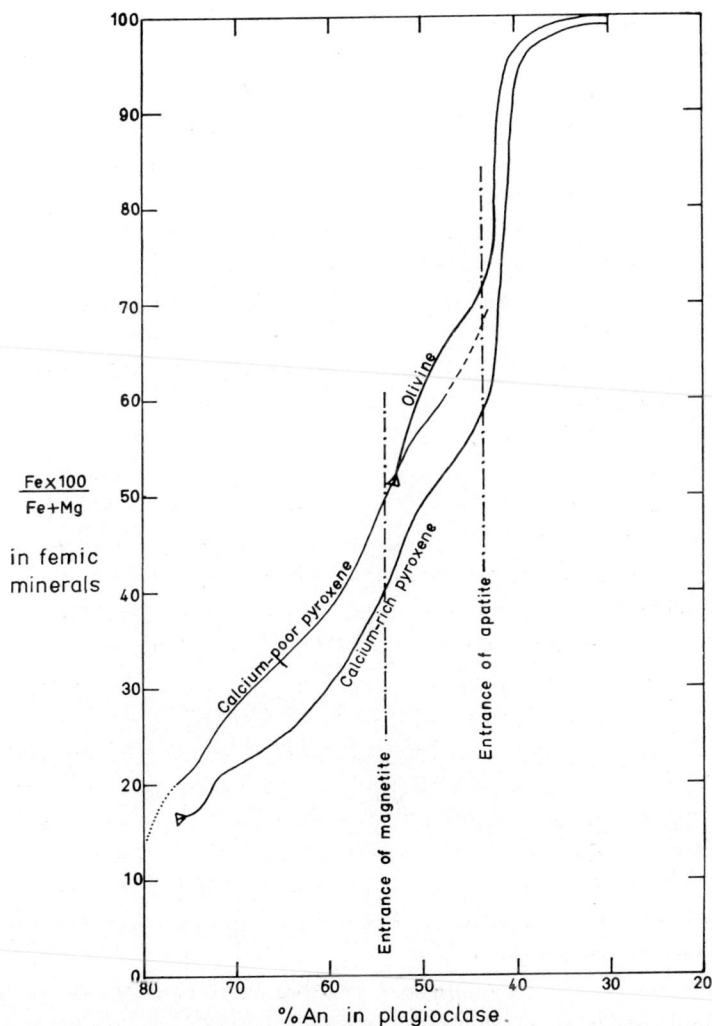

FIG. 216. Plot of An content of Bushveld cumulus plagioclases against Fe ratios of coexisting femic cumulus phases (solid lines). Dashed line refers to intercumulus pigeonite and dotted line to intercumulus plagioclase. The olivines of the Basal Series cannot usefully be plotted, because they ceased to crystallize before plagioclase became a cumulus phase (see Fig. 192).

2 C

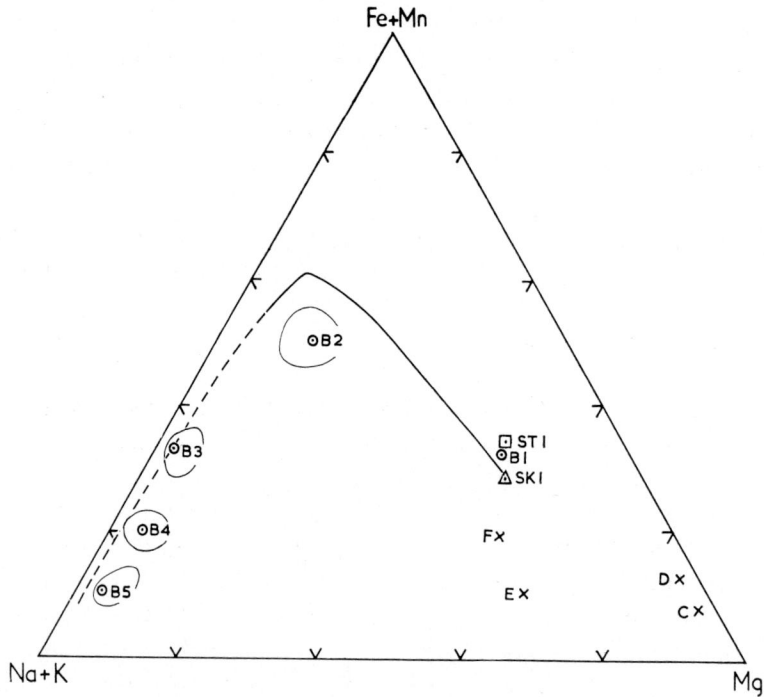

FIG. 217. Plot of analyzed Bushveld rocks from data given by Hall (1932), Boshoff (1942), and Liebenberg (1960).

B1 = ST1, and SK1 = presumed chilled marginal rocks of the Bushveld, Stillwater, and Skaergaard intrusions (Table 26). (Cf. Fig. 132 & 189.)

B2 = average, with scatter area, of five analyses of ferrodiorites, Upper Zone, Tauteshoogte.

B3 = average, with scatter area, of three analyses of ferro-syenodiorites and melanogranophyres, Upper Zone, Tauteshoogte.

B4 = average, with scatter area, of six analyses of acid rocks (felsites and granophyres, often containing fayalite and hedenbergite), Tauteshoogte and Middelburg.

B5 = Bushveld Red Granite, with scatter area including two analyses of leucocratic Rooiberg felsites, Middelburg.

The analyses of layered cumulates from the lower zones do not mean much in view of crystal accumulation, but are plotted here to emphasize this point.

C = average of 9 bronzite-rich cumulates (Basal Series).

D = average of 6 olivine-rich cumulates (Basal Series).

E = average of 6 felspar-rich and pyroxene-rich cumulates (Critical Series).

F = average of 14 gabbroic cumulates (Main Zone).

felsites and late granites, which are not part of the fractionation sequence of the main Bushveld intrusion.

The relevant components are plotted on Fig. 217, and a comparison is drawn with the trend of the calculated successive Skaergaard liquids. The result, taking the trend from the Bushveld 1st liquid, through the ferrodiorites and ferro-syenodiorites to the ferrohedenbergite–fayalite granophyres, is an impressive demonstration of the overall similarity between the Bushveld and the Skaergaard trends. This is to be expected in view of the similar type of cryptic variation, especially shown by the pyroxene trends (Fig. 19 & 214) but also by the trends towards fayalitic olivines, and towards sodic plagioclases and alkali felspars. In fact, the pattern of ferrous iron-enrichment shown by the Skaergaard intrusion, and generally thought of as unique, is a characteristic feature, also, of the Bushveld intrusion. The advantage in the Bushveld is that the earliest crystal accumulates are exposed (broadly equivalent to the Skaergaard Hidden Zone), as well as the later crystal accumulates (broadly equivalent to the Stillwater Upper Hidden Zone).

The similarities between these three large intrusions are, therefore, impressive, from the point of view not only of layering and mineralogy but also of bulk chemistry. In detail, the differences are equally interesting and it has been shown, in the sections dealing with their mineralogy, that the coexisting mineral phases differ slightly but significantly in character and composition. The initial magma compositions are not greatly different (Fig. 217), but it should be remembered that chemical factors such as Ca, Al and Si contents, and Fe^{+++}/Fe^{++} ratios, are not recorded on this type of diagram. Also, the physical conditions of crystallization in these three different-sized magma chambers, the shapes, depths and closures of which are not known in detail, would influence the respective courses of fractionation.

The acid rocks of the Bushveld Complex provide a bewildering array of granites, granophyres and felsites which belong to various phases of igneous activity. Unfortunately they occur together at the level of the roof-zone of the main Bushveld basic intrusion, and include pre-Bushveld lavas, post-Bushveld granites, and possible rheomorphic equivalents of the early, Rooiberg felsite lavas. In addition, it is likely that some of the granophyres, particularly those in the Tauteshoogte and Blood River regions, are acid differentiates of the layered intrusion. These latter, relatively iron-rich granophyres (described as part of the Upper Zone, pp. 376–82) contain fayalite and ferrohedenbergite, and are comparable with similar melanogranophyres of the Skaergaard intrusion (Ch. VI). In contrast, the earlier felsites and later, Red Granite are free from these iron-rich minerals. As demonstrated on Fig. 217 the chemical relationships tend to support this distinction, the Bushveld melanogranophyres plotting in an area that could be viewed as the termination of the trend. In contrast, the Rooiberg felsites and Bushveld Granite are sufficiently extreme in composition as to indicate that the melanogranophyres have a different origin, and not so obviously independent of that of the layered ferrodiorites and ferro-syenodiorites.

SUMMARY OF THE CRYSTALLIZATION HISTORY

The Bushveld Complex as a whole consists chiefly of an assemblage of basic and acid volcanic, hypabyssal, and plutonic rocks, the basic rocks having tholeiitic affinities. The huge Bushveld layered intrusion occupies much of the area of the whole complex, about 25 000 square miles, and is undoubtedly the largest basic intrusion in the world. The most likely way in which it formed is by the injection of a large volume of basaltic magma into the crustal metamorphics, sediments and lavas, entry being gained either by crustal foundering or, more probably, by the upward displacement of crustal material along conical fractures so as to leave a funnel-shaped cavity or cavities. It seems unlikely that the act of magma injection was a single incident, not only in view of the volume involved but also because, in contrast to the much smaller Skaergaard intrusion, little evidence is preserved of a symmetrical Border Group, the marginal relationships being complicated particularly by the extensive occurrence of thin sheets of quickly-cooled basic magma in the form of sills. Nor does it seem likely that the initial magma was emplaced in a single, symmetrical chamber but, rather, that the Bushveld intrusion should be viewed as having been a complex of at least four separate, interconnected magma chambers. Crystallization may have proceeded independently in each chamber, but in view of the lateral persistence of certain distinctive layers it is more likely that this process was interrupted periodically through magma migrating laterally between the chambers, probably at those stages when fresh supplies of basaltic magma entered the complex of chambers from below.

The cooling of the magma would have initiated crystallization and the crystals would have sunk through magma at their temperature, aided by convection currents sweeping crystals from the upper, cooler zones to the deeper levels where the crystals would have accumulated on the floor. This seems a logical, if infrequently accepted hypothesis to account for the origin of the Bushveld layered rocks, and is applied to the following summary of the crystallization history.

The igneous cumulates of the Bushveld intrusion consist chiefly of plagioclase felspar, olivine, pyroxenes, and iron-bearing oxides, with minor amounts of quartz, alkali felspar, apatite, hornblende, biotite, platinoid minerals, and sulphides. The mineralogy is thus the same as found in other layered intrusions consisting chiefly of ultrabasic and basic rocks, where the evidence of a chill-zone is more obviously in favour of crystallization from basaltic magma. The distribution of the minerals follows a precise pattern in relation to the stratigraphic sequence defined by the layering, from ultrabasic assemblages at the exposed base, through basic and then intermediate assemblages, to an acid assemblage near the exposed roof. The gradual compositional changes in the felspars, olivines, pyroxenes and spinels, upwards, is from high-temperature to low-temperature members of their respective solid-solution series, and this cryptic layering, taken together with the graded character of many of the rhythmically layered units, leaves little room for manoeuvre in outlining the crystallization sequence, from the floor upwards.

The earliest minerals to separate from the magma were bronzite and a magnesian olivine, generally concentrated either into pyroxenitic or dunitic layers to form the Basal Series of the layered sequence. Chrome-spinel, present in minor amounts, then became an important cumulus phase and settled periodically to form the economically imported chromitite layers, the lowest of which is taken to mark the base of the Critical Series. As the magma continued to cool, changing composition owing to the separation of abundant olivine, bronzite and chromite, olivine ceased to crystallize and then plagioclase, and later augite, began to crystallize (Fig. 192). The changing composition of the liquid is also reflected in the iron-enrichment of the olivines, pyroxenes, and spinels, and the sodium-enrichment of the felspars, although these overall trends are complicated in the Basal and Critical Series by occasional fluctuations in cumulus compositions with height in the layered sequence. This is probably due to the fact that in the earlier stages of magma-chamber formation, the tectonic conditions were unstable and resulted in periodic flooding of the chambers with fresh supplies of parental basalt magma. The mixing of residual and fresh magmas is a process of magmatic integration rather than magmatic differentiation, and for this reason we have grouped together the Basal and Critical Series as having formed during an Integration Stage, as distinct from the later development of Zones during the more stable Differentiation Stage. Nevertheless, the cooling of the mixed magmas resulted in a distinct, if irregular type of differentiation, and the overall trends show a systematic change in the phase relations throughout the whole of the layered sequence. The main result of temporary magmatic integration would be that the relative bulks of the various cumulates could not be correlated with the proportions of their constituents present in an initial volume of basaltic magma (as we first realized when dealing with the ultrabasic cumulates in Rhum) and this reservation would apply, for example, to the concentrations of chromium in the Critical Series of the Bushveld. Towards the close of the Integration Stage, a magma replenishment occurred, apparently, that was more drastic in its effects and may have been the last to have taken place. The mixing of residual and fresh magmas resulted in the separation of an immiscible liquid from which platinoid and sulphide minerals crystallized, being deposited along with chromite and bronzite at the base of two of the rhythmic units forming the Merensky Reef horizon at the top of the Critical Series. The turbulence associated with the mixing of the residual and fresh magmas gave rise to a characteristic set of layered structures at this horizon, especially channel and scour structures.

The Bushveld intrusion would then seem to have become fairly stable, the subsequent differentiates of the Main Zone being layered but with the cryptic layering more regular and better developed than the rhythmic layering. However, only plagioclase and two pyroxenes are present as cumulus phases in the Main Zone rocks and it may be that (as in the Middle Zone of the Skaergaard) the weak rhythmic layering is more a reflection of the relative densities of the available cumulus phases than of the relative current activity in the magma chamber. A return to the type of markedly contrasted layering found in the Critical Series is found when a dense, iron-rich oxide began to crystallize as a cumulus phase, marking the base of the Upper Zone.

The rocks near the base of the Upper Zone, underlain by almost 20 000 ft of layered cumulates, contain plagioclases and pyroxenes which are well advanced in their trends towards lower-temperature compositions (Fig. 192). The marked iron-enrichment of the magma must also have been accompanied by an increase in oxidation state such that iron-rich oxides were precipitated in abundance, probably as titaniferous magnetites which, at a later stage, were converted to titanomaghemites. The melanocratic layers at this horizon each grade upwards into felspar-rich layers, so the process of crystal accumulation to form rhythmic layering was still operative at this stage. Continued fractionation led to the precipitation of an iron-rich olivine, returning as a cumulus phase after being absent from at least 12 000 ft of cumulates, while the increase in phosphorus content of the magma then led to the precipitation of apatite as a cumulus phase. By this stage the plagioclase was an andesine, the Ca-rich pyroxene a ferro-augite, and the Ca-poor pyroxene a ferropigeonite. Upwards in the ferrodiorite sequence, characterized by the continued presence of rhythmic layering, the plagioclase became more sodic and the olivine and pyroxenes more ferriferous (with the Ca-poor pyroxene eventually ceasing to crystallize) until, at the top of the exposed layered series, the extreme differentiates are found (Fig. 192). Thus the Bushveld layered series is characterized by *extreme* fractionation of the type generally attributed only to the Skaergaard because, apart from a brief reference by Lombaard (1956, p. 73), the information on the ferrodiorites and later differentiates of the Bushveld has not been published outside university theses.

The uppermost layered rocks of the Upper Zone contain, as cumulus phases, plagioclase (An_{30}), olivine (Fo_0), ferrohedenbergite ($Ca_{42}Mg_1Fe_{57}$), up to 5% of apatite, and minor amounts of titaniferous magnetite. In addition, the plagioclases are zoned to about An_{20}, the pyroxenes are partly altered to amphibole, biotite is present in small amounts, and a micrographic intergrowth of quartz and alkali felspar may account for up to 25% of some of the rocks. These rocks have been variously named as grano-diorites, syenodiorites, or perthite-bearing diorites, but their association with the underlying ferrodiorites and their iron-rich character would justify naming them, perhaps, *ferro-syenodiorites*. They are undoubtedly the latest obvious differentiates of the Bushveld layered series, and their constituent olivines and ferrohedenbergites lie on the ends of their respective solid-solution series. As discussed earlier (p. 380) these rocks are occasionally in direct contact with the overlying roof-felsites, but in other areas there are granophyres overlying the ferro-syenodiorites or occurring as sheets or lenses in the roof rocks. These granophyres (containing over 75% quartz+ microperthite) contain minor amounts of plagioclase (zoned from An_{30} to An_{20}), fayalite and ferrohedenbergite, the latter three minerals having, significantly, the same com-positions as in the layered ferro-syenodiorites. It seems logical to assume that the trend of fractionation was extended, by an increase in quartz and alkali felspar, to include the formation of these acid rocks from a final, residual liquid. By analogy with the Skaergaard they are termed melanogranophyres, and in the same way they appear to have been expelled locally as a late-stage, mobile liquid, to form small transgressive sheets in the roof zone.

From this summary of the Bushveld layered sequence it will be apparent that the intrusion displays a remarkably complete sequence of the fractionation products of a tholeiitic basalt magma. In the absence of a clearly-defined chilled zone, marginal and upper border groups, and ideal layered structures such as gravity-stratified layers and trough-banded structures, it falls short in relation to the evidence in the Skaergaard that has led to detailed interpretations of its formation. On the other hand, the Bushveld is the only discovered layered intrusion in which the almost complete assemblage, ranging from ultrabasic to acid differentiates, can be observed directly and, by comparison with certain fragmented patterns in other layered intrusions, can safely be assumed to have formed by crystallization differentiation in place.

CHAPTER XV

SOME OTHER TYPES OF LAYERED BASIC INTRUSIONS

Cuillin, Isle of Skye, Northwest Scotland

The Tertiary igneous centre of the Isle of Skye, in the Scottish Inner Hebrides, includes a large basic intrusive complex forming the spectacular range of mountains known as the Cuillins. Exposure is excellent, revealing a great variety of layered structures and field relationships in or between the ultrabasic, eucritic, and gabbroic parts of the complex, and evidence has emerged for a much more complicated pattern of events than previously envisaged. The present stage of research, based largely on the results presented in three Oxford D.Phil. theses (Carr, 1952; Weedon, 1956; Zinovieff, 1958), has contributed enough information to show that the layered rocks formed by a process of crystal accumulation, and a broad sequence of cryptic layering has been established. The more complex problems relate to the events responsible for the present structural arrangement of the various parts, and these require further work, particularly on some of the difficult mountains.

The first detailed mapping was done by A. Harker for the Geological Survey, as part of his classic work on the intrusive complex of Central Skye (1904). Prior to this, Geikie and Teall (1894) had discussed the nature of the banded gabbro in part of this region (Fig. 3) and had realized that different adjacent bands resulted from different proportions of minerals of essentially the same composition; they were the first to make this important generalization but did not, at that time, attribute it to crystal sorting. Much later, Stewart and Wager (1947) recognized gravity stratification in certain rare layers, and this they attributed to crystal accumulation.

The Cuillin basic intrusion is roughly circular in shape on a map (Fig. 218), and about 5 miles in diameter, although the eastern part has been removed during the later intrusion of a ring-complex of granitic rocks forming part of the separate, Western Red Hills centre (Wager *et al.*, 1965). Elsewhere, the intrusion is in contact with the Tertiary basalt series which was erupted prior to the central intrusive episode. According to Harker's interpretation, the Cuillin complex consisted essentially of a laccolith of ultrabasic rocks intruded by a later gabbroic laccolith. The 'banding' which generally dips towards Druim Hain (Fig. 219 & 220), the centre of the roughly circular basic mass before the displacement of the northeastern part by granites, was attributed by Harker to flow of heterogeneous magma. A minor, but striking aspect of the Cuillin complex is the profusion of cone sheets which dip at about the same angle, and to the same centre, as the banding, and probably led to the idea that the banded basic rocks might also be due to repeated injections of magma. Bailey (1945, p. 183), for instance, viewed the Cuillin gabbros as a confluent cone-sheet complex.

Despite the detail of Harker's work, which included the recognition of different gabbro types, our subsequent work in the region, together with further work on the ultrabasic rocks by Weedon (1961, 1965), has shown the need for a different inter- pretation of the relationships within the intrusion. The fine-scale rhythmic layering and the cryptic layering of the ultrabasic, eucritic and gabbroic rocks, have led us to the conclusion that the various parts of the complex belong, broadly, to a single episode of crystal accumulation within a basic magma chamber situated in the general position of the present complex. Relative movements of large masses of the early-formed cumulates, however, have taken place during and after the settling of the crystals, giving the complicated, present disposition of the layered rocks.

a) *Structural sub-divisions*

The arcuate ultrabasic mass of Sgurr Dubh (Fig. 219) has different contact relation- ships on its two sides. The nature of the inner contact was well described by Harker (1904) but the general structural relations, especially at the outer contact, are far from clear. Probably the outer, convex contact is in part a structural discontinuity, as

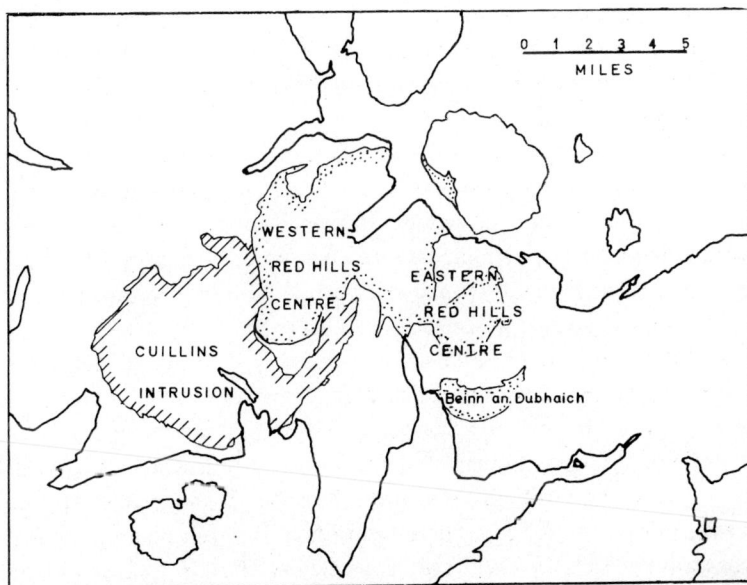

FIG. 218. Sketch-map of the Tertiary central intrusive complex, Isle of Skye (from Moorbath & Bell, 1965) showing the position of the Cuillin basic layered complex in relation to the Red Hills granitic complexes. (See also Fig. 133.)

suggested by Weedon's work (1955, 1961, 1965). The non-layered Outer Gabbros near the ultrabasic rocks show a wide zone of felspar-clouding, attributed by Weedon (1961, fig. 1) to thermal metamorphism by the ultrabasic rocks. It will later be suggested that a hot mass of ultrabasic material, perhaps as a crystal mush, was pushed upwards into its present position against the Outer Gabbros and that heat from the mass produced the clouding of the felspars. The inner contact against the eucrites is totally different from the outer. The layered inner eucrites contain a profusion of ultrabasic blocks, derived from the Sgurr Dubh ultrabasic mass, which decrease in size and amount away from the irregular contact (Weedon, 1961, pp. 199–200) and apparently these were laid down with the crystal precipitate forming the eucrites, after the emplacement of the ultrabasic rocks into their present position.

The Outer Gabbros are not considered further in this account, except in relation to the sequence of events in the formation of the complex as a whole. Weedon (1961, fig. 1) has described the southern part of this large outer complex, which to the north forms most of the higher peaks of the Cuillins range and has yet to be studied in detail. The greater part of the southern outcrops consists of sheets of a rather fine-grained gabbro, with glassy mesostasis, interleaved with basalts and termed the Gars-bheinn gabbro. Inwards this gabbro, containing the characteristically clouded felspars, is coarser grained and rather more basic in composition than the sheet gabbros, and may represent a different intrusion. Later, both gabbro types have been intruded by ring eucrites, one of which forms the southwestern boundary between the whole Cuillin complex and the basalts and Torridonian sediments. Finally, in this 'Outer Gabbro' complex, have been intruded a set of ultrabasic dykes (Bowen, 1928) and a thick, differentiated, ultrabasic sill with a dyke feeder (Weedon, 1960), described briefly on pp. 536–7.

The main arcuate outcrop of Sgurr Dubh ultrabasic rocks (Fig. 219) consists of layered peridotites and dunites with the unusual kind of transgressive contact with the Outer Gabbros mentioned above. The ultrabasic rocks are not now viewed as an independent intrusion but, rather, as the earliest crystal cumulates formed as part of an extensive sequence leading on from peridotites to allivalites and, later, to the inner group of layered eucrites and gabbros. Weedon (1965) has described the ultrabasic assemblage in detail, and recognized a fairly regular change, upwards, from olivine cumulates (dunites) to rocks with both cumulus olivine and plagioclase. Later than Weedon's initial work (1955) we recognized felspathic, layered, ultrabasic allivalites (resembling those of the neighbouring island of Rhum) on Sgurr nan Gillean, well to the north but approximately along the line of strike of the Sgurr Dubh peridotites. Zinovieff's work (1958) in this northern region has now established the existence of a great thickness of layered ultrabasic, plagioclase–olivine cumulates (Allivalite Series) which are now viewed as having formed, together with the Sgurr Dubh olivine-rich cumulates, during the early stages of crystal accumulation in the Cuillin magma chamber. In the central part of the western region, between Zinovieff's and Weedon's mapped

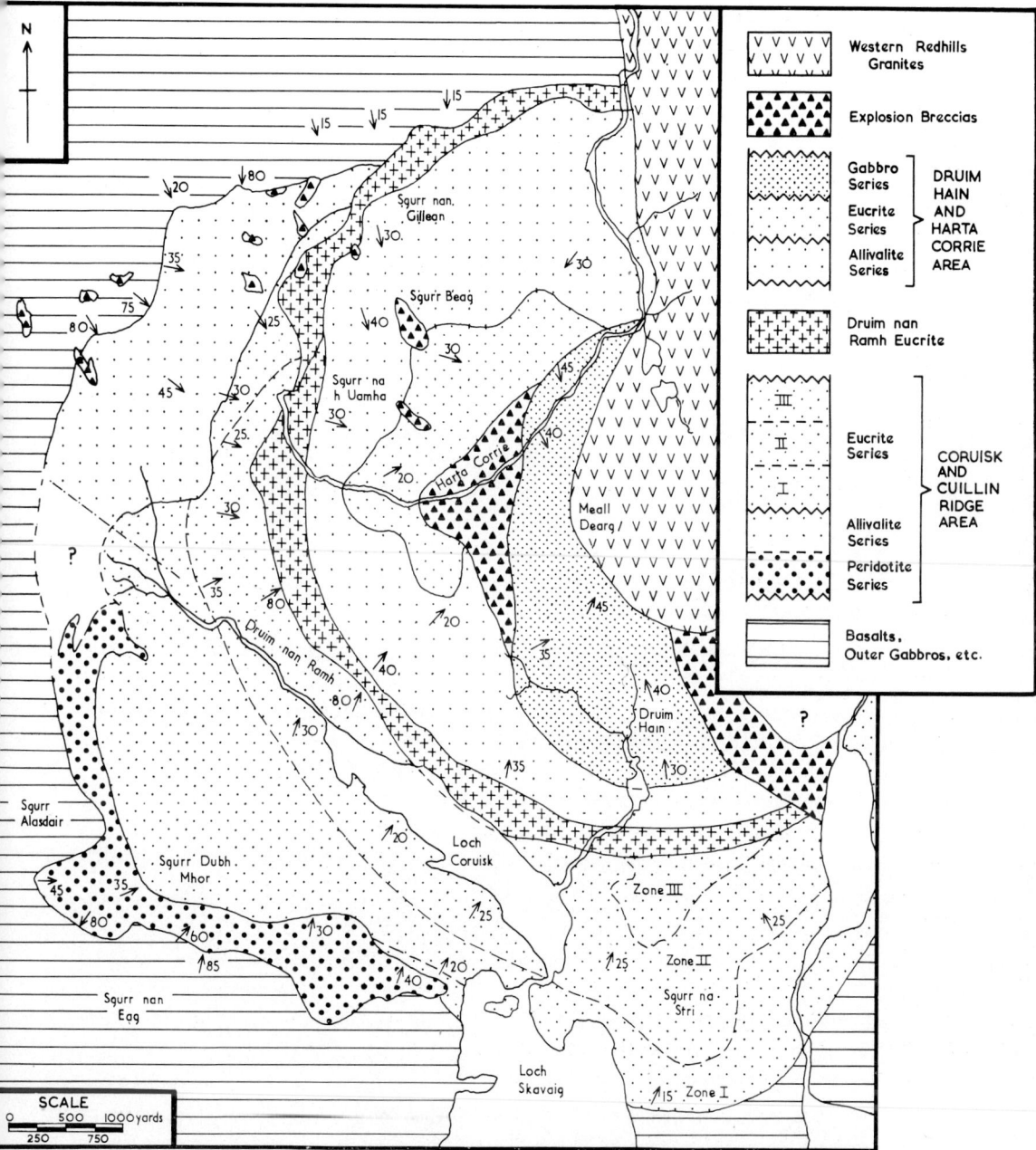

FIG. 219. Geological map of the Cuillin intrusion, compiled from our data and those of Carr (1952), Weedon (1956), and Zinovieff (1958). Un-mapped areas are shown by a query-mark.

areas, Hutchison (1964) has shown that there is a continuation of the Allivalite Series. In places, an outer contact can be seen against the Outer Ring Eucrite (personal communication), allivalitic rocks probably continue into what Weedon has called the Ghrunnda Eucrite. Allivalites, then, appear to lie on the convex side of the Sgurr Dubh peridotites, according to Hutchison's work, and the area north of Sgurr Alisdair must be more complex than shown on Fig. 219.

The eucrites and gabbros, enclosing ultrabasic blocks and, therefore, crystallized after the ultrabasic cumulates, form much of the inner part of the Cuillin layered complex. The term ' Inner Gabbros ', found useful in the earlier stages of mapping, has no real use and the rocks will be referred to as the layered eucrites and gabbros of the intrusion. An unconformity in the layered rocks exists between the ultrabasic rocks on Meall na Cuilce (at the northwest edge of Loch Skavaig) and the outermost, lower zones of eucrite. On the eastern side of Loch Skavaig (Fig. 219), also, there is a non-sequence, a marginal facies of the eucrites being in direct contact with the basalts. In the latter, Sgurr na Stri region, Carr (1954) studied the eucrites in great detail and was able to establish, from the cryptic layering, a layered sequence which could be sub-divided, according to mineralogical and internal structural criteria, into three main zones. Between these and the well-known gabbroic cumulates of Druim Hain (Fig. 219 & 220) there is another structural and compositional break.

The layered rocks range from ultrabasic cumulates, through eucritic cumulates to gabbroic cumulates. Each group is described separately because of the structural breaks between them, although they are finally grouped together and considered as part of a single episode of crystal fractionation. Certain undifferentiated eucrites with ring-shaped outcrop are also present and are apparently related to the pattern of contemporaneous faulting within the intrusion. These arcuate structures, and the dips of the layered rocks, together give the impression of an intrusive complex with a centre in the region of Druim Hain (Fig. 219 & 220), where the latest exposed differentiates are found in close association with volcanic agglomerates. Here the basic complex was subsequently invaded by the Meall Dearg epigranite which forms part of the large, granitic ring-complex centred to the east of the Cuillins, and known as the Western Red Hills complex.

b) *The ultrabasic cumulates*

Sgurr Dubh peridotites. These rocks occur in an arcuate belt between the head of Loch Skavaig and the Coruisk River (Fig. 219 & Pl. XIII). The outer margin, against the earlier Outer Gabbros, has been studied in some detail by Weedon (1965, fig. 1 and pp. 56–61) and shown to be almost vertical. In addition, the almost-vertical, banded rocks within a few feet of the margin frequently have granulated and partly recrystallized textures, suggesting movement relative to the Outer Gabbros. The inner margin against the later-formed layered eucrites is also unusual. In the Meall na Cuilce area, as described by Weedon (1961, figs. 1 & 2; pp. 198–200), the ultrabasic

PLATE XIII

View of the Cuillins Mountains of Skye, looking northwest across the head of Loch Skavaig (see map, Fig. 219). The deep valley behind the small islets is the glaciated basin of Loch Coruisk. Rocks in the foreground are layered eucrites, with Outer Gabbros and peridotites of Meall na Cuilce on the mountain to the left. The fairly high ground at the extreme right is the Druim nan Ramh plateau (eucrites), leading to the higher, northern part of the main Cuillins ridge (background) of allivalites and Outer Gabbros.

FIG. 220. Aerial photograph of the Druim Hain region (north at the top), showing the layered features of the gabbro series, dipping towards a centre close to Meall Dearg (see Fig. 219 for localities). The outcrops of Meall Dearg epigranite (pale coloured, top-right) and of explosion breccias (pale coloured, bottom-right) are clearly visible, together with their sharp, arcuate contacts against the earlier gabbros. (*Royal Air Force Photograph, Crown Copyright Reserved.*)

rocks show clear evidence of having been invaded by basic magma and of having been broken into large blocks which became detached and strewn throughout a fairly extensive and thick layer of the inner layered eucrites. The outer and inner marginal features of the peridotite mass of Sgurr Dubh indicate that there was probably an upward movement of the peridotites (probably while still a hot mush of crystals) relative to the Outer Gabbros, and a collapse downwards along a fracture further in, producing a local arcuate ridge of ultrabasic rocks projecting into the magma. Erosion of the ridge provided blocks of peridotite which became widely distributed along a certain horizon of the inner layered rocks. Whatever the relative movement, it is likely that the present exposure of dunites and peridotites in this area represents only an isolated portion of an ultrabasic layered sequence which probably underlies the layered allivalites and eucrites over an extensive area.

The exposed thickness of ultrabasic rocks in this area has not been estimated by Weedon, because of difficulties experienced in lateral correlation of the more massive and unlayered parts, and in dealing with the brecciated zones. However, the more felspathic layered rocks (Fig. 221 & 222), which dip to the northeast at about 30°, are at least 2000 ft in thickness and the total exposed thickness may be in the region of 4000 ft. The series has been subdivided into three zones by Weedon (1965, p. 44); the lowest consists of dunites and peridotites, the next of felspathic peridotites and allivalites, and the uppermost, of the brecciated and blocky zone. The brecciated zone is a local division based on structural features, while we link Weedon's second zone directly with the thick allivalite series to the north.

In describing these rocks, which we believe to be clear examples of igneous cumulates, the cumulate nomenclature (Wager *et al.*, 1960) is more advantageous than dividing them into dunites, allivalites, etc., particularly if consideration is to be given to the textural evidence for types of intercumulus growth. The lower rocks are olivine–spinel cumulates which may contain 96% or more of olivine (dunites), the only layering being in the form of thin, persistent layers of chrome-spinel (Cf. Fig. 222). These thin layers, which range from a few millimetres to an inch in thickness, may be traced laterally for up to 50 yards (Weedon, 1965, p. 53). More commonly they form irregular, contorted seams which are attributed to movements at the crystal mush stage. The almost monomineralic, olivine-rich rocks are olivine (–spinel) adcumulates, and in their relative abundance in Skye they form a contrast with Rhum (Ch. XI) where olivine adcumulates are not found among the olivine-rich rocks at the base of the layered macro-units, whereas plagioclase adcumulates are present at the top of many of those units.

Above the olivine (–spinel) adcumulates, Weedon (1965, p. 45) refers to a group of peridotites in which plagioclase and augite are found as large plates poikilitically enclosing cumulus olivine and spinel crystals; these are apparently olivine (–spinel) heteradcumulates. Upwards in the sequence, plagioclase enters as a cumulus phase in gradually increasing amounts, leading to rocks which Weedon calls felspathic peridotites and allivalites. These are olivine–plagioclase heteradcumulates and plagioclase–olivine heteradcumulates, so that the ultrabasic sequence as a whole, in this area, shows a regular upward increase in the ratio of cumulus plagioclase to cumulus olivine—a feature also observed within each macro-unit in Rhum (Ch. X).

The cumulus olivines of these rocks show a slight, but significant iron-enrichment upwards in the sequence, from Fo_{87} to Fo_{84} (Weedon, 1965, pp. 47 & 51). These determinations, from optical measurements, are believed more dependable than the estimated normative olivine compositions (Fo_{93}–Fo_{90}), from whole rock analyses, chosen for discussion by Weedon (*op. cit.*, p. 55). Where cumulus, the plagioclases are determined as An_{88} in composition (Fig. 224) while the intercumulus plagioclase of the lower, olivine-rich cumulates is determined as An_{85}. The difference in composition is only slight, bearing in mind the difficulty of optical measurements on the large poikilitic crystals; there seems little doubt that the intercumulus plagioclases have grown by the adcumulus process and that the rocks containing them are heteradcumulates. Again, the optical determinations of the plagioclase compositions are preferable to the norma-

FIG. 221. Layering in the Sgurr Dubh ultrabasic rocks, below the rock barrier of An Garbh-choire, accentuated by the resistant character of the felspathic, relative to the olivine-rich layers. (From Weedon, 1956.)

tively calculated compositions which are, in some cases, An_{45} and improbably low (Weedon, 1965, tables 1–4). Slight zoning around some of the cumulus plagioclase has been recorded (Weedon, 1965, p. 50), the range being from An_{88} (core) to An_{74} (rim). This means that although the majority of these ultrabasic rocks are either adcumulates or heteradcumulates, a few are strictly mesocumulates, although outer zones with anorthite percentages lower than An_{74} have not so far been recorded.

Allivalite series. A large part of the northern Cuillins consists of about 6000 ft of allivalitic rocks which are relatively rich in cumulus calcic bytownite, with subsidiary, cumulus olivine and augite. Similar rocks have been found to the south and west by Hutchison (1964). These rocks continue the trend of felspar enrichment described above for the olivine-rich rocks of Sgurr Dubh and are considered, therefore, to be

FIG. 222. Polished hand-specimen of an ultrabasic layered rock from the locality shown in Fig. 221, showing that the contacts between layers (differing greatly in persistance to weathering) are not so sharp as suggested by the weathered surfaces. The thin, black layer 2 inches from the base is a chromite-rich cumulate. (Specimen 8 inches tall.) (From Zinovieff, 1958.)

later-formed crystal accumulates.   Mapped by Harker as ' gabbro ', they occupy the
main ridge of the northern and western Cuillins, and the subsidiary ridge of Druim
na Ramh (Fig. 219).   At the northern end of the ridge on the commonly-followed
tourist path up Sgurr nan Gillean, good fine-scale layering, very similar to that found in
the Rhum allivalites (Ch. XI) can be seen.   Zinovieff (1958) sub-divided this great
thickness of felspar–olivine cumulates (his ' Allivalite Series ') into five major rhythmic
units, while from the Bruach na Frith region, northwest of Sgurr nan Gillean, he has
described an unlayered, Border Zone facies.   From Zinovieff's observations the latter
consists, in part, of a fine-grained marginal gabbro adjacent to the earlier, Gars-bheinn
gabbros, and it is attributed to chilling at the border of the Cuillin basic magma chamber.
Inwards, the chilled facies passes into coarser grained rocks which include a variety with
elongate, apparently crescumulate olivines, suggestive of inward growth within a tranquil
division of the marginal border zone.   This rather inaccessible, but important border
zone needs more detailed study before we can be sure of the relationships.   Hutchison
(1964), for example, has found that the Allivalite Series further to the south cuts across
the Outer Ring Eucrite.   The lower allivalite units are apparently in contact with
border-zone rocks, whereas the peridotites at their southeastern limit are in contact
with the gabbroic country rocks of Gars-bheinn, and allivalites are absent.   These
relationships are attributed to uplift of the southern peridotites to their present position,
as discussed above, the uplift apparently dying out northwards where the structur-
ally-higher allivalite layers pass into an undisturbed border zone.   There may be
a radial fault separating these southern and northern areas (shown hypothetically
on Fig. 219), coinciding with the distinct topographical feature of the Coruisk River
valley.

The allivalites in the region northwest of Druim nan Ramh, as shown on Fig. 219,
dip inwards but are overlain, unconformably, by the layered eucrites.   This relationship
is very similar to that observed between peridotites and eucrites to the south.   If the
ultrabasic allivalites and their border zone are in their original position it is apparent that
there must have been a downward collapse of layered rocks to the east, as also occurred to
the east of the peridotites of Sgurr Dubh, with uprise of the contemporary magma and
subsequent deposition from it of layered eucrites on, or banked up against, the earlier
ultrabasic cumulates.

The largest outcrop of layered allivalites lies to the east of an extensive ring-dyke of
undifferentiated eucrite (Druim nan Ramh eucrite), and the resulting structural complexities have not yet been resolved.   Zinovieff (1958) suggested that the allivalites to
the east of the ring-eucrite, with some overlying eucrite, had been faulted upwards into
their present position, and that subsequently, continued deposition of layered eucrites
and gabbros took place.   This is difficult to envisage, because the western allivalites are
partly overlapped, unconformably, by layered eucrites.   As seen on the map (Fig. 219),
the magma chamber in which the eucrites of the Coruisk–Cuillin Ridge area formed
would have extended into the area now occupied by the eastern block of allivalites.   If

2 D

such late-faulting has occurred, an enormous central *uplift* would have been involved, which is difficult to reconcile with the presence of later layered eucrite and layered gabbros in the central, Druim Hain–Harta Corrie area.

The difficulty in explaining the relationships in this part of the complex justifies alternative hypotheses. One is that the Druim nan Ramh eucrite marked the beginning of another layered intrusion, after a further collapse of the earlier layered rocks in the central region. The outer margin of the eucrite (Fig. 219) is such that it cuts across most of the previous layered rocks and even cuts through basalts to the north. Within the new magma chamber, a new series of allivalites (Units 3, 4, & 5 of Zinovieff's nomenclature), and then eucrites and gabbros, formed. The radial dips as mapped by Zinovieff, however, which run counter to the contacts between the layered units, indicate the need for more detailed work in the area, and any hypothesis to explain the arrangement of the rocks must be tentative. On the map here presented (Fig. 219) the Druim nan Ramh eucrite is treated tentatively as a ring dyke, as Zinovieff suggested, and not as the earliest facies of a second intrusive centre.

The allivalites are grouped by Zinovieff into five major rhythmic units (each *c.* 1000 ft thick) each of which has a felspar–augite cumulate, without conspicuous rhythmic layering, in the lower parts, giving place upwards to rhythmically layered, felspar–olivine cumulates. There is a gradual increase in the degree of rhythmic layering upwards in each unit, but cryptic variation of the cumulus minerals within a single unit could not be detected (Zinovieff, 1958). It seems that the base of each macro-unit is a felspar–augite cumulate rather than an olivine-bearing cumulate, and at the moment no acceptable explanation can be offered to account for this sequence, which is the reverse of that found in Rhum and other layered ultrabasic sequences, and particularly unexpected in view of the abundance of cumulus olivine in the Peridotite Series, believed to be the earliest of the Skye layered rocks. If, despite the previously mentioned reservations, we take the five units as a whole, there are gradual compositional changes suggesting that they can be viewed as parts of a single, thick series of crystal accumulates. Thus there is a gradual decrease in the average amount of cumulus olivine from about 40% in Unit 1 to about 20% in Unit 5, while the cumulus felspars change in composition from An_{87} to An_{82}, and the cumulus olivines from Fo_{84} to Fo_{81} (Fig. 224). Zoning is rarely observed, and the rocks of this series are either adcumulates or heteradcumulates.

c) *The eucritic cumulates*

These layered rocks are characterized by the presence of a cumulus plagioclase which is less calcic than in the ultrabasic layered rocks, and they are continuously exposed on Sgurr na Stri and northwards towards the vicinity of Druim Hain. Carr (1952) sub-divided them into an outer (marginal) zone (not shown on Fig. 219) and three layered zones; Zones I–III have a total thickness of about 5000 ft, while Zinovieff (1958) has recognized a further zone, in the Harta Corrie area (Fig. 219), about 1500 ft thick. The strike of the layering is approximately arcuate, and the dips vary from about $15°$ in the outer to $35°$ in the innermost exposures.

The outer margin of the eucrites is variable in character. In the Sgurr na Stri area, Carr (1952) found a Marginal Zone of unbanded eucritic rocks in contact with basalts and sediments; a similar zone of coarse, unlaminated rocks with a $70°$ dip was found in contact with the Sgurr Dubh peridotites—the Allt Beag Zone of Weedon (1961, fig. 2). At about the position of the eucritic Zones I and II, peridotite blocks, discussed above, are particularly abundant. They occur in the layered eucrites as far east as Loch na Creithach (eastern edge of Fig. 219) while Zinovieff (1958) has found essentially similar blocks, but of allivalite, on Druim nan Ramh to the north, and has ascribed Carr's 'olivine-eucrite phase blockstuff' in the eucrites of Sgurr na Stri to a similar origin. While this interesting phenomenon is most conspicuous at about the junction between Zones I and II, a few peridotite blocks have been found lower in Zone I, and in Zone III.

The eucritic cumulates of Zone I are, on the whole, medium-grained orthocumulates, whereas those of Zones II and III are well-laminated adcumulates in which rhythmic layering is better developed than in Zone I. Other distinctions are based on the character of the cumulus phases; all contain cumulus plagioclase, olivine and clino-pyroxene, but Zone I also contains cumulus magnetite and inverted pigeonite, Zone II does not contain cumulus magnetite but has more olivine than the other zones, and Zone III again contains cumulus magnetite. The plagioclase composition changes upwards through these eucrite zones, with some slight irregularities, from An_{75} to An_{67}, while the olivine changes from Fo_{74} to Fo_{67}. An augite analyzed from Zone II has the composition $Ca_{41.7}Mg_{46}Fe_{12.3}$*(Fig. 224).

The 'calcic-phase phenocryst' felspars are a special feature in all layers, but especially upwards from the base of Zone II. These are large plagioclases of a relatively constant composition (An_{85}), regardless of the felspar composition of the host rocks in which they are found. Weedon (1961) has discussed the possibility of a special magmatic event occurring between the deposition of Zones I and II. He draws attention to the fact that the junction between Zones I and II is characterized not only by the advent of 'calcic phase' felspars in abundance but also by the marked increase in igneous lamination and in the proportion of ultrabasic blocks. The latter factors suggest a marked increase in current activity, and perhaps the incoming of the abundant calcic phenocrysts indicates the addition of a new surge of basalt magma of the Hebridean Porphyritic-Central type. The reason for the persistence of 'calcic phase' pheno-crysts, through a great thickness of layered rocks (Fig. 224), remains a problem. Unless they are continually added to the magma, perhaps through magma replenishments, it would be expected that they would quickly sink out with the other felspars and be characteristic of only a limited thickness of the layered series.

* In contrast, an augite analyzed from the allivalitic cumulates (Zinovieff, 1958) has the composition $Ca_{44.6}Mg_{46.7}Fe_{8.7}$.

d) *The gabbroic cumulates*

This series of well-layered rocks is about 2000 ft thick and forms much of the plateau known as Druim Hain (Fig. 219 & 220). Its position within the intrusion, together with the general dip of all the layering towards Druim Hain, suggests that the gabbros can be viewed as an upward continuation of the layered sequence from the Zone III eucrite; if this is so, the gabbroic cumulates, truncated by the later granites of Meall Dearg, represent the latest fractions of the Cuillin intrusion to be preserved. In fact, the zone of gabbros (Fig. 219) is separated from layered eucrites and allivalites by a 50 to 100 ft-wide zone of crushed and mylonitized rocks. Between the allivalites outside and the gabbros inside the crush belt, a considerable thickness of layered rocks is missing, the whole of the eucrites being locally absent. This suggests downward faulting of the gabbros along a concentric fracture, bringing later layered rocks against earlier ones.

Fine-scale rhythmic layering is strikingly developed in the gabbros of Druim Hain

FIG. 223.   Fine-scale rhythmic layering in the Gabbro Series, Druim Hain. View looking west, along strike, with the main Cullins Ridge (just north of Sgurr Alisdair) in the background (see Fig. 219 for localities). (From Zinovieff, 1958.)

(Fig. 223). Rare gravity-stratification is also seen (Fig. 3) in which there is generally a marked concentration of magnetites at the base and an upward grading through pyroxene-rich, to felspar-rich rocks. The dip of the layering is steep, increasing from $30°$ in the marginal, less well-layered parts to almost $45°$ inwards; this feature, like that in the allivalites, is attributed to the centripetal tilting of the layered rocks during subsidence of the central region.

The gabbroic composition of these cumulates contrasts with the ultrabasic and eucritic composition of the bulk of the Cuillin rocks. Harker's nomenclature led to the concept of the whole intrusion being largely of gabbros, with subordinate peridotites and dunites, whereas the recent work has demonstrated the much wider extent, and greater thickness, of allivalites and eucrites. The layered gabbros are mainly ortho-cumulates with good igneous lamination of the plagioclase felspars. Cumulus felspar, augite, and subsidiary olivine, together with a few 'calcic phase' phenocrysts, are the same cumulus phases as found in the lower, eucritic cumulates. On the other hand, the advent of cumulus apatite and abundant cumulus magnetite represents a significant advance in the degree of fractionation. Upwards in the gabbro layers, the cumulus plagioclase felspar changes in composition from An_{69} to An_{58} and the olivine from Fo_{66} to Fo_{58}. The augite composition (derived from optics) changes from about Fe_{16} to Fe_{21}, indicating appreciable enrichment in iron relative to that in the augites of the allivalites and eucrites (Fig. 224). Relative to the compositions of the cumulus felspar and olivine, the magnetite and apatite enter as a cumulus phase earlier than in the Skaergaard intrusion; in the case of the iron oxide, this probably results from a higher degree of oxidation in the initial Cuillin basalt magma.

e) *The major events in the development of the complex*

The Cuillin layered intrusion of Skye consists of a series of igneous cumulates with an apparent thickness of at least 18 000 ft. The overall change is from ultrabasic peridotites and allivalites (olivine-rich cumulates and plagioclase–olivine cumulates) near the western margin, through eucritic cumulates to gabbroic cumulates, with in-creasingly sodic plagioclases and ferriferous olivines and pyroxenes towards the centre. Contemporary movement, often in the form of faulting which caused relative vertical displacement, has considerably disrupted part of the sequence. There are also faulting movements which, apparently, post-dated the development of the layered series, notably that which brought the gabbroic cumulates of Druim Hain to their present position. The suggested sequence of events is summarized in the following paragraphs.

The intrusion of the outer masses of unlayered Gars-bheinn gabbros and outer ring eucrites, under the basalt cover, came first and was followed by the formation of an extensive pool of basic magma from which the layered rocks developed. First to be deposited was the Sgurr Dubh Peridotite Series, formed by the accumulation of olivine and chrome-spinel in the earliest stages and of these phases, together with cumulus plagioclase, later. North of Sgurr Dubh and perhaps originally, also, in the Sgurr

Dubh region, a sequence of predominantly felspar–augite and felspar–olivine cumulates (the Allivalite Series) was deposited. Most of the ultrabasic rocks of these episodes are adcumulates or heteradcumulates, which are rhythmically layered on both a large and a small scale; the degree of fine-scale layering and the average felspar to olivine ratio increased at successively later stages. At a distance of about $\frac{1}{4}$ to $\frac{1}{2}$ mile inwards from the western margin, collapse of the layered rocks then changed the form of the magma chamber. Over a north–south distance of a mile or two, a steep fault-scarp of ultrabasic layered rocks was produced which appears to have been progressively eroded to give blocks of the ultrabasic rocks, to be spread by the circulating magma and deposited together with the cumulus minerals forming the eucrites. At this stage the ultrabasic rocks subjected to erosion were probably the allivalites, but later the underlying Peridotite Series was eroded, apparently as a result of an uplift such that they projected into the magma chamber. The uplift gave the Sgurr Dubh sector its present relationship to the surrounding rocks, the amount of uplift dying out northwards where the Allivalite Series is in its original position relative to the Outer Gabbro Series, since it exhibits a chilled Border Zone. The clouding in the felspars of the outer, Gars-bheinn gabbros is probably to be attributed to this emplacement of the still hot, but semi-solid mass of the Sgurr Dubh peridotites.

The eucritic cumulates were being laid down, during these movements, against peridotites or allivalites, while in other parts of the intrusion (e.g. Sgurr na Stri) they were banked directly adjacent to a eucritic marginal zone. They are more laminated than the ultrabasic cumulates, show good rhythmic layering, and are predominantly of orthocumulate or mesocumulate, rather than adcumulate type. There is a cryptic variation throughout the zones of the Eucrite Series, which is continued into the gabbroic orthocumulates of the uppermost layered series, although minor reversals occur and may be due to periodic, minor influxes of fresh parental magma. The gabbroic cumulates probably formed above the eucritic cumulates, but they are at present only preserved, due to collapse, in the central area.

Despite the extreme structural complexity which has delayed our understanding of the Cuillin as a layered intrusion, a cryptic variation in the various units of cumulates is clearly demonstrable and seems to tie them into a single series. The plagioclase changes composition, broadly, from An_{88} to An_{58}, the olivine from Fo_{87} to Fo_{58}, and the augite from Fe_9 to Fe_{21}, whilst the incoming of cumulus apatite and magnetite in abundance, in the uppermost layers, bears further testimony to the progressive nature of the fractionation in this intrusion.

At the present stage of knowledge there are, doubtless, various ways of piecing together the complex, jigsaw puzzle of rock units which now form the central and eastern part of the mountainous Cuillins. The conclusion we present in Fig. 219 is based mainly on the field and petrological work by Carr (1952), Weedon (1956) and Zinovieff (1958), and is essentially that reached by one of us (G.M.B.) through collaborating with Zinovieff in an attempt to produce a synthesis of events. The main principle which has guided this recent work on the complex has been the recognition of an order of formation of the rocks by means of the cryptic variation, chiefly deduced from the evidence of the changing anorthite content of the cumulus plagioclases. Abrupt changes in the expected regularity, sometimes associated with recognizable structural features, have been used to interpret the series of disruptive events in the sequence of crystallization within the magma chambers, together with certain post-consolidation movements such as the down-faulting which has preserved the gabbro series of Druim Hain. An overall cryptic variation appears to link the peridotites,

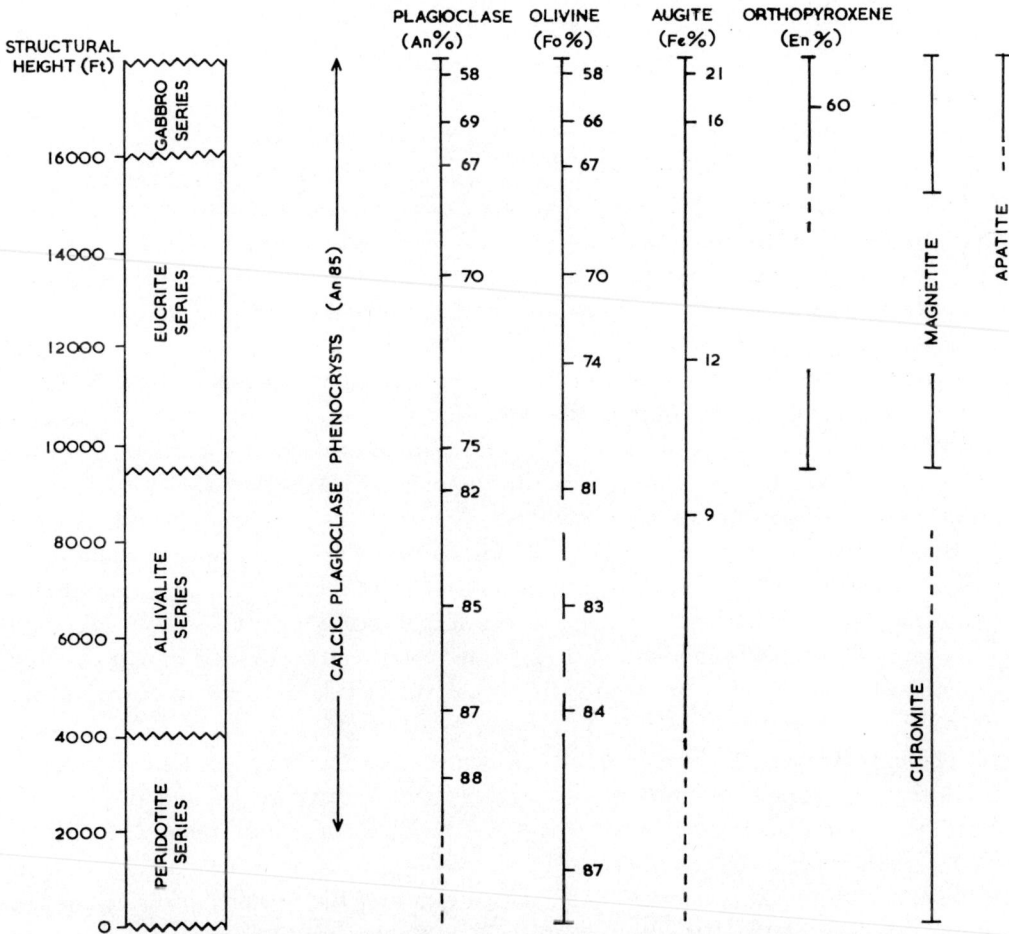

FIG. 224. Generalized stratigraphic section of the Cullin layered series, showing the cryptic variation of the cumulus mineral phases (intercumulus phases as broken lines). Compositions obtained from optical measurements except for the chemically analyzed augites (see text). In view of the extensive faulting, the series of rock types is shown as a broken sequence.

allivalites, eucrites, and gabbros. From geophysical evidence, Tuson (1959) has shown the probable existence of another large basic intrusion underlying the post-Cuillin, Western and Eastern Red Hills granites. However, these granites are not now regarded as a fractionation product of basic magma, but as derived from magmas generated from those country rocks which were partially melted by the basic magma body (Brown, 1963; Wager *et al*, 1965; Moorbath & Bell, 1965).

Whether the fractionation consequent to the production of the Cuillin rocks produced any differentiates later than the gabbros, cannot be observed. The only granitic intrusion recognized as belonging, according to its age relations, to the Cuillin centre (i.e. pre-cone sheets) is the Coire Uaigneich epigranite, but it seems that this acid magma originated from the melting of Torridonian sediments rather than by the fractionation of basic magma (Brown, 1963).

KAP EDVARD HOLM, EAST GREENLAND

Four layered intrusions have been discovered in the relatively small area of the East Greenland coast in the immediate vicinity of the large and prominent Kangerdlugssuaq Fjord (Fig. 4), and have been described in this book. The Skaergaard basic intrusion (Part I) lies on the eastern flanks of the mouth of the fjord, the Kangerdlugssuaq alkaline intrusion (Ch. XVI) on the west–central flanks, and the Kaerven basic intrusion (Ch. XV) just to the north of the latter. The Kap Edvard Holm basic intrusion is situated on the western side of the mouth of the fjord, approximately 10 miles from the Skaergaard intrusion and due south of the Kangerdlugssuaq intrusion. First discovered by one of us (L.R.W.) in 1930, during a visit to the region by a party of the British Arctic Air Route Expedition, it was later included in a reconnaissance study of the area, in the company of Professor W. A. Deer, on the British East Greenland Expedition in 1935–6. In 1953, three members of the East Greenland Geological Expedition, under the leadership of Professor Deer, mapped the Kap Edvard Holm Complex. Accounts of the layered basic intrusion have been presented by Abbott (1962) and in part by Deer and Abbott (1965), and of the associated, later syenite ring-dyke intrusions by Lwin (1960).

The exposed part of the intrusion occupies an area of about 300 km² (Fig. 225), although the mapping indicates that it once extended eastward beyond the present coastline, and that the lower part of the layered series lies beneath the broad Hutchinson glacier to the north. Much of the mapped area is covered by glaciers and small ice fields, but it has been possible to define the position of the western intrusive contact against gneisses of the basement complex and, further south, against the Tertiary basalts. East of the contact, the complex consists predominantly of layered basic rocks dipping radially towards a centre now occupied by a later syenite intrusion at Kap Boswell. The dips decrease from about 50°–70° near the margins to about 15–25° towards the centre, in the Low Point area, although local irregularities are to be found.

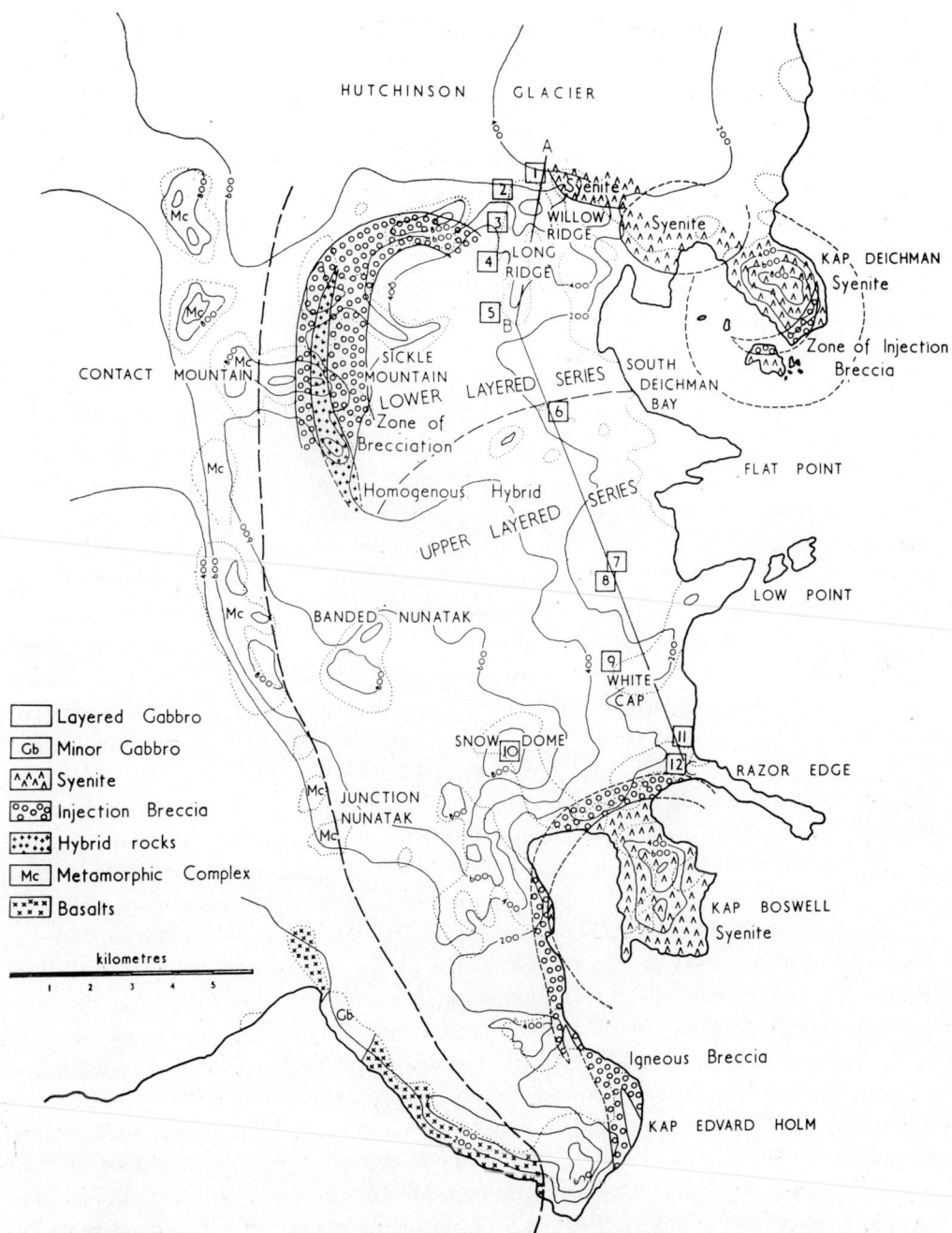

FIG. 225. Geological map of the Kap Edvard Holm Complex. (From Deer & Abbott, 1965. The numbers in squares refer to the localities of rocks from which pyroxenes have been analyzed: see Fig. 226.)

The regional crustal flexure (pp. 13–16) is doubtless responsible for the high dips, and when a correction is made for this superimposed eastward tilting of the intrusion subsequent to consolidation, the layers assume the saucer-shaped forms, with dips increasing gradually towards the margins, analogous to those of the Skaergaard intrusion. The coastal flexuring has resulted in displacement of large blocks of the layered series so that continuity along strike is frequently destroyed, while dykes of the coastal swarm, intruded at the time of monoclinal folding, form a belt in *en echelon* arrangement, across the centre of the complex from southwest to northeast. Subsequent to dyke intrusion, further disruption of the layered basic rocks accompanied the intrusion of syenites and resulted in local veining, thermal metamorphism, and brecciation.

a) *The Layered Series*

The total exposed thickness of layered rocks is about 7500 m, neither a base nor a top being exposed. Although the position of the margin can be inferred, no marginal border group rocks are exposed and either an ice-covered region or a later marginal gabbro separates country rocks from the lowest members of the layered series. The lower 6300 m are plagioclase–augite–olivine cumulates, and the upper 1200 m are plagioclase–augite cumulates. However, the layered series has been sub-divided broadly into two parts by Abbott (1962, fig. 11) and by Deer and Abbott (1964, fig. 2), not according to this change in cumulus phases but on the basis of an abrupt change in cumulus mineral compositions which occurs 3800 m above the lowest exposed layers. Thus the Lower Layered Series is of plagioclase–augite–olivine cumulates and the Upper Layered Series of more of these plus the overlying, plagioclase–augite cumulates (Fig. 226). The abrupt change, at the base of the Upper Layered Series, to cumulus minerals which are higher-temperature solid solutions than those in the layers directly below, is taken to imply a stage of injection of a fresh supply of magma to the differentiation chamber, and similar but smaller-scale injections are believed to have taken place at other periods in the intrusion's history, and form the basis for extra sub-division of the layered series.

The Lower Layered Series. The lowest members, structurally, of the Lower Layered Series are well exposed at the northern edge of the intrusion (Fig. 225) at Willow Ridge, where they can be traced along strike for about 5 km. Fine-scale rhythmic layering is well developed, although the more common cumulus assemblages in these layers consists either of almost equal proportions of plagioclase, olivine and augite, or of plagioclase-rich layers, the scarcity of melanocratic layers implying an overall relative abundance of cumulus plagioclase. Extreme examples, containing 94% plagioclase, are fairly abundant as thin layers, and at about 500 m above the base a 50 m thick layer of plagioclase cumulates occurs, termed, by Abbott, the Anorthosite Zone. The compositions of the cumulus minerals vary, throughout the 500 m (average) thickness of cumulates up to the top of the Anorthosite Zone, from higher to lower temperature solid-solutions, the layers being termed, collectively, the Willow Ridge Unit. The plagioclase changes from An_{79} to An_{67}, the olivine from Fo_{77} to Fo_{68}, and the augite

METRES

LOWER LAYERED SERIES — UPPER LAYERED SERIES

PLAG. AUG. CUMULATES

CUMULATES

PLAGIOCLASE AUGITE OLIVINE

7500 — $6241\ Ca_{42.7}Mg_{37.4}Fe_{19.9}$ — An_{51}—An_{42}

7000 — $6559\ Ca_{43.0}Mg_{38.1}Fe_{18.9}$

6500 — $6618\ Ca_{42.2}Mg_{40.5}Fe_{17.3}$ — An_{66}—An_{49}

6000

5500 — $6278\ Ca_{41.4}Mg_{41.8}Fe_{16.8}$ — An_{61}—An_{63} — Fo_{66}
$6335\ Ca_{42.0}Mg_{41.8}Fe_{16.2}$ — Fo_{66}

5000

4500 — An_{82}—An_{75} — Fo_{77}

4000 — $3245\ Ca_{42.7}Mg_{46.0}Fe_{11.3}$ — An_{82} — Fo_{85}

3500 — An_{77}—An_{65} — Fo_{70}

3000 — An_{80}—An_{62} — Fo_{67}

2500 — An_{73}—An_{57} — Fo_{68}

2000

1500 — $6517\ Ca_{42.6}Mg_{41.1}Fe_{16.3}$ — An_{67}—An_{54} — Fo_{67}
$6520\ Ca_{42.9}Mg_{42.2}Fe_{14.9}$ — An_{66}—An_{62} — Fo_{68}

1000 — $6525\ Ca_{42.9}Mg_{42.2}Fe_{14.9}$ — An_{81}—An_{65} — Fo_{67}

An_{70}—An_{65}—An_{76}—An_{64}

500 — $6537\ Ca_{43.1}Mg_{43.1}Fe_{13.5}$ — An_{67}—An_{71}—An_{64} — Fo_{68}

0 — $6067\ Ca_{43.1}Mg_{44.1}Fe_{12.8}$ — An_{79}—An_{76} — Fo_{77}

PLAGIOCLASE

OLIVINE

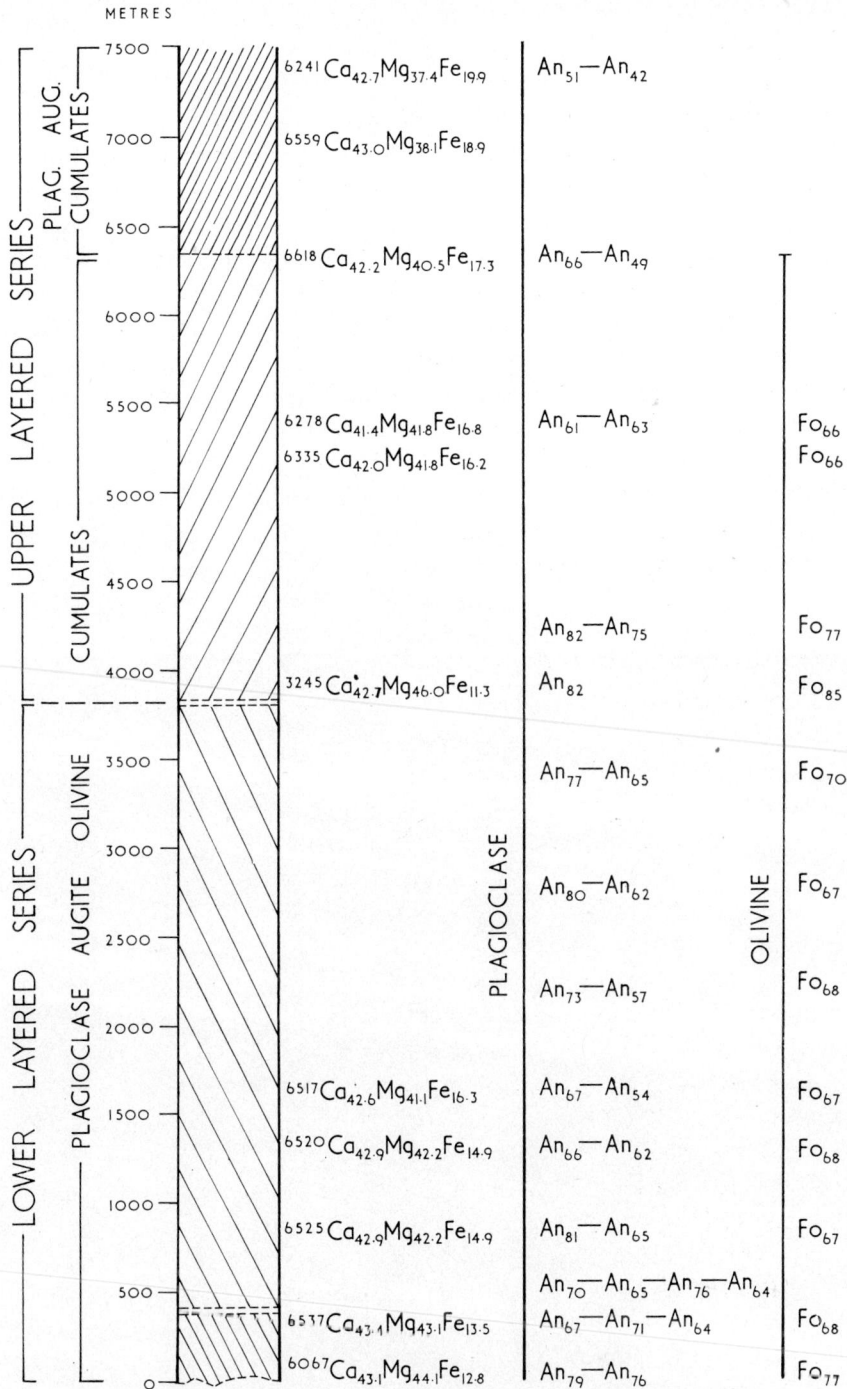

FIG. 226. The relative heights of rocks of the lower and upper layered series of Kap Edvard Holm (from Deer & Abbott, 1965). The pyroxenes were chemically analyzed, and the values for plagioclases and olivines obtained from optical properties and d_{130} spacings, respectively.

from $Ca_{43.1}Mg_{44.1}Fe_{12.8}$ to $Ca_{43.4}Mg_{43.1}Fe_{13.5}$. Zoning is slight, and most of these cumulates are of adcumulate or heteradcumulate type (although a few mesocumulates occur), the extreme plagioclase adcumulates being comparable with those of Rhum and Stillwater.

The sub-zone of plagioclase adcumulates is overlain by a series of coarser-grained, plagioclase–augite–olivine cumulates, about 3300 m in thickness, which comprise the rest of the Lower Layered Series and are grouped together as the Long Ridge Unit, being well exposed along the length of this extensive ridge. Fine-scale rhythmic layering is particularly well developed here (Fig. 227), resulting chiefly from the relative concentration of cumulus plagioclase and augite. Olivine rarely exceeds 8% of any rock and there is some suggestion of a lateral variation in the amount of cumulus olivine, it being absent over 150 m thickness to the east, in the South Deichmann Bay area. Iron ore, although rarely exceeding 5% of any rock, is significantly of cumulus status, in contrast to the Willow Ridge Unit. The cumulus plagioclase at the base of

FIG. 227. Rhythmic layering in the plagioclase–augite–olivine cumulates of the Lower Layered Series, Long Ridge, Kap Edvard Holm. (From Abbott, 1962.)

this unit is slightly more calcic than in the underlying rocks, and shows oscillatory zoning with cores of An_{70} and a median zone as calcic as An_{76} (Fig. 226). Upwards, the plagioclase and olivine cumulus phases show irregular variations in composition which have not yet been satisfactorily explained. Abbott (1962) has suggested that the variations may reflect minor pulses of fresh magma, or variations in water-vapour pressure (also correlated with plagioclase oscillatory zoning), but the problem needs further study, particularly as the augites show a regular trend of compositional change (Deer & Abbott, 1965, fig. 3). The cumulates of the Long Ridge Unit differ from those below in being predominantly mesocumulates, with an appreciable proportion of pore material in the form of relatively sodic rims to the plagioclases (Fig. 226), and small amounts of a hastingsite amphibole, biotite, quartz, apatite, K-felspar, and even allanite (Abbott, 1962, p. 120). These features are indicative of appreciable trapping of pore liquid, and some of the irregularities in the supposed cumulus mineral compositions may, in fact, be due in part to making-over of the original cumulus phases, by reaction with this liquid, as discussed in connection with the Stillwater cumulates (Ch. XIII), especially as the olivines in the Long Ridge cumulates are characteristically rounded and show signs of resorption.

The Upper Layered Series. Although there is no change in the cumulus assemblage (plag–aug–ol cumulates), there is an abrupt change in the mineral compositions at the 3800 m level (Fig. 226), and this is taken as the base of the Upper Layered Series. The cumulus minerals at the base of this series are the highest temperature solid-solutions found in the Kap Edvard Holm intrusion, with plagioclase An_{82}, olivine Fo_{85}, and augite $Ca_{42.7}Mg_{46.0}Fe_{11.3}$. Clearly, this suggests the influx of fresh magma at this stage, subsequent to the previous magma having fractionated to the extent of precipitating An_{77} and Fo_{70}. There is a thickness of 3200 m of rhythmically layered cumulates in which plagioclase and augite are preponderant but in which olivine is a more abundant cumulus phase than in the upper part of the Lower Layered Series, again suggestive of a slightly different magma chemistry for the Upper Layered Series. Plagioclase-rich cumulates occur, with up to 94% modal plagioclase in extreme cases, while melanocratic layers are relatively common at the base of each minor rhythmic unit, one such specimen from the 4500 m level containing 64% olivine, 22% augite, and only 3% plagioclase (Abbott, 1962, tables 24 & 26). Iron ore becomes an important cumulus phase upwards in this series, constituting 18% of a cumulate at the 5300 m level. The extreme cumulates are of adcumulate or heteradcumulate type, but the more uniform cumulates are usually mesocumulates with zoned felspars and small amounts of amphibole, biotite, and sometimes hypersthene. Throughout the plagioclase–augite–olivine cumulates of the Upper Layered Series, the cumulus phases show a regular pattern of cryptic variation: i.e. plagioclase An_{82} to An_{61}, olivine Fo_{85} to Fo_{66}, and augite $Ca_{42.7}Mg_{46.0}Fe_{11.3}$ to $Ca_{41.4}Mg_{41.8}Fe_{16.8}$ (Fig. 226).

At the 6300 m level, olivine ceases to crystallize and the rest of the exposed Upper Layered Series rocks are plagioclase–augite cumulates. Significantly, however, there is a return to cumulus phases of slightly higher temperature solid-solutions at the base of these cumulates (Fig. 226), so again there may have been a fresh influx of magma at this stage, rather than a change in phase relations as the result of fractionation in a closed system. Thereafter, the plagioclase changes from An_{66} to An_{51} and the augite from $Ca_{42.2}Mg_{40.5}Fe_{17.3}$ to $Ca_{42.7}Mg_{37.4}Fe_{19.9}$, the uppermost exposed rocks (7500 m level) being the most advanced fractions found in the Kap Edvard Holm intrusion. The rocks do not show as much rhythmic layering and as many extreme cumulates as at lower horizons, and there is no record of the incoming of iron ore as an important cumulus phase, as was noted in the lower, olivine-bearing cumulates of this Upper Layered Series, again suggesting some discontinuity in the series as a whole. The rocks are chiefly uniform mesocumulates, the felspars being zoned to compositions as sodic as An_{42}. The pore material includes apatite and quartz, although the rocks have been partly altered by the later syenite intrusions and these minerals, together with calcite and clinozoisite, have been attributed by Abbott to soaking by, and introduction of materials from, the syenite magmas.

b) *Summary*

The layered series of the Kap Edvard Holm intrusion show several features indicative of a formation by bottom accumulation of crystals separating from a gabbroic magma, including rhythmic layering, igneous lamination, and cryptic layering, as well as the textural features of igneous cumulates. The paucity of exposure means, however, that the marginal envelope and the lowest and uppermost layered rocks cannot be observed, and the exposed rocks, like those in Rhum and some other layered intrusions, represent only a part of the original intrusion.

The exposed layered rocks are 7500 m in thickness, and considering the limited extent of cryptic variation in the exposed series then, viewed as a closed system of products from a fractionated basic magma, the total Kap Edvard Holm intrusion must have been of impressively large dimensions. However, it is clear from several lines of evidence that only half, almost exactly, of the exposed layered series formed from what was probably the initial magma, supplemented periodically by minor influxes of fresh magma and, presumably, by extrusion of magma from the crystallization chamber. At the stage when the precipitating felspars were An_{77}, and the olivines Fo_{70}, a major influx of fresh magma occurred such that the phases then precipitated were, respectively, An_{82} and Fo_{85}. Further evidence for this major change in conditions is afforded by the two distinctive augite trends, differing in Ca-contents and appertaining to the Lower and Upper Layered Series, obtained by analysis of twelve specimens (Deer & Abbott, 1965, fig. 3). The cumulus phases at the base of the Upper Layered Series, which are of higher temperature solid-solutions than elsewhere in the intrusion, indicate a drastic clearance of residual magma from the chamber before replenishment with the new magma. It is particularly significant, in this respect, that the Upper Layered Series is characterized by a profusion of xenoliths of layered cumulates, completely absent from the Lower Layered Series. Replenishment with fresh magma has been proposed for Rhum, Bushveld and some other intrusions, but it has always been a matter of some surprise that such an event was not accompanied by the partial disruption of the existing cumulates in the magma chamber. In the Kap Edvard Holm intrusion, however, the event appears to have left its mark in the form of this and other evidence.

The Upper Layered Series, generally, shows cryptic variation of a regular pattern, and it would appear that by then the magma chamber had ceased to be disturbed as much as during the formation of the Lower Layered Series, where compositional irregularities are commonplace. The regular change in plagioclase compositions from An_{82} to An_{51}, over about 3500 m in thickness, indicates appreciable fractionation, and the absence of exposure above this level invites analogy with the Stillwater case, except that the absence of evidence for the composition of any of the Kap Edvard Holm magmas precludes an estimate of the likely composition of an Upper Hidden Zone. The Kap Edvard Holm magmas were probably rich in normative felspar because even at the base of the Upper Layered Series, calcic plagioclase is the dominant cumulus mineral and ultramafic cumulates are absent, even though the olivine (Fo_{85}) and augite ($Ca_{43}Mg_{46}Fe_{11}$) are equivalent in composition to such phases found in the very thick ultramafic zones of other layered intrusions. It is likely that extreme fractionation of the Upper Layered Series magma could have given rise to late differentiates rich in sodic plagioclases and iron-rich pyroxenes, and Deer and Abbott (1965) have analyzed a ferroaugite from an iron-rich granophyre which is associated with the upper layered rocks and may represent a filter-pressed, late differentiate. However, the relatively early separation of magnetite and the common occurrence of interstitial amphibole, biotite, and other hydrous minerals in the Kap Edvard Holm cumulates, suggest that the magmas were more hydrous than, say, the Skaergaard magma, and the effect of higher oxygen partial pressures would be to produce late differentiates showing less extreme iron-enrichment than in the Skaergaard case.

KAERVEN, EAST GREENLAND

This basic intrusion lies about half-way up the western side of the Kangerdlugssuaq fjord in East Greenland, about 30 km northwest of the Skaergaard intrusion (Fig. 4). It was found by one of us (L.R.W.) in 1930–1 and re-visited in 1932, and together with W. A. Deer in 1935–6 (see Wager, 1934, for preliminary mention). In 1953 the present authors made further collections, and based on this and previous collections a petrological study has been made (Ojha, 1960).

The present outcrop has the shape of part of an annulus (Fig. 228) with a fairly constant width of about 1 km and length of about 8 km. The eastern, outer margin is in contact with Basement Complex gneisses against which the basic magma has chilled, whilst the inner margin is defined by a later syenite intrusion. It appears that the Kaerven intrusion was originally almost circular in shape, with a diameter of about 10 km, the centre being now occupied by the Notch Mountain syenite.

The intrusion is broadly divisible into a Border Group and a Layered Series. The contact with the gneiss varies from vertical to an 80° dip inwards, and the best exposed part of the Border Group in the southeast is about 50–60 ft wide. The layered rocks total about 1600 ft in thickness and are generally characterized by rhythmic layering dipping radially inwards at about 30° (although locally exceeded, probably through post-consolidation movement). The Layered Series shows marked cryptic layering and, on the basis of the appearance of certain cumulus phases at specific horizons, has been sub-divided into four zones.

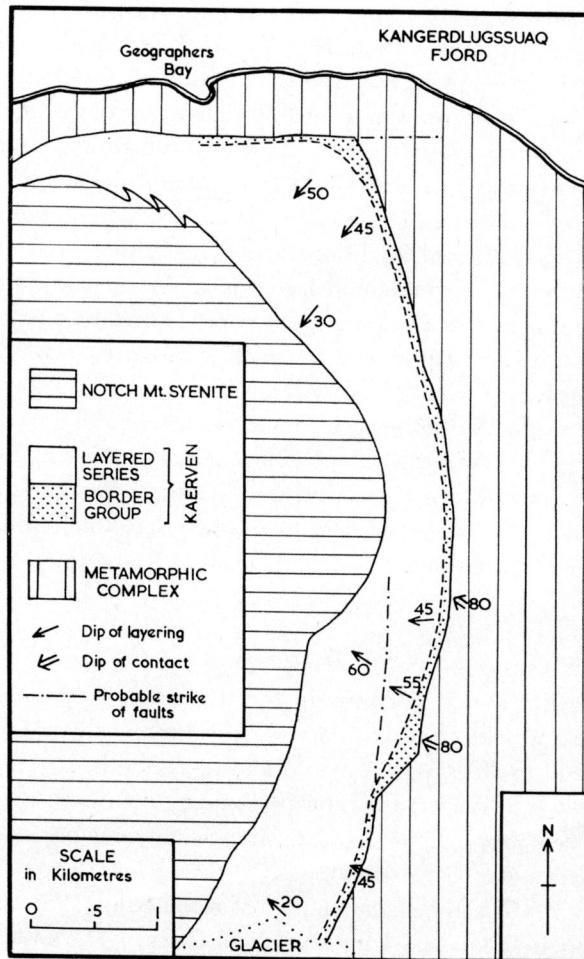

FIG. 228. Geological map of the Kaerven region, showing the present outcrop of the layered basic intrusion. (See Fig. 4 for general locality.)

a) *The Border Group*

A feature common to all examined parts of the margin (except on the extreme south-east) is the occurrence of an olivine-free, fine-grained hypersthene gabbro which is believed to represent the chilled facies of the Kaerven parental magma, probably somewhat modified by contamination. It is structureless at the margin, and a few feet inwards it becomes coarser-grained and rich in inclusions of acid gneiss. Here there is an abundance of granitic veins and of biotite, which may have been produced from the gneiss inclusions.

An analyzed specimen of the chilled gabbro contains about 46% plagioclase (An_{50}), 36% orthopyroxene (Of_{38}) and clinopyroxene ($Ca_{37}Mg_{38}Fe_{25}$), and 12% iron oxides. The Fe_2O_3 content is unusually high (5·7%) and the SiO_2 relatively low (47·8%) compared with other tholeiitic basic magmas, and for this reason the chilled rock is believed to have suffered from marginal oxidation, perhaps by reaction with water from the country rocks.

The southeastern part of the Border Group differs from the rest in that the contact with the gneiss is inwards at a low angle (about 45°). Furthermore, the zone is here characterized by the occurrence of a ' plutonic breccia ' consisting of rounded and sub-angular dunite blocks embedded in an olivine-rich, eucritic matrix (Fig. 229). This provides a striking analogy with the zone of picrite blocks in the northern Marginal Border Group of the nearby Skaergaard intrusion where, again, a gently-dipping low angle contact with country rocks may be observed. The Kaerven dunite blocks contain 99% olivine (Fo_{83}) and, presumably, represent a near-marginal accumulation of the earliest precipitated olivines of this intrusion, which have formed an extreme type of adcumulate. The matrix can be described broadly as olivine eucrite, containing cumulus plagioclase (An_{72}) and olivine (Fo_{77}), and intercumulus orthopyroxene (Of_{21}) and clinopyroxene. The textures of the matrix material are difficult to interpret because of cataclastic deformation, and much clouding and alteration, of the minerals. By analogy with the better-exposed Skaergaard example, it is likely that the matrix to the blocks represents slow crystallization of the Kaerven magma under tranquil conditions in a marginal envelope, while the dunite blocks are part of the earliest crystal fractions to accumulate in the marginal zone.

b) *The Layered Series*

The rocks of the Layered Series have the overall character of igneous cumulates, and have many features in common with those of other basic layered intrusions. They have tholeiitic, rather than alkali basalt affinities in that they contain orthopyroxene, show marked iron-enrichment with fractionation, and have produced, in some cases, patches of a granophyric late residuum. Conspicuous layering, of both rhythmic and cryptic types, together with igneous lamination (Fig. 230) and other textures, impart to the whole of the layered rocks the characteristic features of a layered intrusion, broadly of Skaergaard type.

2 E

FIG. 229. 'Plutonic breccia' from the Marginal Border Group, Kaerven. Rounded blocks of dunite occur in a matrix of olivine eucrite. ×$\frac{3}{4}$.

FIG. 230. Plagioclase–pyroxene–magnetite cumulate, showing good igneous lamination. Zone D of the Kaerven layered series. ×$\frac{3}{4}$.

The rock types are composed largely of plagioclase, olivine, orthopyroxene, clino-pyroxene, and iron ore. In addition, small quantities of accessory minerals, in the following maximum amounts in any rock, are present: biotite, 7%; quartz, 4%; apatite, 1%; amphibole and sulphides (pyrite and pyrrhotite), trace. On the basis of cryptic layering, the Layered Series has been sub-divided into Zones A–D (Fig. 231). In the lowest zone A (250 ft thick), olivine and plagioclase are the only cumulus phases; in zone B (200 ft) olivine has disappeared and bronzite joins plagioclase as a cumulus phase; in zone C (600 ft), cumulus augite joins plagioclase and hypersthene; in zone D (600 ft), magnetite and a small amount of apatite join plagioclase, hypersthene and augite.

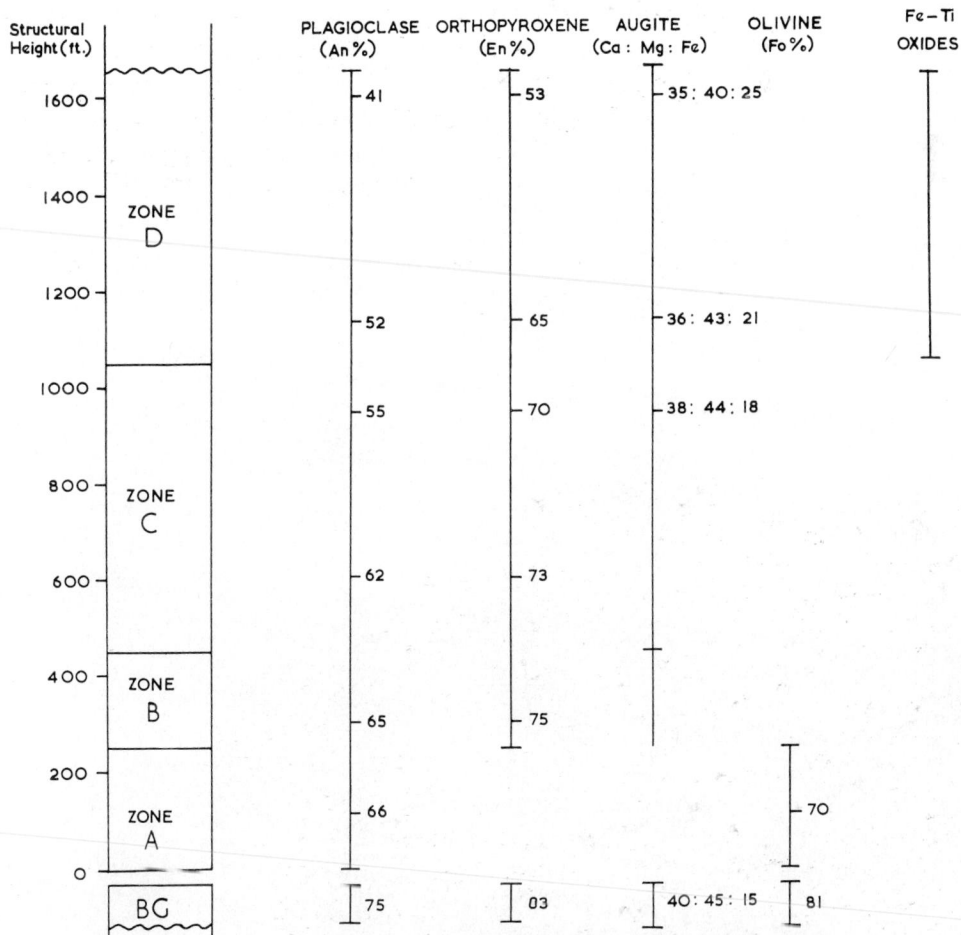

FIG. 231. Cryptic variation and distribution of the cumulus-phase minerals in the layered series of the Kaerven intrusion. The Border Group (BG) data are for the picrites of cumulate origin. (Compositions estimated from optical properties.)

The intrusion displays, in only 1700 ft of layered rocks, relatively marked cryptic variation of the plagioclase felspars and pyroxenes, and several phase changes (Fig. 231). In the early disappearance of cumulus olivine and the late incoming of magnetite and apatite, the Kaerven layered series shows close analogy with the Skaergaard layered series. However, the olivine disappears earlier in the fractionation stages of Kaerven (as defined by plagioclase or olivine compositions) and hypersthene is relatively abundant. In addition, the calcium-poor pyroxene is a primary hypersthene even at compositions as ferriferous as Fs_{46}; in the Skaergaard, Bushveld and Stillwater intrusions, pigeonite (generally inverted to hypersthene) is the calcium-poor phase at compositions more ferriferous than Fs_{30}. This may be due to a slightly higher water content in the Kaerven magma, which could be responsible for depressing the pyroxene crystallization temperatures in relation to sub-solidus inversion temperatures (see Fig. 23) and thus extending the field of hypersthene crystallization: the relatively high biotite content of the Kaerven rocks may also be attributed to the higher water content of this magma.

In the field, several good examples of fine-scale rhythmic layering are to be seen,

FIG. 232. Fine-scale rhythmic layering in plagioclase–pyroxene (–magnetite) cumulates, Zone D of the Kaerven layered series. The layered basic rocks are cut by leucocratic veins of the later, Notch Mountain syenite.

particularly in Zone D (Fig. 232). The rocks, however, are quick weathering and it is not possible to trace individual layers for more than a few yards. The rhythmic layering gives rise to thin, contrasted layers of rock of which two extreme examples, of adjacent rocks from Zone D, contain 22% plagioclase (with 26% ilmenite–magnetite) and 95% plagioclase, respectively. Felspar-rich and melanocratic cumulates also occur in the other zones, and several cases in which the amount of plagioclase exceeds 90% are found. Such extreme rocks are generally adcumulates, although the rocks of Zone A are dominantly orthocumulates and upwards in the succession there is a progressive increase in the proportion of mesocumulates, heteradcumulates, and adcumulates.

NORTHEAST SCOTLAND

a) *Introduction*

Gabbroic and ultrabasic rocks of Caledonian age outcrop in a large area of Aberdeen-shire and Banffshire, and were first described in any detail by Read (1919, 1923). Within an area measuring about 30×30 miles, seven isolated bodies occur, known as the

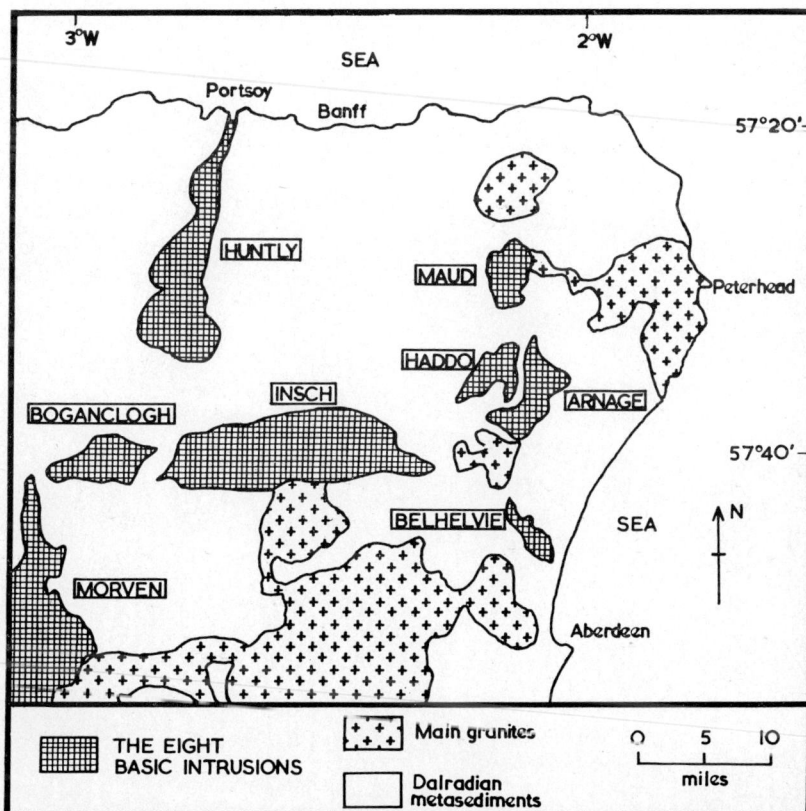

FIG. 233. Sketch-map showing the eight associated basic intrusions of North-east Scotland. (After Wadsworth *et al.*, 1966.)

Arnage, Belhelvie, Boganclogh, Haddo House, Huntly, Insch, and Maud masses, while more recently another, the Morven–Cabrach, has been shown to belong to this suite (Fig. 233).   The bodies are intrusive into Dalradian metasediments, and it is now generally believed likely (Stewart & Johnson, 1960) that they once formed part of a large, single intrusion which post-dated the early Caledonian folding, but pre-dated the later folding in this region.   The later folding has resulted in several structural complexities amongst the basic intrusions, not least of which is the tilting of the original layering to high angles (Fig. 234).   Sometimes this results in layering that is vertical or, occasionally, overturned (Shackleton, 1948).   The existence of some gravity-stratified layering and, in certain areas, of near-horizontal layering has permitted these conclusions to be drawn, while recent petrological studies have shown the isolated, generally poorly-exposed masses to have many features in common.

The separate bodies vary greatly in size and character, and it is not our intention to describe each one separately or in detail.   In regard to the exhibited layering and cryptic

FIG. 234.   Steeply-inclined layering in gabbroic rocks of the Huntly mass, Sinsharnie Quarry.   (Photo. by F. H. Stewart.)

variation, the most important are the Belhelvie, Insch, and Huntly masses. Those of Arnage and Haddo House are unlayered and appreciably contaminated, while Maud, Bogancloch, and Morven–Cabrach have not yet been studied to the same extent as the others. The extensive, early work of Read is referred to in the more recent papers discussed here, and it is in the latter, including papers by Read and co-workers, that the data relevant to our subject are mostly presented. Work on the Huntly mass is still in progress and only the few available, relevant facts are mentioned in the account given here of the Belhelvie and Insch intrusions. While each is referred to as an intrusion, they are all to be viewed as having once probably formed part of a single, much larger layered intrusion.

b) *Belhelvie*

This intrusion lies to the southeast of the other bodies (Fig. 233) and is exposed over an area of 5×2 miles. Stewart (1946) described the layered sequence, grading from ultrabasic rocks at the western side, through troctolites, olivine norites, hypersthene gabbros and olivine gabbros, to amphibolitized gabbros in the east. Fine-scale layering dips at 80–90° and strikes approximately north–south, and on the basis of recognized gravity-stratified layers, Stewart proposed post-consolidation movement of a layered intrusion into a nearly vertical position, with the base to the west. Igneous lamination is well developed in the rocks, and the cryptic variation observed for the orthopyroxenes has supported the general hypothesis.

More recently, Wadsworth, Stewart and Rothstein (1966) have published an account of cryptic layering in the Belhelvie intrusion, and they kindly allowed us to read and refer to the manuscript prior to publication. The nomenclature of the various main units has been revised, the intrusion being sub-divided into the following units (with the earlier-named groups in parentheses):

> Unit C (olivine gabbro and hypersthene-gabbro group)
> Unit B (troctolite group)
> Unit A (ultrabasic group).

The thickness of the exposed cumulates is difficult to estimate, but it has been suggested that the total thickness is at least 6000 ft, with Unit A, 2500 ft; Unit B, 1000 ft; and Unit C, 2500 ft.

The cumulates of unit A contain mostly olivine as the cumulus phase, with cumulus augite, orthopyroxene and spinel occasionally present in small amounts. The pyroxenes usually, and the plagioclase always, are of intercumulus status and are unzoned, the rocks being mostly olivine (–spinel) heteradcumulates. Plagioclase enters as a cumulus phase in Unit B and rhythmic layering is well developed, olivine-rich or plagioclase-rich layers being fairly common. Both augite and orthopyroxene are present as important cumulus phases in Unit C, together with cumulus plagioclase and olivine, and rhythmic layering is relatively common. Zoning of the intercumulus material is rare in Belhelvie, and most of the rocks are either adcumulates or heteradcumulates.

Cryptic layering has been shown by Wadsworth *et al.* (1966) to be a prominent feature of this intrusion. Total or partial chemical analyses of two orthopyroxenes, two augites, and four plagioclases were supplemented by optical or X-ray measurements of the same minerals, and olivines, from other rocks. Taking the three units as a whole, the cumulus minerals generally show a systematic change in compositions, upwards, towards lower-temperature members of their solid solution series. The olivine changes from Fo_{86} to Fo_{80}, orthopyroxene from En_{85} to En_{79}, plagioclase from An_{81} to An_{74}, and augite (only two measurements) from $Ca_{42}Mg_{51}Fe_7$ to $Ca_{41}Mg_{48}Fe_{11}$.

This evidence is a convincing demonstration that depite the present high dips, the Belhelvie rocks must originally have formed as a layered series of crystal accumulates. The original, once almost-horizontal floor is not exposed, but the cryptic layering, and the gravity-stratification observed by Stewart (1946), together indicate that it lay in the direction of the most westerly-exposed outcrops. The observed layered rocks represent only a limited amount of the fractionation sequence to be expected from a differentiated basic intrusion, and one may suggest either that the residual magma was expelled from this part of the magma chamber (Wadsworth et al., 1966) or that later cumulates formed and were subsequently removed, owing to folding and local erosion (a roof-zone not being recorded). The exposed, early-stage cumulates of the Huntly mass (plagioclase, An_{85} to An_{80}) and of the Insch mass (plagioclase, An_{84} to An_{77}; olivine, Fo_{82} to Fo_{78}) are similar to those of Belhelvie, but in the case of Insch and, probably, of Huntly (Wadsworth et al., 1966), later differentiates are preserved. This, and the concept of each mass once forming part of a single intrusion, favours the hypothesis that the sequence exposed in Belhelvie is only part of an originally more extensive sequence of layered rocks.

c) *Insch*

All the work so far published on this intrusion is by H. H. Read and his co-workers, including the first petrographic account (Read, 1923), a description of layering in the ultrabasic rocks (Sadashivaiah, 1954), and a study of the trace element contents of the rocks (Read & Haq, 1963). More recently, Read, Sadashivaiah and Haq (1965) have presented chemical analyses of the hypersthene gabbros, suggesting that they belong to a later, separate intrusion, and originated by the contamination of basic magma at depth. For the purposes of our discussion, however, the most important accounts are presented by Read, Sadashivaiah and Haq (1961), and by P. D. Clarke in an unpublished Ph.D. thesis (University of Edinburgh, 1965). We are indebted to Dr Clarke, during examination by one of us (G.M.B.), for permission to reproduce parts of his unpublished data in this account.

Read et al. (1961) sub-divided the rocks of the Insch basic mass into the following groups, in order of formation:

3. Hypersthene gabbros.
2. Olivine gabbro, syenogabbro, and syenite series.
1. Dunite–troctolite series.

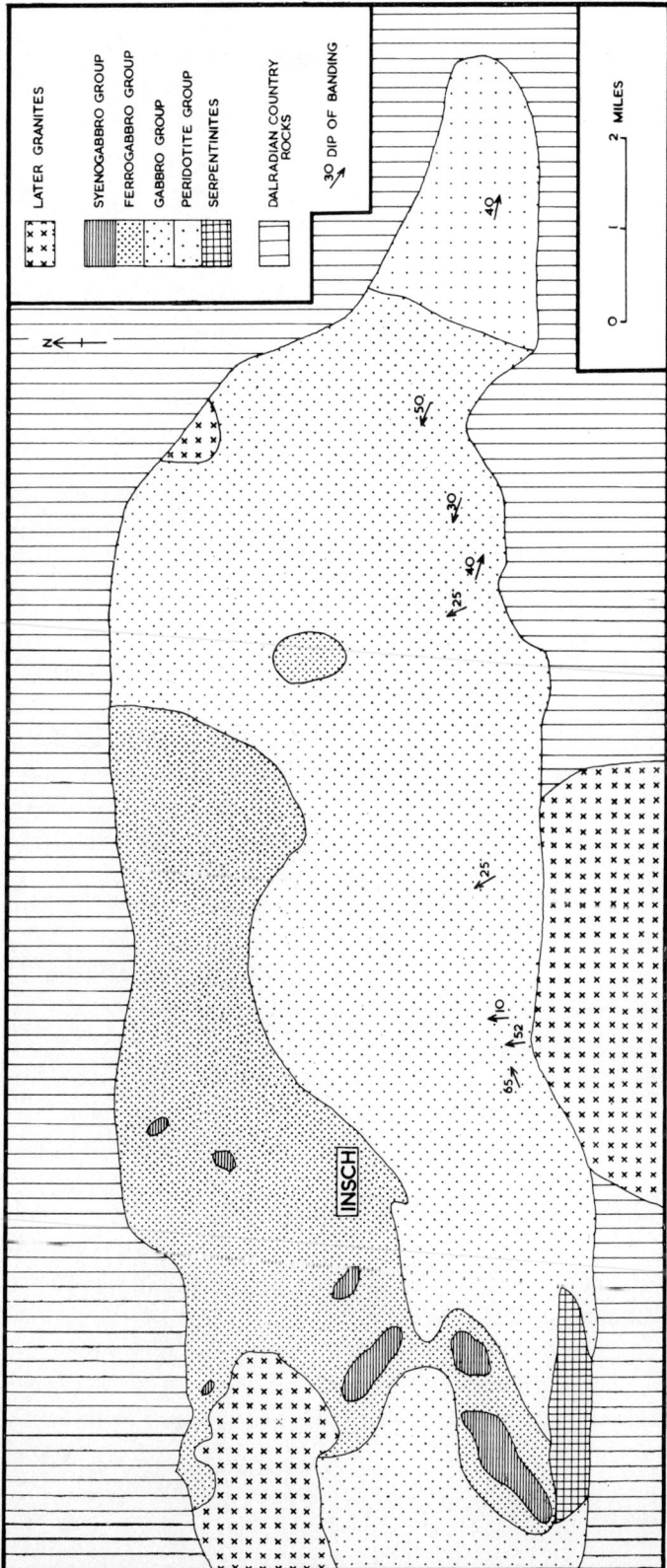

FIG. 235

Geological map of the Insch intrusion, Aberdeenshire. (After Clarke, 1965.)

Chemical analyses of 38 specimens were presented, and it was shown that fractional crystallization of a basaltic ('gabbroic') magma could account for the sequence from ultrabasic dunites and peridotites, through troctolites and olivine gabbros, to syeno-gabbros and syenites, but it was suggested that the hypersthene gabbros, as mentioned above, do not belong to this fractionation sequence. Clarke (1965) does not accept the latter reservation, and has presented further chemical and mineralogical data from which an impressive story of cryptic variation within the Insch intrusion has emerged.

The exposed part of the intrusion is about 18 miles long and 5 miles wide, and occupies the south–central part of the whole complex (Fig. 235). Clarke has sub-divided the intrusion into the following four groups, belonging to a single fractionation sequence (the sub-divisions of Read *et al.*, 1961, being given in parentheses):

4. Syenogabbro group (Syenogabbro and syenite).
3. Ferrogabbro group (Olivine gabbro).
3. Gabbro group (Hypersthene gabbro).
1. Peridotite group (Bourtie Dunite–Troctolite series).

The peridotite group is confined to an area in the extreme eastern part of the intrusion, but may include an outcrop of serpentinites in the southwestern corner (Fig. 235). The gabbros cover the largest area, and outcrop in the greater part of the eastern and southern parts of the intrusion. The northwestern part consists largely of ferro-gabbros, with syenogabbros outcropping on small, isolated hills. The Insch rocks rarely show rhythmic layering, and so it is difficult to estimate the true thickness of each rock group or the stratigraphic height in the sequence. Some rather irregular, streaky layering is present in the peridotite and gabbro groups; a crude size-sorting in the gabbro group; and, particularly, igneous lamination in the gabbro group.

In general, the sequence proposed by Clarke is based on a regular cryptic variation exhibited by the plagioclase felspars, which can then be taken as a parameter against which the compositions of the other, less extensively distributed cumulus phases can be plotted. Chemical variations involving both major and trace elements can then be plotted against established variables such as An%, Ca%, iron ratios, etc. Despite the absence of a scale depicting stratigraphic heights or relative thicknesses, the chemical and mineralogical variations exhibited within the Insch intrusion are remarkably regular and the whole rock sequence is clearly to be viewed as a single, cryptically layered sequence.

The cumulus phases, and their compositions, are shown in Fig. 236, the most signifi-cant variations being as follows. The plagioclase felspars become progressively more albitic and later, in the syenogabbro group, are joined by an alkali felspar. Olivine disappears as a cumulus phase at Fo_{62}, taken as the base of the gabbro group, and reappears at Fo_{42}, the base of the ferrogabbro group, thereafter showing marked iron-enrichment to Fo_4 in the syenogabbros and being absent from the syenites. Ortho-pyroxene may locally be absent because of rhythmic layering in the peridotite group,

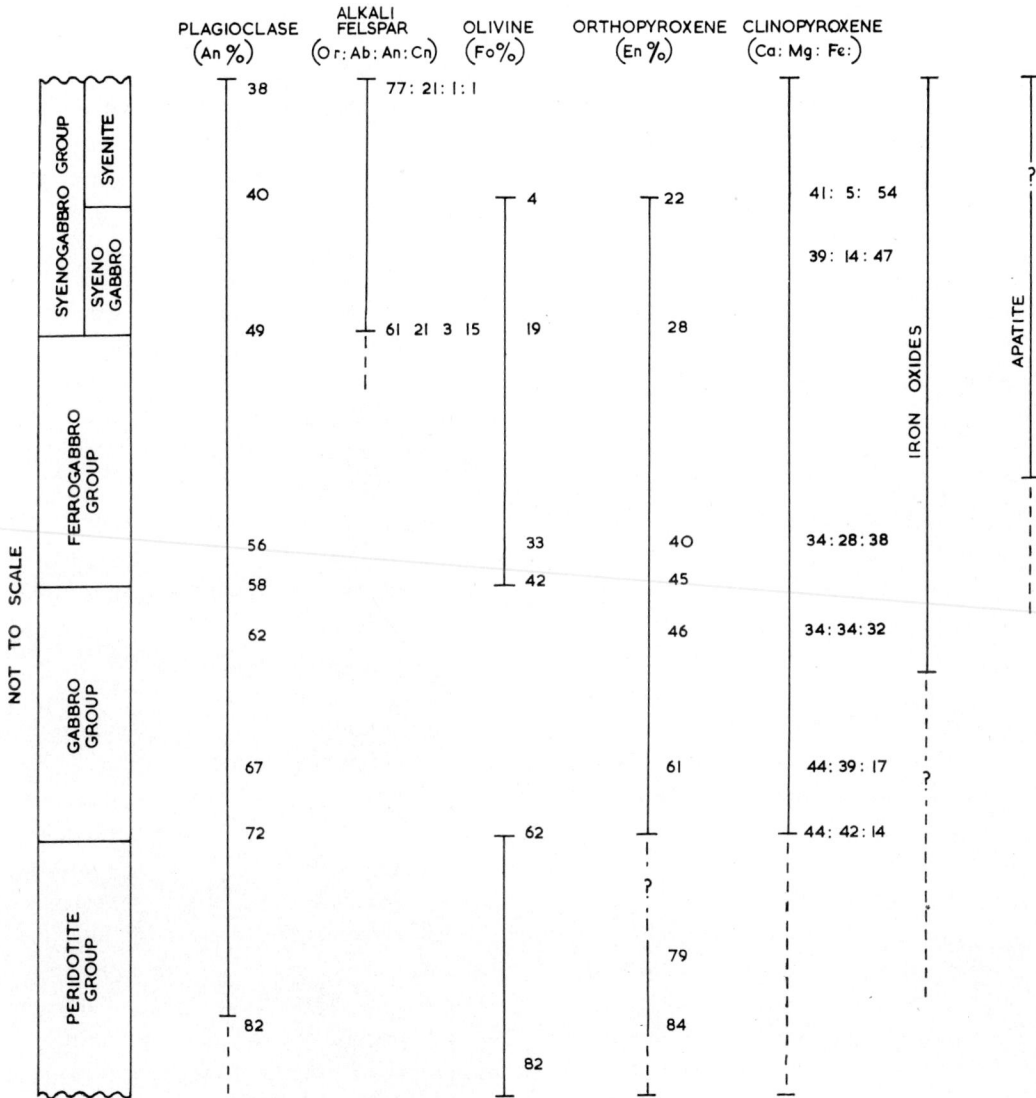

FIG. 236. Vertical column of the main sub-divisions of the Insch intrusion. The true thicknesses are not known, owing to poor exposure and rarity of layered structures. The distribution and compositions of the cumulus phases are shown, the broken lines indicating occurrence as an intercumulus phase and the queries referring to rocks in which the status of the phase is not clear. (After Clarke, 1965.)

but is generally present throughout the whole series, becoming progressively more iron-rich up to En_{22} beyond which, in the syenites, it is absent. The orthopyroxenes from En_{84} to En_{56} are bronzites or hypersthenes crystallized directly from the magma, whereas those ferriferous orthopyroxenes within the range En_{55} to En_{32} have apparently inverted from pigeonites and ferropigeonites. Ca-rich pyroxenes, ranging from augites to ferroaugites, are present as cumulus phases from the base of the gabbros upwards. Marked iron enrichment is the characteristic feature of this trend, also, and the latest Ca-rich pyroxene has the composition $Ca_{41}Mg_5Fe_{54}$. Iron oxides enter as a cumulus phase towards the middle of the gabbro group, and cumulus apatite at the base of the ferrogabbro group.

In many respects, the cryptic variation of the Insch rocks is comparable with that shown in the large, tholeiitic layered basic intrusions such as the Skaergaard and Bushveld. This is particularly true of the way in which olivine temporarily ceases to crystallize, of the Ca-rich and Ca-poor pyroxene crystallization trends, and of the relative stages at which cumulus iron oxides and apatite first appear. In detail, however, the cryptic variation at Insch shows several interesting and significant differences. The most important difference is in the production of late-stage syenitic, rather than granitic, differentiates. The general absence of quartz (occasionally present, but $<1\%$, in the syenites) at first suggests that although direct evidence of the original magma composition is unobtainable, it may have been of alkali olivine-basalt, rather than of tholeiitic basalt affinities. However, felspathoids are also absent from the syenites, and the Ca-rich pyroxene trend, together with the presence of a Ca-poor pyroxene throughout, weigh strongly against this hypothesis. Also, the temporary cessation of olivine crystallization is a feature of a tholeiitic, rather than an alkali olivine-basalt fractionation series. The Insch fractionation sequence is analogous in many respects to that found at Okonjeje (Ch. XVII), as pointed out by Clarke (1965). More significant, however, is a comparison with the Kiglapait intrusion (Fig. 245). There are slight differences, but this is found in a detailed comparison of any two layered intrusions. Using the plagioclase compositions as a parameter, the stages at which there is an exit and reappearance of olivine, for example, vary subtly from one intrusion to another, as do the stages at which pigeonite and other cumulus phases first crystallize or cease to crystallize. The main conclusion to be drawn from Insch and Kiglapait is that certain magmas are of such a composition, or crystallize under such conditions, that although the majority of the cumulates contain, broadly, a tholeiitic mineral assemblage (particularly in regard to the pyroxene trends which afford a sensitive indicator of the main magma-types) they give rise to late differentiates of syenitic, rather than granitic composition. Hitherto, the main distinction has been between the fractionation products of basaltic magmas with either tholeiitic or alkali olivine-basalt affinities. Within the tholeiitic group, in addition to variation in alumina-contents, a distinction has also been drawn between the products of the more and less oxidized magmas. Whereas oxidized magmas result in abundant quartzose differentiates, strongly reducing conditions have still given rise to a small amount of quartz-bearing late differentiates (e.g. Skaergaard). The recent descriptions of the Insch and the Kiglapait intrusions, unfortunately yet to be published but both made available to us by the generosity of the respective authors, will doubtless result in a consideration by petrologists of the significance of the particular type of differentiation involved in the production of syenites as late differentiates of tholeiitic basalt magmas.

DULUTH, MINNESOTA, U.S.A.

The classic studies by Grout (1918*a, b, c, d*) resulted in the early establishment of the Duluth gabbro as the standard example of a lopolithic, differentiated intrusion in which the striking rhythmic layering was attributed to crystal settling and sorting by the agency of convection currents in the magma. Since then, the complex has received little attention apart from the important descriptions by Taylor (1956), abridged from a more extensive account in his Ph.D. thesis (Taylor, 1955), and by Goldich, Taylor and Lucia (1956). The complex is of particular interest in that although made up of several distinct intrusions, closely related in space and time, the rock types of the separate intrusions show a compositional relationship which can be expressed in terms of strong fractionation of basic magma. Thus the intrusive bodies include a layered gabbro grading into syenogabbro; ferrodiorites (or ferrogranodiorites); and basic ferrohedenbergite granophyres, the series (taken as a whole) showing several similarities to the trend of iron-enrichment showed by the Skaergaard and Bushveld intrusions.

The complex forms an immense, sill-like mass extending for 150 miles northeast from Duluth (Fig. 237). It is of Middle Keweenawan age, being intruded chiefly among Keweenawan lavas and sediments of Later Precambrian age. The earliest member of the complex is a coarse, anorthositic gabbro from which apophyses intrude the lavas without evidence of chilling, and in which recrystallized basalt xenoliths are abundant. In many respects this member is reminiscent of the massive anorthosites of the Beaver Bay area to the north, and of the Adirondacks, except that igneous lamination is frequently well-developed. The felspar laths range in size from 0·2 to 6 inches in length, and comprise 75–90% of the rocks, the rest being composed of poikilitic titanaugite and accessory Fe–Ti oxides, apatite, biotite, and amphibole. The plagioclase is a labradorite, generally between An_{58} and An_{70} in composition.

The layered gabbro intrusion, which is the better-known member of the Duluth complex, shows intrusive contacts and chilling against the basalts and anorthositic gabbro (Taylor, 1956, pp. 50–1). The composition of the chilled rock is suspect because of possible contamination, but nevertheless is close in composition to that of the basic dykes which constitute the latest igneous episode of the region. In the rather remote Ely region, 2 miles southeast of Old Babbitt, the northern base of the intrusion is chilled against country rock, and at about 100 ft above the base, amongst gabbros dipping at 25–30° southeast, there occurs a thick (150 ft) sulphide-bearing zone (Anderson,

1956). The chief sulphides are chalcopyrite, cubanite, pyrrhotite and pentlandite, and may represent an early separation of immiscible sulphide-rich liquid, similar to that giving rise to the sulphide zone found near the base of the Stillwater intrusion.

The layered gabbro intrusion is about 15 000 ft thick and is composed chiefly of troctolite, olivine gabbro, felspathic gabbro, and syenogabbro. The series shows well-developed rhythmic layering on a fine scale, particularly on Skyline Drive (city of Duluth) near Bardon's Peak (Fig. 238), together with gravity-stratified layering and marked igneous lamination. The chief cumulus minerals are olivine, plagioclase and augite, although titaniferous magnetite and pigeonite (inverted to hypersthene) are invariably present and may be of cumulus status. There is little evidence of cryptic

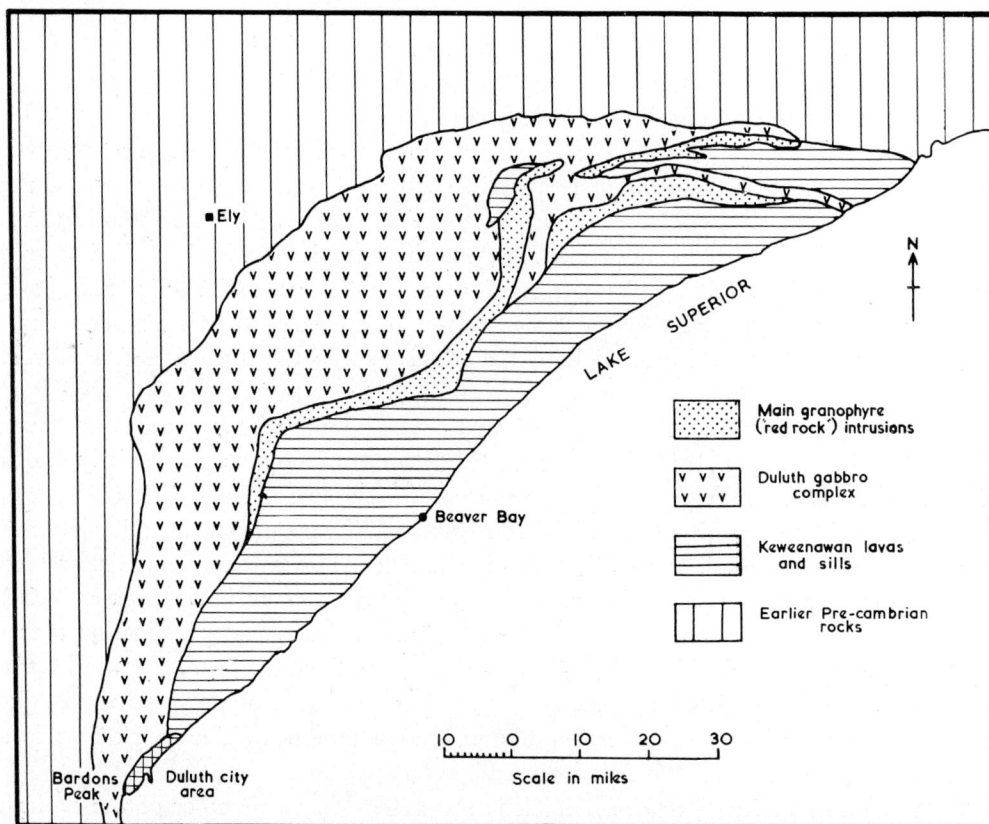

FIG. 237. Geological map of the Duluth area, Minnesota. The Duluth gabbro complex consists of several separate, but inter-related intrusions of layered gabbro, anorthosite, syenogabbro, ferrogranodiorite, and ferrogranophyre (see text). Towards the roof of the complex, along the southeastern edge of the exposed area, intrusive granophyres ('red rock') are abundant and are shown separately. The roof consists largely of Keweenawan lava flows and differentiated sills, such as the Endion sill (northeastern edge of Duluth city) and Beaver Bay diabase. (After Schwartz, 1965, fig. 1.)

variation throughout the greater thickness of the layered series, the plagioclase varying from An_{65} to An_{59} and the olivine from Fo_{61} to Fo_{53}, but irregularly with height. This, and local irregularity in the layering and signs of autobrecciation, led Taylor (1956) to suggest that the layered series crystallized from successive pulses of basic magma.   However, towards the top of the exposed layered series, about 14 000 ft from the base, the plagioclase is An_{53} in composition, olivine has disappeared, and quartz and iron ore are present in appreciable amounts (up to 12% titaniferous magnetite). These latter rocks have been termed syenogabbros in view of the presence of about 6% of quartz plus potash felspar.

At Bardon's Peak the layered gabbros are intruded by a separate body of coarser olivine gabbro, the Bardon's Peak intrusion, which is nowhere chilled against the earlier layered intrusion.   This later intrusion can only be traced laterally for about $1\frac{1}{2}$ miles and is tabular in shape, with a thickness of about 200 ft.   It is, however, well

FIG. 238.   Rhythmic layering in olivine–plagioclase cumulates, Skyline Drive, Bardon's Peak, Duluth.   (From Taylor, 1956.)

layered, containing a peridotite layer 50 ft thick and other finer scale, mafic and felspathic layers. Several gabbro pegmatites, which intrude the main Duluth layered intrusion and which are zoned and contain large titanaugite crystals up to 6 inches in length, arranged perpendicular to the walls, are attributed by Taylor to material squeezed off from the Bardon's Peak intrusion.

Several bodies of dark, pyroxene granodiorite intrude the main layered series, particularly in the Enger Tower area (Goldich *et al.*, 1956). These interesting rocks show fine-scale layering (*op. cit.*, fig. 11) and consist essentially of andesine plagioclase (An_{47} to An_{42}), ferroaugite, hornblende, magnetite, and up to 30% of quartz plus K-felspar. They are comparable, in some respects, with the ferrodiorites of the Skaergaard except for the high quartz, K-felspar, and hornblende contents, and the absence of iron-rich olivine. The term ferrogranodiorite (*op. cit.*, p. 82) is useful in preserving

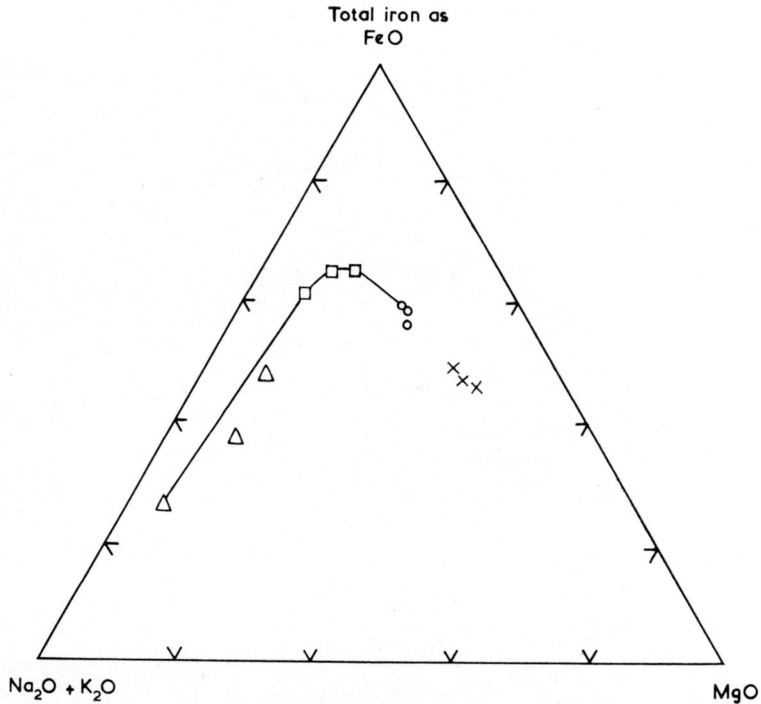

FIG. 239. Trend shown from chemical analyses of rocks of the Duluth complex. The rock groups belong to separate intrusions (see text) but suggest an overall genetic relationship. (After Taylor, 1956, fig. 7.) The analyses used for the plotted points have not all been published, so the parameters given are as used in Taylor's paper. Hence the trend position can only be compared broadly with those shown in similar diagrams in this book, where atomic ratios are plotted (e.g. Fig. 132, 189, & 217).

Circles: basalt and diabase dykes (probably similar to initial Duluth magma); crosses: gabbroic cumulates; square: ferrogranodiorites; triangles: ferrogranophyres.

this distinction. However, one example listed (*op. cit.*, table 3, no. 13) contains only 6% of quartz+K-felspar+hornblende but 5% of olivine (presumably iron-rich), and is truly analogous to the Skaergaard ferrodiorites. One of the ferrogranodiorite intrusions shows a gradation towards rocks termed adamellites, in which the plagioclase is more sodic (An_{40}) and the K-felspar contents more abundant (25%). These rocks also contain about 10% quartz, 15% ferroaugite, and 10% magnetite, and are clearly a further, more advanced stage in the trend towards iron-enriched granitic rocks. Finally, there are the separate intrusions of granophyre, which Grout (1918*b*) termed ' red rock ' and which contain an even more sodic plagioclase (from An_{40} to An_{32}) and more K-felspar (40%), together with the other constituents of the ferro-adamellites. These granophyres (often miarolitic) are clearly iron-rich, the pyroxenes approaching ferro-hedenbergite in composition ($2V\gamma = 60°$; $\beta = 1·729$).

Several chemical analyses of the rocks are published by Goldich *et al.* (1956) and the data are plotted on an $MgO-(Na_2O+K_2O)-(FeO+Fe_2O_3)$ diagram by Taylor (1956, fig. 7). As shown in Fig. 239, these data indicate that if this impressive series of Duluth complex rocks is taken as a whole, they illustrate a trend of iron-enrichment, followed by alkali enrichment, comparable with that of the Skaergaard. However, the significant difference lies in the fact that the Duluth series does not show iron-enrichment to quite the extreme shown by the Skaergaard, while the amount of quartz (not shown on this diagram) and K-felspar is higher at the intermediate stages of fractionation in Duluth. These features, taken together with the presence of amphibole and earlier precipitation of magnetite, suggest that in Duluth we are dealing with a more oxidized magma than in Skaergaard.

Despite the fact that the Duluth complex consists of a group of separate intrusions, it contains 15 000 ft of one layered gabbroic intrusion and an impressive array of other intrusions which apparently show close genetic inter-relationships. So far, the detailed mapping has been confined to the southern sector, in the neighbourhood of the city of Duluth, but it may be that further work will show whether the iron-rich granodiorites and granophyres are filter-pressed differentiates from the basic layered intrusion or whether the whole assemblage, ranging from peridotites to granophyres, is due to crystallization from successive magmas drawn from a differentiating chamber at deeper levels. The former may be the more likely hypothesis, in view of the observed relative ages and also because the related sills to the north, such as the Endion sill (Ernst, 1960) and the Beaver Bay diabase (Muir, 1954) show evidence of *in situ* differentiation towards iron-rich granophyres, comparable with the overall trend shown by the huge Duluth complex.

BAY OF ISLANDS, NEWFOUNDLAND, CANADA

The Bay of Islands complex consists essentially of four large, now isolated bodies of ultrabasic and basic layered igneous rocks (the Table Mountain, North Arm Mountain, Blow me Down, and Lewis Hill masses) and subsidiary masses of variable size including the Mount St. Gregory mass (Fig. 240 & 241). Those north of the Bay of Islands were

2 F

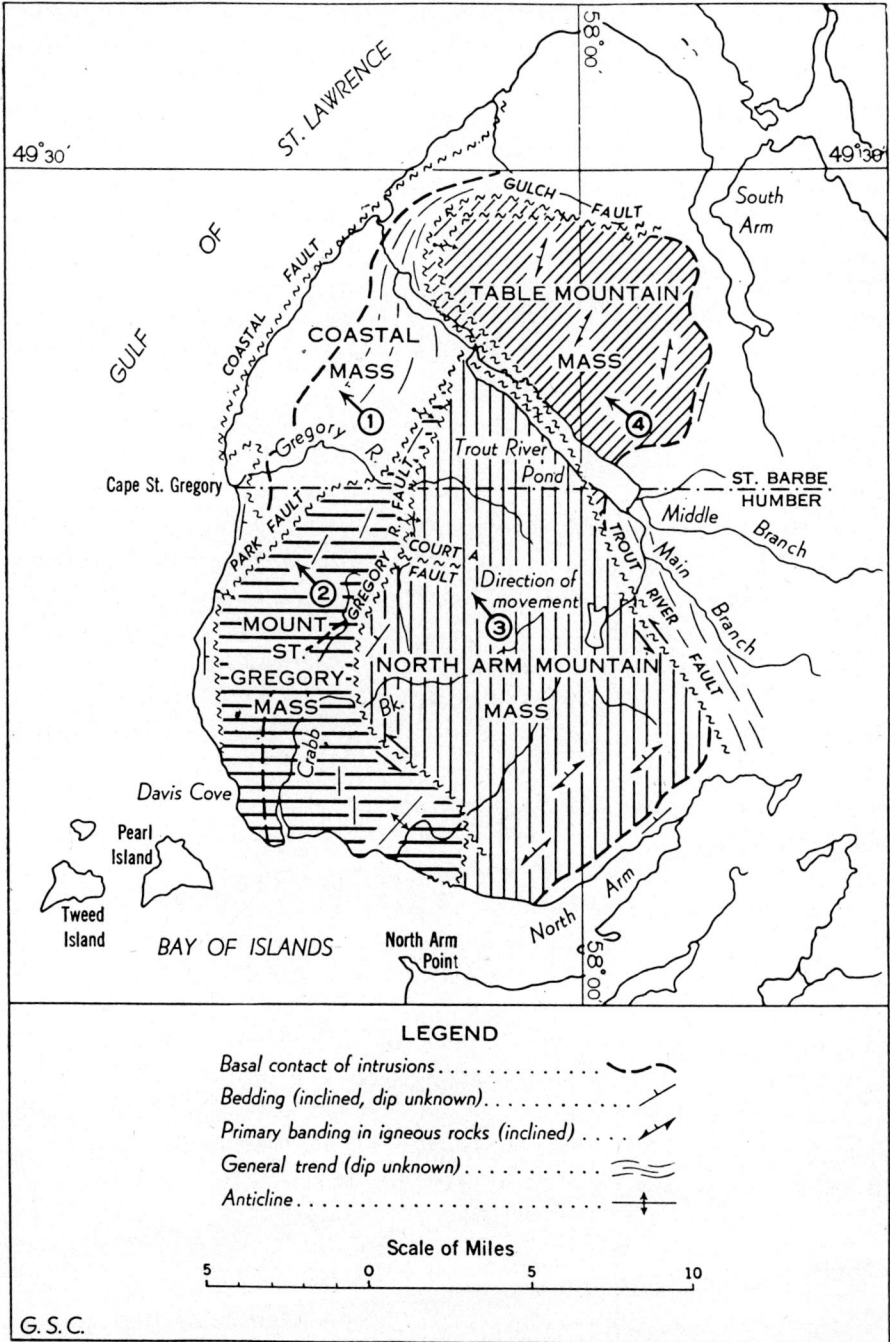

FIG. 240. The structural units of the northern half of the Bay of Islands
          Igneous Complex. (From Smith, 1958, fig. 8.)

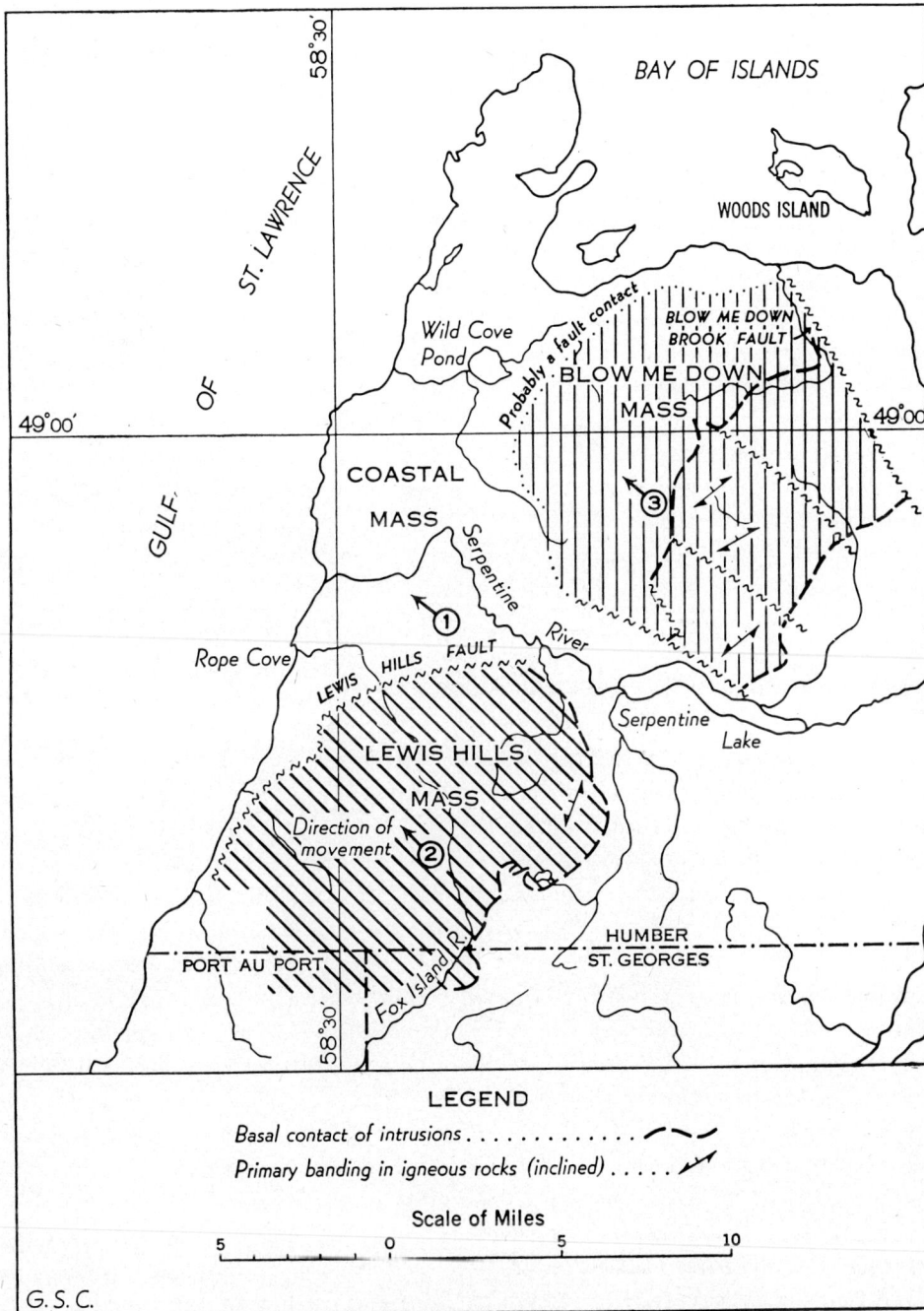

FIG. 241. The structural units of the southern half of the Bay of Islands Igneous Complex. (From Smith, 1958, fig. 11.)

described by Ingerson (1935) and those to the south by Cooper (1936), while Smith (1958) has written an account of the whole complex.

The complex, outcropping in a belt 65 miles long and 10 miles wide, has intruded Middle to Late Ordovician sediments and lavas of a typical eugeosynclinal association, subsequent to which extensive uplift, folding and faulting, of doubtful orogenic age, have resulted in the present, complex relationships between the intrusive masses. Ingerson (1935) believed that the Table Mountain and North Arm Mountain masses represented separate laccoliths, and defended this argument (1937) in reply to a proposal by Buddington and Hess (1937) that these two masses, and possibly the Blow me Down and Lewis Hills masses to the south, are parts of an original single lopolith that had been broken up by faulting. Cooper (1936, p. 7) favours the latter hypothesis in introducing the comprehensive name for the whole complex, while Smith (1958, p. 85), although less commital as to the form of the ' primary pluton ', believes that thrust faulting and associated transverse faulting have been responsible for destroying the original continuity of one intrusion and producing the present exposed, isolated masses.

The extensive structural dislocations preclude the construction of a suitable vertical section, showing the layered units representative of the intrusion as a whole, but from the data given by Smith (e.g. fig. 13) a generalized section is given in Fig. 242. This does not represent the sequence in any selected member of the masses or the complete sequence likely to have been present in the original intrusion but, rather, is an attempt to present a synthesis of the cumulate sequence recorded in the several reports. Only the minimum of details about which there is general agreement is given, additional minor variations being at present in need of further investigation. Despite extensive mapping of the complex, there is little available information on the detailed characteristics of the layering, the rock textures, and the mineral compositions. It is felt that Smith's (1958) interpretation of the relationships, and the descriptions now available of similar associations from other regions, will now open the way to a detailed study of this very thick sequence of what appear to be igneous cumulates. From the evidence available at present, the main point to be stressed is the association of a thick sequence of layered ultramafic rocks with an overlying mass of weakly layered basic rocks, the significance of which is mentioned in connection with the Great Dyke (pp. 465–8).

The four plutons differ in the amount of uplift, tilting and faulting, which has resulted in variable proportions of ultramafic and gabbroic rocks being exposed (Smith, 1958, fig. 13). The North Arm pluton, with the overall structure of a gravity-stratified sheet, is perhaps the closest to an example of the idealized sequence. The ultramafic zone is underlain by a narrow thermal aureole of metamorphosed sedimentary and volcanic rocks, the lower contact being sheared and serpentinized. The ultramafic rocks have been sub-divided by Smith into a lower sequence of enstatite peridotites, interlayered both on a large scale (hundreds of feet) and small scale (*op. cit.*, pl. 6) with dunites, and an upper sequence of dunites together with thin layers rich in chromite or enstatite. The enstatite peridotites are described as containing $>75\%$ olivine crystals,

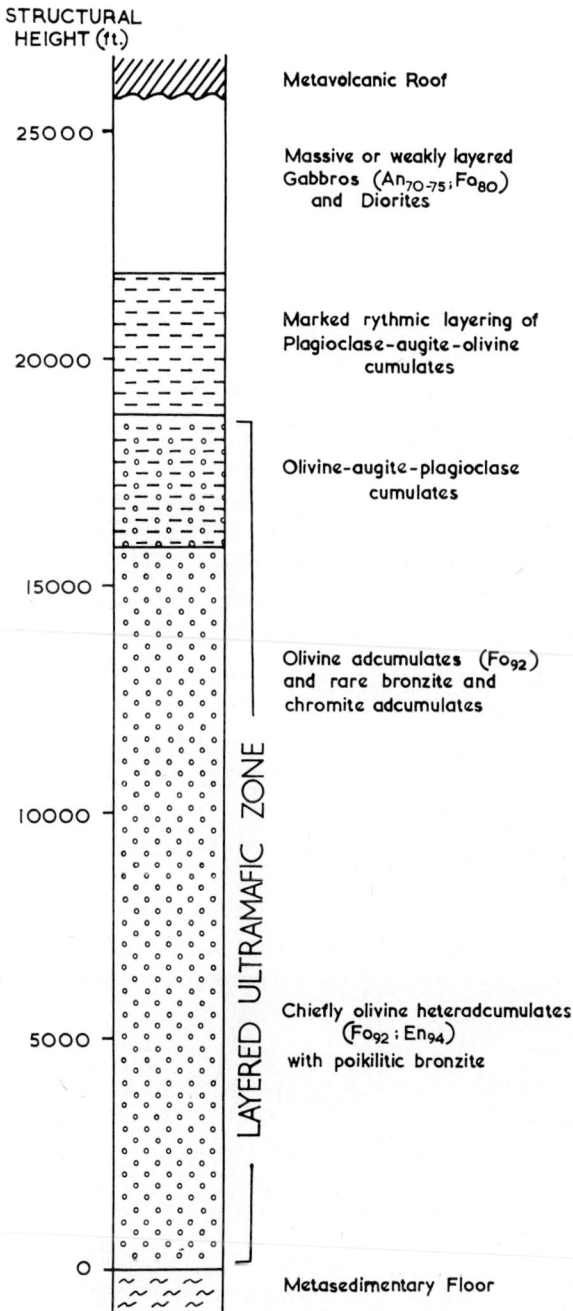

STRUCTURAL HEIGHT (ft.)

Metavolcanic Roof

Massive or weakly layered Gabbros (An_{70-75}; Fo_{80}) and Diorites

Marked rythmic layering of Plagioclase-augite-olivine cumulates

Olivine-augite-plagioclase cumulates

Olivine adcumulates (Fo_{92}) and rare bronzite and chromite adcumulates

Chiefly olivine heteradcumulates (Fo_{92}; En_{94}) with poikilitic bronzite

Metasedimentary Floor

LAYERED ULTRAMAFIC ZONE

enclosed by large enstatite plates, and it appears that the lower sequence consists generally of olivine alone as a cumulus phase (Fo_{92}), the layers being either of olivine adcumulates (dunites) or of olivine heteradcumulates (enstatite peridotites). Occasional cumulus enstatites (En_{94}) occur, but apparently not enough to justify Smith stressing that the sequence is unusual in the concentration of enstatite, relative to olivine, at the base of the sequence. The upper part of the ultramafic sequence shows a preponderance of olivine adcumulates, together with thin layers rich in cumulus chromite or cumulus enstatite. Throughout this great thickness of well-layered rocks ($c.$ 15 000 ft) no cryptic variation has so far been detected, although the recorded data are scarce. Towards the top of the Ultramafic Zone (Fig. 242), augite and plagioclase enter as cumulus phases (Smith, 1958, fig. 2). The entry of these two phases as cumulus minerals,

FIG. 242

Generalized vertical section for the series of rocks exposed in the Bay of Islands complex. The sequences are incomplete in individual masses, but from the data given by Smith (1958) and Cooper (1936), the overall pattern of variation amongst the cumulates is close to that shown.

subsequent to the deposition of a great thickness of olivine and orthopyroxene cumulates, is a feature characteristic of the cryptic layering shown by the Bushveld, Stillwater, and Great Dyke intrusions.

Ingerson (1935), Cooper (1936), and Smith (1958) have drawn attention to a zone of extremely well-layered rocks which, in each mass, lies between the Ultramafic Zone and the Gabbro Zone, and which they called the ' Critical Zone '.   The individual layers have been called gabbro, anorthositic gabbro, anorthosite, clinopyroxenite, olivine gabbro, troctolite, and felspathic dunite.   These layers appear simply to be cumulates consisting of variable proportions of cumulus plagioclase, augite, and olivine, and differ from the underlying layers in the greater overall proportion of cumulus plagioclase (and possibly augite) and in the finer scale of the layering.   This zone is overlain by relatively massive gabbros (with Ni–Cu sulphide concentrations) in which layering is rarely developed, intruded by or, rarely, overlain by diorites and quartz diorites believed to be hybrid in origin.

The majority of published opinion is that the Bay of Islands complex crystallized at a greater depth and has since been uplifted and dissected by later orogenic movements. Smith (1958, p. 84) stresses that the ultrabasic rocks form part of a single intrusion, and that isolated ultrabasic layered masses in orogenic regions may once have formed parts of similar intrusions, certain small, isolated ultramafic parts of the Bay of Islands complex now being divorced from the once-associated basic rocks.   The absence of cryptic variation in the Ultramafic Zone is a problem, but it may be that the very slight variations in mineral compositions found in other layered ultrabasic bodies (e.g. Great Dyke and Rhum) may ultimately be detected here.   On the other hand, periodic influxes of parental magma can stabilize temperature for a long time (as in Rhum), and the Bay of Islands may have behaved as this type of open system.   Cooper (1936) suggested that the intrusion was of this type, and also that magma may have been erupted periodically at the surface.   Some cryptic variation is evidenced by the entry of cumulus augite and plagioclase towards the top of the Ultramafic Zone.   In the case of the Great Dyke, the cryptic variation supports continuity between the Ultramafic and Basic Zones, although even here a fresh influx of magma is suggested by certain features (p. 466).   In the Bay of Islands, the distinctive fine-scale layering in the ' Critical Zone ' may be due to a fresh influx of magma setting up temperature gradients and turbulence in the body of mixed magmas.   Also, the Gabbro Zone may represent yet another magma influx in view of some evidence for transgressive and interdigitating contacts between these and the earlier rocks (Smith, 1958, fig. 12).   In the absence of sufficient mineralogical data to correlate the various units of the complex, it is at present impossible to tell whether the full sequence is regular or interrupted.   However, it does seem likely that the whole assemblage once formed in one large magma chamber by crystal accumulation, and that the formation of the Ultramafic Zone and of the gabbroic rocks were both a part of this single process.

Kiglapait, Labrador, Canada

The Kiglapait intrusion is a Precambrian body of layered rocks exhibiting marked cryptic variation. Situated on the Labrador coast, it is adjacent to the Nain anorthosite mass and was viewed as a part of it by Wheeler (1942) and Douglas (1953). An extensive, very detailed study by S. A. Morse (1961) has shown Kiglapait to be a separate layered intrusion, the excellent exposures permitting a clear picture to be obtained of the form and internal relationships. The results of Morse's work are to be presented in a memoir now in preparation, but he has kindly allowed us to read his unpublished work on which this account is based and from which Fig. 243–5 are reproduced. This account is necessarily brief, with the omission of much pertinent data, because further work in progress (personal communication, Morse, 1965) is aimed at clarifying the chemical and mineralogical relationships, and this may also result in a modification of some of the information presented here.

The intrusion is elliptical in plan, measuring about 32×27 km. It has the form of a lopolith or funnel (Fig. 244), the dip of the floor not being known with certainty but suggesting conformity with the layering which has a moderate dip ($40°$) near the outer margins, decreasing towards the centre except in the southeastern sector (Fig. 243). A total thickness of about 7800 m is exposed, on average, of which 700 m is a lower border zone (OBZ and IBZ), 400 m an upper border zone, and the remaining 6700 m consists of a layered series of igneous cumulates. The layered rocks have been divided into a Lower Zone of olivine–plagioclase cumulates (troctolites) and an Upper Zone in which augite enters as a cumulus phase. The Upper Zone shows extreme cryptic variation and has been sub-divided into UZa, UZb, UZc, and UZd according to the character of the cumulus phases (Fig. 245).

No evidence of a chilled phase of the intrusion is found in the Outer Border Zone, which is present only against gneiss and metasediments. Since the OBZ is absent against the Nain anorthosite mass, it has been suggested that the latter may still have been hot at the time of the Kiglapait intrusion. The Outer Border Zone, where present, averages about 500 m in thickness. The rocks are finely-layered gabbros, olivine gabbros, norites and hornblende gabbros, believed by Morse to be orthocumulates involving plagioclase, olivine, orthopyroxene, brown hornblende, augite, and iron ore as cumulus phases. Xenoliths and effects of contamination are common, while some of the rocks have a granulitic texture suggestive of displacement while in a hot, largely crystalline condition. The zone is relatively mafic but the mineral compositions denote a composition which could represent relatively undifferentiated parental material (Fig. 245). These layered rocks appear to grade into the Inner Border Zone consisting of massive, unlayered olivine gabbros, about 200 m in thickness. The rocks are viewed as orthocumulates, with plagioclase and subsidiary olivine as the only cumulus phases. The mineral compositions (Fig. 245) suggest less under-cooling than in the Outer Border Zone, rather than a succession of cumulates overlying

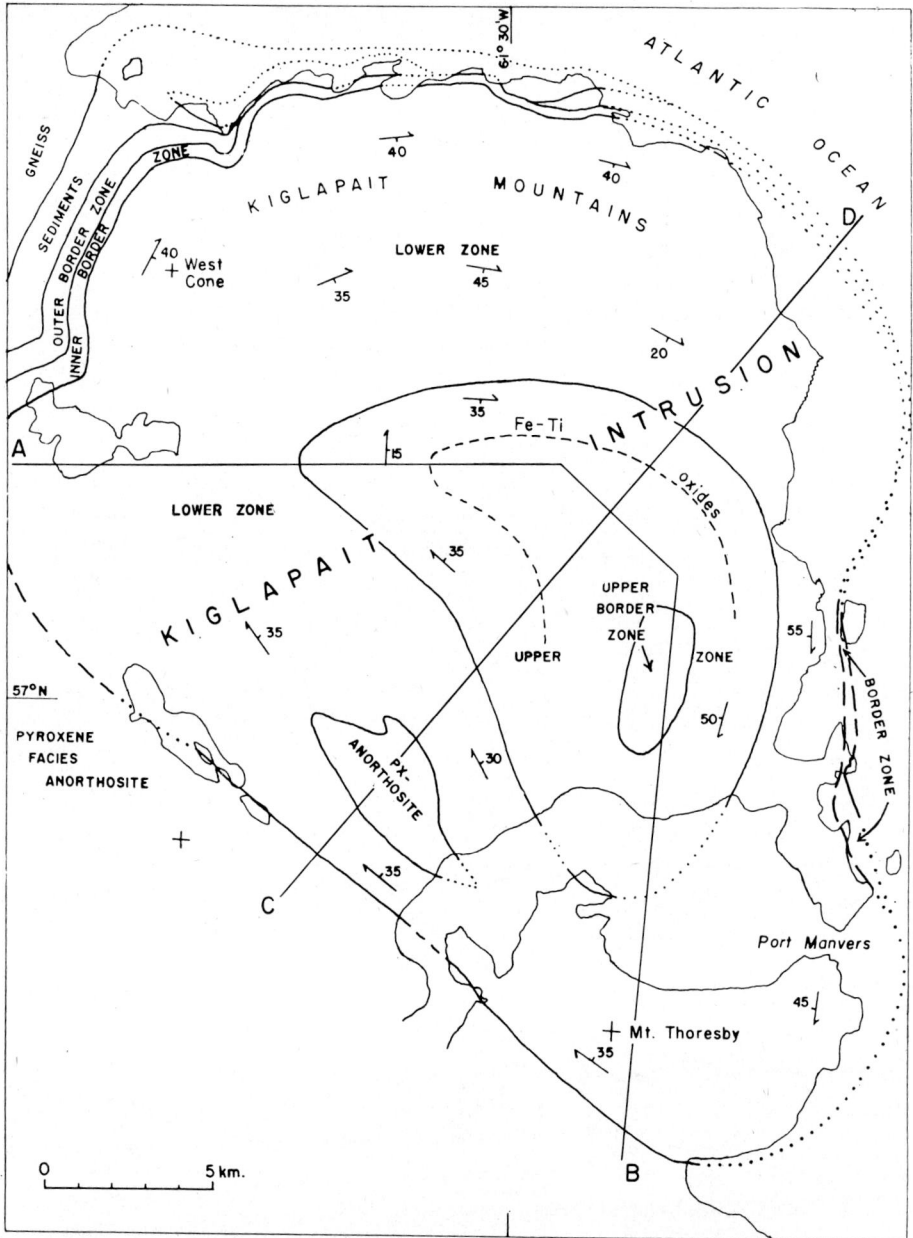

FIG. 243. Geological sketch-map of the Kiglapait intrusion, showing the Border Zones and the Lower and Upper Zones of the Layered Series (Fig. 243–245 reproduced by courtesy of S. A. Morse).

the latter, although Morse feels that the values in parentheses require further confirmation.

The layered series is characterized by a great thickness of rocks showing magnificent rhythmic layering and other structural features of igneous cumulates. Igneous lamination is common, while 'graded layering' (sometimes the reverse of gravity stratification), slump structures and scour structures are recorded. Unlayered rocks alternating with sets of well-layered rocks are common in the Lower Zone. Olivine and plagioclase are the only cumulus phases here, and the rocks are mainly adcumulates amongst which extreme examples (anorthosites and dunites) occur. In the Upper Zone, augite and titaniferous magnetite are important cumulus phases, as well as apatite and alkali felspar (Fig. 245). The layering results in the occasional development of layers rich in iron-ore, one such example, a few metres thick, being traceable for about 17 km along strike (Fig. 243). The Upper Zone rocks are chiefly mesocumulates, a few orthocumulates and adcumulates being recorded from UZd.

Cryptic layering is very pronounced in the Kiglapait intrusion (Fig. 245), and is comparable in extent with that shown by the Skaergaard. Ultrabasic cumulates are not found in the layered series, the most calcic plagioclase being An_{62} and the most magnesian olivine, Fo_{72}. The plagioclase becomes very sodic (An_{28}) in UZc, giving place to a mesoperthite in UZd in which the Na-phase is $Or_2Ab_{96}An_2$, the K-phase about Or_{97}, and the total about Or_{60}. The olivine persists as a cumulus phase throughout the layered series (to Fo_4) and in this respect the intrusion shows affinities with an alkali-basalt fractionation sequence rather than, as in the Skaergaard, a tholeiitic sequence in which silica enrichment results in the temporary cessation of olivine crystallization. Also orthopyroxene seems to be absent as a cumulus phase, although it is not uncommon as an intercumulus phase and occurs as exsolution lamellae in the augites. Augite becomes progressively richer in iron upwards in the layered series until ferroaugites crystallize in UZc and UZd. Further evidence of the regular cryptic

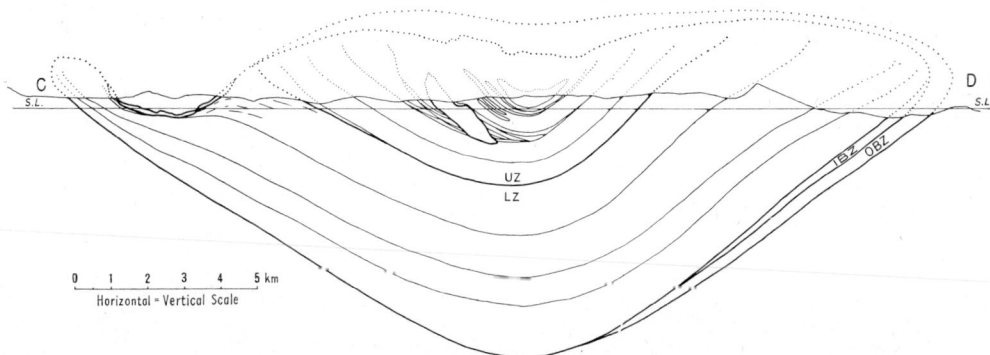

FIG. 244. Horizontal sketch-section of Kiglapait (C–D, Fig. 243), showing the form of the intrusion and disposition of the layering.

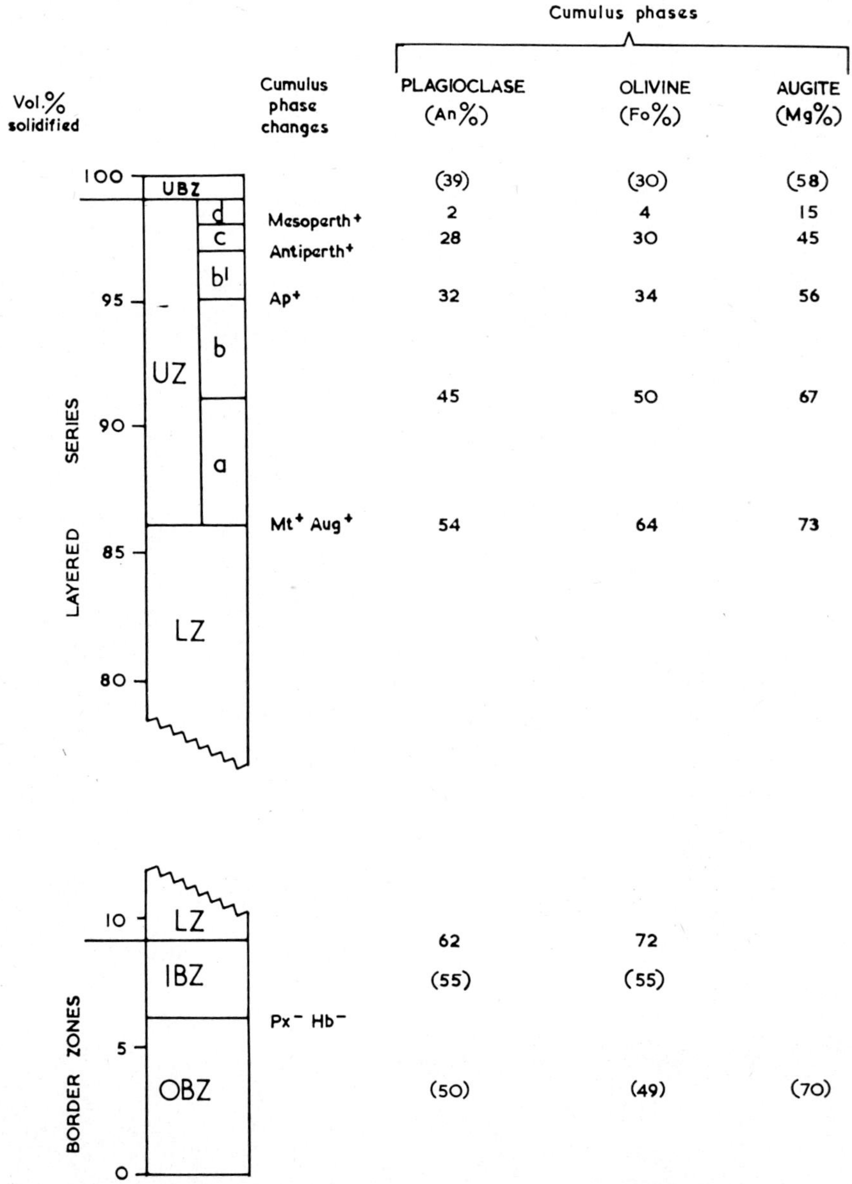

FIG. 245. Cryptic variation of cumulus minerals in the layered series, and mineral compositions in the Outer and Inner Border Zones, of the Kiglapait intrusion. (Values in parentheses require further verification.) Compositions estimated from optical measurements, three chemical analyses of plagioclases and three of olivines being used as a control. The estimates of volume percentage solidified are at present only tentative, and S. A. Morse (pers. comm., 1965) is working on a revision of these values.

variation is shown by the incoming of cumulus magnetite at the base of UZa, and of cumulus apatite at the base of UZb.

An Upper Border Zone of unlayered fine-grained gabbros, with variable modes, lies above UZd (Fig. 243). The mineral compositions are close to those in UZb and lead Morse to suggest that earlier crystals were abstracted from the magma even as it cooled relatively quickly, and that a true chill zone may once have overlain the Upper Border Zone.

The latest fractionation products exposed in the Kiglapait intrusion contain a K-rich felspar, fayalite, ferroaugite, and subordinate amounts of apatite and iron oxide. Morse calls these rocks larvikites rather than alkali syenites, because of a suspicion that they are undersaturated and may be Ne-normative. However, no chemical analyses of rocks are yet available for Kiglapait, so the nature of the original magma and the characteristics of the fractionation trend, in terms of magma-types, are not yet known. In general the persistence of cumulus olivine and the absence of quartz-bearing late differentiates would suggest that the intrusion has alkaline affinities. On the other hand, the sporadic presence of orthopyroxcne in the layered series and the absence of felspathoid minerals in the late differentiates suggest tholeiitic affinities. In the absence of a chilled border rock providing information on the composition of the parental magma, the Ca-rich clinopyroxenes of layered intrusions, in terms of their Ca-contents (Poldervaart & Hess, 1951) and also in terms of their Al and Ti contents (summary in Brown, 1967) provide the best means of distinguishing between the fractionation products of alkali-basalt and tholeiitic basalt magmas. Chemical analyses of these minerals are not yet available but are being made (personal communication, Morse, 1965). The available petrological evidence would favour tholeiitic affinities, if choice needs to be made only between the two main basalt magma-types. However, it is clear that the physico-chemical conditions in the Kiglapait intrusion prevented a build-up in silica content sufficient either to cause a gap in the crystallization sequence of olivine or to allow the crystallization of quartz in the late fractionation stages. In other words, the trend of successive liquids followed a course such that despite the abundant separation of olivine, the plane of silica saturation (Yoder & Tilley, 1962, fig. 1) was not crossed. Highly reduced conditions may favour such a tendency (Osborn, 1959) although they would not be expected to yield the amount of magnetite found in Kiglapait. Alternatively, crystallization under a relatively high pressure would extent the pyroxene stability field such that for certain compositions, olivine removal would not lead to the crystallization of a silica phase. In these respects, amongst many, the Kiglapait intrusion is clearly one which is going to be of very great petrogenetic significance. Other intrusions with tholeiitic affinities, but with orthopyroxene more abundant than in Kiglapait, and as a cumulus phase, show the formation of saturated syenites as late differentiates (e.g. Insch, p. 443; Okonjeje, p. 523) but these are not well layered. The symmetrical and grand scale of the layering at Kiglapait means that the intrusion provides a particularly clear example of a fractionated series of rocks that trends from olivine gabbros to syenites free from quartz or felspathoid minerals.

Great Dyke, (Southern) Rhodesia

This huge, dyke-like intrusion has a total length of about 330 miles and an average width of $3\frac{1}{2}$ miles. The strike is NNE, northern outcrops of the intrusion lying about 30 miles west of Salisbury (Fig. 246). The rocks consist chiefly of a layered ultramafic assemblage of dunites, chromitites and pyroxenites, with lesser amounts of gabbro and quartz gabbro. The Great Dyke has intruded Archaean granites and granitic gneisses and is generally viewed as Precambrian in age. Macgregor (1951, table 1) gives ages of 1900 million years for some of the intruded granites, but estimates that the Great Dyke was probably intruded in Upper Precambrian times ($<$1000 million years). It seems unlikely, as pointed out by Willemse (1964, p. 125) that the Great Dyke, as suggested by several geologists, can be viewed as contemporaneous with, and a lateral extension of, the Bushveld activity.

Tyndale-Biscoe (1949, p. 35), in considering the character of this unique intrusion, agrees that 'whatever it is it is not a dyke'. Since the early, lucid descriptions by Lightfoot (1927) attention has been focussed on the form of the intrusion, and on the means whereby conformable layering could be developed in a body of such elongate shape. Weiss (1940) carried out gravimetric traverses in the Hartley area and showed a sharp positive anomaly of 50 milligals, indicative of an extension of the ultramafic rocks to an appreciable depth below the level of exposure, so that in this sense the intrusion is a dyke. Hess (1950, fig. 2) advanced the theory that the Great Dyke acted as a feeder to a vast, sheet-like intrusion, extending about 60 miles each side of the present axis, which has since largely been removed by erosion. Since then, B.G. Worst has conducted a remarkably detailed study of the field relations of the Great Dyke, described at length in an unpublished thesis (1956), the critical information from which has since been published (1958). The structures of the intrusion are complex (Worst, 1958, pl. 50), correlation between layers necessitating a detailed consideration of faults, dips, and variable pitch along the whole length of exposures. From this, Worst agrees that the Great Dyke is not a dyke, but suggests that it is the remains of four lopoliths which were formed along a straight line and may have inter-sected one another. Subsequent faulting has preserved the parts of the four complexes in a graben, and only superficially can the intrusion be viewed, according to the present outcrops, as a dyke-like body. The original lateral extent of the complex as a whole has not been estimated, but even if elongate it is likely to have been of Bushveld dimensions. In fact, it has been suggested (pp. 345, 348) that the Bushveld consisted of four separate but interconnected magma chambers, in which case the four lopoliths of the Great Dyke complex, measuring 195, 60, 50 and 27 miles in length, respectively, probably constitute a similarly related assemblage.

a) The ultramafic layered rocks

The general orientation of the Great Dyke, and the positions of the four separate centres defined by Worst, are shown in Fig. 246. The reader is referred to Worst

(1958, pl. 50) for detailed maps showing the structures of the several parts of the intrusion. The layers have a synclinal structure along the length of the dyke, the dips decreasing gradually from the margins towards the axis, from about 25° to 5° on average. The undulatory pitch of the layers results in outcrops which form arcs of concentric circles; the pyroxenite layers support more vegetation than the serpentinized dunites,

and the layering, faults, and marginal contacts are clearly discernible on aerial photographs.

A total exposed thickness of 10,000 ft of layered rocks has been recorded from the Hartley area, 6000 ft from the Selukwe area, and 3000 ft from the Wedza area to the south (Worst, 1958, fig. 1). In the latter area, a 5000 ft borehole has increased the section to 6000 ft and provided a collection of exceptionally fresh dunites free from the extensive serpentinization found in surface exposures. The thick Hartley section consists of 7000 ft of ultramafic rocks overlain conformably by 3000 ft of gabbros and norites (Fig. 247). Seven thick pyroxenite layers occur in the ultramafic sequence, one of them occurring at the base, while the rest consists of serpentinized dunite and harzburgite layers and of ten thin chromitite seams. The complete sequence is given in Table 31, together with the thicknesses of each layer, and

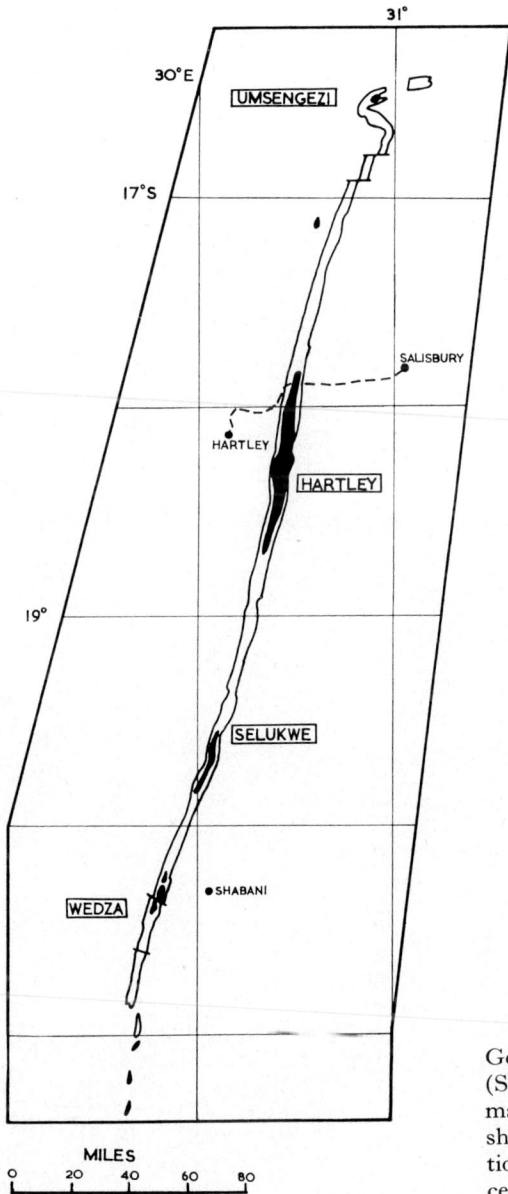

FIG. 246

Generalized map of the Great Dyke, (Southern) Rhodesia. The four main outcrops of basic rocks are shown in black, located at the positions of the four separate, named centres (after Worst, 1956, 1958).

TABLE 31

Thicknesses of layers in the Hartley Area of the Great Dyke, listed from top to base (compiled from Worst, 1958). Six possible macro-rhythmic units are here delineated (I–VI)

Unit	Layer	Thickness
	Gabbro and Norite	3 000 ft
I	Felspathic pyroxenite Thin Pt-bearing pyroxenite Pyroxenite No. 1 Olivine pyroxenite	700 ft
	Picrite Harzburgite	250 ft
	Chromite Seam No. 1	2 in and 4 in
	Harzburgite	80 ft
	Chromite Seam No. 2	14 in
	Harzburgite Dunite	380 ft
II	Pyroxenite No. 2	30 ft
	Harzburgite	60 ft
	Chromite Seam No. 3	2 in and 6 in
	Harzburgite	200 ft
III	Pyroxenite No. 3 Olivine pyroxenite	580 ft
	Harzburgite Dunite	200 ft
IV	Pyroxenite No. 4 Olivine pyroxenite	170 ft
	Dunite	175 ft
V	Pyroxenite No. 5	200 ft
	Harzburgite Dunite	150 ft
	Chromite Seam No. 4	5 in
VI	Pyroxenite No. 6 Olivine pyroxenite	180 ft
	Dunite	490 ft
	Chromite Seam No. 5	4 in
	Dunite	400 ft
	Chromite Seam No. 6	6 in
	Dunite	260 ft
	Chromite Seam No. 7	5 in
	Dunite	400 ft
	Chromite Seam No. 8	6 in
	Dunite	300 ft
	Chromite Seam No. 9	4 in
	Dunite	350 ft
	Chromite Seam No. 10	3 in
	Dunite	1 100 ft
	Pyroxenite No. 7	500 ft
	Total	10 200 ft

Note: (i) All but the uppermost of the Pyroxenites are Bronzitites.
(ii) All the dunites in this area are largely serpentinized.

a diagrammatic sequence given in Fig. 247. Sections from other parts of the complex differ in detail, supporting Worst's hypothesis of four separate centres, although the cumulus mineral assemblage is the same throughout. Of particular interest, however, is the fact that above Pyroxenite No. 2 (numbered from the top downwards) the relationship between the layers is similar in the four centres, Pyroxenite No. 1 retaining a constant thickness of about 700 ft for 140 miles (Worst, 1958, p. 351). This suggests that a stage was probably reached when the four magma chambers were joined, crystal deposition then taking place in one vast magma chamber.

The layered ultramafic rocks are clearly igneous cumulates, consisting of variable proportions of cumulus olivine, bronzite, and chromite. Clinopyroxene enters as a cumulus phase in the uppermost pyroxenite layer, as does plagioclase in lesser amounts, both then persisting throughout the overlying gabbroic rocks. The dunites (usually serpentinized) are chiefly olivine adcumulates, chromite usually being $<5\%$, while the pyroxenites are chiefly bronzite adcumulates (Worst, 1958, tables 4 & 9 and pl. 54). Occasional poikilitic crystals of olivine or bronzite are recorded from some rocks which are probably heteradcumulates, but these are relatively rare, as are the meso-cumulates containing poikilitic crystals of relatively sodic plagioclase (Fig. 247). The distribution of cumulus minerals in the various thick layers has not been studied systematically by Worst, although he records the presence of harzburgites (olivine–bronzite cumulates), olivine pyroxenites, and picrites from the upper half of the ultramafic layered series. As shown in Fig. 247, the Hartley section can be viewed as consisting essentially of about 3000 ft of olivine adcumulates, with five thin layers of chromite adcumulates, overlain by six macro-rhythmic units (totalling a further 4000 ft) similar in scale and mineralogy to the units in the Basal Zone of the Bushveld and in the Ultramafic Zone of the Stillwater intrusion. Each unit usually has an olivine adcumulate at the base, an olivine–bronzite adcumulate in the central part, and a bronzite adcumulate at the top. The thicknesses of each cumulate type, and the modes, are not yet sufficiently well documented to permit a more accurate section being drawn but, clearly, these huge macro-units (averaging 600 ft in thickness) merit further study.

The olivines of the ultramafic rocks reach more magnesian compositions than do any recorded from other basic layered intrusions, the composition varying from Fo_{94} in the lowest layers to Fo_{85} in the layers immediately below the gabbros, at which level olivine ceases to crystallize (Fig. 247). Similarly the orthopyroxenes vary in composition with height in the ultramafic layers, from En_{93} to En_{77}, although this mineral is also found in the overlying gabbros. In the chromites, as in those of the Bushveld, there is a regular decrease in Cr_2O_3 and in Cr:Fe ratio upwards in the layered sequence (Worst, 1958, tables 15 & 16). Hence cryptic layering in the ultramafic units of the Great Dyke is particularly well developed, such variation generally being difficult to detect in the extreme ultramafic parts of most layered intrusions. Apart from the chromites, the mineral compositions are estimated from

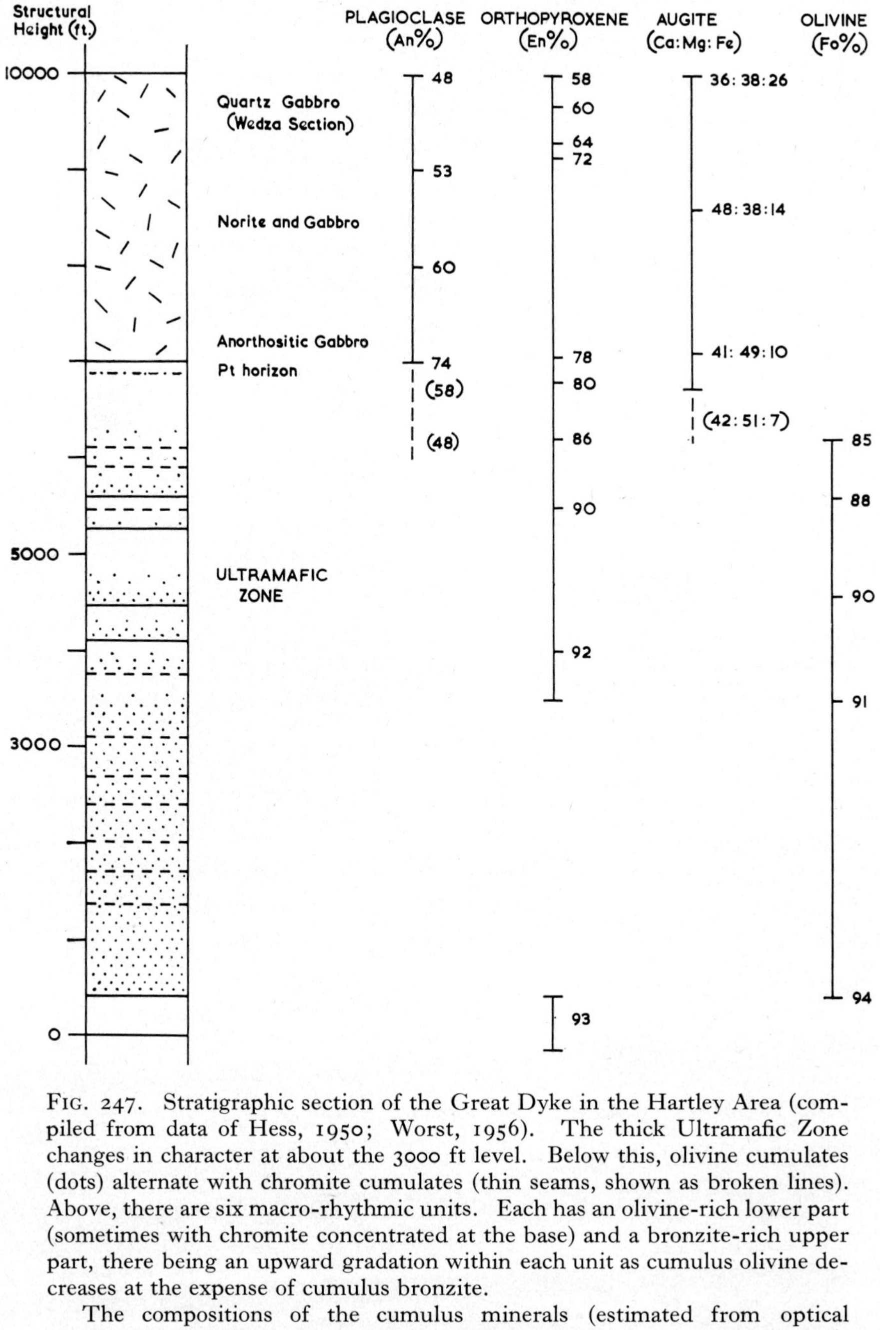

FIG. 247. Stratigraphic section of the Great Dyke in the Hartley Area (compiled from data of Hess, 1950; Worst, 1956). The thick Ultramafic Zone
changes in character at about the 3000 ft level. Below this, olivine cumulates
(dots) alternate with chromite cumulates (thin seams, shown as broken lines).
Above, there are six macro-rhythmic units. Each has an olivine-rich lower part
(sometimes with chromite concentrated at the base) and a bronzite-rich upper
part, there being an upward gradation within each unit as cumulus olivine decreases at the expense of cumulus bronzite.

The compositions of the cumulus minerals (estimated from optical
properties) are shown, and the levels at which each exists as a cumulus phase
or, rarely, as an intercumulus phase (broken lines).

optical properties, Worst providing much data in his unpublished thesis (1956, tables 29 & 30).

b) *The gabbroic rocks*

Basic rocks are now confined to the four areas designated by Worst as the centres, or feeder-areas, of each of the separate lopoliths (Fig. 246). This is attributed by him to excessive sagging in these regions of maximum overload, and preservation of higher layers during subsequent erosion of the region. A maximum thickness of 3000 ft is preserved in the Hartley area (Fig. 247), consisting of about 300 ft of gabbro and 2700 ft of norite. In the southern, Wedza area, 500 ft of gabbro and norite is overlain by 700 ft of quartz gabbro.

No layering has been described from the basic rocks by Worst or by Hess (1950), although igneous lamination is well-developed. The rocks are described as varying slightly in their modal constituents, but generally they contain about 60% plagioclase and 40% pyroxenes, only one layer at the base of the gabbroic series, 100 ft thick, being markedly felspathic (82% plagioclase). Worst (1958, p. 305) and Hess (1950, p. 164) attributed this layer of anorthositic gabbro to crystal sorting, viewing it as related to the underlying Pyroxenite No. 1 layer which becomes slightly felspathic towards the top (Table 31). However, the felspathic pyroxenite contains only *inter-cumulus* plagioclase (8%). Moreover, this intercumulus plagioclase is An_{58} (indicating that mesocumulates occur), whereas the cumulus plagioclase of the overlying anorthositic gabbro is An_{74} (Worst, 1956, table 28).

The constituent cumulus minerals of the basic rocks are plagioclase, augite, and either hypersthene or inverted pigeonite. Iron oxides never exceed 2% of any rock, while 8–9% quartz occurs in the relatively rare, quartz gabbros of the Wedza area. Cryptic variation is well-developed, the felspars and pyroxenes showing a gradual change towards lower-temperature members upward in the succession. The plagioclases change from An_{74} to An_{48} (Fig. 247) and the hypersthenes from En_{78} to En_{58}. Those hypersthenes less magnesian than En_{67} show augite exsolution textures indicating that they have inverted from pigeonites (Hess, 1950, table 1). The compositions of augites are difficult to estimate from optical measurements and the values quoted by Worst (1958, fig. 6) are viewed with some scepticism, particularly in regard to the low Wo contents quoted. Hess's values are probably more correct (1950, table 1), ranging from $Ca_{41}Mg_{49}Fe_{10}$ to $Ca_{36}Mg_{38}Fe_{26}$.

The mineralogical evidence (Fig. 247) suggests that the basic rocks are to be viewed as part of the regular fractionation sequence exhibited by the underlying ultramafic rocks, there being no sharp break in the compositions of the orthopyroxenes, which occur as cumulus minerals throughout the intrusion. There is, however, a marked change in the cumulus assemblage above the uppermost thick pyroxenite layer, despite the incoming of augite and plagioclase towards the top of this layer. Olivine and chromite disappear, while plagioclase and augite become important cumulus phases.

2 G

The data are not sufficiently precise to warrant more than a suggestion, at this stage, that a major, fresh influx of parental basic magma may have been responsible for the sudden change in the cumulus mineral assemblage at the top of the uppermost pyroxenite layer. It is perhaps significant, in this respect, that a thin Pt-rich horizon occurs towards the top of this layer, and that this feature of the Bushveld intrusion (Ch. XIV), the Merensky Reef, has been correlated with a fresh magma influx and the subsequent deposition of the relatively non-layered gabbroic rocks of the Bushveld Main Zone.

c) *General considerations*

A maximum thickness of 10 000 ft of layered rocks is exposed in the Great Dyke intrusive complex. The rocks range in composition from dunites (Fo_{94}) at the base to quartz gabbros (An_{48}) at the top, the compositional variation being regular and attributable to differentiation through crystal accumulation in position. Crystallization probably took place in four large magma chambers of appreciable lateral extent, which may have been inter-connected during parts of their crystallization history but probably only during the formation of the upper third of the layered sequence. The present elongate, dyke-like outcrop is attributed to extensive faulting and the preservation of a narrow strip of the original lopolithic intrusions as a graben. The marginal contacts are steep-sided, in contrast to the $25°$ inward dip of the layered rocks, and are to be viewed as fault-lines rather than intrusive margins. Hence no estimate can be made of the original magma composition, from chilled margins. Hess (1950) collected two fine-grained dolerites, overlying the gabbroic rocks in the Hartley area, which he believed to represent the original parental magma chilled against the roof. Analyses were given later (Hess, 1960, table 48) and shown to be close in composition to Bushveld chilled rocks. The rocks are noritic in comparison, with normative plagioclase An_{52-55}, normative hypersthene En_{50-55}, and normative olivine absent. Worst does not comment on these outcrops, and it may be that he views them either as xenoliths (common in this area) or as one of the later doleritic intrusions commonly found associated with the Great Dyke. Whatever their origin, we feel that these rocks cannot seriously be considered as representative of the parental magma of the Great Dyke complex. They are close in composition to the norites which constitute the latest observed differentiates (Worst, 1958, table 17, No. G396), and if they are to represent the overall parental magma then great difficulties would arise in accounting for the enormous thickness of ultramafic differentiates. It is unlikely that the latter were restricted to the present narrow axial region, the early crystal phases being more likely to have accumulated evenly across the entire floor of the original magma chambers. Also it is impossible, on chemical grounds, for an olivine-free noritic magma to precipitate such a huge bulk of olivine cumulates without giving rise to an equally great bulk of quartzose late differentiates. By postulating that the fine-grained norites represent a chilled roof phase, immediately overlying quartz-free gabbros and norites in the Hartley area, Hess precludes the possibility that later differentiates existed, and so the chemical and field evidence are mutually prohibitive. The alternatives are either that the original magma was some kind of basalt and that late differentiates once overlying the gabbros have been removed by erosion; that the original magma was picritic; or that the complex crystallized as an open system in the earlier stages, the great thickness of ultramafic rocks being built up with the aid of replenishments of parental magma (see Ch. XII, XIV). The latter hypothesis is not in accord with the regularity shown in the cryptic layering, no obvious reversals in mineral compositions being recorded. In any case, the earliest cumulus minerals of the Great Dyke are unusually high-temperature members of their solid solution series. The olivines (Fo_{94}) and the bronzites (En_{93}) are more magnesian than in any other basic layered intrusion, but similar to the minerals found in certain layered ultramafic complexes (Brown, 1967, table 3). Also, the lowest chromites are very rich in Cr_2O_3, the uppermost layers in the Great Dyke corresponding in composition to the lowest in the Bushveld. Until more is known of the physical properties of picritic magmas the subject will not be pursued further here. However, the Great Dyke contains an assemblage of rocks which are of especial significance in this connection. The ultra-mafic layers correspond in composition to those found in other layered ultramafic complexes but are particularly significant in that they display a marked, regular cryptic

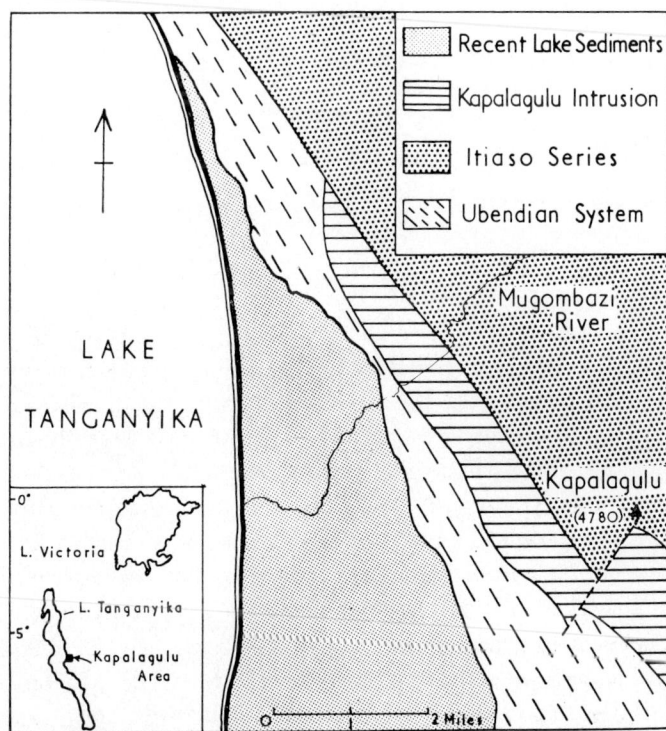

FIG. 248. Geological sketch-map showing the position of the investigated part of the Kapalagulu intrusion. (From Wadsworth, 1963, fig. 1.)

variation and grade upwards, in composition, into a thick gabbroic series through which the cryptic variation continues without a sharp break.

KAPALAGULU, TANZANIA (TANGANYIKA)

The Kapalagulu intrusion has been described by van Zyl (1959) and Wadsworth (1963), and shown to consist of layered basic and ultrabasic rocks in which cryptic layering is well developed. The intrusion lies close to the eastern shore of Lake Tanganyika (Fig. 248), about 70 miles south of Kigoma. It is exposed as a long, narrow belt, approximately 9 miles by 1 mile, trending northwest–southeast. The northern,

TABLE 32

The sub-divisions of the Kapalagulu intrusion

Zones	Cumulate assemblage and sub-zones	Thickness (ft)
Main	(e) Plagioclase–augite–magnetite	700
	(d) Plagioclase–augite–orthopyroxene	1300
	(c) Plagioclase–augite–orthopyroxene	500
	(b) Plagioclase	300
	(a) Plagioclase–augite–orthopyroxene	550
Intermediate	Plagioclase–olivine–augite–orthopyroxene	750
Basal	Olivine–chrome-spinel	400
		4500

western, and eastern margins terminate abruptly against the country rocks owing to faulting, while the southern part of the intrusion is not exposed. The age is not known but it is likely to be post-Bukoban system which, in turn, is possibly of Palaeozoic age (Wadsworth, 1963, pp. 108 & 115). The steep dip of the layered structures is approximately 80–85° NE (van Zyl, 1959, fig. 1), while the contacts are equally steep and extensively faulted and sheared. The tectonic events leading to the present disposition of the intrusion have not been investigated but it is clear, from the present structures and cryptic layering, that a once-larger layered intrusion has been tilted to the northeast and faulted, subsequent to consolidation.

A maximum thickness of 4500 ft of layered rocks is exposed, the greatest thickness being preserved in the region of the Mguje river (Wadsworth, 1963, fig. 2) close to the 4780 ft-high Kapalagulu mountain. The layered series was divided into three zones by van Zyl (1959, table 1): a Basal Zone, Intermediate Zone, and Main Zone.

Wadsworth has retained this nomenclature but subdivided the Main Zone into five sub-zones. The thickness and cumulate assemblage of each zone are given in Table 32 and the distribution and compositions of the cumulus minerals in Fig. 249.

The Kapalagulu layered rocks show a regular pattern of cryptic variation. The lower, Basal Zone layers contain only olivine and minor chrome-spinel as cumulus phases, bronzite and calcic plagioclase joining olivine as cumulus phases at the base of the Intermediate Zone. Towards the top of the Intermediate Zone, augite enters as a cumulus phase while olivine disappears at the top. Cumulus plagioclase, orthopyroxene, and augite persist throughout the Main Zone, except where a 300 ft-thick layer of plagioclase cumulates (anorthosites) occurs. Towards the top of the Main Zone, the Ca-poor pyroxene is an inverted pigeonite while magnetite, apatite, and micro-pegmatite are important mineral phases. A 20–30-ft layer relatively rich in sulphides (pyrrhotite, with minor pyrite, chalcopyrite, sphalerite, and pentlandite) occurs in the

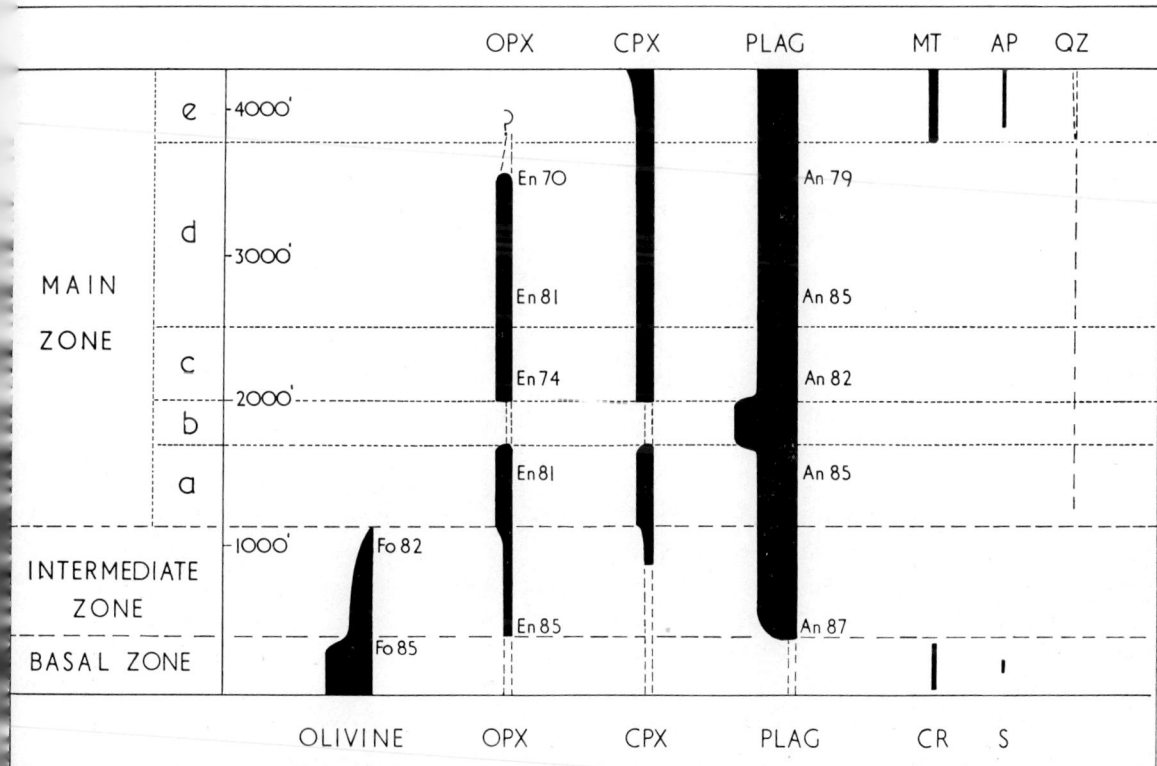

FIG. 249. Summary of the cryptic layering in the Kapalagulu layered series, and the sub-division into zones and subzones based on the cumulus mineral assemblage (CR, chromite; S, sulphides). The dotted lines indicate where minerals occur only with intercumulus status. Major modal variations are shown by changes in the thicknesses of the individual columns. (From Wadsworth, 1963, fig. 3.)

middle of the Basal Zone, the sulphides being interstitial to the cumulus olivine crystals (van Zyl, 1959, pl. 1).

Wadsworth has described the textures briefly, drawing attention to the preponderance of adcumulates and heteradcumulates which characterize the whole of the Basal and Intermediate Zone rocks. In the Main Zone, interstitial quartz and micropegmatite are relatively common in many rocks (up to 10%) but zoning of the cumulus minerals remains inconspicuous and the rocks are viewed as mesocumulates.

The olivines, pyroxenes, and plagioclase felspars each change gradually in composition towards lower melting-temperature members upwards in the layered series (Fig. 249). No chemical analyses of minerals or rocks are yet available, however, the variation being estimated from optical determinations (Wadsworth, 1963, fig. 4). The earliest cumulus assemblage (Fo_{85}, En_{85}, and An_{87}) is similar to that found in most other layered intrusions, and likely to have been precipitated from a basaltic magma. The uppermost exposed differentiates, however, are not likely to be the latest formed by fractionation of this magma although it is impossible to say whether a further sequence once existed, which has been removed by faulting and erosion, or whether the residual magmas were expelled from the magma chamber. On the other hand, magnetite, apatite, and quartz appear relatively early in the fractionation sequence as defined by the plagioclase composition (An_{79}), suggesting that the physico-chemical conditions were rather different, in Kapalagulu, from most of the other intrusions where such phases generally appear when the plagioclase is about An_{60} in composition. Wadsworth has detected a significant reversal in the cryptic variation of the Main Zone (Fig. 249) at a level used, on this basis, to define the base of sub-zone (d). He was unable to confirm an additional reversal within the Basal Zone, associated by van Zyl (1959, p. 22) with the precipitation of the sulphide-rich layer. Both authors attribute the reversals to the influx of fresh supplies of undifferentiated magma, analogous to the mechanism discussed here for similar features of the Bushveld (Ch. XIV).

Rhythmic layering is particularly well developed in the Intermediate Zone, but is also found in sub-zones (a) and (c) of the Main Zone. Igneous lamination, and slumped and current-bedding structures, are also mentioned by Wadsworth (1963, p. 112). The rhythmic layering led van Zyl (1959, pp. 24–28 and fig. 7) to propose, and consider in some detail, the process of undercooling and rhythmic nucleation to account for the fine-scale layering (cf., Fig. 84). Van Zyl cites the literature relating to the Liesegang phenomenon and other aspects of metastable crystallization, and discusses the effect of undercooling in relation both to rhythmic layering and to reversed zoning in isomorphous mineral groups.

CHAPTER XVI

LAYERING IN GRANITES, QUARTZ SYENITES, AND FELSPATHOIDAL SYENITES

GRANITES

Most of the banding and parallelism of platy minerals in granitic rocks seems to be due either to flow of a mush of crystals and liquid or to heterogeneity inherited from earlier rocks, the partial melting and metasomatism of which produced the granite. In a few granites, however, there is evidence of crystal settling, and in these we must postulate that at one time the granite was a fluid magma containing sporadic crystals which, in favourable circumstances, were able to sink.

The most convincing evidence for the sinking of crystals in granitic magma has been found by Harry and Emeleus (1960) in certain fluorite-bearing granites of South Greenland. At the head of Tigssaluk fjord, 25 km north of Ivigtut, there are two granite stocks, each about 5 km across, in which layering is indicated by the distribution of the dark-coloured minerals. The granites are made up mainly of microcline–microperthite, oligoclase, biotite, and quartz; early accessory minerals are sphene, apatite, allanite, and rare zircon; fluorite, as a late mineral, is ubiquitous. The layering tends to occur within a few hundred metres of both the margin and roof, and is due to a concentration of rather well-shaped crystals of biotite and the accessory minerals. The dark layers, usually a few centimetres thick, grade laterally and vertically into normal granites (Fig. 250), and often the layers show gravity stratification (Fig. 251). The layers are not very persistent and in some places show false bedding and in others, trough banding structures; cases suggesting slumping of a crystal mush have also been described.

The textural features to be seen in thin sections of the rocks showing layering are described in a second paper by Emeleus (1963). In the relatively mafic layers the minerals fall into three groups on the basis of shapes and textural relations: (1) euhedral crystals, (2) euhedral crystals modified by further growth, and (3) crystals of interstitial or poikilitic habit. Biotite and most of the accessory minerals belong to the first group. Plagioclase and the larger crystals of alkali felspar are of the second group, while quartz, smaller crystals of microcline–microperthite, an interstitial plagioclase, and fluorite are

FIG. 250.　Layering in the Tigssaluk granite, Ivigtut region. The mafic layers are of biotite and accessory minerals. (From Emeleus, 1963.)

FIG. 251.　Gravity-stratified layering in the Tigssaluk granite. (From Emeleus, 1963.)

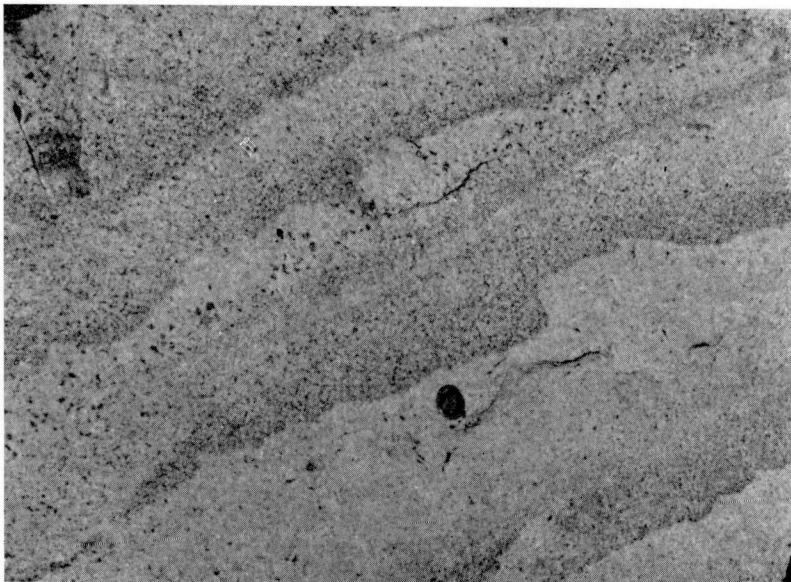

of the third group. Emeleus considered that all the minerals, with the exception of fluorite, were available from an early stage of crystallization. The crystallization of biotite, sphene, allanite, zircon, apatite, and the opaque minerals were completed fairly early, crystallization of the felspars extended over a longer period, while most of the quartz and fluorite completed crystallization relatively late. A factor in the development of these textural features is, no doubt, the relative abundance in the melt of the elements forming the different minerals. Even if the crystallization of the accessory minerals continued until late in the crystallization sequence, their form would not be modified greatly from their euhedral shape because so little further material would be deposited from the interstitial liquid. On the other hand there would be enough potential plagioclase, alkali felspar and quartz in the liquid to produce obvious mantling, or poikilitic textures. Emeleus concludes that there was some crystal settling of such minerals as biotite, accessories, and larger crystals of felspar, and that minerals of the second and third textural groups, on the whole, crystallized from intercumulus liquid around the settled crystals. However, he points out that the layered rocks are not simply orthocumulates, because the proportions of quartz, alkali felspar, and plagioclase in the poikilitic and interstitial material are not, apparently, the same as in the normal granite.

The other acid pluton showing layering is the hornblende granite of Alangorssuaq, 40 km south of Ivigtut, in which dark layers, one to twenty centimetres thick, consist of concentrations of the darker minerals of the granite with some quartz and felspar crystals which are usually smaller than those in the more average parts (Harry & Emeleus, 1960, p. 178). Both margins of the layers may be sharp, or one may be sharp and the other gradational into the normal granite. Two-dimensional exposures often show gracefully curving, bifurcating layers, thinning towards their ends (*op. cit.*, fig. 10). While individual layers thin and disappear beyond a few metres, the unit of the granite containing the evidence for layering persists for two kilometres. The granite described by Harry and Emeleus may be a continuation of the Helene granite which Harry and Pulvertaft have shown in a later paper (1963) to exhibit perhaps even better layering. Harry and Emeleus (1960) ascribe the various examples of layering to the local accumulation of dark minerals during the crystallization of a granite magma, the concentration of them being governed largely by density contrast. Some of the features of the layers simulate sedimentary structures and others suggest currents in the magma. They suggest that the granites showing layering crystallized from magma of unusual fluidity.

In high level epigranites such as those of the British Tertiary Province, no evidence of significant settling of crystals has been seen. No doubt the viscosity of the magma was high, and cooling relatively rapid, so that the time available for sinking to take place was relatively short. The evidence from the South Greenland granites suggests that mafic bands in certain granites are most likely to be layers formed by crystal sinking in these particular granite magmas. The phenomenon seems to be restricted to intrusions formed from granitic magmas of particularly low viscosity (cf. Shaw,

1965), and is not likely to be a petrogenetic factor of general importance for explaining variations within granitic intrusions. However, when layering can be demonstrated, a study of the layered structures and of the rock textures and mineralogy provides evidence of complete liquidity and low viscosity, and will probably help with the complex problems of the physico-chemical properties of granitic magmas.

KÛNGNÂT AND NUNARSSUIT QUARTZ SYENITES, SOUTH GREENLAND

In quartz syenites, and still more in nepheline syenites, the evidence for crystal settling is more abundant than in granites, and in some intrusive complexes of agpaitic affinities, layering due to settling of crystals is as striking as in the basic layered intrusions.

The Kûngnât intrusion, some 15 km west of Ivigtut in South Greenland, exhibits layering in an unambiguous way. It was investigated by B. G. J. Upton while a research student at Oxford, the field work and publication being done under the auspices of the Greenland Geological Survey. The rocks were intruded at the close of the Gardar period, about 1240 million years ago. Upton (1960) has divided the complex into: (1) the southwestern marginal syenite—the earliest, but not well exposed part of the intrusion; (2) the western syenite, divided into a lower and upper layered series; and (3) the eastern layered series and border group (see map, Fig. 253). Later than all these layered rocks is a remarkably fine example of a ring dyke of gabbroic composition. Of the layered syenites, the lower and upper layered series of the western syenites will be described in some detail, and the rest only cursorily.

A sequence of 1800 m of western layered syenites is seen in the western half of Kûngnât (Fig. 252). The layering and the very common igneous lamination dip, on the whole, towards a centre to the east of the existing part of the intrusion. The three lower units, forming the lower layered series, are separated from the upper layered

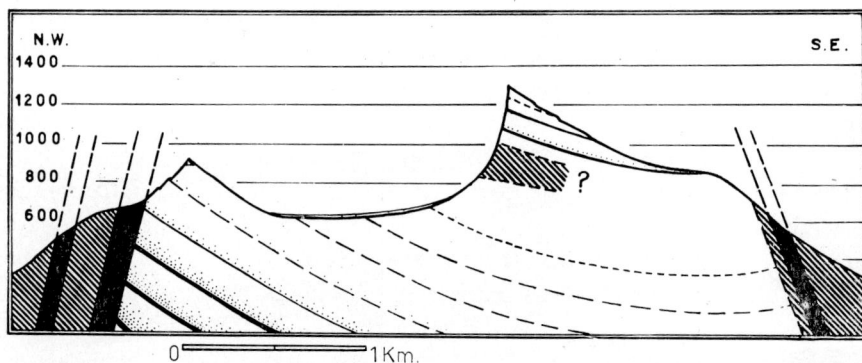

FIG. 252. Section across the western layered series, Kûngnât. (From NW corner of Fig. 253 to Røverborg.) Line-shading: gneiss; solid: ring-dyke rocks. The density of the layering in the syenite is shown diagrammatically. (From Upton, 1960.)

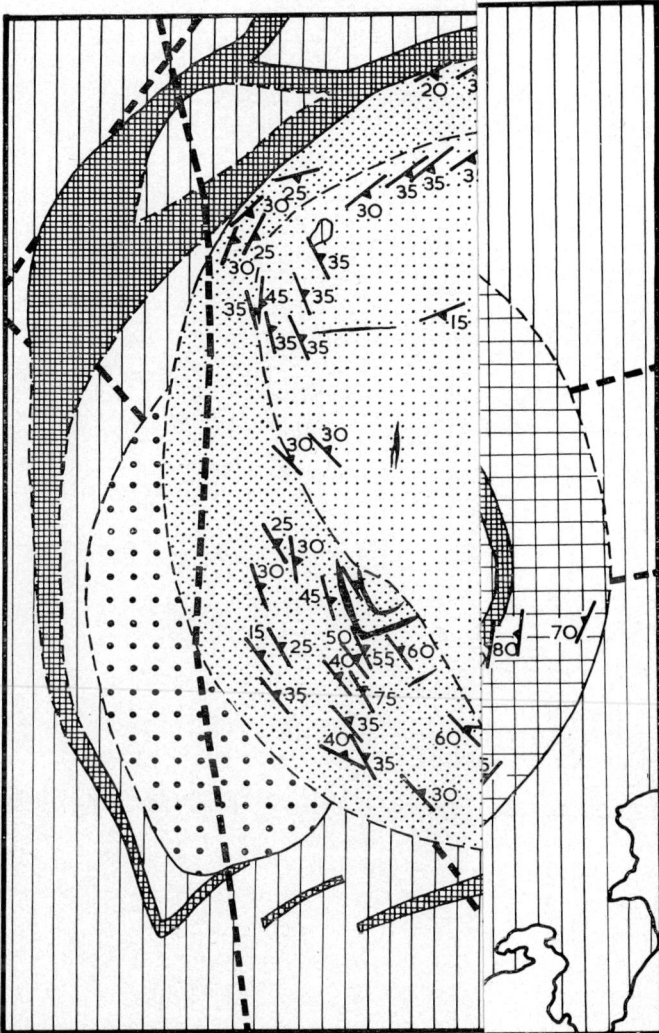

Legend:

Gneiss ... ~~~ es

Dolerite Dykes

Attitude of Layered Str~~~

SCALE IN KILOMETRES

1 0 .... 5 10

FIGURE 253

Geological map of the Kûngnât quartz-syenite c~~~
(After colour-map of Upton, 1960.)

[face p. 474]

series by a zone of shadowy blocks of grey gneiss. The lowermost rocks are dark olive-green syenites, and mafic layers form a large proportion. Upwards, the layering becomes less common and almost dies out below the zone of inclusions, but above this zone, strongly layered rocks occur again.

FIG. 254. Rhythmic layering in syenites, western lower layered series, Kûngnât. (From Upton, 1960.)

FIG. 255. Small trough-band structure in syenites, western lower layered series, Kûngnât. (From Upton, 1960.)

Rhythmic layering is well developed (Fig. 254) and trough banding is not uncommon (Fig. 255). The axial plane of the trough-band sets is usually vertical and directed approximately at right angles to the nearby margin, that is, towards the same centre as the layering. The trough banding dips downwards toward the centre of the intrusion, at the same angle as the rest of the layering. The larger trough bands are 10 or 12 m across and some trough-band sets are made up of thirty units, all showing strong gravity stratification. In some cases, the axis of each successive trough is offset from the one beneath, and always in the same direction. Sometimes a higher trough-band set cuts across the one beneath, producing a false-bedding effect.

The eastern syenites consist of a border group, and the eastern layered series forming Kûngnât Peak itself. The broad border group has abundant gneiss inclusions and pegmatites near the margin, while the main mass shows feeble fluxion banding, normally vertical or dipping into the intrusion at steep angles (Fig. 253). This banding is cut by the abundant pegmatites, which are clearly later. A thickness of 800 m of layered rocks is exposed in the eastern layered series, the rhythmic layering and the igneous lamination, where developed, dipping towards the centre of the mass; trough banding is occasionally seen (Upton, 1960, fig. 19 & 20). The basic ring dyke, later than all the syenites, consists of a variety of gabbroic rocks among which are types showing a remarkable similarity to the perpendicular-felspar rock of the Skaergaard intrusion (Upton, 1960, fig. 27).

Confining our attention to the well layered western syenites, which illustrate the chief features of all parts of the intrusion, we find a systematic variation of the minerals, as summarized in Upton's diagram reproduced here (Fig. 256). From the diagram it is clear that cryptic layering is present, the olivine varying from Fo_{20} to Fo_0, the augite from $Ca_{42}Mg_{25}Fe_{33}$ to almost pure hedenbergite, and the alkali felspars from

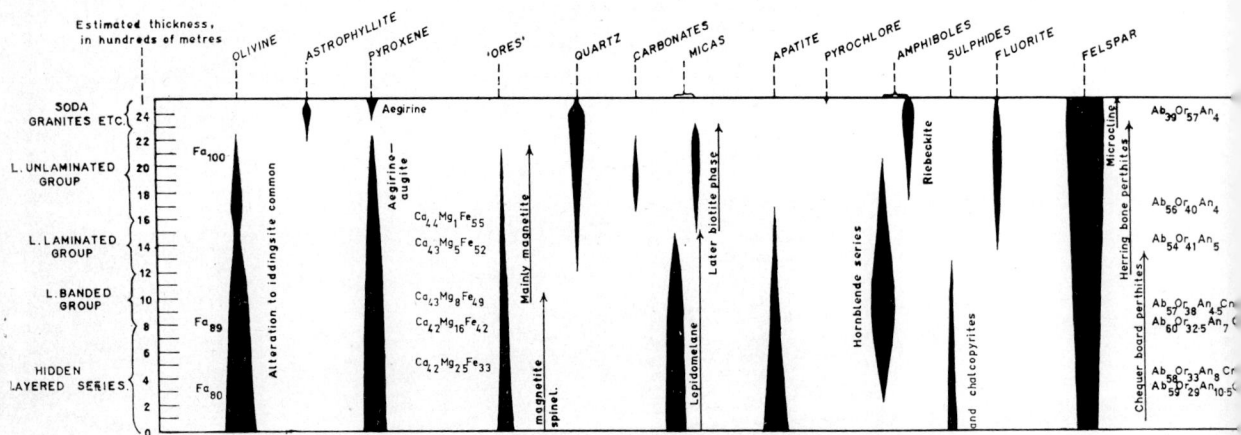

FIG. 256. Modal and cryptic variation in the layered syenites of Kûngnât. Certain assumptions were made by Upton in adding the Hidden Layered Series and granites to the sequence. (From Upton, 1960.)

$Ab_{59}Or_{29}An_{10.5}Cn_{1.5}$ to $Ab_{56}Or_{40}An_4$. Some of the minerals of the rocks, especially the alkali felspar (now perthite), the iron-rich olivine, the iron-rich pyroxene, the titaniferous magnetite and the apatite, often show textural features suggesting that they are cumulus minerals, e.g. the pyroxene and olivine of a mafic layer (Fig. 257).

Upton (1960) considers that crystal settling has taken place in the western syenite and, to some extent, in the eastern syenites, and that the former existence of magmatic currents is indicated by the cross bedding, trough banding, and igneous lamination. Crystal sorting under the influence of currents of variable velocity is considered, in a general way, to be the cause of the rhythmic layering. A decrease in the average vigour

FIG. 257. Micro-drawing of textural relations in the mafic part of a syenitic rhythmic unit of the western lower layered series, Kûngnât. Cumulus fayalite (open stipple) and ferroaugite (close stipple), accompanied by alkali felspar, apatite, biotite and ilmenomagnetite. ×20. (From Upton, 1960.)

of the convection currents is considered to have taken place with time. A later paper by Upton (1961, p. 16) makes the suggestion that in most of the rocks there was a considerable amount of trapped liquid and that they are, in fact, ortho- or mesocumulates. During the crystal fractionation resulting from the settling of early crystals, the rocks remained over-saturated in silica and became somewhat richer in silica as the fractionation proceeded.

The Nunarssuit Complex, described by Harry and Pulvertaft (1963), is much larger than Kûngnât but of about the same age and only 50 km to the south. It shows good rhythmic layering in parts of the extensive quartz syenite, which in many respects is similar to that of Kûngnât. It is worth separate mention as another case of layering in quartz syenites and also because, as pointed out by Ferguson and Pulvertaft (1963, pp. 13–14), erosional structures are frequent. After discussing and illustrating mafic layers, which may or may not show gravity stratification, they describe discordant

Fig. 258. Fine-scale rhythmic layering in the Nunarssuit syenite, with slight discordance shown in the lower part of the photograph. (From Harry & Pulvertaft, 1963.)

Fig. 259. Erosion channel in the Nunarssuit syenite. Regular layers (right-hand side) terminate against the trough structure, their disposition indicating that the scoured channel was at least 3–4 ft deep (note hammer for scale). Within the channel, conditions ultimately gave rise to crystal deposition in regular, almost horizontal layers (top centre). (From Harry & Pulvertaft, 1963.)

structures. In some cases an upper layer cuts across underlying layers in such a way that the discordance must be interpreted as erosion of some of the layered sediment of crystals before deposition of the overlying layer (Fig. 258). In other cases, like that illustrated in Fig. 259, the discordance is on the flank of a trough-band structure. (Note, particularly, regular layering at the right-hand side of the photograph.) The structure is called, by them, an outwash channel produced by erosion. It appears, also, that the trough structure is due to an erosion channel and then, in it, trough-shaped bands were deposited. Occasional erosion effects have been noted in the Skaergaard trough banding but not on this scale. They also comment on the presence of slump structures and slump breccias. From the map presented by Harry and Pulvertaft (1963, pl. 3), the conspicuous layering is apparently near the margin of the syenite intrusion and it is in a similar situation, in the cross-bedded belt, that erosion is most conspicuous in the Skaergaard layered series.

Kangerdlugssuaq Quartz Syenites–Felspathoidal Syenites, East Greenland

The evidence for crystal settling in the Kangerdlugssuaq alkaline intrusion is not so clear as in the quartz syenites of Kûngnât. The intrusion has been investigated intermittently since 1930 by one of the authors (LRW) with W. A. Deer, and re-examined by both of us in 1953, but only a preliminary account has so far been written (Wager, 1965) of which the following is an abridgement.

The intrusion is one of the Tertiary igneous intrusions of East Greenland and is only some 20 km northwest of the Skaergaard intrusion (Fig. 4). It is approximately circular, about 33 km in diameter, and the distribution of the rock types is annular as shown on the map. The outer, nordmarkite ring is in contact on the northeast side with earlier intrusions of layered basic rocks and of fayalite–quartz syenite, the layered basic intrusion (Kaerven) having been studied by one of us (G.M.B. with D. N. Ojha) and described in Chapter XV. On the southwest margin, basic rocks and syenites seem to form part of another early intrusion (the January Nunatak complex) which has been cut across and largely replaced by the Kangerdlugssuaq intrusion. On the east side of the intrusion, various granitic rocks (the Snout Complex) were emplaced subsequently.

The small-scale map (Fig. 260) shows the nordmarkite ring to be 11 km across on the west and narrowing to only 2 km adjacent to the Kaerven intrusion. The outer contact is outward dipping or vertical and the adjacent gneisses of the basement complex are only slightly metamorphosed by the intrusion. The inner rings of pulaskite and the central mass of foyaite are, together, 14 km in diameter and are also somewhat asymmetric. The various rock types grade into each other and the boundaries between the successive rings, as given on the map, mark only the general positions of gradational boundaries.

The bulk of all the rocks consists of perthitic alkali felspars, which become more sodic towards the centre of the intrusion. The earliest and most abundant rock type is the grey to fawn-coloured, fairly coarse-grained *nordmarkite* of the outer ring. The outer part of the nordmarkite ring has about 10% of modal quartz and this decreases steadily inwards to a vanishing point, which is taken as the inner limit of the nordmarkite. Besides the normal alkali felspar there are sporadic, large, dark cryptoperthite crystals in some of the rocks. They contain small inclusions of augite and hornblende that are not the alkaline types properly belonging to the rock. Large cryptoperthites with similar inclusions are seen to have developed as porphyroblasts in many of the altered basalt inclusions and it is virtually certain that the scattered, large cryptoperthites found in many parts of the nordmarkite, particularly in the inner part, have resulted from the complete disintegration of metasomatized basalt inclusions. The chief ferromagnesian mineral of the nordmarkites is an alkali amphibole. There is sometimes subsidiary aegirine and biotite, the latter being noticeably commoner near the inner limit of the nordmarkite ring. Iron-ore, apatite and, rarely, sphene are the chief accessories.

Inwards from the inner limit of the nordmarkite there is a ring of pulaskite which varies steadily in character and composition centripetally. The outermost part, described as the *transitional pulaskite*, is a felspathoid-free syenite and, with the incoming of a trace of nepheline, this passes gradually into the *main pulaskite*. The amount of nepheline increases inwards, and at about 5% nepheline by volume we have drawn the arbitrary limit of the pulaskite; the rock further inwards is a nepheline–sodalite syenite which we have called *foyaite*.

The different rock types shown on the map and used for descriptive purposes are simply arbitrary divisions of a systematically varying assemblage of minerals. Thus the quartz in the nordmarkite becomes gradually reduced in amount until where, say, close scrutiny of a square yard of a glaciated surface reveals only one or two interstitial quartz grains. A little further inwards, none is found and, for a distance of a kilometre or so, the rock is free from both quartz and felspathoids. The incoming of felspathoids is equally gradual ; inwards from the felspathoid-free syenite, interstitial nepheline or its alteration products make their first appearance in sporadic traces, and from there inwards they increase steadily in amount.

The foyaite area is largely ice-covered but the amounts of nepheline and sodalite can be seen to increase inwards, the typical foyaite having 30 or 40% of felspathoids. Sodalite tends to occur as small, well-shaped crystals. Nepheline is in larger units than in the pulaskites and is more idiomorphic in habit. The dominant ferromagnesian mineral is zoned aegirine but melanite, astrophyllite, and alkali amphibole, the latter being rather different in type from that in the rocks of the outer rings, also occur. The foyaites are heterogeneous, due largely to the development of indefinite, pegmatite streaks.

The evidence that the development of the Kangerdlugssuaq intrusion has involved some bottom accumulation of crystals comes from a consideration of certain basalt inclusions and from the platy parallelism of the rather tabular felspars. The basalt inclusions occur in clusters in the quartz nordmarkite and, to a lesser extent, in the transitional pulaskite. They are clearly derived from the thick spread of Thulean basalts which is still preserved to the northwest, north, east, and southwest of the intrusion (see map in Wager, 1947, pl. 6). Although the Kangerdlugssuaq intrusion, at the present level of erosion, is nowhere in contact with the basalts but only with the basement metamorphic complex and other Tertiary intrusions, it is apparent that the intrusion penetrated, by piecemeal stoping, a considerable way into the originally overlying lavas. The inclusions, varying in size from hundreds of metres to small fragments (Fig. 261), tend to occur in zones where basalt is more abundant than the surrounding syenite, although many isolated masses are also found. The zones of inclusions dip inwards at 40 to 50°, dips of some of the better defined zones being shown on Fig. 260. This peculiarity is interpreted as the result of the inclusions sinking and coming to rest on solidified syenite, the temporary top surface of which had an inward dip of about 40 to 50°.

Although the majority of the rocks of the intrusion show no obvious directional structures, strong platy parallelism of tabular felspars is found at certain places,

FIG. 261. Zone of basalt inclusions in the nordmarkites close to the northwest edge of the Kangerdlugssuaq intrusion (as given in Wager, 1965).

2 H

especially in the northeastern sector. This feature, which we call igneous lamination, dips inwards at about 30–60° (see Fig. 260) and thus has a spatial disposition similar to that of the zones of basalt xenoliths. The parallel arrangement of the tabular felspars is believed to have resulted from the deposition of the crystals on the successive, inwardly inclined top surfaces of already solid material, by flowing of magma parallel to those surfaces.

It would seem likely that the transitional junctions between the groups of rocks shown on the map have the same disposition as both the igneous lamination and the zones of basalt inclusions, and they are shown with such a spatial arrangement on the section attached to Fig. 260. From a wider view of the geology of the area, the relationship of the intrusion and its component parts to the surrounding metamorphic complex and overlying basalts is believed to be as shown diagrammatically in Fig. 262. Such an arrangement of the different units of the complex implies solidification of the nordmarkites first, followed by the pulaskites, and ending with the foyaites.

It has been remarked that the basalt xenoliths must have accumulated on inward-dipping surfaces of already solidified material, and that the igneous lamination equally implies a similarly inclined, solid–liquid interface. Since cooling must be due to loss of heat upwards it would be expected, in the absence of convection, that solidification would be from the top downwards and not as indicated by Fig. 262. However, until the time of formation of the main pulaskite, the evidence from the occurrence of basalt blocks shows that the magma was stoping upwards; it would not, at the same time, be solidifying against the roof. The arrangement of the rock types suggests that gentle convection currents existed which aided stoping and, for a time, prevented solidification of magma against the roof.

Evidence for the sequence of solidification of the different parts of the intrusion comes

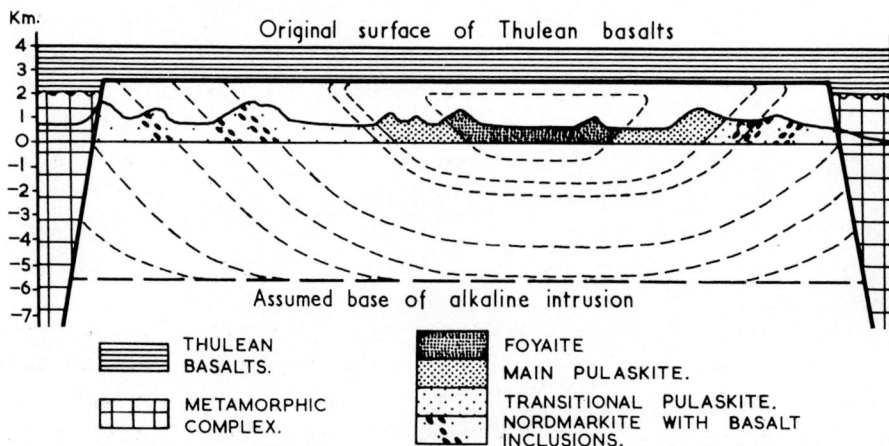

FIG. 262. Hypothetical section of the Kangerdlugssuaq intrusion. (See text and Wager, 1965, where this diagram is discussed.)

not only from the general disposition of the parts but also from the distribution of the aplitic and pegmatitic veins; thus the nordmarkites are cut by nepheline-bearing pegmatites, presumably derived from the pulaskite and foyaite stages of the magma, and so the nordmarkite must have been formed while the more alkaline material was liquid. On the other hand the inner, nepheline-bearing syenites are not intruded by quartz-bearing aplites and pegmatites, presumably because they were not present as solid rocks when the nordmarkite magma was in existence.

The three-dimensional picture of the intrusion makes it clear that the volume of the nordmarkites must be much greater than that of the quartz-free syenites. The relative proportions will depend upon the thickness of the intrusion. If we make the assumption shown in Fig. 262, the total volume would be 6470 km³ and the proportions of the rock types are estimated as approximately:

Nordmarkite	5700 km³ or 88% of the whole
Transitional pulaskite	250 km³ or 3·9% of the whole
Main pulaskite	430 km³ or 6·7% of the whole
Foyaite	90 km³ or 1·4% of the whole

These proportions suggest that the bulk composition of the Kungerdlugssuaq intrusion corresponds to a quartz syenite. The initial magma would have had this composition unless gains or losses of constituents took place during solidification—a possibility mentioned below.

The structure and sequence of events in the Kangerdlugssuaq intrusion imply a continuous development, from an initial quartz syenite magma, of nordmarkites, syenites without quartz or felspathoids, syenites with increasing nepheline, and a small quantity of foyaite. Comparison with other alkaline rocks and with experimental data suggests that this sequence of rocks cannot be solely the result of crystal fractionation. The later part may be, but the change from a quartz-bearing rock to a nepheline-bearing rock seems at present to require addition or subtraction of elements to or from the magma, with fractionation taking place at the same time. Possible causes for the change from a quartz syenite to a felspathoidal syenite are desilicification due to incorporation of basalt inclusions, loss of silica through the roof, or addition of alkalis and alumina from below. This difficult petrogenetic problem is not, however, our present concern, which is to consider the general mechanism by which the intrusion developed, and in particular, to what extent Kangerdlugssuaq can be considered a layered intrusion.

The intrusion has the internal structure of a succession of saucers of decreasing size, packed one within the other; the outer saucer was the first formed and then, successively, those lying within. The intrusion, in fact, shares with layered intrusions a fundamental characteristic: namely, solidification inwards from the sides and upwards from the bottom. The most important characteristic of layered intrusions is the accumulation of crystal fractions at the bottom of the liquid and not where the heat loss must mainly have taken place, at the top. Had the magma solidified without movement of the liquid or the crystals, then the solidification would have taken place largely from above downwards. That this was not the manner of cooling of the Kangerdlugssuaq intrusion seems to imply that crystals, formed near the roof, sank to the floor which, near the margins, sloped inwards. More probably, a convective circulation developed in the intrusion, both sinking and convection currents being involved in the development of the solid rocks at the bottom of the liquid. In this way crystals were available to produce a crystal mush, at the base of the liquid, which had a top surface conforming to a saucer-shape throughout the long period of solidification. The only parts of the saucers to be seen are the edges, which dip inwards at about 45°, the floor not being exposed.

In the case of most intrusions that are claimed to be layered, the chief evidence is of a structural kind such as rhythmic layering and igneous lamination. In many cases there is the additional, highly significant evidence of cryptic layering and often there is useful supporting evidence from the microscopic textures. In the Kangerdlugssuaq intrusion there is practically no rhythmic layering although, rarely, thin layers rich in sphene, alkali pyroxene or melanite are found, which are attributed to accumulation in special abundance of these particular crystals and are considered to be rhythmic layering; in a few places there is, also, strong platy parallelism of the tabular felspars which is regarded as igneous lamination. However, the chief structural feature giving a clue to the form of the successive top surfaces of the crystal accumulate are the zones of basalt inclusions.

If it be accepted from the structural features that the intrusion is layered, then the steadily changing composition of the minerals and rocks must be considered as cryptic layering, although not in a closed system. The microscopic textures, while not providing evidence against the crystal accumulation hypothesis, can at present give little useful support because the proper interpretation of the textures is not clear. In the nordmarkite stage the only mineral considered to have settled is the alkali felspar, and it is likely that this had only a slight tendency to sink, its density probably being close to that of the magma. At the pulaskite stages, in addition to the alkali felspar, it is probable that sphene and alkali pyroxene were cumulus minerals. At the foyaite stage, nepheline, sodalite, and melanite are likely to have been additional cumulus minerals. There was probably much trapped liquid in all the rocks since the cumulus felspars probably sank and accumulated very gently. If, in the nordmarkites and early pulaskites, the only cumulus mineral was alkali felspar, then the quartz, alkali amphibole, alkali pyroxene, and mica must have formed from the intercumulus liquid, and the rocks are orthocumulates. The later pulaskites and foyaites, also, are probably orthocumulates. In the later pulaskites, nepheline has the textural relations of an intercumulus mineral and in some rocks, sodalite seems to have formed from the intercumulus liquid. In the later foyaite, both these minerals appear to be of cumulus status. Strong zoning in the pyroxene of the foyaites also suggests the crystallization of much trapped liquid around settled pyroxene crystals. Where, however, settling of crystals is gentle, as seems to have been the case in the Kangerdlugssuaq intrusion, there is an approach to the cooling conditions usual in granites. In the nordmarkite the intercumulus liquid was, no doubt, large in amount and it appears to have crystallized much as a stationary granite magma would have crystallized.

ILIMAUSSAQ FELSPATHOIDAL SYENITES, SOUTH GREENLAND

The fine description of this agpaitic intrusion by Ussing (1912) made it clear that many of the rocks showed layering on a grand scale. In addition, it appeared that in some places a border group analogous to that since found in the Skaergaard was present, and that the saucer-shaped layering was banked up against it. Ussing concluded that the mechanism involved was crystallization differentiation, combined with gravitative effects. He explained the upper naujaite as due to flotation of the early-formed and light sodalite crystals, combined with a tendency for the sodalite, which is an early mineral, to form '. . . as a result of convection or diffusion where cooling was taking place . . .' (Ussing, 1912, pp. 349–50). The lujavrite, which underlies the naujaite, was thought to be poor in sodalite because this mineral had floated upwards. The banded kakortokites, which are closely related in composition to the lujavrite, are different in structure, being in the form of markedly gravity-stratified sheets (Ussing, 1912, pp. 355–62). The hypothesis which Ussing adopted to explain the banded kakortokites was: first, recurrent crystallization due to outside effects causing repeated reductions and increases in pressure ; and second, sinking of the crystals so formed, under the influence of gravity, at various rates. He pointed out (1912, pp. 361–2), as we have done for the Skaergaard intrusion, that the magma responsible for the Ilimaussaq intrusion was, probably, unusually mobile. In recent years further work has been done on the intrusion, mainly by Sørenson (1958), Hamilton (1964) and Ferguson (1964), and the following account is based on the new investigations, together with Ussing's original monograph.

The Ilimaussaq is intruded into more-or-less horizontal sandstones and lavas which are resting on the Julianehaab granite and other Pre-Gardar basement rocks (Fig. 263). A peralkaline magma stoped its way into these rocks and differentiated to give the present complex, the roof of lavas being still partly preserved. Between the country rocks and the layered rocks there is an impersistent, marginal and upper border group of augite syenite, usually with a little nepheline, best seen round the southern half of the intrusion and under the roof of lavas in the northern part. The marginal parts of the augite syenite were chilled against the country rocks, and then consolidated from the margin inwards. The relationship between the early augite syenite magma of the border group and the agpaitic magma from which the layered rocks were formed is not clear, but it does not seem safe to conclude that the augite syenite represents the original magma of the whole intrusion. After the formation of the border group, a complication occurred in the form of an alkali granite which intruded the roof region and tended to hybridize with the augite syenites of the border group, giving various heterogenous syenites (Hamilton, 1964).

In describing the layered alkaline rocks, the synthesis arrived at by the recent investigators, and summarized diagrammatically by Ferguson (1964) in two figures

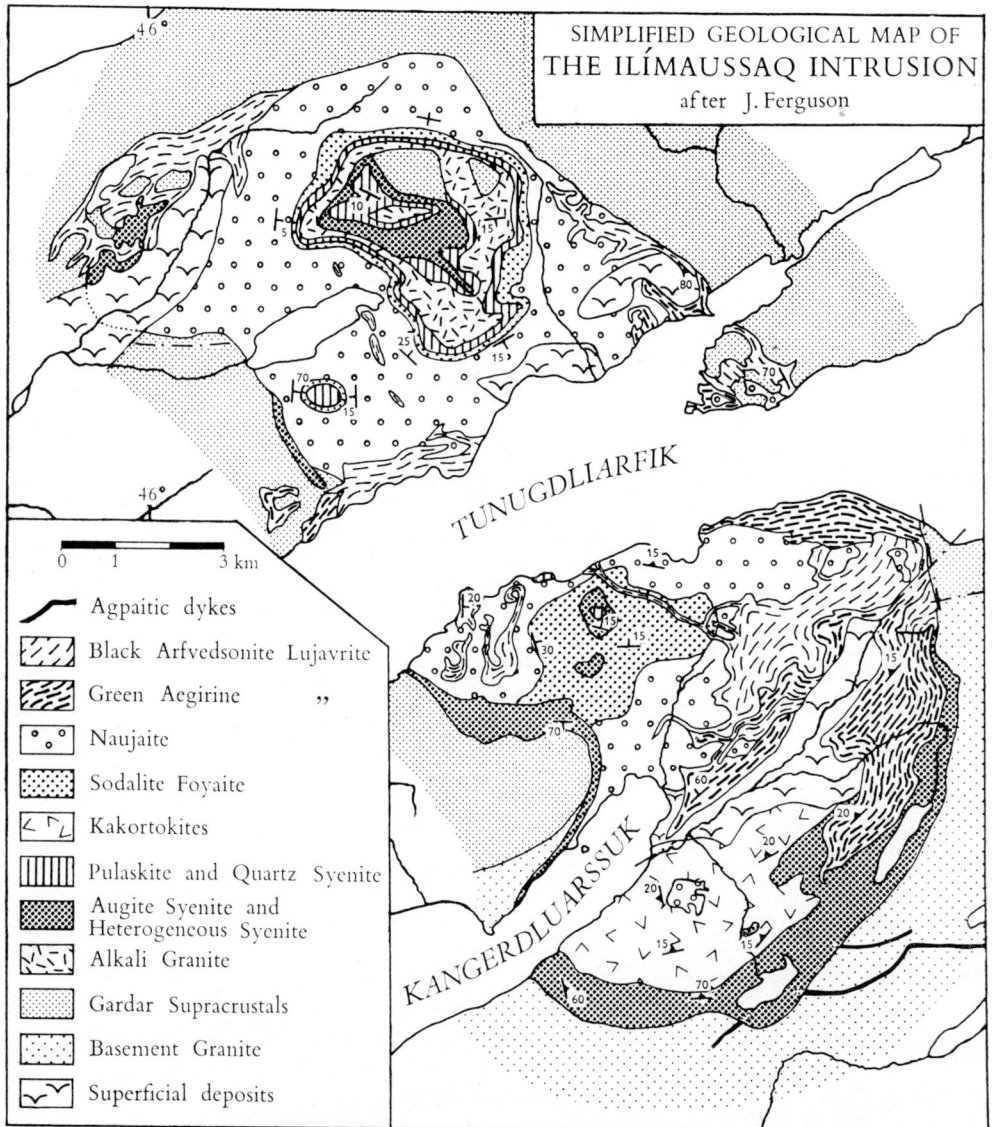

SIMPLIFIED GEOLOGICAL MAP OF
THE ILÍMAUSSAQ INTRUSION
after J. Ferguson

TUNUGDLIARFIK

KANGERDLUARSSUK

0 1 3 km

Agpaitic dykes
Black Arfvedsonite Lujavrite
Green Aegirine   ,,
Naujaite
Sodalite Foyaite
Kakortokites
Pulaskite and Quartz Syenite
Augite Syenite and Heterogeneous Syenite
Alkali Granite
Gardar Supracrustals
Basement Granite
Superficial deposits

FIG. 263. Simplified geological map of the Ilimaussaq intrusion. (Provided by H. Sørenson.)

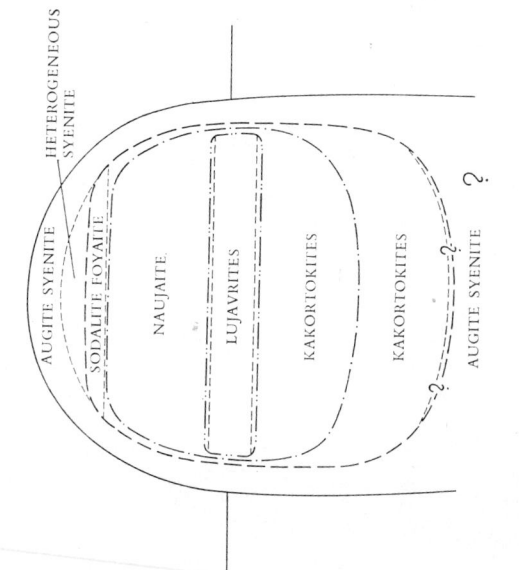

OUTER BORDERS OF SUCCESSIVE
AGPAITE DIFFERENTIATES

Third phase

Second phase

First phase

Boundaries of rock units

Alkali Granite omitted as it is not part of in situ differentiation

SCALE

Vertical ⊢————⊣ 1 km
0

Horizontal ⊢————⊣ 1 km
0

FIG. 265. Simplified version of Fig. 264, showing the
disposition and relative volumes of the rock types.
(From Ferguson, 1964.)

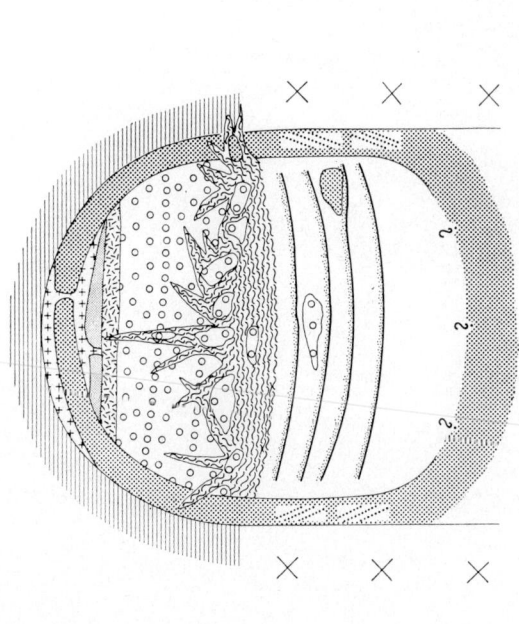

COUNTRY ROCK

Gardar continental series

Julianehåb Granite

INTRUSIVE

Alkali Granite

SCALE

Vertical ⊢————⊣ 1 km
0

Horizontal ⊢————⊣ 1 km
0

BORDERGROUP

Heterogeneous Syenite

Augite Syenite

AGPAITES

Lujavrites

Naujaite

Sodalite Foyalite

Kakortokite

FIG. 264. Broad form and relationships between the
rock types of Ilimaussaq. (From Ferguson, 1964.)

reproduced as Fig. 264 and 265, will be given first. The uppermost layered rock is a sodalite foyaite and this passes downwards into the very coarse-grained naujaite, about a kilometre thick. The naujaite is composed of well-shaped sodalite crystals surrounded by large, poikilitic crystals of aegirine, arfvedsonite, alkali felspar and eudialyte, each poikilitic crystal containing hundreds of sodalite crystals (Fig. 266). The naujaites are considered to have formed concomitantly with the kakortokites. The latter are superbly layered rocks (Fig. 267 & 268) consisting of varying proportions of perthite, arfvedsonite, eudialyte, nepheline and aegirine, and there seems no doubt that they are layered and were formed by the settling of crystals. Finally, a residual lujavrite liquid resembling, in the manner of its origin, the Sandwich Horizon of the Skaergaard intrusion, but of relatively much greater thickness, was left between the downward-growing naujaite and the upward-growing kakortokite. This liquid is partly in the position where it originated,

Fig. 266. Poikilitic texture of naujaite, Ilimaussaq. Sodalite crystals (2–3 mm diameter) are enclosed by large, poikilitic patches of aegirine, arfvedsonite, alkali felspar, and eudialyte. Approx. scale $\times \frac{1}{2}$. (From Ferguson, 1964.)

FIG. 267. Layered kakortokites of Ilimaussaq (foreground). The dark mountain to the left is of sandstone strata and diabase sills, while that to the right of centre (below the snow-covered mountains) is of naujaite overlain by sodalite foyaite. (From Sørenson, 1958.)

FIG. 268. Well-layered kakortokites of Kringlerne, Ilimaussaq. In the centre of the cliff-face a mass of altered naujaite is enclosed by the kakortokites, the layers passing comformably over the inclusion. (From Sørenson, 1958.)

but much of it has been injected into the overlying naujaite and even into the border group and the country rocks. The general picture is thus closely similar to that of the Skaergaard, despite the completely different rock-types involved.

The kakortokites are exposed through a thickness of 1000 m, and the depth to which they extend below this is not known. Ussing (1912) distinguished 25 to 30 rhythmic units, each made up of 'black, red, and white kakortokites'. An average unit consists of black kakortokite with a fairly well defined base, about 1·5 m thick, overlain by red kakortokite, usually thinner than the black variety and sometimes rather inconspicuous, but always there; finally there is white kakortokite, about 12 m thick, after which the rhythms of black, red, and white layers begin again. The units of layering are persistent and do not vary much in thickness. Ferguson and Pulvertaft (1963, p. 16) refer to a large inclusion which has compressed underlying layers to two-thirds of their usual thickness and to a depth of 15 m, which they consider as indicating the depth of the unconsolidated crystal mush at any one time. More significantly, they record that the overlying layers pass conformably over the top of the inclusion without thinning at the crests (see Fig. 268).

There is a marked igneous lamination produced by tabular felspars in the black kakortokite (Upton, 1963, fig. 3), while there is not much parallelism of the felspars in the white variety. Discordance in layering, and trough banding, are not reported by Ferguson and Pulvertaft (1963) but were occasionally observed by Upton, and considered as evidence for the previous presence of some magma currents. In the black, lower part of each unit there are no conspicuous poikilitic crystals and the cumulus minerals, in order of abundance, are believed to be arfvedsonite (black), perthite, nepheline, aegirine, and eudialyte. The dominant cumulus minerals of the red part, in order of abundance, are perthite, eudialyte (red), and nepheline; arfvedsonite and aegirine are usually poikilitic, and are regarded as usually of intercumulus origin. In the white part, perthite, nepheline, and sometimes eudialyte are cumulus minerals, while arfvedsonite, aegirine, and sometimes eudialyte are sub-poikilitic and believed to be intercumulus minerals (Upton, 1961, fig. 16). Zoning is occasionally seen in the eudialyte, but is not otherwise conspicuous.

Cryptic layering in the kakortokites apparently exists in that the $KAlSiO_4$ content of the nepheline diminishes upwards from about 19% to 10%. Perhaps other cumulus minerals will show slight changes in composition due to cryptic layering, but such variations have not so far been reported.

Ussing (1912, p. 356) eliminated successive injections as the cause of the layered kakortokites. He mentions supersaturation as being a possible cause of the sequence of the precipitated minerals (*op. cit.* p. 358) but thinks this unlikely. He then points out that the sequence of black, red, and white kakortokites form an often-repeated unit which could have been produced by separation under gravity, since the black kakortokite (S.G. = 3·1) is characterized by especial abundance of the denser minerals; the red kakortokite (S.G. = 2·9) includes more of the less dense minerals, and the white kakortokite (S.G. = 2·7) is dominantly of alkali felspar and nepheline. He argued that separation under gravity will not, by itself, produce the repetition of units, but that in addition there must have been some repeated outside factor causing recurrent crystallization. He states that it is not improbable that ' a volcanic outburst should influence the conditions of the subjacent magma bodies ', and suggests that the volcanic outburst may have produced periodicity in crystallization by decreasing the pressure, by causing loss of heat, and by causing currents in the magma. He states: '. . . it must be admitted as possible that volcanic outbursts under favourable conditions may leave a permanent record upon the structure of abyssal rocks, and the characteristic feature of such marks must be that they are recurrent '. He believed that to explain the layering, the ' simplest supposition is perhaps that the recurrent layers have originated in consequence of repeated variations in pressure. Each reduction in the pressure may have

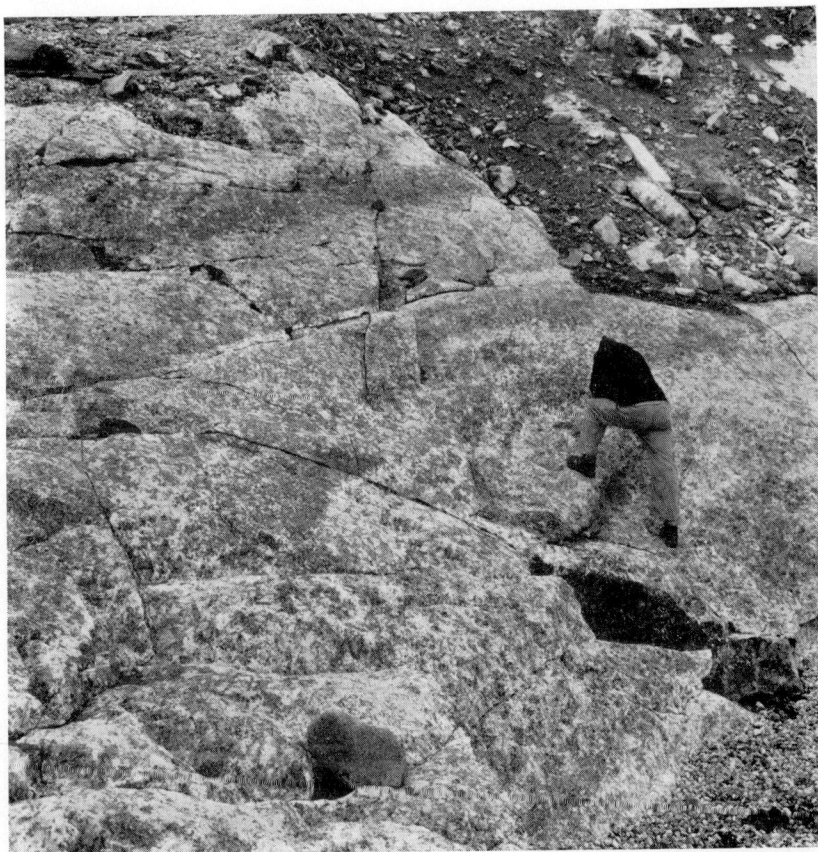

FIG. 269. Rather indistinct layering in the naujaite, Ilimaussaq. The dark layers (rich in aegirine, arfvedsonite, or occasional eudialyte) vary from 60 cm to 1 m in width, alternating with felspar-rich layers. (From Ferguson, 1964.)

caused the dissociation of a certain quantity of volatile matter from the magma, and this process in its turn may have caused the crystallization of a certain quantity of the magma' (*op. cit.* p. 361). With this latter hypothesis, Ferguson (1964, p. 67) records his general agreement. Periodic, slight reductions of water pressure are most likely if the magma reservoir were functioning as the source of lavas erupting periodically from an overlying volcano (as postulated for Rhum, Ch. XII). The construction of detailed models of the processes involved, such as the nucleation and growth of the crystals, the manner of heat loss to permit the growth of the crystals, etc., has not so far been attempted for the Ilimaussaq intrusion. Also the hypothesis of successive convective overturns of the magma, to account for the rhythmic layering, should be explored for this intrusion. Upton (1961, p. 28) has observed some evidence of currents in the magma and it should also be remembered that whether there are currents or not, the crystals have to sink either through stationary or flowing magma. Quite gentle currents may be important for the mechanism and yet not leave evidence so definite as trough banding or washouts in the layering.

FIG. 270. Closer view of the rather indistinct but undoubted layering in the naujaites of Ilimaussaq. In the field, the red, white, and black layers (as in the kakortokites) are more easily discerned. (From Sørenson, 1958.)

The naujaite, which overlies the sandwich layer of lujavrites, shows rather indefinite layering throughout (Fig. 269 & 270). It is caused primarily by variation in the abundance of sodalite which may form up to 80% of the rock although, more usually, about 40%. The sodalite occurs in well-shaped crystals, 2-3 mm across, surrounded by poikilitic patches of aegirine, arfvedsonite, alkali felspar (often up to 10 mm across), and eudialyte (Fig. 266). The sodalite is, apparently, a cumulus mineral and the poikilitic minerals are intercumulus (Hamilton, 1964, fig. 23). The less sodalite-rich layers, according to Ferguson (1964, p. 58), have a non-poikilitic texture and may represent the result of the crystallization of agpaitic magma without addition of cumulus sodalite.

Ussing and most subsequent writers have proposed that many of the sodalite crystals ascended in the magma, forming a flotation cumulate. Hamilton (1964) has shown that some of the sodalite is replacing nepheline, but whether the nepheline also floated upwards is not clear and seems unlikely, since it occurs as settled crystals in the

FIG. 271. Fissile lujavrite contorted around an inclusion of naujaite, Ilimaussaq. (From Ferguson, 1964.)

kakortokites. In places, arfvedsonite and aegirine also seem to be cumulus minerals in the naujaites, but they certainly cannot have floated upwards; as phases denser than the liquid, they may have been carried up by a central, upward convection current, as suggested for certain plagioclases of the Skaergaard upper border group. The foyaites and naujaites, like the Skaergaard upper border group rocks, are pictured as having grown downwards. As in the Skaergaard case, these events were much more complex than those giving rise to the layered series. Contamination, pegmatite formation and, probably, filter-pressing were apparently involved, as shown by Hamilton (1964).

The appearance of the lujavrite, lying between the kakortokites and the naujaites, is dominated by 40% of small, acicular aegirine and arfvedsonite crystals, which are arranged roughly parallel to each other and impart a fissility to the rocks. The other minerals occurring in fairly well-shaped crystals are microcline, albite, nepheline, sodalite, and eudialyte. The lujavrite forms a 200-m sandwich layer, which includes

FIG. 272. Rhythmic layering in the lujavrites, Ilimaussaq. The dark layers are rich in aegirine or arfvedsonite, alternating with felspar-rich layers. Gravity stratification can be discerned in the dark layers. (From Ferguson, 1964.)

abundant blocks of naujaite fallen from above; it also injects the overlying rocks in irregular, and often wide veins. The flow banding, picked out by the acicular aegirines and arfvedsonites, is shown wrapping round a block of naujaite in Fig. 271. The sandwich layer tends to be aegirine lujavrite in the lower part and darker, arfvedsonite lujavrite in the upper, between which is a 40-m zone of alternating layers of the two types, together with some layers relatively rich in felspars (Fig. 272). The mush of crystals forming the lujavrites seems to have had a relatively high water content and remained fairly mobile, in view of the lujavrite vein-formation. The formation of the lujavrites, apart from certain late-stage veining by natrolite and analcite, was the latest episode in the development of the Ilimaussaq intrusion.

So far, there is only a limited amount of evidence for cryptic layering in the settled cumulates of the Ilimaussaq layered series, and none has been recorded in the foyaites and naujaites to confirm the suggestion that they developed by downward crystallization. In the upper border group of the Ilimaussaq intrusion, the most obvious primocryst mineral is sodalite. It does not have a cumulus status in the kakortokites, presumably because all the sodalite primocrysts floated upwards and contributed to the development of the upper border group. In the Skaergaard intrusion, primocrysts were caught in the upper border group by a congelation process and it may be that in the Ilimaussaq intrusion, apart from the sodalite which could have floated, some of the other primocryst minerals such as nepheline, aegirine, and arfvedsonite were occasionally carried up by convection currents and caught near the roof by some sort of congelation process.

GRØNNEDAL-IKA NEPHELINE SYENITES AND CARBONATITE, SOUTH GREENLAND

In this complex, where nepheline syenites are associated with carbonatites, the relationship between the many different rock types is highly complex, partly because there was a number of successive intrusions and partly because the complex has been torn by a succession of transcurrent faults. Some of the rock types of the complex have been known for a long time, but the first thorough mapping and synthesis of the field data into a coherent story are due to Emeleus (1964). The detailed petrological account, however, has not yet been published. Layering is shown by the nepheline syenites which preceded the carbonatites, the latter being flow banded but not layered. The layering of the nepheline syenites is characteristically inclined at about 45°, and Emeleus gives a useful discussion of how such high-angle layering may have developed.

The complex, which is situated about 7 km east of Ivigtut (Emeleus, 1964, pl. 1), stretches for 8 km in a NNW direction. The reconstruction of the intrusion before the faulting (Fig. 273) shows it to have been oval in plan, the long axis being about 6 km in a northwest direction, which indicates the considerable effect of the tear faulting which causes it now to cover 8 km. To simplify the discussion, the reconstructed map of the intrusion will be used.

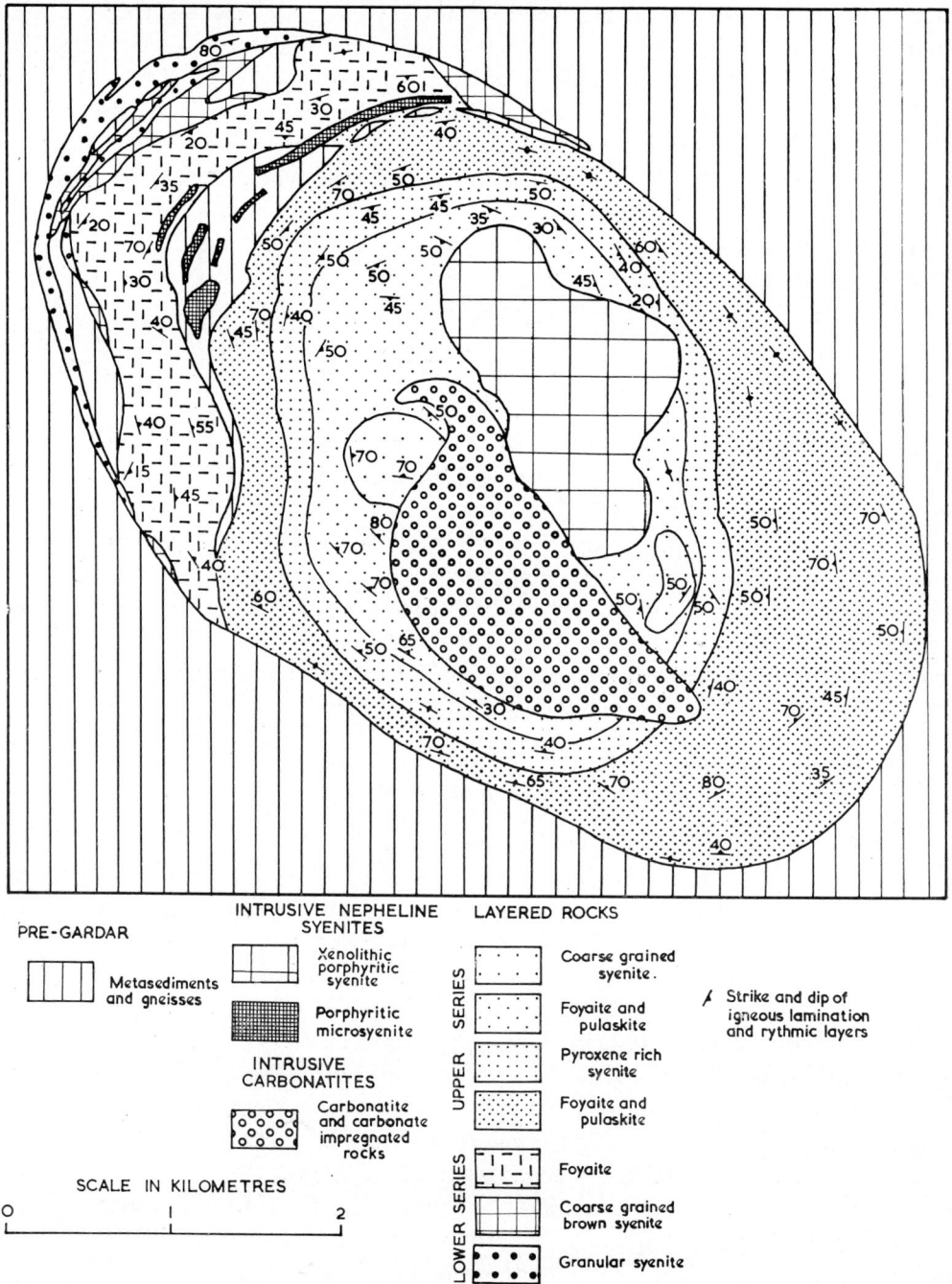

FIG. 273. Geological map of the Grønnedal–Ika complex, reconstructed to represent the form prior to transcurrent faulting. (Provided by C. H. Emeleus.)

The layered nepheline-syenite rocks are divided into a lower and upper series. The division is marked by a thick and rather continuous, sheet-like inclusion of the country rocks (Pre-Gardar gneisses, and metasediments). Most of the syenites show igneous lamination, and rhythmic layering is well developed in the pyroxene-rich syenites of the upper series. The igneous lamination, the result of parallel arrangement of tabular felspars, dips with considerable regularity at between 30 and 60° towards the centre of the intrusion (Fig. 273). A section across the northwestern half of the intrusion is given in Fig. 274, the pattern indicating, diagrammatically, the igneous lamination.

The lower layered series is probably the older of the two. The two outer members (syenites) are without igneous lamination, but the foyaite shows extremely good parallelism of the thin tablets of alkali felspar, usually about 10×7 mm on the flat surface and 1-2 mm thick. The rock also has prisms of nepheline, and poikilitic aegirine, biotite, and cancrinite. Felspar and nepheline are, apparently, cumulus minerals and the others have formed from intercumulus liquid (Fig. 275). Sharp variations in mineral proportions take place across the strike, giving rather inconspicuous rhythmic layering.

The upper layered series is of foyaites and pulaskites, and perthitic alkali felspar and nepheline are the chief cumulus minerals. In the pyroxene-rich syenite, the aegirine–augite occurs as well formed crystals and is, presumably, a cumulus mineral (Upton, 1961, fig. 15). Changes in the proportions of the cumulus pyroxene crystals gives layering, as shown in Fig. 276 (see, also, Emeleus, 1964, fig. 4). Extensive, poikilitic alkali amphibole is found, especially in the more melanocratic layers, and has formed from the intercumulus liquid.

Microporphyritic syenite occurs as intrusive sheets in the lower series, and a con-

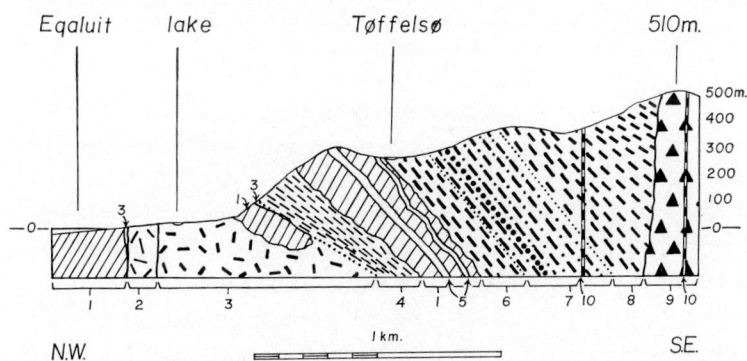

FIG. 274. Cross-section of Grønnedal–Ika, from the northwest corner to the patch of xenolithic porphyritic syenite (see Fig. 273). Disposition of igneous lamination and layering shown diagrammatically. 1, gneiss; 2, granular syenite; 3, coarse-grained brown syenite; 4, foyaite of the Lower Series; 5, microporphyritic syenite; 6, foyaite of the Upper Series; 7, pyroxene-rich syenite of the Upper Series; 8, foyaite and pulaskite of the Upper Series; 9, xenolithic porphyritic syenite; 10, alkaline dykes. (From Emeleus, 1964.)

2 I

siderable intrusion of xenolithic, porphyritic syenite occupies a northeast–central position.   Finally, intrusive carbonatite magma has invaded the complex, mainly in a southwest–central area, and has often altered the adjacent syenite to give carbonate-impregnated rocks.   The carbonatite is flow banded, and this is seen best on weathered surfaces (Emeleus, 1964, fig. 12).

The whole group of layered rocks is believed by Emeleus to be the result of crystal settling, and the age of formation, from the stratigraphy, must decrease upwards; apparently no field evidence conflicts with this view.   The coarse brown syenite is believed to be the first of the syenites to have formed and the granular syenite is seen

Fig. 275.   Micro-drawing of laminated nepheline syenite, Grønnedal–Ika, showing cumulus crystals of alkali felspar and of nepheline (centre, altered partly to cancrinite).   ×15.   (From Upton, 1961.)

in places to have intruded it.   The microporphyritic and xenolithic syenites, which are seen cutting the upper series, are late in the sequence of events but pre-date the carbonatites.

The dip of the igneous lamination and rhythmic layering is generally between 30 and 60°.   Since the direction of dip boxes the compass it is argued that the high dips are not due to any regional tilting of the area.   They are, in fact, believed to be the original dips resulting from crystal settling and Emeleus (1964) discusses the implication of this.   In most basic layered intrusions the evidence suggests that layers due to crystal settling were close to the horizontal at the time of formation.   This is true for the central part of the Skaergaard intrusion, although in the cross-bedded belt, towards the margin of the layered series, the dips reached 10 or 20°; even so, slumping is rare.   The layering in the eastern part of Rhum seems not to have exceeded 10 or 15° and slumping, attributed to instability of the crystal mush, occurs sporadically (Brown, 1956).   On the other hand, in the Skye ultrabasic–basic layered intrusion the layering dips at 45°; there is some doubt, however, whether this is the original dip or whether it has been induced by central subsidence (see Ch. XV).

In the highly alkaline intrusion of Ilimaussaq, the lower layered rocks are still quasi-horizontal and the same is true of the Lovozero complex (Ch. XVII). Among the quartz syenites showing layering, such as at Kûngnât and Nunarssuit, the dip of the layering is usually around 45° or more. The quartz syenite–felspathoidal syenite intrusion of Kangerdlugssuaq also provides evidence that the surface of accumulation of the crystals had a dip of 45°, and the same is true of the nepheline syenites of the Grønnedal-Ika complex.

Emeleus (1964) emphasizes the factors liable to affect the stability of steeply inclined layers of settled crystals, i.e. (1) the viscosity of the magma, (2) the density differences between the magma and the cumulus phases, and (3) the rate of deposition of the cumulus crystals (that is, the rapidity with which the intercumulus liquid crystallized

FIG. 276. Layered and laminated syenite from the pyroxene-rich syenites of the Upper Series, Grønnedal–Ika. (From Emeleus, 1964.)

to cement the crystal mush). Considering these factors *seriatim*, we would agree that where it might be expected that the magma had a relatively low viscosity, as in the case of basic and highly alkaline magmas, the layering tends to be almost horizontal. We suspect, however, that the viscosity of the magma is only important in determining whether or not vigorous convection takes place. If there were no convection currents but only direct sinking of crystals to the floor (probably a rare condition in major intrusions), the settled crystals should still lie parallel to the floor, whether this be horizontal or inclined, except that above a certain angle, slumping would occur and gradually produce a less-inclined floor. If, however, convection currents were established they would tend to be more vigorous in the more fluid magmas, and such currents would be able to flow horizontally across considerable stretches of the floor before turning upwards at the centre of the convection cell. Probably sluggish currents, in magmas of high viscosity, would not have long distances of horizontal flow and there would be a lesser development of semi-horizontal layering.

The density difference between crystal mush and liquid is, presumably, a highly significant factor in limiting the angle of slope of a pile of cumulus crystals. Clearly, if the crystals had the same density as the liquid there would be no tendency for the pile to slump, but with cumulus crystals of greater density than the liquid, slumping of masses of crystals with their enclosed liquid would take place if the slope exceeded a critical value. No doubt, the shape of the crystals and degree of compaction of the sediment are other contributory factors. Unfortunately, little is known of the density of natural magmas. However, it would seem likely that a quartz-syenite magma, such as that giving rise to the Kangerdlugssuaq intrusion, would be only slightly less dense than alkali felspar, the chief cumulus mineral during most of the period of solidification, and it is perhaps for this reason that the angle of slope of the top surface of the pile of cumulus crystals usually seems to attain 45° or more before becoming unstable.

Emeleus (1964) points out that if crystallization of the intercumulus liquid more-or-less kept pace with crystal accumulation, as in adcumulates or heteradcumulates, this would greatly increase the angle of stability of the interface between the pile of crystals and the overlying liquid. In Rhum, one of us (Brown, 1956) considered this relationship between slumping and the rate of solidification of the crystal mush, and observed the tendency for the felspathic adcumulates to break into blocks, rather than to slump. At present it is not known whether the conditions of crystallization of the syenites in the Kangerdlugssuaq and Grønnedal-Ika intrusions approximated to those of adcumulates or heteradcumulates.

In considering the formation of the layered syenites of the Grønnedal-Ika complex, Emeleus suggests that interrupted convection currents may have occurred during the formation of the rhythmically layered, pyroxene-rich syenites, and that a steady convection current, which we would suggest was relatively slow, existed during the formation of the main part of the layered rocks. No evidence has so far been given of any cryptic variation, but this may be forthcoming from detailed mineralogical studies. Emeleus does, however, suggest that the extensive raft of country rocks is due to subsidence of overlying gneisses and schist material, which may have been initiated by a new injection of magma into the chamber where the layered intrusion was forming. If this happened, at least once, the cryptic layering is unlikely to be regular from bottom to top of the layered series and may exhibit one or more major fluctuations.

GENERALIZATIONS

Among the vast amount of granite that has been examined by petrologists there is, so far, practically no evidence for any settling of crystals. Apart from slight variations in An–Ab–Or–Qz ratios and potential ferromagnesian contents, granite magmas tend to be close to a ternary minimum composition. Thus on cooling there is little flexibility for the order in which the minerals begin crystallization, most of the chief minerals being capable of crystallization within a narrow temperature range, from the onset of crystallization. The available evidence suggests that granite magmas are highly viscous and that any sinking or floating of crystals would be slow or would not take place at all. Thus a mass of granite magma, not under tectonic stresses causing flow, would probably be stationary and if the crystals which form in it are also stationary, the cooling of the mass would be by simple conduction. In such cases the imaginary surface, marking the condition required for the beginning of crystallization, would move steadily inwards as cooling proceeded. Crystal phases would nucleate and grow as heat was conducted away and, by hypothesis, without movement of the crystals or the liquid. The production of nuclei of the various minerals of granite must be an extremely uniform process which resulted in an even distribution of the nuclei of the various minerals, since the resulting texture of the rock is relatively uniform.

In a few intrusions, however, sinking of crystals in a granite magma seems to have taken place. All the cases known to the authors have been mentioned in the first section of this chapter, and the fact that there are so few is, presumably, an indication of the rarity of the phenomenon. The granites which show slight evidence of crystal sinking also show structures suggesting flow, probably of the convection-current kind (see Fig. 250). It is, perhaps, likely that convection currents, involving cooler and warmer parts of the granite magma body, may be a more frequent factor than sinking of crystals, providing that the viscosity is reduced by the presence of water, fluorine, or other volatile constituents. The effects of slow convection currents would not obviously be different from flow movements connected with the emplacement of the granite. In a theoretical study of the possibilities of convection in granite magma, Shaw (1965) gives it as his opinion that ' natural convection must be reckoned with in any large granite pluton, at any rate in the early stages of crystallization '. It may be that the different types of granite, often with roughly annular outcrops as mapped in a granitic complex, may not result from separate intrusions but from convective over-turns bringing slightly different parts of a magma body to levels at which cooling results in solidification. Although clear evidence of the sinking of crystals and of convective flow in granite magmas is rare, certain features of granitic complexes may ultimately prove to be due to the variable effects of these processes.

Sinking of crystals in a quartz-syenite magma have evidently taken place in the Kûngnât intrusion described above. Geologists working in South Greenland have also found layering, presumably due to crystal settling, in the Nunarssuit pyroxene–fayalite syenite (Harry & Pulvertaft, 1963) and in the Klokken larvikite intrusion (Ferguson & Pulvertaft, 1963, p. 12). The Kangerdlugssuaq intrusion has been treated separately from Kûngnât because it passes from a quartz syenite to a nepheline–sodalite syenite, without any apparent break in some uniform process of genesis. Some degree of bottom accumulation of crystals and, probably, some convective circulation seem to be indicated by the structural and compositional features of this intrusion.

The felspathoidal rocks of the Ilimaussaq intrusion provide good evidence of crystal sinking and, probably, of fairly vigorous convection which maintained reasonable homogeneity in the successive residual magmas. Presumably, the original magmas of this and, probably the Lovozero intrusion (Ch. XVII) had abundant water, chlorine, and other volatile components which resulted in their having relatively low viscosities, thus allowing the development of convection currents and layering comparable with those of basic layered intrusions. Nepheline syenites associated with carbonatites, such as Spitzkop and Grønnedal-Ika, often seem to show little order in the structural arrangement of the various rock types but in the Grønnedal-Ika complex, at least, there was crystal accumulation giving rise to layering in the main nepheline-syenite intrusion, which immediately preceded the injection of carbonatite magma.

CHAPTER XVII

ADDITIONAL LIST OF LAYERED OR FRACTIONATED INTRUSIONS

BRITISH TERTIARY EXAMPLES

a) *Ben Buie, Isle of Mull, Scotland*

Layered eucrites in the Ben Buie intrusion, emplaced during the sequence of the Early Caldera in Mull, were first described briefly by Bailey *et al.* (1924). The intrusion is situated on the southwestern periphery of the Tertiary central igneous complex (Fig. 133), and is exposed over about $6\frac{1}{2}$ square miles. The form and mode of explacement have been discussed by Lobjoit (1959). A petrological study by the same author has not yet been published, but the thesis account (Lobjoit, 1957) provides a detailed description of the layered structures, textures, and mineralogy. Analyses of chilled marginal gabbros show the parental magma to be close to an alumina-rich tholeiitic basalt. The exposed layered series consists of igneous cumulates in which rhythmic layering, cryptic layering, and slump structures are developed. Broadly, the rock types range from peridotites and allivalites (cumulus plagioclase about An_{90}) through eucrites to gabbros. The discovery of layered ultrabasic rocks in Mull emphasizes the overall similarities between the Hebridean Tertiary volcanic centres, those of Rhum, Skye, and Mull being similar in the rock associations and differing, chiefly, only in the relative proportions of each rock type.

b) *Ardnamurchan, Scotland*

Ardnamurchan is a Tertiary volcanic centre on a peninsula of Northwest Scotland (Fig. 133), similar in size and character to those centres which, on the nearby islands of Skye, Rhum, Mull, and St. Kilda, contain layered basic igneous rocks and are described in this book. Three small, separate centres at Ardnamurchan are defined by the disposition of cone-sheets and ring-dykes (Richey, 1932, 1948). Centre 2 includes an arcuate outcrop of hypersthene gabbros (Wells, 1954) in which rhythmic layering is well developed, some lower layers being troctolitic. Richey believed Centre 3 to consist of a series of concentric ring-dykes, four being eucritic of which the largest is called the Great Eucrite. A detailed study by Bradshaw (1961) has shown the eucrites to belong to a single intrusion cut by later, relatively narrow ring-dykes. Rhythmic layering is present but weakly developed, while regular cryptic layering could not be detected. The presence of 'calcic-phase phenocrysts' of plagioclase (cf. Skye, Ch. XV) throughout the layered sequence, together with the absence of cryptic layering, suggests periodic influxes of a parental magma containing bytownite phenocrysts (i.e. the Porphyritic Central-Type Magma of Bailey et al., 1924). Two pyroxenes crystallized in both Centre 2 and Centre 3 basic intrusions, indicating tholeiitic affinities. The form of the Centre 3 eucrite is still in doubt, but the shape prior to dissection by later, narrow ring-dykes, and the dip of the layering, suggest that it may have been a large lopolith or funnel-intrusion rather than a ring-dyke or a series of ring-dykes. It would be unfortunate if this now-classic example of a ring-dyke complex turned out to be of the form suggested here, but to the best of our knowledge the evidence from the contacts, also, are not convincingly in favour of a ring-dyke form. Also, ring-dyke fractures require the foundering of crustal blocks into an underlying magma chamber, and we would expect that where this is appreciable, as for very thick ring-dykes, such a process would be associated with acid, rather than basic intrusions.

c) *Carlingford, Louth, Eire*

The Carlingford complex, on the east coast of Ireland (3 miles south of the border between the Irish Free State and Northern Ireland), is a Tertiary basic complex similar to those in Northwest Scotland, and is adjacent to the Tertiary granitic ring-dyke complexes of Slieve Gullion and the Mourne Mountains. A gravity survey (Cook & Murphy, 1952) has shown the likely presence of a large basic magma reservoir beneath the Carlingford and Slieve Gullion complexes, and the layered basic rocks of Carlingford have been described by Le Bas (1960). The layered complex, believed to be a lopolith, consists of four major rhythmic units, in each of which the ratio of cumulus olivine to cumulus plagioclase decreases upwards. Cryptic variation in the compositions of olivines, plagioclases, augites, and orthopyroxenes is confined to each major unit, within which the compositional variations are quite large, considering that the units vary only between 200 ft and 500 ft in thickness. The deposition of each unit is probably attributable to a fresh pulse of parental magma, Carlingford providing a good example of the repetition in cryptic layering rarely found in those other intrusions where, presumably, the fresh magma pulses comprised a smaller proportion of the total bulk in the magma chamber. Le Bas suggests that the parental magma was an alumina-rich tholeiitic basalt, and it is interesting that a similar magma-type has been proposed for the other basic layered intrusions of the Thulean Tertiary Province (Skaergaard, Rhum, Skye, Mull, and Ardnamurchan).

d) *St. Kilda, North Atlantic*

A mass of eucrite, which is part of an annular mass that appears to be the earliest intrusion now exposed of the St. Kilda Tertiary centre, was described by Cockburn (1935). It has been visited by one of us (L.R.W.) and has subsequently been mapped and described by Harding (1961). Rhythmic layering is not conspicuous, igneous lamination is only moderately developed, and cryptic variation in the plagioclases has not been observed with certainty, in the main mass. Nevertheless, textural features of the rocks suggest that sedimentation of the crystals has occurred, and if the large rafts of gabbroic rocks found embedded in later, more acid intrusions form an upper part of the eucrite mass, it would seem that here there is preserved part of a layered intrusion of the kind better formed in Skye.

OTHER EUROPEAN EXAMPLES

a) *Lizard, Cornwall, England*

The Lizard complex, about 30 square miles in extent, is an ultramafic mass in an orogenic environment. In a recent account, Green (1964) shows it to be similar in many respects to the Tinaquillo peridotite in Venezuela (MacKenzie, 1950) and the Mount Albert ultramafic complex in Quebec (Smith & MacGregor, 1960). A primary mineral assemblage of olivine, aluminous enstatite, aluminous diopside, and green aluminous spinel has, in part, recrystallized to olivine, enstatite and diopside with normal alumina contents, plagioclase, and brown chromite. Green attributes the banding to recrystallization, and envisages the complex as formed from a peridotite magma or hot crystalline diapir, rising from the mantle during the development of non-hydrostatic, stress-fields under orogenic conditions of major folding and thrusting. The complex is included here because others (e.g. Challis, 1965) disagree and feel that 'Alpine-type' peridotites (Thayer, 1960) could have formed by crystal accumulation from basic magmas. The present authors have observed layering, in the Lizard peridotite, suggestive of crystal sorting and settling, and do not agree that the olivines (Fo_{90}) of the Lizard (Green, 1964, fig. 6) indicate that they, together with those of peridotite nodules (Fo_{89-92}), need have different origins from those formed by crystal accumulation in basic magmas. Several layered intrusions (e.g. Great Dyke and Bay of Islands, Ch. XV) contain more magnesian olivines. In addition, a troctolite intrusion occurs in the Lizard area and this suggests further analogy with the intrusions mentioned, and with the argument developed by Challis (see p. 525) that under orogenic conditions, the ultramafic and basic components may become separated through faulting. However, certain aspects of the Lizard mineralogy, such as the highly aluminous pyroxenes, are unusual for basaltic differentiates and at present the origin of the complex must remain in some doubt.

b) *Dawros, Connemara, Eire*

This small ultramafic intrusion, $\frac{1}{2} \times \frac{1}{4}$ mile in size, has been described in detail by Rothstein (1957). It provides a good example of an intrusion in which, despite the subsequent dynamo-thermal metamorphism that has recrystallized part of the rocks, primary layered structures and textures can still be recognized. In many respects it is comparable with the Lizard complex where, as we have remarked, primary layered structures appear to be preserved in a few places, also observed in the Lizard by Rothstein (1957, p. 3). At Dawros, approximately 500 ft of layered cumulates are found, the general sequence of cumulus phases being as follows: olivine; olivine+orthopyroxene; olivine+chrome-spinel; olivine+diopsidic augite. The phase relations are discussed in a separate paper by Rothstein (1961). Fine-scale rhythmic layering results in rocks consisting of only one of the cumulus phases available at a particular horizon, while a slight but significant cryptic variation was detected. Thus the Mg:Fe ratio of the orthopyroxenes changes upwards from 10·6 to 6·4, and of the augites from 9·8 to 6·5 (Rothstein, 1958). No plagioclase is found in these rocks, which resemble more the ultramafic layers in intrusions such as the Great Dyke and the Bay of Islands, rather than the ultrabasic layers of Rhum, Bushveld, and Stillwater. A gabbro found associated with the Dawros intrusion is believed by Leake (1964) to be transitional into the ultramafic rocks, but Rothstein's belief (1964) that it is an independent, if genetically related intrusion is more in accord with similar occurrences in the Lizard complex and the Bay of Islands. The subsequent recrystallization of some of the Dawros rocks is similar to that observed in the Lizard, resulting in the formation of cataclastic textures and in recrystallization.

c) *Northern Norway*

Ultramafic and mafic layered rocks outcrop over a considerable area in the province of Finnmark, and particularly on the islands of Seiland, Stjernöy, and Söröy. A detailed account of a large complex with its focus on Stjernöy (Oosterom, 1963) serves to illustrate the general relationships of the layered rocks in this region of Norway, and provides references to earlier literature on the region. Rhythmic layering is well developed and generally takes the form of alternations of leucocratic and melanocratic layers with sharp boundaries, there being few cases of a gradation between adjacent layers. The overall sequence of layered rocks is from peridotite, through olivine melagabbro, to gabbro. The cumulus plagioclases vary in composition from about An_{90} to An_{70}, and the olivines from about Fo_{90} to Fo_{75}. However, there is no systematic variation in compositions with structural height, and the pyroxenes fail to show iron enrichment. At present, it is difficult to reconcile the regular layered structures with the anomalous compositions of certain mineral phases, either in relation to position within the layered sequence or in relation to coexisting phases. Oosterom presents several criteria for distinguishing between the layered igneous rocks and the banded gneisses with which they are intimately associated. However, his hypothesis for '. . . the emplacement of the layered gabbro and ultramafite on Stjernöy involving differential anatexis with local palingenesis in a deep-seated zone of the Caledonian orogen ', intended to explain the origin of the parental basic magma, adds a further complication to the understanding of the environment in which the layered rocks developed.

d) *South Greenland*

Igneous layering is developed in nearly all the larger intrusions of the Gardar Province of South Greenland, in rocks ranging from gabbro to granite in composition. This remarkable variety of layered intrusions is described in general terms by Ferguson and Pulvertaft (1963), with references to the work completed or in progress. The greater proportion of the intrusions are syenitic, and examples, together with reference to layering in the granites, have been described in Chapter XVI.

Layered gabbros occur in the northwestern part of the Nunarssuit complex, the central and greater part of the complex being of layered pyroxene–fayalite syenites and the layered Helene granite (Harry & Pulvertaft, 1963). At present, the only published information (Ferguson & Pulvertaft, 1963, p. 11) indicates that rhythmic layering and igneous lamination are well developed in the gabbroic rocks, layers of pyroxene adcumulates and olivine adcumulates alternating with plagioclase-rich layers.

The remarkable, giant dykes of Tugtutoq (Upton, 1961, 1962, 1965), west of the main Ilimaussaq intrusion, are up to 700 m wide. Within the greater part of the dykes, synformal layering is very well developed and is attributed to crystal accumulation from a parental, alkali olivine-basalt magma. The margins of the dykes crystallized under tranquil conditions, and in some cases there are vertical zones of fluxion banding and of perpendicular-felspar rock in which elongate, curved plagioclases, standing perpendicular to the cooling walls, have their convex surfaces uppermost (Upton, 1962, p. 22). These marginal relations are reminiscent of those found in the tranquil and banded divisions of the Skaergaard marginal border group (Ch. V). The central parts of the dykes are rhythmically layered, the dip decreasing, in general, from about 45° near the marginal zones to near-horizontal in the centres. The mafic zones are well laminated and rich in cumulus olivine and ilmenomagnetite, with subsidiary apatite and plagioclase, while augite, biotite, and microperthite occur as intercumulus phases, indicating that the rocks are chiefly orthocumulates. Layers of average rock, free from igneous lamination, alternate with the mafic layers and are believed to have accumulated under still conditions, in contrast to the conditions of current activity responsible for the winnowing and lamination of the mafic layers. Here again, an analogy exists with the Skaergaard in which currents of variable velocity are believed responsible for the alternation between well-sorted and average layers (Ch. VIII). The Tugtutoq dykes are remarkable in the interesting structures exhibited for intrusions of this type, and it is to be expected that in view of their limited size and excellent exposure, further studies will contribute much to the understanding of cooling and crystallization processes in igneous intrusions. The elongate shape of the dykes would preclude the development of a single, symmetrical convection system as in the Skaergaard, and Upton has recognized, from the layered structures, the previous existence of several independent convection cells. A similar giant dyke, 400 m wide, occurs on the northeastern margin of the Nunarssuit intrusion and is called the Eqaloqarfia dyke (Pulvertaft, 1965). Here, a perpendicular-felspar layer occurs near the floor of well-layered trough structures, which characterize the lower exposed part of the dyke and are even more impressive than those at Tugtutoq.   The layering is due chiefly to alternations of olivine-rich cumulates and ' felspar-phyric gabbro ', the absence of igneous lamination, graded layering, and textural evidence of tight packing of cumulus crystals together leading Pulvertaft to propose that crystallization (from a tranquil magma) was chiefly from the floor upwards, with only occasional, preferential sinking of olivines.

Similar macro-dykes have been observed in central Sweden (Krokström, 1930, 1936), South-West Africa (Simpson & Otto, 1960) and East Greenland (Douglas, 1964), but the layered structures have not yet been described in detail.   The macro-dykes as a whole are broadly analagous to the Great Dyke of Southern Rhodesia, except that they are much smaller in size, show marginal contacts, and appear to be true vertical-sided dykes rather than parts of larger intrusions preserved in a graben.

e) Troodos, Cyprus

This large plutonic complex, described by Gass and Masson-Smith (1963), occupies an area of about 200 square miles in the region of Mt. Olympus in the Troodos range, southwest of Nicosia.   It is composed of a suite of rocks ranging in composition from dunite and peridotite, through pyroxenite and olivine-gabbro, to gabbro and grano-phyre, all of which are thought to be differentiates of one parent magma.   The gradations in composition occur from the core of the complex outwards, and layering or banding dips at high angles, approaching the vertical.   Hence, if this is a layered intrusion it has been tilted and domed.   Gass and Masson-Smith view the mass as emplaced by under-thrusting during the Alpine orogeny, but have not considered the relationship of the present form to that of the original, differentiated intrusion.   They consider that the suite originated from a magma generated by partial fusion of mantle material which, from the gravity data, appears to form a huge, near-surface slice, at least 7 miles thick, beneath Cyprus.   The abundant pillow-lavas, which overly the plutonic complex unconformably, contain olivine phenocrysts of composition Fo_{90-92} (Gass, 1958) in a groundmass or glass of basaltic composition.   Gass and Masson-Smith suggest that the basalt lavas were derived from the mantle by partial fusion, the olivines being of mantle composition.   They consider that the plutonic rocks originated from a peridotitic magma, consisting of a crystal mush of olivines and interstitial basaltic liquid, which was derived, also, from the mantle material.   This material seems the most likely source for the interesting suite of rocks, although the character of the magma from which the plutonic rocks formed by crystal differentiation need not necessarily be ultrabasic. For example, Gass and Masson-Smith (1963, p. 433) consider that the peridotite olivines (Fo_{90-92}) are too magnesian to be in equilibrium with basaltic magma, yet some of the basaltic pillow-lavas provide evidence that this is not necessarily so.   In fact, in view of what has been said elsewhere in this book about the problems of associating very magnesian olivines with basaltic magmas (e.g. Great Dyke, Bay of Islands, and Lizard), the Troodos lavas may be significant in showing that equilibrium is possible.

North American Examples

a) *Maine, U.S.A.*

Two occurrences of layered basic intrusions in Maine have been commented upon, briefly, in abstracts of papers read at the Annual Meetings of the Geological Society of America.

The Moxie pluton in west-central Maine (Visher, 1961, p. 165A) measures about 25×5 miles and is probably a funnel-shaped intrusion. Dunite, troctolite, olivine norite, anorthosite, and ferriferous olivine-norite are interlayered, and the mineral compositions show a wide range: orthopyroxene, En_{91} to En_{34}; olivine, Fo_{90} to Fo_{17}; and plagioclase, An_{90} to An_{56}. The nature of the rhythmic and cryptic layering were not discussed in this account.

A gabbroic body in the Pleasant Bay area of the southwestern coastal region has been described briefly by Bickford (1962, p. 17A). Rhythmic layering and igneous lamination have developed, and the sequence of rock types, although complicated by extensive faulting, is suggestive of cryptic layering. The rock types include olivine gabbro, gabbro, quartz gabbro, and quartz ferrogabbro, within which sequence the plagioclase compositions become progressively more sodic. Initial soda and silica enrichment, followed in the later stages by iron enrichment, is attributed to the existence of different sets of conditions, in regard to total composition and Po_2 during the formation of the olivine gabbro–quartz gabbro series (constant or increasing Po_2) and the quartz ferrogabbros (constant total composition).

b) *Baltimore, Maryland, U.S.A.*

This is a funnel-shaped intrusion, about 50 square miles in extent, in which rhythmic layering is well developed (Herz, 1951). The mineral variation within the altered gabbros is not continuous with that in the ultramafic layered series, and it is envisaged that a time interval separated the intrusion of magmas responsible for the two groups. In the ultramafic series, the mineral compositional range is: olivine, Fo_{86-90}; orthopyroxene, En_{83-90}; and plagioclase, An_{78-86}. In the gabbroic series, the range is: olivine, Fo_{68-80}; orthopyroxene, En_{77-80}; and plagioclase, An_{78-83}. (A more sodic plagioclase, An_{64-70}, is probably an intercumulus phase in certain rocks rather than, as suggested by Herz, a separate, coexisting phase.) The compositional variation within each series is not correlated with structural height within the layered sequence, but as the rocks are well layered and are believed by Herz to be crystal accumulates from basic magma, the compositional variation within the Baltimore intrusion can probably be taken to imply significant cryptic variation. The late-stage trend is towards rocks with primary hornblende (bojites) and quartz, and in view of the absence of extreme iron-enrichment and the evidence for appreciable late-stage hydrothermal alteration the magma was, presumably, relatively hydrous. Emplacement during the Appalachian orogenic movements has resulted in the development of deformation structures within the complex, producing irregularities in the layering.

c) *Wisconsin, U.S.A.*

A gabbro–granophyre complex near Ashland, on the southern limb of the Lake Superior geosyncline, has been described by Leighton (1954). The gabbroic rocks are discussed relatively briefly, but are believed to exhibit rhythmic layering in addition to a superimposed fluxion structure. The olivines range from Fo_{73} to Fo_{48}, the ortho-pyroxenes from En_{72} to En_{54}, the augites from Fe_{13} to Fe_{32}, and the plagioclase (cores) from An_{82} to An_{52}. These values are not given in relation to structural height within the layered series, so cannot at present be discussed in terms of cryptic variation, but Leighton shows a close correlation between the compositions of coexisting olivines and orthopyroxenes. The granophyre ('Red rock ') is a separate, slightly later intrusion, believed to have originated at deeper levels either by crustal fusion or by fractionation of basic magma.

d) *Willow Lake, Oregon, U.S.A.*

This very small intrusion, $1 \times \frac{1}{2}$ mile in size, is situated on the southern edge of the Bald Mountain tonalite batholith and is one of the earlier units of the batholith, probably early Cretaceous in age, to have been emplaced. The significance of the basic intrusion far outweighs its size, in view of the remarkable type of layering that has been described in detail by Taubeneck and Poldervaart (1960). The intrusion is funnel-shaped, with contacts inclined inwards at 60-70°. The rocks are broadly described as metanorites and hornblende metagabbros, the constituent minerals (plagioclase, orthopyroxene, augite, olivine, and magnetite) showing evidence, in some cases, of metamorphism by the later tonalite. A wide variation in mineral compositions is recorded but this cannot be related to cryptic variation, especially in view of the size of the intrusion and the special processes involved in its crystallization.

The layering is of two types, called ' Willow Lake type ' and ' Skaergaard type ' by Taubeneck and Poldervaart. All the layered rocks are inclined at high angles of 50-90° in the main funnel and 45-80° in the western extension. The Willow Lake type is found chiefly at the outer periphery of the intrusion, and inwards this type gives place to layering predominantly of ' Skaergaard type ', and then to weakly layered rocks at the centre of the intrusion. The so-called Skaergaard type of layering consists chiefly of alternating, thin layers rich in either plagioclase (An_{40-58}) or orthopyroxene (En_{65-70}) and is similar to the fine-scale rhythmic layering of other basic intrusions, attributed to crystal accumulation, except that in Willow Lake the layering dips at high angles.

The special type of Willow Lake layering is quite different from the other, better-known type. It is characterized by a profusion of very thin layers ($\frac{1}{4}$ in or less in thickness) within which certain minerals have their long axes inclined at 60-90° to the plane of layering. The elongated minerals are plagioclase and, less commonly, hornblende, orthopyroxene and augite, the crystals often showing curved and feathery forms. Taubeneck and Poldervaart explain the rhythmic character of this layering as due to undercooling near the margins, followed by a slight rise in temperature owing to the latent heat of crystallization and, then, by further undercooling. Such a process, resulting in a large number of alternating layers of contrasted mineralogy, would also account for the elongate, branching form of the crystals growing into the magma from the walls formed by the preceding layer. Taubeneck and Poldervaart also draw attention to the markedly similar rhythmic banding that has developed around the circumference of xenoliths of country rock, acting as local bodies against which under-cooling of the magma could temporarily have occurred. Inwards from the region of Willow Lake layering the more usual type of layering is found, due to precipitation of crystals in equilibrium with the magma and formed when the heat loss had ceased to be so drastic.

The Willow Lake type of layering is analogous in character and origin to that found in the tranquil division of the marginal border group of the Skaergaard intrusion, where the perpendicular-felspar rock is attributed to the inward growth of elongate plagioclases during periods of supercooling of the tranquil magma (Ch. V), and to the harrisitic layers in Rhum, where elongate olivines have grown upwards from the floor into tranquil supercooled magma (Ch. XI). We have given the general name of 'crescumulate' to this type of rock, and the Willow Lake type of layering is of similar type. Comparable structures are also seen near the margins of certain dykes, however, where crystal accumulation has not occurred. The term 'Willow Lake type layering' or, more logically, 'banding', if used to describe the products of supercooled magma near the margins of minor intrusions, is therefore a useful term. In the larger intrusions, how-ever, where layers due to supercooling may alternate or be closely associated with crystal accumulates deposited under equilibrium conditions, the term crescumulate layers can usefully be employed to describe the former, in contrast to the other types of cumulates in adjacent layers.

e) *Guadalupe, California, U.S.A.*

This complex, over 60 square miles in extent, includes a series of layered gabbroic rocks (Best, 1963). Rhythmic layering and igneous lamination are well developed, and cryptic variation is exhibited by the plagioclases (An_{80} to An_{28}), Ca-rich pyroxenes (Fe_8 to Fe_{28}), and Ca-poor pyroxenes (Fe_{13} to Fe_{57}). Iron oxides, hornblende, and biotite are important constituents of the upper layered basic rocks, and Best has pro-posed that the Guadalupe magma was rich in water and crystallized under conditions in which a relatively high Po_2 was maintained. These conditions prohibited extreme iron-enrichment in the pyroxenes and led to quartzose late-differentiates, including quartz monzonite and leucogranophyre.

f) *Southeastern Alaska, U.S.A.*

At least thirty-five ultramafic bodies associated with gabbro, occurring in a belt 30 miles wide and 400 miles long in Southeastern Alaska, have been discovered and are considered in general by Taylor and Noble (1960). They are approximately of

Lower Cretaceous age and pre-date the Coast Range batholiths. Hornblende pyroxenites are the dominant rocks in most of the complexes, but eight of the more obviously layered examples contain peridotites and related rocks. They all have enough features in common to suggest that they originated by the same mechanism, and it appears that ultramafic magma was involved. Two well-described examples are considered here.

The Duke Island complex (Irvine, 1963) consists of a somewhat disturbed intrusion of olivine-pyroxenites which was succeeded by a layered peridotite and dunite mass. Earlier gabbroic rocks, with plagioclase predominantly An_{75} to An_{50}, show a little rhythmic layering but no cryptic variation has been established. The ultramafic rocks outcrop in two areas, believed to be connected in depth and totalling 9 square miles in area. Olivine, diopsidic augite, and hornblende, with a little chromite and magnetite, make up virtually the whole of the ultramafic rocks, orthopyroxene and plagioclase being notably absent. The dominant rock types are olivine-pyroxenite, with from 15-30% olivine, and hornblende-pyroxenite. Dunite, peridotite, and hornblende–olivine-pyroxenite are present in small amounts. The olivine ranges from Fo_{85} in the dunites to Fo_{78} in the hornblende–olivine-pyroxenites, and the magnesium content of the diopsidic augites shows a similar decrease. Veins and dykes, believed genetically related to the ultramafic rocks, are of mafic pegmatite consisting of hornblende and plagioclase (An_{90-100}) in large crystals, locally attaining 100 cm in length.

The olivine–augite rocks show abundant rhythmic layering, due to the different sizes of the accumulated crystals or clusters of crystals, rather than to different proportions of the minerals (Irvine, 1963, fig. 8 & 9). Most layers are 10 or 20 cm thick, but may attain a metre. Commonly, the bottom of each layer has grains of olivine and pyroxene about 5 mm across while towards the top, the grains are from 2 mm to 0·2 mm. The top of the rhythmic unit is often a finely laminated, olivine-rich rock. Both pyroxene and olivine are involved in this size grading, but there is apparently a tendency for the olivine to be in smaller crystals and, therefore, to be more abundant towards the top of a graded layer. To the best of our knowledge, no other case of graded layering due to size-sorting has been described so convincingly as for Duke Island. Irvine has also described, from the more easterly part of the complex, a common type of layer with a base of upward-growing augites succeeded by a fine-grained, dunitic layer. This seems to be crescumulate layering similar in character to that in Rhum (see Ch. XI) except that at Duke Island the upward-growing crystals are diopsidic augites whereas in Rhum they are elongate olivines. In both cases each crescumulate layer is overlain by a layer of small, cumulus olivines.

Irvine considers that two successive pulses of ultramafic magma were forced into the earlier gabbroic rocks and there crystallized with bottom accumulation of crystals, probably under the influence of magmatic currents. The sequence and spatial relations of the rocks are complex, presumably due to contemporary movements. Blocks of layered ultramafic rock occur embedded in later, layered ultramafic rock, and slump structures are also found. After intrusion and solidification of gabbroic magma there was, apparently, an injection of ultramafic magma which solidified to give the layered olivine-pyroxenite rock and this was followed by a second injection of more ultramafic magma, to give the layered peridotite–dunite series of rocks.

Irvine's suggestion, that the layered rocks were formed from an ultramafic magma is also made for the other ultramafic bodies of the region by Noble and Taylor (1960). From the relative areas and compositions of the exposed rocks, the ultramafic magma is estimated to have had a normative composition as follows: diopsidic augite, 46%; magnesian olivine, 27%; anorthite, 11%; magnetite, 6%; and minor amounts of silica-deficient silicates. From this, diopsidic augite and olivine crystals are thought to have formed and, by crystal settling, produced the layered rocks. The absence of any interstitial plagioclase, and usually of magnetite, suggests that the rocks are adcumulates. The unusual, ultrabasic pegmatites of anorthite and hornblende are considered to have developed from a water-rich residual liquid. As crystallization proceeded, there should have been enrichment in iron relative to magnesium in both clinopyroxene and olivine, and more detailed studies might detect this as a slight

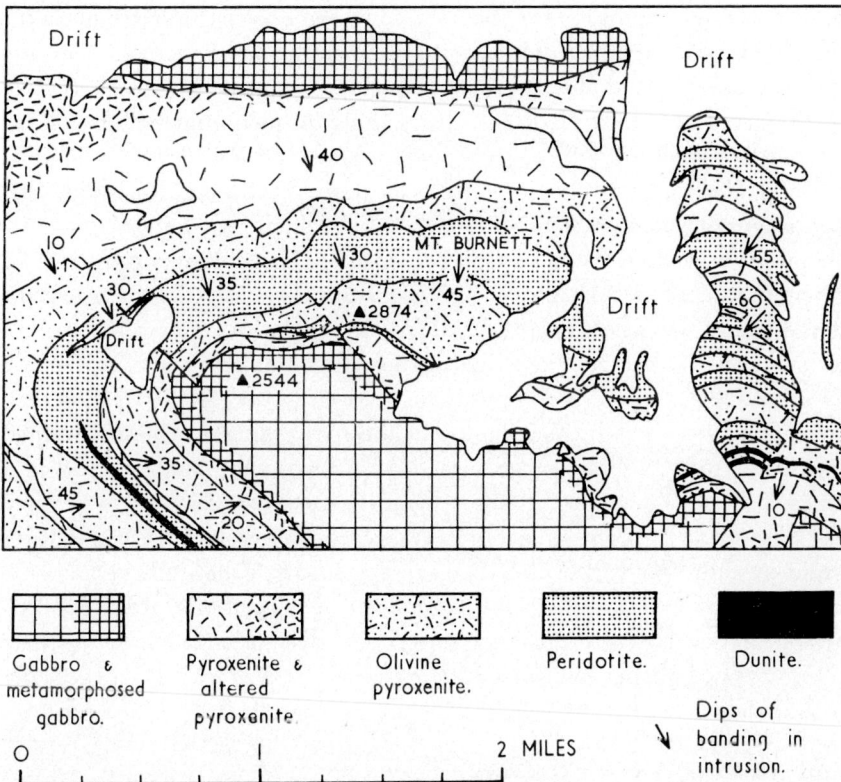

FIG. 277. Geological map of part of the Union Bay ultramafic complex (after coloured map of Ruckmick & Noble, 1959, pl. 1).

2 K

cryptic variation. The fact that crystals of olivine and diopsidic augite could sink through the magma, and become size-sorted during the process, provides evidence of the former existence of liquid with few or no crystals suspended in it, there being little likelihood that the injected material was an ultramafic crystal mush.

The Union Bay complex (Ruckmick & Noble, 1959) is composed of sheets of various ultramafic rocks, except for the eastern region where there appears to be a dunite plug which has been forcibly emplaced during the last stages of formation of the sheet complex. As at Duke Island, there was an earlier gabbroic mass into which the ultra-mafic rocks were injected, hornblende pyroxenite developing at the contact between the two. The sheet complex consists mainly of cumulus diposidic augite, olivine and, perhaps, magnetite. The map (Fig. 277) shows that the succession of layers of pyroxenite, olivine-pyroxenite, peridotite, and dunite have the general form of a shallow basin, with dips of up to 45°.

The Union Bay complex is considered by Ruckmick and Noble to have formed from a magma having a composition corresponding to 80% diopside ($Di_{75}Hed_{25}$), 15% magnetite, and 5% magnesian olivine ($Fo_{80}Fa_{20}$). They consider that successive injection of different magmas is the main cause of the sheet structures, but admit that minor layering is probably the result of bottom accumulation of crystals. One argument used by them (*op. cit.*, p. 1006) against the layering being due to crystal accumulation, referring to Wager and Deer (1939), is that essentially monomineralic rocks exist, i.e. rocks without the expected 20% or so of material from an intercumulus liquid. Since ultrabasic and ultramafic monomineralic rocks occur in Rhum, Bushveld, Stillwater and other layered intrusions, and can now be explained as having formed from an accumulation of crystals extended by post-depositional, adcumulus growth (Ch. XI & XIII), their argument is not tenable. While successive injections may account for some of the sheets and the disturbances of their regularity, it may be that much of the sheet structure is true layering, due to the settling of crystals as at Duke Island. Whether subsequent work will show this to be the case or not, it again seems likely that ultramafic magma was involved and that the pyroxenites, peridotites, and dunites are not derived from gabbroic magma as they probably are, for instance, in Rhum, Stillwater and Bushveld. On the other hand, there are several cases of very thick ultramafic layered assemblages (e.g. Great Dyke and Bay of Islands, Ch. XV) overlain by gabbroic layered rocks with which they are, apparently, genetically related. Whether the parental magma was basic in these latter intrusions or, more probably, ultrabasic, it would appear that an early precipitation of very magnesian olivines and pyroxenes can result in cryptic layering and (especially in the case of the Great Dyke) gradation into gabbroic cumulates.

g) *Muskox, Northwest Territories, Canada*

The Muskox intrusion, which crosses the Arctic Circle in the Coppermine River area, was discovered by the Canadian Nickel Company in 1956. The intrusion was mapped for the Geological Survey of Canada by C. H. Smith in 1959 and 1960, and brief accounts have been published (Smith, 1962; Smith & Kapp, 1963). Attention was focussed on this intrusion as part of Canadian studies related to the International Upper Mantle Project, and in 1963 the Geological Survey of Canada drilled three separate, vertical holes of 4000 ft, 3500 ft, and 2500 ft depth (Findlay & Smith, 1965). At present, the intrusion and the rocks collected from the surface and the drill-cores are being investigated systematically and in great detail by Smith and his collaborators (e.g. Bhattacharji & Smith, 1964; Jambor & Smith, 1964; Agterberg, 1964). For this reason we have confined our discussion to this brief, preliminary mention, although it is already apparent that when the present studies have been completed the Muskox intrusion will be of great interest amongst layered intrusions. The intrusion is Precambrian in age, the country-rocks having been dated as 1765 to 1720 million years old.

The intrusion is dyke-like in plan and funnel-shaped in cross-section, and as now exposed it appears as a narrow dyke, 35 miles long, which widens towards the north into an intrusion about 40 miles long and, at a maximum, about 7 miles wide (Smith & Kapp, 1963, fig. 2). The latter part has a synclinal form, the eastern and western edges dipping inwards at 5 to 20°, with a slight northward plunge of about 4°.

Smith (1962) has divided the intrusion into four main units:

1. A nearly vertical feeder-dyke, 500–2000 ft wide, consisting of bands of bronzite gabbro and picrite parallel to its walls.
2. A marginal zone, 200–1200 ft thick, parallel to the walls of the intrusion which dip inward at 23–57°. This zone grades inwards from fine-grained bronzite gabbro, through picrite and peridotite, to dunite in places.
3. A central layered series, 6500 ft thick, consisting of about 35 main layers which vary in individual thickness from 10 to 1100 ft. The lower layers are cumulates with orthopyroxene, olivine, and rare chromite as the cumulus minerals, while cumulus augite and plagioclase enter at higher levels.
4. An upper border zone, less than 200 ft thick, consisting of gabbro with interstitial micropegmatite, grading upwards into granophyre.

The feeder-dyke, at its southern end, consists of bronzite gabbro. Traced northwards, pods of picrite are found which gradually coalesce to become a single, central band of picrite. Bhattacharji and Smith (1964) have proposed a mechanism to explain this, discussed below.

The marginal zone shows a decrease in plagioclase and orthopyroxene, and an increase in olivine, from the margins inward. Simultaneously, the olivine becomes more forsteritic and the plagioclase more anorthitic. The central layered series apparently exhibits the result of sedimentation of crystals, the layering dying out near the margins. The rock types are reported as dunite, peridotite, felspathic peridotite, picrite, olivinite, clinopyroxenite, websterite, orthopyroxenite, troctolite, olivine gabbro, norite, and anorthositic norite. More critically expressed, the rocks are igneous cumulates with orthopyroxene, olivine and, less commonly, chromite as cumulus phases in the lower layered rocks, joined by cumulus augite and, finally, by cumulus plagioclase higher in the layered sequence. The approximate proportions of the different rock types in the whole intrusion indicate an overall ultramafic character, the average amount of olivine being about 58%.

The compositions of the olivine in different parts of the intrusion have been described by Smith and Kapp (1963, fig. 4). The 'central core' of the intrusion has olivine which is Fo_{85-80} and this is said to become less magnesian towards the margins, to Fo_{70-60}. This is taken to mean a grouping together of marginal and layered rocks, rather than a lateral variation within individual layers. Apparently the top 2000 ft of the central layered series shows cryptic variation in the olivine from Fo_{80} to Fo_{60}, and with the beginning of the cryptic variation in the olivine, plagioclase enters as a cumulus phase. In the lower 4000 ft or so, the olivine only varies within the limits Fo_{85} to Fo_{80}.

Until more information is available on the compositions of each cumulus phase in the Muskox layered series, on the relationship between the marginal and layered rocks, and on the inward sequences in the marginal zones, it would be premature for us to give more than this brief summary. The layered series shows evidence for cryptic variation of the type found in other ultrabasic–basic layered intrusions, such as the iron-enrichment of the olivines and the late entry of cumulus augite and plagioclase (c.f. the Bushveld, Stillwater, and Great Dyke). The presence of cumulus orthopyroxene indicates, further, that the intrusion has tholeiitic, rather than alkali-basalt affinities. In regard to the feeder-dyke, marginal zone, and upper zone rocks, however, the situation is much more complicated. The character of the chilled rock, and the tendency for a change from orthopyroxene-bearing to olivine-bearing rocks inward from the margins, are suggestive of supercooling of a hypersthene-normative magma near the margins; in that case, slower cooling to give the olivine-rich rocks should give rise to abundant quartz-bearing differentiates in the upper part of the layered series. In fact, layered rocks containing magnesian olivine (Fo_{60-70}) are overlain directly by a 200 ft-thick zone of quartz-bearing gabbros and granophyres, called the 'Upper Border Zone'. If this is, indeed, a border zone (presumably crystallized against country-rocks), then the gabbro must represent the initial magma cooled against, and contaminated with, overlying siliceous country rocks, seeing that the amount of granophyre increases upwards rather than downwards. To postulate that the quartz-bearing rocks are late differentiates of the layered series would present greater problems, since one would then need to consider a sudden change from abundant ultrabasic rocks to rare, quartz-bearing gabbros and granophyres, and the absence of an upper border zone would also need to be explained. Whatever the ultimate explanation, it would seem to us that the importance of Muskox is chiefly as an ultrabasic layered intrusion with a well-preserved feeder, showing more than the usual amount of cryptic variation (in comparison with, say, the Great Dyke and Southeastern Alaska ultrabasic intrusions), rather than as one showing the extreme variations found in basic layered intrusions of the Bushveld, Skaergaard, Stillwater, and Kiglapait type.

A particular feature of special significance in the Muskox intrusion has been considered by Bhattacharji and Smith (1964) where they suggest, from model experiments, that magma flowage has resulted in the differentiated nature of the feeder-dyke. Flow of magma up the feeder is thought to have resulted in the concentration of the earlier, magnesian olivines at the centre of the dyke, successive surges precipitating less magnesian olivines outwards. This relationship is also related to the pattern in the funnel intrusion above the dyke, where the olivines are more magnesian in the ' core ' (i.e. the inner marginal zone and the lower layered series). Although this is an interesting hypothesis, we find it difficult to accept as an important generalization. The range of olivine compositions involved is high, i.e. from Fo_{85} to Fo_{60}, and this must represent a large range in temperature of crystallization. This, in turn, means that the feeding fissure and the marginal zone need to be open continuously at temperatures down to those when Fo_{60} is in equilibrium with the magma, i.e. well after the deposition of about 5000 ft of layered rocks directly above the fissure (Smith & Kapp, 1963, fig. 4). It is difficult to accept these conditions, especially in the narrow feeder-dyke where extremely tall, thin, vertical sheets of successively-crystallized material would need to stand while magma poured past on each side without congealing at the walls, and while it found its way always between the thick cumulates of the overlying funnel-intrusion and its walls. At present we feel that further consideration might be given to the possibility that in the dyke and marginal zone, the inward increase in the Fo-contents of the olivines and the increase in olivine relative to orthopyroxene are due to a gradual decrease, inwards, in the degree of supercooling of the magma (i.e. in relation to the systems Fo–Fa and Fo–SiO_2, respectively).

h) *Caribou Lake, Ontario, Canada*

This basic intrusive body, situated in the Grenville Province of the Canadian Shield, covers an area of about 7 square miles (Friedman, 1957). Rhythmic layering helps to define two synclinal areas in the western part, but the cryptic variation suggests a differentiation path originating in the extreme southeast of the complex, where the only ultrabasic rocks are found. The olivine (Fo_{80}) is restricted in occurrence, whereas plagioclase, orthopyroxene and augite show systematic changes in composition which are suggestive of cryptic variation. The orthopyroxene range is from En_{88} (picrites) to En_{60} (norites) and the plagioclase range from An_{68} (picrites) to An_{50} (norites), the relatively sodic felspars in the picrites probably indicating that the measured plagioclase is of intercumulus rather than cumulus status. The compositions of ' coexisting ' mineral phases are plotted against one another, and compared with similar data from other layered basic intrusions. We feel that certain anomalies observed by Friedman may well be due to the distinction not being drawn between cumulus and intercumulus phases.

i) *Sudbury, Ontario, Canada*

It should be stated at the outset that we do not believe there is satisfactory evidence, at the time of writing, to indicate that crystal settling was an important process in the famous Sudbury intrusion. However, Coleman (1905), who provided the first good geological map of the Sudbury area, advanced the view that the sulphide ores were of magmatic origin and genetically related to the norite. In the latest major work on the Sudbury ores and their origin, Hawley (1962) is firmly in favour of the view that the sulphide masses were produced by gravitative separation of an immiscible sulphide liquid from the norite magma. Coleman also considered that the thick layer of micropegmatite over-lying the norite was produced by differentiation of some single, initial magma in place, but other hypotheses have since been proposed and no strong concensus of opinion has yet been reached.

In view of the importance of the Sudbury intrusion in igneous petrology, the amount of work that has been done on the intrusive history and nickel deposits, and the fact that Sudbury has often been linked with layered basic intrusions, we decided that a brief summary was desirable in this book. Recent summaries of the geology are provided by Hawley (1962) and Thomson (1957), and the map presented here (Fig. 278) is based largely on their data. In the literature, the so-called ' Nickel Irruptive ' is taken as consisting, from the top downwards, of:

> Micropegmatite
> Transition or Hybrid Zone
> Norite and quartz diorite with associated sulphides.

Above the micropegmatite is a thick layer of acid tuffs and breccias (volume estimated as 300 cubic miles) which Williams (1957) considers to be the product of glowing avalanches from a group of volcanic vents. He also described rhyolitic intrusions near or at the base of the tuffs, i.e. immediately above the micropegmatite. The term Nickel Irruptive, covering the norite and micropegmatite, will be avoided in the following discussion; it tends to imply a single intrusion of magma and suggests, too readily, the hypothesis that the micropegmatite and norite were produced by differentiation in place, from a single intruded mass of magma.

The norite intrusion is dated as about 1700 million years by radiometric methods, the pre-norite basement being chiefly of Huronian igneous and sedimentary rocks. Inside the Sudbury Basin, the thick acid tuffs and breccias (the Onaping volcanics) are overlain by Onwatin Slates and Chelmsford Sandstones (arkose and greywacke) which are also considered to be pre-norite in age. These inner rocks are surprisingly little metamorphosed or deformed, and seem to have overlain the basement of Huronian and Pre-Huronian rocks unconformably at the time of the intrusion of the norite.

Before the intrusion of the Sudbury norite, events of great magnitude, probably related to the development of the Sudbury Basin, produced the so-called common Sudbury breccias (recently described in detail by Speers, 1957). These rocks are

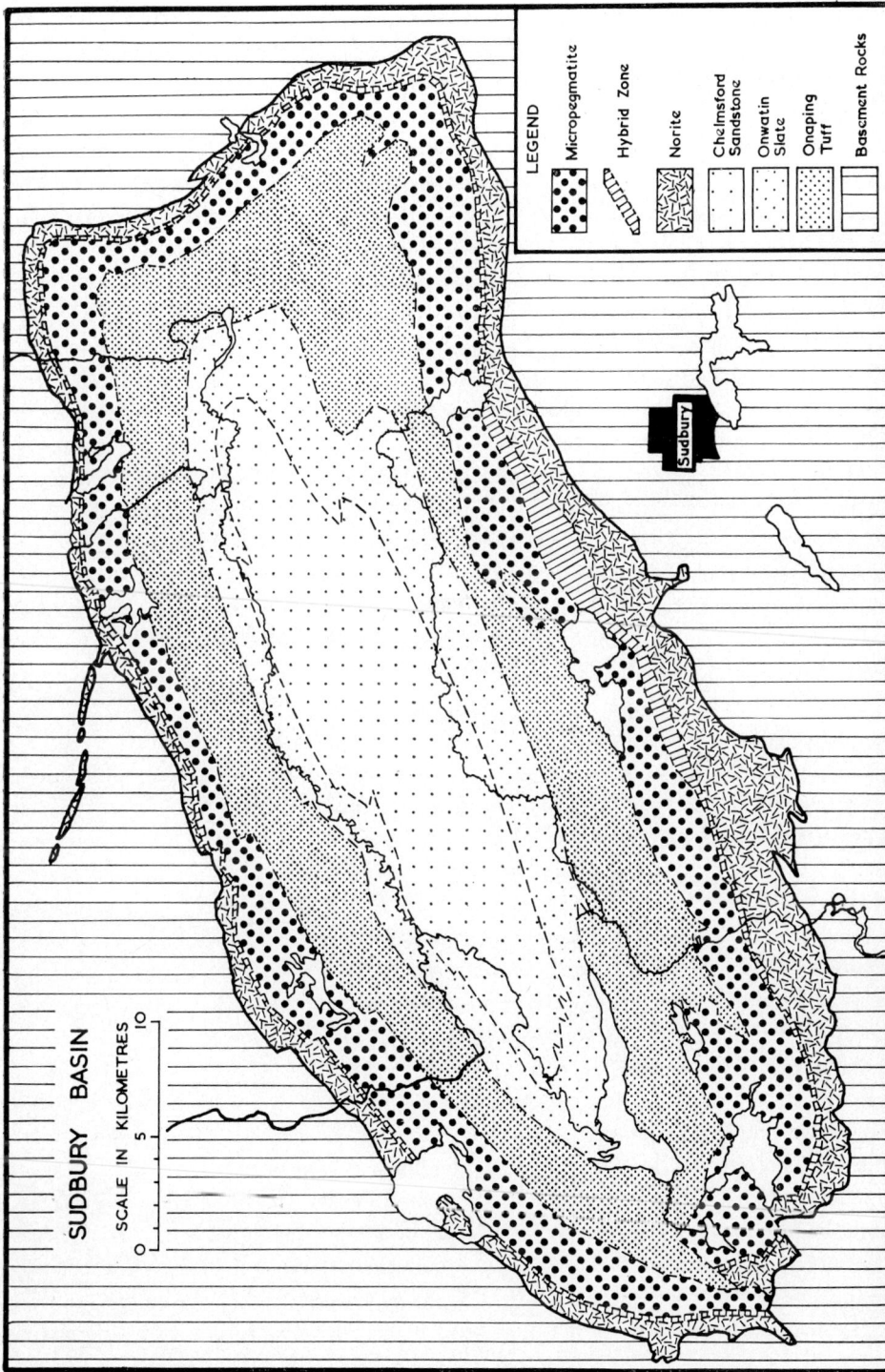

FIG. 273. Simplified geological map of the Sudbury Basin, after the coloured map in the Sudbury Area Guidebook (6th Commonwealth Mining and Metallurgical Congress, Canada, 1957), and by reference to Thomson (1956, Map No. 1956–1), Speers (1957, fig. 2), Hawley (1962, fig. 1), and Stevenson (1963, fig. 1).

distributed around the periphery of the whole Sudbury Basin, so a connection has been postulated between them and the formation of the rocks of the basin. The breccias are not found among the rocks within the basin, nor do the inner rocks show any other evidence of the tectonic and explosive events which would seem to have been required to produce the breccias. Thus their formation must have preceded the Onaping tuffs, the Onwatin Slates and the Chelmsford Sandstones, as well as the Sudbury norite. The significant distribution of the breccias seems, nevertheless, to imply a distinct connection with the special igneous and tectonic events which gave rise to the Sudbury Basin.

The Onaping tuffs and breccias have been considered recently in some detail by Williams (1957). The formation is made up of andesitic and rhyolitic tuffs, heavily loaded with lithic fragments that are chiefly of rhyolite but also notably of quartzite, amphibolite and granite, ranging in size from minute particles to gigantic blocks. In their gross textures, the coarse tuffs and breccias resemble the chaotic, unstratified deposits of Peléan glowing avalanches. Williams considers that Coleman's estimate of their thickness is conservative and he suggests a total thickness of 5000 ft, believing that the mass resulted from the rapid extrusion of froths of andesitic and rhyolitic magmas. Some masses of rhyolite are interpreted as intrusions and others, occurring as blocks in the breccias, as having been derived from early intrusions. Williams believes that the formation of the 300 cubic miles of tuffs and breccias caused a large-scale subsidence which contributed to the formation of the Sudbury Basin. The Onaping volcanics apparently pass upwards into the Onwatin–Chelmsford formation by the gradual incoming of water-lain sediments.

The norite mass was intruded after the eruption of the common Sudbury breccias and the Onaping tuffs and breccias, and after the deposition of the succeeding sediments. In form, the norite is presumably basin- or funnel-shaped, although the deeper parts are unknown from direct observation. Early workers have postulated that the norite was intruded as a flat-lying sheet which was subsequently folded into its present form. After reviewing the evidence, Hawley (1962) considers that the norite was probably intruded essentially in the form it now has, although there has been a limited amount of subsequent thrusting, faulting, and compression.

The general impression gained from descriptions of the norite intrusion and examination of the rock-types is that it is a relatively uniform norite with little conspicuous layering. This impression was confirmed by a brief visit by one of us (L.R.W.) to the area in 1958. Despite this, it would be of great interest to decide whether the Sudbury norite shows any cryptic variation, although this is not an inviting study because the rocks tend to be altered. The plagioclase compositions could be determined in detail, particularly in vertical sections such as that obtained from drilling at the Creighton Mine where, apparently, a mile of the norite was penetrated. It is clear that if the norite is layered and the successive cumulates have a shallower inward dip than the foot-wall of the norite intrusion (see Hawley, 1962, p. 24), traverses across the body at the present level of erosion would tend to remain at approximately the same layered horizon; it is for this reason that the vertical drill-cores would provide a more fruitful study. A significant observation, mentioned by Wilson (1956), is that near the top of the norite there is a well-defined zone rich in magnetite, which may correspond to the entry of cumulus magnetite as a result of crystal settling and fractionation. In places, the norite passes gradually into quartz diorite. This rock-type usually occurs near the foot-wall and is often associated with the sulphide masses; it does not seem to have the sort of distribution appropriate to a later differentiate of a layered intrusion.

The original magma which gave rise to the norite is considered by most workers 'to have been essentially saturated with sulphides so that before any appreciable crystallization occurred, immiscible sulphide liquids were able to settle vertically downwards towards the foot-wall contact and into offset fractures to give segregations of massive sulphides which must have displaced any silicate liquid originally there. Above these ores, in places, are found the mixed silicate–sulphide ores and disseminated sulphide blebs in quartz diorite or norite . . . especially along more gently dipping contacts which have not been controlled by faulting'. (Hawley, 1962.) Hawley describes sulphide (mainly pyrrhotite with minor chalcopyrite and pentlandite) as forming rounded blebs in norite, or as massive ores due, apparently, to the coalescence of masses of immiscible sulphide liquid. In some cases, as a counterpart, the silicate occurs as rounded blebs and seems to have crystallized from immiscible silicate droplets in a considerable volume of sulphide liquid (Hawley, *op. cit.*, p. 34). Sometimes the sulphide masses occur along the foot-wall of the intrusion and in other cases they form offset deposits, with associated norite or diorite, which seem to have resulted from the silicate magma, together with settled sulphide liquid, being pushed into the underlying country rocks.

The relationships between the norite and the overlying micropegmatite have interested petrologists since the Sudbury complex was first described. Coleman (1905), followed by many others, considered that a great bulk of magma had differentiated in place, by fractional crystallization under the influence of gravity, to produce the more mafic norite and a residual, overlying liquid which later solidified to give the micro-pegmatite. The main alternative to this hypothesis has been presented in some detail by Phemister (1937) and again by Thomson (1957). Phemister noted that the 'transitional' zone between the norite and the micropegmatite is only a few hundred feet thick, and very constant in position. He demonstrated, from mineralogical and chemical evidence, that the zone begins rather abruptly above the norite and, after two or three hundred feet, gives place equally abruptly to the overlying micropegmatite. Phemister considered that the zone resulted from the micropegmatite having been intruded as a separate and distinct magma on top of the partly solidified mass of the basic member, with the formation of hybrids along the junction (Phemister, 1937, p. 42). This hypothesis also has its difficulties, one being the improbability of the acid magma being intruded so uniformly against the whole body of norite, with rather constant thickness and without showing any transgressive relationships.

A third alternative hypothesis was being discussed in 1958, at the time one of us (L.R.W.) was accompanied on a visit to the area by Mr F. Langford (who was mapping the region around the Levack and Fecunnis mines). Briefly, this supposes that the micropegmatite was formed from part of the Onaping tuffs at the time of intrusion of the norite magma. The magma, intruded along the junction between what are now the inner rocks of the basin and the basement rocks outside, is postulated to have metamorphosed and metasomatized the tuffs, thus forming the micropegmatite. On this hypothesis, the hybrid zone represents the product of mixing of the tuffs, or their metasomatized equivalent, with the norite. The thickness of the micropegmatite is fairly constant, and often there seems to be a gradual transition from patchy micropegmatite to the Onaping tuffs and breccias. Near the base of the Onaping tuffs, and thus at about the top of the micropegmatite, occurs the Trout River conglomerate which Stevenson (1963) has described in some detail and shown to be a quartzite breccia. He believes that this is a sedimentary horizon but whether this is so or not, its highly siliceous composition is such that it would not be converted to micropegmatite* so easily as would the tuffs. Thus the approximate coincidence of this horizon with the top of the micropegmatite may be a result of the difficulty of altering the quartzites to micropegmatites near the limit of the major metamorphic effect. The association of acid and basic rocks is a general problem, and pertinent to a study of layered intrusions where, in most cases, extreme fractionation appears to have given rise only to minor amounts of late-stage acid rocks. In Skye, Brown (1963) has proposed that the intrusive acid rocks are due to the eventual melting of acid basement rocks by basic intrusions, and Sudbury seems to provide evidence of a similar process *in situ*.

Whether or not bottom accumulation of crystals contributed to the formation of the Sudbury norite has to be left an open question at the present time. The most effective way of deciding the matter would, as we have stressed above, be through investigating whether there is a cryptic variation in the silicate minerals of the intrusion. The Sudbury norite displays a kind of layering due to the separation and sinking of a relatively dense, immiscible sulphide liquid, but the authors do not believe that the 'micropegmatite' at Sudbury is directly the consequence of the settling of crystals, under the influence of gravity, to give a lower layer of norite and a residual, upper layer of granitic liquid.

AFRICAN AND AUSTRALASIAN EXAMPLES

a) *Freetown, Sierra Leone, West Africa*

The petrology of this intrusion has been described in detail by Wells (1962), and the results of a gravity survey by Baker and Bott (1962). The intrusion forms a crescentic outcrop in the coastal region, 30 miles N–S × 9 miles E–W. The layers dip westward

---

* The name 'micropegmatite' is generally synonymous with the more widely used name 'granophyre'. We feel that the latter is preferable for such a rock, the origin and grain-size of which have no connection with those of pegmatites.

towards a centre situated, according to the gravity anomalies, in the Atlantic (8-12 miles WSW of York). The dip of the layering increases westward from 10-20° to about 40°, and is approximately parallel to the dip of the floor. The thickness of exposed layered rocks is about 20 000 ft, and Wells has sub-divided the layered series into four very thick zones. In general, each zone shows a progressive increase, upwards, in the ratio of cumulus plagioclase to cumulus olivine. Within each zone, both major rhythmic layering (units averaging a few hundred feet in thickness) and fine-scale rhythmic layering occur. Despite the great thickness of layered rocks, and the convincing structural evidence of formation by bottom accumulation of crystals (e.g. Wells & Baker, 1956, fig. 3), the Freetown layered series is remarkable in the absence of any cryptic layering. This feature has often been encountered in layered ultrabasic rocks, but not to such an extent in layered basic rocks. The mineral compositions (from optical properties) vary irregularly, with height, from An_{59} to An_{64} (plagioclase), Fo_{58} to Fo_{67} (olivine), and En_{65} to En_{72} (orthopyroxene, the more ferriferous being inverted from pigeonite). Apart from this limited compositional range, a relationship between phases of particular compositions, and a cryptic variation within a zone or a macro-unit, have not been detected. Wells suggests periodic influxes of an olivine-gabbro magma to account for the absence of cryptic layering, but proposes a majority of ancillary processes, including diffusion of volatiles, flow of magma rich in suspended crystals, and periodic magma chamber expansion, to account for the rhythmic layering.

b) *Kunene, Angola, West Africa*

This elongate Precambrian intrusion, striking N–S and measuring 300 × 25 km, has been described by Stone and Brown (1958) in relation to the cryptic variation exhibited by rocks collected from a traverse across its width. The layers dip inwards, the olivine:plagioclase ratio decreasing towards the centre. The plagioclase changes gradually from An_{80} to An_{50} and the olivine from Fo_{84} to Fo_{65}, apatite and magnetite becoming increasingly common in the central, higher layers. Simpson and Otto (1960), as part of a detailed study in progress, have described an extension of the complex to the south of the Kunene river. Olivine–plagioclase cumulates show large-scale layering and cryptic layering, and are underlain by massive anorthosites containing sodalite and contorted, calcareous chert bands. Genetically related gabbro intrusions west of Otjijanjasemo are layered, and comparable in structure to the Great Dyke (*op. cit.*, p. 226).

c) *Okonjeje, South-West Africa*

The Okonjeje complex is one of a series of related post-Karroo intrusives in the Damaraland region, including the Messum complex (Martin *et. al.*, 1960). Described extensively by Simpson (1954*b*), Okonjeje is exposed over a circular area of about 20 km² and consists essentially of a lopolithic complex of tholeiitic olivine gabbros and ferro-gabbros, associated closely in space and time with a ring-dyke complex of alkali olivine-gabbros, essexites, and pulaskites. The 'differentiated group' of the tholeiitic series, including gabbros and ferrogabbros, displays a remarkably regular mineralogical and chemical variation (*op. cit.*, fig. 2) from An_{80} to An_{30} (plagioclase), En_{78} to En_{27} (orthopyroxene), Fo_{73} to Fo_{20} (olivine), and Fe_{17} to Fe_{32} (augite). However, this variation is from the *top to the base* of the lopolithic, stratified series, and from this and other evidence Simpson (1954*b*, p. 141) envisages the emplacement of successive injections of thin sheets, from the top downwards. The concise map and sections given by Simpson (1954*b*, pl. 29) leave little doubt that the complex as a whole is probably due to separate injections of contrasted magmas, although some fine-scale rhythmic layering has developed within the sheet of alkali olivine-gabbro. Hence the complex is not a layered intrusion in the sense defined in this book, but Okonjeje may well serve as a type-example of the effects of successive tapping of a differentiating body at depth, the resultant rock series showing the type of mineralogical and chemical variation found within layered intrusions of the Skaergaard type.

d) *Messum, South-West Africa*

The description of this complex by Korn and Martin (1954) mentions the existence of a basic lopolith, consisting of 6000 ft of layered gabbros and anorthosites. Since then, however, Mathias (1956, pp. 8 & 15) has concluded that it is a sheet complex. However, graded layering occurs between olivine eucrites and anorthosites, although mineralogical and textural data are not yet available as an aid to interpretation of the layering. The separate basic intrusions may be related chemically, as at Okonjeje, but this has not yet been investigated.

e) *Somalia, Northeast Africa*

More than twenty small, layered basic intrusions have been mapped within the Precambrian to Lower Palaeozoic metamorphic fold-belt lying south of the Gulf of Aden (Daniels, 1958). Several show good rhythmic layering on a coarse and fine scale, and some show cryptic layering, although the majority have been deformed and metamorphosed by later orogeny. The most recent account (Daniels *et al.*, 1965) gives more petrographic information than in the papers dealing specifically with Sr–An relationships in the plagioclases (Butler & Skiba, 1962; Skiba & Butler, 1963). One of the bigger intrusions, Radasafaka (12 square miles), consists of 13 000 ft of layered rocks divisible into five major units, each characterized by an olivine-rich base grading upwards into a plagioclase-rich top. The plagioclase composition becomes more sodic upwards in some units (e.g. An_{79} to An_{65} in the 4000-ft lowest unit), and more sodic upwards in the series as a whole. Although ultrabasic layers are mentioned from several intrusions, the most calcic plagioclase recorded is An_{79} and the olivines range from Fo_{79} to Fo_{68}, so the intrusions appear chiefly to be basic layered bodies, with overall enrichment in the mafic cumulus minerals towards the base of each. They are not thought to have once formed part of a single, large intrusion, though particular groups of them may once have been connected in this way.

f) *Insizwa, South Africa*

The strata of the Karroo basin are riddled with dolerite dykes and sheets, intruded contemporaneously with, or immediately following, the late-Triassic (Stormberg), Drakensberg basalts (Walker & Poldervaart, 1949). Some of the sheets are so thick as to assume the form of major intrusions, of which Insizwa (East Griqualand, Natal) is the thickest (3000 ft). The neighbouring mountain masses of Ingeli, Tonti, and Tabankulu probably form part of the same huge, undulating sheet with an original area of about 700 square miles. The intrusion is best-known for the extensive descriptions of the Cu–Ni sulphide minerals found in the Basal Zone (Scholtz, 1936). The petrographic account by Scholtz has been followed by a more detailed description of a vertical traverse by Bruynzeel (1957). A lower chill phase is succeeded upwards by picrites, olivine gabbros, and quartz–hypersthene gabbros. Rhythmic layering and cryptic layering occur, the olivine changing from Fo_{89} to Fo_{66} and the plagioclase from An_{90} to An_{70} (though complicated by zoning and resorption). Both pyroxenes show iron-enrichment, and the exsolution and inversion textures have been described in detail.

g) *Dun Mountain, New Zealand*

This is the better-known of the large masses of layered ultramafic rocks associated with the Permian volcanics of South Island. The petrology of the group as a whole has been described by Challis (1965) with particular reference to the Red Hills intrusion. The petrography and layering in the Dun Mountain intrusion ($2 \times 2\frac{1}{2}$ miles) has been described by Lauder (1965). The intrusions consist of well-layered cumulates in which olivine, bronzite, diopsidic augite and chromite are the cumulus phases. The compositional ranges are limited, with olivine Fo_{94-89} and bronzite En_{93-89} throughout a thickness of about 4000 ft, no regular cryptic variation being observed. A few occurrences of gabbros are recorded, the association with ultramafic rocks being reminiscent of that in the Great Dyke and Bay of Islands (Ch. XV). Challis (1965, pp. 357-61) favours an origin through crystal accumulation from tholeiitic basalt magma, agreeing with Smith (1958; see, also, p. 454) that subsequent dislocation of the ultramafic cumulates, as has occurred in New Zealand, gives the impression that ‘ Alpine-type masses ’ differ in origin from stratiform cumulates derived from basic magmas.

h) *Giles Complex, Central Australia*

This complex consists of at least fourteen isolated, Precambrian layered masses extending over a 100 mile area in the northwest corner of South Australia (Nesbitt & Kleeman, 1964). Work is in progress, but R. W. Nesbitt (personal communication, 1965) reports that fine-scale rhythmic layering, gravity stratification, slump structures, and cryptic layering have been observed in the Mount Davies (Tomkinson Range) intrusion. This intrusion is 10 $\times$ 3 miles in size, and consists of 14 000 ft of well-layered peridotites, gabbros and norites. The plagioclase changes upwards from An_{87}

to An_{67}, the olivine from Fo_{95} to Fo_{82}, the orthopyroxene from En_{94} to En_{74}, and the augite from Fe_4 to Fe_{13}. The dip of the layering is now high (80°), due to subsequent tilting and folding.

SOVIET RUSSIAN EXAMPLES

There are doubtless several examples of layered intrusions in the U.S.S.R. and other parts of the Eurasian Continent of which we are unaware. Owing to language difficulties, also, we have not been able to do justice to the published literature on the well-known examples, nor enjoyed access to the field localities and to photographs of layered structures in the same way, for example, as in the case of comparable Greenland examples. Nevertheless, this book would be incomplete without drawing the reader's attention to the Lovozero alkali complex, comparable in many respects with the Ilimaussaq of South Greenland, and to the extensive ultramafic bodies in the Ural Mountains, both having engaged the attention of eminent Soviet petrologists in recent years.

a) *Lovozero, U.S.S.R.*

The Lovozero and Khibina alkaline complexes are situated in the Kola peninsula (about 100 miles southeast and south of Murmansk), and were first described in some detail by Fersman (1937). Since then, aspects of the layered Lovozero complex have been described, for example, by Vorobyeva (1940), Yeliseyev (1941), and Vlasov *et al.* (1959). An English translation of the last-mentioned book was published too late for our use, in 1966. The writer is indebted to Dr. J. B. Dawson (St. Andrews University) for drawing his attention to a paper by Atamanov *et al.* (1962) in which a summary of recent studies is presented, and on which the present account is based. The latter paper is particularly instructive in that it shows the extent to which certain Soviet geologists disagree on interpretations of the petrogenesis of Lovozero, particularly where criticizing the work by Vlasov *et al.* (1959) to which users of the English language will tend to refer, almost exclusively.

The Lovozero complex, about 6 km in diameter, has intruded Precambrian granites, granitic gneisses, schists, and basic and ultrabasic rocks. Radiometric determinations indicate that the intrusive complex is about 260 million years old. The form of intrusion is close to a funnel-shape (Atamanov *et al.*, 1962, fig. 4), the walls having a steep, inward dip. Similarly, the nearby Khibina complex is thought to have been injected in the form of inverted cones. An early intrusive phase of fine-grained nepheline syenite is restricted to the margins of the Lovozero complex and outcrops, on the north-west side, as a continuous zone 150–200 m wide and about 10 km long. Xenoliths of nepheline syenite are common in the layered series, but the fact that the plane of undisturbed contact between the marginal and layered groups is only 15° suggests that the nepheline syenite may be a marginal facies of the layered series. The layered rocks are said to ' flow round ' the xenoliths (a feature noted in the Skaergaard, where xeno-liths of the marginal group were transported to the floor and deposited with layers of crystal accumulates; see Fig. 46), indicating a close temporal relationship between the two. The major part of the Lovozero complex consists of a layered series of lujavrites, foyaites, juvites, and urtites; two of these rock names are derived from the region (Lujaur-Urt, Kola Peninsula). The third major component has been called 'the eudialyte complex', the rocks being relatively melanocratic lujavrites rich in euhedral eudialyte, sphene, apatite, amphibole, and certain rare minerals. The relationship between this and the underlying layered series is complicated by the presence of a wide zone of 'porphyritic lujavrites' occurring as veins and lenses, and attributed to a fourth stage of activity. However, the eudialyte-rich rocks are weakly layered and may turn out to be an upper part of the layered series.

The exposed part of the layered series is about 1700 m thick, the dip being generally 8–12° but increasing to 40–50° near the margins. Layers up to 10 cm in thickness can be traced laterally for several kilometres without any apparent transgressive character-istics. Rhythmic layering is remarkably well developed at Lovozero, involving alter-nations between melanocratic layers rich in aegirine (lujavrites) or amphibole, and leucocratic layers rich in nepheline (urtites) or alkali felspar (juvites). Less extreme cumulates are termed foyaites and leucocratic lujavrites. Atamanov *et al.* (1962, fig. 2) have subdivided the layered rocks into five 'series', each having a lower 'zone' con-sisting of numerous leucocratic and mesocratic alternations and an upper 'zone', pre-dominantly of leucocratic rocks, also showing rhythmic layering. For consistency, and in order to avoid confusion in nomenclature, we would rather refer to the whole as a *layered series*, and to the five subdivisions as *zones* if cryptic variation is involved, or as *major rhythmic units* if cumulus mineral proportions alone are involved. A suggestion of cryptic variation is implied by the observation that felspar-rich rocks (juvites) are common in the lower 'series', whereas nepheline-rich rocks (urtites) are commoner in the upper 'series'. Hence it may ultimately be desirable to refer to the Lovozero layered series according to a zonal nomenclature, each zone being perhaps divisible into subzones *a*, *b*, etc. If this is so, then equally the fine-grained nepheline syenite may turn out to be a marginal border group, and the eudialyte-rich layered rocks to be an upper zone of the layered series. Clearly we are not in a position to do more than make these tentative suggestions, offered in order to see whether Lovozero can be compared more closely with other well-described layered intrusions rather than as an attempt to achieve a rigorous conformity. If the relationships were as suggested here, and the mineralogical and chemical variations, and textural relations in terms of crystal accumula-tion, investigated with this possibility in mind, there is little doubt that the remarkable assemblage of layered rocks and the unusual mineral assemblage at Lovozero would add much to our present knowledge of layering in alkaline intrusions.

b) *Urals, U.S.S.R.*

The Hercynian mountain chain of the Urals extends N–S for about 2000 km and has an average width of about 200 km. The core of this folded complex of schists and sediments, together with lavas and plutonic rocks, contains an impressive array of ultramafic and basic igneous rocks. We are not aware of any published descriptions of the latter which indicate a layered origin, and they appear to belong to the orogenic or 'Alpine-type' of ultramafic assemblage, such as we have described from South-eastern Alaska, Cornwall, and Cyprus. However, layering has been observed in the Lizard of Cornwall and in certain Alaskan ultramafic rocks, the latter having been compared generally with the Urals (Taylor & Noble, 1960).

The fine series of coloured maps by Duparc and Tikonowitch (1920) show broad strips of basic plutonic rocks running parallel to the strike of the Urals tectonic structures, the assemblage including a wide variety of gabbros, olivine gabbros, norites, and diorites. The ultramafic rocks occur as a chain of isolated small bodies (average diameter 10 km), each characterized by having a core of dunite or peridotite and a sheath of harzburgite, pyroxenite, or hornblendite. The ultramafic bodies are famous for their platinum content (Duparc & Tikonowitch, 1920), the platinoid mineral assemblage being very similar to that found in the Bushveld and associated dunite pipes (Ch. XIV). The importance of the region is shown by the extensive Russian literature on the igneous rocks and mineral deposits. (See recent volumes of *Mineralogical Abstracts*). It is clear that this book is not the appropriate place for further discussion, except to emphasize that when more is known of the origin of the often-distinctive layered or banded structures to be seen in ultramafic and ultrabasic rocks of orogenic regions, then the Urals, together with the Appalachian and Alaskan examples, will be of particular genetic significance.

c) *Taimyr, U.S.S.R.*

An interesting reference to a layered intrusion in the Taimyr peninsula of Siberian Russia (Ravich & Chaika, 1956) has been given by S. I. Tomkeieff (*Mineralogical Abstracts*, 1957, **31**, p. 392). Although this region is one of folding, possibly as a branch of the Urals Hercynian orogenesis, the layered body is aparently related to the Permian phase of igneous activity, i.e. the 'Siberian traps' (King, 1962, pp. 427 & 402). The large sheet-intrusion is about 900 m thick and is exposed over 20 km of outcrop. Divisible into five units, the uppermost unit is granitic while the lower four units are gabbroic. The four gabbroic units are well layered, and each has a melanocratic base and a leucocratic top. Of particular significance, also, is the preservation of a lower-most chill zone consisting of olivine dolerite. While the layering is attributed to crystallization differentiation, the hypothesis that this took place while the magma was intruding in a horizontal direction would be difficult to reconcile with the apparent conformity and disposition of the rhythmic layering and with the development of the granitic unit near the top, rather than the central part of the sheet. Crystallization from the bottom upwards is suggestive, rather, of a convective system resulting in transference of heat within the magma body.

CHAPTER XVIII

DIFFERENTIATED SILLS

INTRODUCTION

The phenomenon of crystal settling in magmas has been discussed in Chapter I, and shown to be a likely process wherever the conditions permitted enough time for the sinking of crystals of appropriate size and density. These conditions are rarely realized in lava flows or in thin intrusive sheets, but are to be expected in the thicker basic sills. In most sills, however, crystal accumulation appears to have played a minor rôle in their differentiation, and is generally confined to the earlier stages of crystallization, involving the accumulation of olivine crystals in a single, relatively thick layer near the base. Examples of this type are the Palisades sill, the Shiant sill, and the Elephant's Head dyke (a relatively broad, differentiated body associated with the New Amalfi sill). The Black Jack sill is fairly rich in olivine in the lower parts, but owing to some irregularities in olivine distribution the rôle of crystal settling has been doubted. The tholeiitic sills of Tasmania and Antarctica are relatively rich in pyroxene in their lower parts and this appears to be due to crystal settling, especially in the thicker Antarctic sills.

In most of the well-differentiated sills, the contrasts between the rock types are attributable to processes other than crystal sorting and accumulation in successive layers. Rhythmic layering and igneous lamination are rare or absent and the rock textures are not usually of the type found in igneous cumulates. The often regular variations in mineralogical and chemical compositions cannot be related to regular changes in *cumulus* mineral compositions, and cannot therefore be described as cryptic layering. Hence, in view of the general absence of rhythmic layering, cryptic layering, igneous lamination and other textural features of igneous cumulates, differentiated sills should not be called layered intrusions. On the other hand, the evidence in some sills of subordinate crystal accumulation merits their inclusion in a section of this book. The thicker the sill, the greater the likelihood that conditions of crystallization approached those of the larger, layered intrusions. For example, the Insizwa intrusion (p. 525) is a thicker variety of the Karroo dolerite sills and exhibits layering not found in the New Amalfi and thinner sills of this province. The size of an intrusion is not the only factor influencing the rate of cooling, however, and the intrusion of a relatively small body at some depth, or into a hot environment (for example, the Basistoppen sill (Ch. VI) and the Willow Lake intrusion (Ch. XVII)) may result in conditions approaching those found in larger intrusions.

The cooling history of sills, and the mechanism of differentiation, are outside the scope of this book. However, certain aspects are of interest in relation to the conditions of crystallization at the margins of the larger intrusions, sills generally providing an ideal means of studying the effects of magma crystallization where heat loss to the walls was relatively rapid. Supercooling probably played an important rôle during the crystallization of the greater part of the thinner sills, and the occasional banding in some sills, and the characteristic variolitic textures in others, are probably to be explained according to the order in which the crystals nucleated from supercooled magma, rather than to processes of crystal accumulation under gravity.

The Palisades sill and the Shiant sill are taken as two examples exhibiting many of the features outlined above. Not only do each contain an olivine-rich layer near the base, but they provide evidence of extensive fractionation of two contrasted magma types, the tholeiitic basalt and the (alkali) olivine-basalt. Other sills are discussed together, only those features relevant to the present context, or of other general significance, being discussed.

PALISADES SILL, NEW JERSEY, U.S.A.

This well-known sill is about 1000 ft thick and has been described in detail by Walker (1940). It is a classic example of a minor basic intrusion in which the relationships between the chilled margins and the various differentiates are well exhibited. The original magma had the overall composition of a tholeiitic basalt, and differentiated in the body of the sill to produce olivine dolerites near the base and quartz dolerites near the top. The plagioclases become progressively more sodic towards the top of the sill, while the pyroxenes become more iron-rich and exhibit a series of phase changes similar to those observed in the pyroxenes of the thick Skaergaard layered series. The late-stage differentiates occur in a pegmatitic zone approximately two-thirds of the way up the sill, and have higher contents of silica, iron, and alkalis than found in the other rocks of the sill (Walker, 1953).

Although the whole differentiation sequence is impressive, the feature of the Palisades sill which has attracted the greatest attention is the olivine-rich layer (Fig. 2). The layer occurs at about 50 ft above the lower contact, averages about 20 ft in thickness, and contains 20-25% of olivine. Olivine occurs in the chilled marginal dolerite (c. 2%) and below the olivine layer (<5%), but is absent throughout the rest of the sill. It is generally accepted that in this respect the Palisades sill provides a striking example of crystal accumulation, the olivine crystals having sunk towards the floor and, by effective separation, resulted in the production of quartz-rich late differentiates. The composition of the initial magma was, apparently, sufficiently rich in silica that magnesian olivines were in equilibrium with the liquid only for a short period of the cooling history, and was sufficiently oxidized that iron-rich olivines failed to crystallize at some later fractionation stage.

Walker's contention (1940) that both the large and small olivines of the olivine-rich layer nucleated in the magma after intrusion, and then sank to the level of the floor (50 ft above the base of the sill by then, because of inward crystallization from the margins) has since been questioned. Jaeger and Joplin (1956, p. 445) suggest that the olivine layer consists of an accumulation of phenocrysts present in the magma at the time of intrusion; calculations of the cooling rate of the sill and of the settling velocities of the crystalline phases would suggest that only fair-sized phenocrysts would have time to settle before the magma became too viscous to allow further appreciable crystal settling. It is of interest, in regard to this hypothesis, to note that the olivine pheno- crysts in the chilled dolerite have the same composition as the large olivines in the olivine-rich layer (Walker, 1940, p. 1068). Also, the 2% of olivine observed in the chilled dolerite and, therefore, once distributed through 1000 ft thickness of magma could, theoretically, form a 20-ft layer of olivine by accumulation. The rocks of the 20-ft thick, olivine-rich layer contain a maximum of only 25% olivine so that clearly the olivine phenocrysts alone could have contributed to the formation of the olivine- rich layer. Other, smaller olivine phenocrysts probably reacted with the magma to form orthopyroxene—as evidenced by the pyroxene reaction rims around certain olivine phenocrysts in the chilled dolerite. A feature of the olivine-rich layer which is not explained by this hypothesis is the fact that small olivine crystals exist together with large crystals, and that the former are more iron-rich than the latter. We would suggest that the smaller crystals probably nucleated from supercooled magma just above the chilled floor, resulting in their smaller size and more iron-rich composition than those in equilibrium with the magma. This zone of supercooled liquid, relatively viscous and containing only a few small olivine crystals, apparently extended for about 70 ft above the floor, the larger olivines being able to sink only through the uppermost 20 ft to mingle with the small olivines and thus produce the olivine-rich layer.

These olivine relationships cannot be so well explained if we accept Hess's hypothesis (1956) that there was no crystal settling in the Palisades sill. Hess believes that crystal settling would result in igneous lamination, notably absent from the Palisades sill. Clearly this reasoning could be applied to the olivine-rich layer only if the olivines had a tabular habit, which they have not. In regard to other minerals, such as tabular plagioclases, we believe that igneous lamination requires the aid of horizontally moving currents, unaided settling in relatively viscous magmas probably tending to make the tabular crystals drift to the floor in a haphazard arrangement. However, Hess points out that small crystals would sink much more slowly than larger pockets of cool magma, and hence that convective currents are more likely than unaided crystal settling. We accept this view, and on that basis would agree that if convection was operative, the absence of igneous lamination amongst the Palisades felspar crystals makes it unlikely that these crystals are of cumulative origin.   The textures of the olivine dolerite, however, do not preclude crystal accumulation of the olivines, the felspars and pyroxenes being poikilitic and clearly crystallized from the intercumulus liquid.

Hess's hypothesis (1956) is that the rocks of the Palisades sill crystallized from the margins inwards, the fractionation being due to the convecting magma feeding the growing crystals and removing heat from the solid–liquid interface.   The selective diffusion of material towards and away from the growing crystals would deplete the magma in the components of the higher-temperature crystalline phases (e.g. Mg and Ca) and enrich it in components such as silicon, iron, and alkalis.   This process seems the most likely one to account for the regular fractionation sequence above the olivine-rich layer, and for the relatively calcic plagioclase and magnesian pyroxene present as poikilitic crystals within that layer.   Also, it would seem to be the most likely process to have operated in most sills in which crystallization differentiation, without evidence of crystal settling, has occurred.

SHIANT SILL, NORTH SCOTLAND

The Shiant isles, 20 miles north of the Isle of Skye, consist almost entirely of post-Jurassic differentiated sills (Walker, 1930).   A 400-ft thick sill is present on Garbh Eilean, and extends to the neighbouring island of Eilean an Tighe.   The upper part of a sill on Eilean Mhuire is now believed to be a down-faulted portion of the main sill, so all the exposures are referred to as a single, Shiant sill.   Walker sub-divided the sill into a lower picrite portion (37 ft), a picrodolerite (120 ft), and an upper crinanite (250 ft).

The picrite layer contains 61–66% olivine, the picrodolerite about 45% olivine, and the crinanite about 20% olivine.   Although there is a sharp drop in olivine content at the top of the picrite layer, a graphical representation of olivine content *vs* height in the sill shows the variation to be in the form of a smooth curve with a marked change of slope, rather than a broken sequence (Walker, 1930).   Drever (1953) and Drever and Johnston (1959) view the picrite as a separate intrusion, and Dr Drever is at present supervising a drilling programme aimed at elucidating the character of the sill from top to bottom.   From the present evidence, however, the Shiant sill appears to provide an excellent example of a sill in which olivine settling has taken place.   In fact, the top of the picrite layer does not appear so sharply defined, according to olivine content, as the Palisades olivine layer, and added to that there is the gradual decrease in olivine content in the picrodolerite to crinanite sequence.   The continued crystallization of olivine is due to the Shiant sill being of (alkali) olivine-basalt magma affinities.   Such liquids, if they lie to the nepheline side of the critical plane of undersaturation (Yoder & Tilley, 1962, fig. 13a, stippled area) would, as is found in Shiant, give late-stage differentiates bearing felspathoidal minerals rather than free silica.

The compositions of the Shiant olivines have been studied in detail by Johnston

(1953) who has shown that the unzoned olivines of the picrite layer are Fo_{84}, and that the olivine phenocrysts in the chilled rock at the base have a similar composition (*op. cit.*, p. 168). Upwards in the sill, the olivines become increasingly more zoned (*op. cit.*, fig. 2), extreme examples having rims of Fo_{10}. Johnston's diagram, however, shows a feature of especial interest which has not, so far, been discussed in regard to its petrological implications. The olivine compositions are such that the cores remain close to Fo_{80-84} throughout the whole sill, only the *zoning* being responsible for the iron-enrichment upwards. This can only mean that all the olivine cores crystallized at about the same temperature, and that although olivine settling took place initially, it later ceased, so that some of the magnesian olivines were prevented from reaching lower levels of the sill. The fact that the olivine phenocrysts of the chilled rock have a similar composition to the largely-settled crystals suggests analogy with the Palisades sill, the settled crystals being viewed as an accumulation of phenocrysts present in the magma at the time of intrusion. Clearly, the Shiant magma must have been richer in olivine phenocrysts than the Palisades magma, as to be expected in alkali olivine-basalt magmas. The fact that olivine settling appears to have been retarded after the initial stages of picrite formation lends support to the view that this retardation may also have happened in the Palisades sill, there preventing the settling of crystals other than the olivine phenocrysts.

The felspars of the Shiant sill, from the data given by Johnston (1953, pp. 162–4) lend further support to the above hypothesis. They are An_{80} in the picrites, zoned from An_{80} to An_{55} in the picrodolerites, and zoned from An_{80} to An_{35} in the crinanites. Again, incomplete settling of the An_{80} plagioclases is suggested, although the plagioclases in the picrites are generally poikilitic and may only be cumulus at the centres of each crystal. The unzoned nature of the olivines and felspars in the picrites, and the increased zoning upwards, indicates that apart from crystal settling the greater part of the intrusion crystallized from the bottom upwards, the diffusion process resulting in the progressive enrichment of the residual liquids in iron and sodium.

The textures of the picrodolerites and crinanites are not of the type found in igneous cumulates, in contrast to the textures of the picrites. Characteristically they are ophitic and sub-ophitic, and suggest that apart from the suspended phenocrysts, the bulk of the successive liquid fractions crystallized fairly rapidly. An exception is found near the top of the sill on Eilean Mhuire (Drever, 1957) where rhythmic layering, gravity stratification, and igneous lamination are present. This is very unusual in sills and may be due to a local, high volatile concentration which lowered the crystallization temperatures such that crystallization here took place over a greater time interval. The liquid would thus be less viscous, also, and permit local crystal settling. The fine-scale layering may be due to crystal sorting and settling, or to variations in water-vapour pressure (Drever, 1957), although the presence of convection currents in the thin, upper part of a sill seems unlikely, as do the roof conditions which would permit both build-up of water vapour on the one hand, and loss on the other. Alternatively, it would seem more likely that rhythmic layering in thin sills can be attributed to rhythmic nucleation from a supercooled liquid, according to the process discussed in Chapter VIII.

OTHER SILLS

In view of what has been said above, it seems likely that differentiation in sills can broadly be attributed to two major processes. Firstly, the settling of early formed crystals, shown above to be confined largely to phenocrysts already present in the magmas at the time of emplacement, and secondly, to crystallization differentiation by crystal growth from the margins, but chiefly from the floor upwards. In regard to the latter point, the latest differentiates are never found in the lower half of the sills but invariably at about $\frac{2}{3}$ to $\frac{3}{4}$ of the way up the sills. A subsidiary, attendant process would be the migration of volatile constituents, chiefly water, to a level immediately below the downward crystallized part of the sills where the latest, often pegmatitic differentiates are to be found.

The Dillsburg sill (Hotz, 1953) has a chilled margin similar in composition to the Palisades sill, and with similar olivine phenocrysts. The drill-core section (370 ft thick) passes, beyond the upper chilled rock and normal dolerite, into pegmatite dolerite, transitional zone, granophyre, and transitional zone again. This is abundant proof that differentiation has taken place largely through crystallization from the margins inward, and Dillsburg is probably the best intrusion for studying the downward-crystallized portion. The mineral compositions display a symmetry which suggests that had the drill-core gone deeper, normal diabase would then have been found, succeeded by magnesian rocks. By analogy with the Palisades sill, and in view of the character of the chilled marginal rock and the thickness of granophyre, a thick olivine layer would be expected near the base of the Dillsburg sill.

Olivine settling appears to have taken place in the Elephant's Head dyke of South Africa (Poldervaart, 1944) to form picritic cumulates in irregular basins near the floor. This dyke is broad and therefore similar, in general, to the macro-dykes of South Greenland (p. 507), and is included here because it is clearly associated with the less differentiated, higher level, New Amalfi sill, the crystallization history of each taking place in a similar environment. Poldervaart suggests that the dyke was a feeder to the sill and it may well be that the settled olivines were present in the magma at the time of emplacement, thus explaining their absence from the overlying sheet which, according to Poldervaart, was formed from residual liquid drawn from the dyke.

The Black Jack sill near Gunnedah, Australia (Wilkinson, 1958) is of alkali olivine-basalt type and displays several of the features found in the Shiant sill, such as the persistence of olivine throughout its crystallization range (Wilkinson, 1956). However, Wilkinson is of the opinion that olivine settling did not occur here. The olivine content decreases, generally, from about 20% near the base to about 4% near the top, changing in composition from Fo_{79} to Fo_{40}. Wilkinson believes that this can be explained according to crystallization from the floor upwards, because the olivine content of basaltic magmas decreases with fractionation. He suggests (1958, p. 31) that the most olivine-rich teschenite near the base (23% olivine) fulfils '. . . the compositional requirements of a parental magma . . .', and views each successive rock in the upward sequence as compositionally equivalent to successive liquids. We find this hypothesis untenable, because one cannot equate a rock containing the most magnesian olivine, in a differentiated sequence, with the parental magma which must normatively be more iron-rich than the earliest ferromagnesian minerals to separate from it. If the Black Jack initial magma were of this composition it would also have contained 23% olivine phenocrysts (Fo_{79}) in suspension, and therefore approximately one-quarter of the sill should consist of olivine of this composition. Furthermore, all these olivines would need to have been effectively concentrated away from the body of magma in order for it to progress to the precipitation of olivines as iron-rich as Fo_{40}. We believe that an alkali olivine-basalt magma would contain olivine phenocrysts on intrusion, but not the large amount found in the lowest rock, and that in any case the absence of such olivines (Fo_{79}) from the upper parts of the sill must mean that they settled to give the rocks rich in minerals of this composition near the base. The absence of olivines from upper rocks cannot be accounted for by reaction with liquid, in the case of a teschenite magma, so that if crystal settling were ruled out then the initial magma would need to have been free from olivine phenocrysts and, hence, even further removed from the composition suggested by Wilkinson.

Several differentiated sills are free from olivine or olivine replaced by orthopyroxene and, therefore, in bulk composition need to be not only quartz normative, but with sufficient silica to prevent the early crystallization and separation of magnesian olivine which is possible in moderately oversaturated basalt magmas. The Tasmanian dolerite sills are generally of this type, although sporadic olivines occur in a few, and Edwards (1942) gave a remarkably fine account of the variation in pyroxene mineralogy throughout some of these sills. He suggested that the differentiation was due to gravitative accumulation of pyroxene, but no modal analyses are given, and from the absence of a reference to pyroxene-rich layers it may be that the differentiation was due to crystallization from the margins and, chiefly, from the floor upwards. McDougall (1961) has also studied the pyroxene mineralogy of a dyke-like extension of a Tasmanian dolerite sill in detail, but in this case it is more obvious that the differentiation is due to fractional crystallization from the walls inwards. The Ferrar dolerites of Antarctica are, as pointed out by Edwards, similar in composition to those of Tasmania. Gunn (1962, p. 832) has shown, however, that these sills are often enriched in modal pyroxene in their lower parts. In the 800-ft Mount Egerton sill (Gunn, 1963) coarse-scale rhythmic layering occurs, individual melanocratic layers containing up to 70% of magnesian pyroxene, and leucocratic layers, 84% of calcic plagioclase. Marked cryptic variation is exhibited by this intrusion and it is one of the clearest examples of a sill, so far described, in which crystal accumulation has been the chief factor in its differentiation. This example seems to show that even in a sill of only moderate thickness, crystal settling can occur throughout the greater part of its cooling history, and that pyroxenes and plagioclases can settle to the floor, given the right conditions, in the same way as olivine settling took place in several other sills.

Most of the sills described so far in this chapter can be said to exhibit certain features in common. They are sills in which crystal settling has probably taken place, although in most cases this is confined to the earliest stages of crystallization and involves the minerals, usually olivine, present in the magma at the time of emplacement. Further crystallization *in situ* would be relatively rapid in most cases, and the textural evidence suggests that crystal settling was then negligible and that consolidation took place by crystallization from the floor upwards, and to a lesser extent from the roof downwards. The crystallization differentiation at this stage probably took place through the mechanism advocated by Hess (1956) and briefly discussed earlier in this chapter.

Many sills, however, show no evidence of crystal settling and yet they are frequently heterogeneous and banded, the separate bands having either a contrasted mineralogy or, in many cases, merely a differing grain-size. The more common banding is due to an alternation of average and pegmatitic facies. The usual explanation for a pegmatitic band is that it was relatively hydrous, but this imposes difficulties associated either with the problem of liquid immiscibility or with the rhythmic increase and decrease of water content in a shallow-level, crystallizing magma body. One of the most interesting sills so far described, of this banded type, is the Gars-Bheinn sill of Skye (Weedon, 1960). Bands rich in olivine (Fo_{88}) alternate with those rich in plagioclase (An_{79}) and augite, there being a slight upward change in olivine and plagioclase compositions towards lower melting-temperature members. The olivines are usually small, discrete crystals, often tabular and exhibiting igneous lamination but occasionally skeletal in habit. The plagioclases are poikilitic in the olivine bands, but in the felspar-rich bands they occur as large crystals elongated perpendicular to the banding, and often radiate from centres near the top of each felspathic band. Weedon points out that from the evidence of the chilled marginal rock, the magma was aphyric when emplaced. He proposes that each period of olivine settling resulted in a change in magma composition such that felspars crystallized at the top of each olivine-rich band, and that local volatile concentration took place at each of these stages, resulting in pegmatitic felspar growth. We do not feel convinced that volatiles could be concentrated locally and frequently in this way, and favour an alternative hypothesis. Briefly, this is that steady cooling of the aphyric liquid would result in supersaturation and, first, in olivine nucleation, the often-skeletal olivines growing and sinking to form an olivine-rich layer. This process would leave the liquid free from suspended crystals, and plagioclase which nucleated between the settled olivine crystals could then grow upwards into this liquid, in a manner analogous to the crescumulate growth in Rhum. However,

Weedon has shown that in many cases the elongated felspar crystals radiate from centres near the top of the felspar-rich bands, rather than from the base. The liquid immediately adjacent to an olivine-rich band, because of continuous crystallization at that level, would be slightly hotter than that further away, so it is possible that plagioclase would nucleate a short distance above the base of the liquid and sprout downwards. The Gars-Bheinn sill thus provides an example in which banding is probably produced by rhythmic nucleation of separate mineral phases from stationary, supercooled magma. Certain mineral phases, under these conditions, may tend to grow large and in an elongate form, perpendicular to the plane of banding. As a general concept this may explain much of the coarse banding usually attributed to pegmatitic growth from volatile-enriched parts of the magma, though not necessarily in the case of those pegmatitic bands rich in iron and alkalis (Walker, 1953).

CHAPTER XIX

SOME GENERALIZATIONS ON
LAYERED INTRUSIONS

The title chosen for this book serves to emphasize the most significant general feature of the large number of igneous intrusions that have been described. Apart from the property of layering, most of the rocks have certain other characteristics that developed as a result of their depositional origin and which are diagnostic in those cases where layering is locally absent or inconspicuously developed. One of the main aims of the authors has been to demonstrate in detail that for one intrusion, the Skaergaard, the layering is due to a process of crystal accumulation from basaltic magma and that the assemblage of igneous rocks enclosed within the marginal envelope of chilled parental magma is due to the slow fractional crystallization of that magma. The greater part of the intrusion, fortuitously opened by erosion and kept clear of debris and overgrowth by glacial action, is laid bare for internal diagnosis like a carefully dissected organism. Within the enclosing skin, the walls and the internal parts can be examined in minute detail in relation to their structural, textural, and chemical characteristics. In this way the observations can be used, according to petrological practice, to reconstruct a picture of the evolutionary processes responsible for the growth of each part, and for the development of the whole igneous body of interconnected parts, from a known parent. The analysis of the Skaergaard intrusion, presented in Part I, necessitated a development and discussion of most of the principles involved in the formation of layered igneous rocks and, as pointed out in the Preface, could not profitably be repeated in this chapter. Turning from the Skaergaard to the igneous bodies described in Part II, the analyst is presented with a wide variety of cases, each generally presenting a fragmentary picture. Layered structures are generally present, but these and associated features show obvious or subtle differences from those found in the Skaergaard, and thus provide further evidence for the origin of layered igneous rocks. If the crystallization history of the one intrusion can be understood, so far as is possible within the limits imposed on geologists, we are then in a better position to learn more of the varying crystallization histories of the other, broadly comparable but less complete intrusions. Much remains to be done along these lines, and it is to be hoped that the available information drawn together in this book will serve as a basis for further work. We have assumed that the emphasis given to the analysis and interpretation of the Skaergaard intrusion will be viewed as the establishment of a basis for comparison, rather than of a nucleus of ideas in relation to which other ideas need be restricted in their orbits. If generalizations are to be made and yet, at the same time, useful distinctions drawn, this can best be achieved at the outset by avoiding terms such as the ' Scottish Bushveld' or the ' Canadian Skaergaard ' in referring to other specific layered intrusions !

The layering of igneous rocks is a remarkably widespread feature amongst certain rock-types, particularly those of gabbroic affinities. As the list of discovered layered intrusions has grown, it is understandable that the geologist has begun to wonder whether unlayered gabbros are purely a local product of an otherwise layered body or, if not, whether they pose a special problem. A body of basic magma with a limited thickness, such as the 1000 ft or so of the thicker sills, has generally crystallized without layering in the sense defined in this book, but there the doleritic texture is a further indication of the relatively quick rate of cooling. Given a coarser-grained gabbroic texture, the likelihood is that cooling conditions would normally have resulted in appreciable crystal settling and layering. The form of intrusion is expected to influence the development of layering, however, and a narrow, steep-sided body intruded into a hot environment may crystallize slowly to give gabbroic textures and yet fail to form by crystal deposition from the bottom upwards.

We have described cases of layering from practically all the better-documented sources of information, so it is now possible to make certain generalizations in regard to the *rock-assemblages* with which layering is associated. The feature is confined chiefly to major intrusions with a basaltic parentage and may in part be characteristic of all such intrusions. Alternatively, some of these intrusions may not show layering but may have other features suggestive of crystal accumulation, such as igneous lamination or igneous cumulate textures. Certain syenites and felspathoidal syenites show well-developed layering, and the same is true of a significant, but minute proportion of known granites. The majority of granites, and of intrusive bodies solely of intermediate composition such as diorites and syenites, are devoid of layering (as distinct from certain banded structures of a different origin). The layering of ultrabasic and ultramafic rocks presents one of the greatest problems. In intrusions such as the Bushveld and Stillwater, the ultramafic and ultrabasic (felspar-bearing) rocks clearly form part of a layered sequence of basaltic differentiates leading to layered gabbros and diorites. Certain examples consisting of layered ultrabasic rocks alone, such as Rhum, can also be shown to have developed by crystal accumulation from a basaltic magma, though since displaced from the original site of formation. The latter example may be viewed as a link with the sort of ultrabasic and ultramafic bodies which show well-developed layered structures and yet no apparent link with a parental basic magma, such as the New Zealand and Alaskan intrusions. On the other hand, the highly magnesian compositions of the olivines and pyroxenes in the latter intrusions suggest possible derivation from magmas more picritic than basalts, such as those which probably gave rise to the Great Dyke and Bay of Islands intrusions. The great thicknesses of ultramafic rocks in the latter intrusions are highly magnesian, although the rocks show a distinct sequence of rhythmic and cryptic layering and are overlain by a small differentiate of gabbroic cumulates. The sedimentation structures in the Duke Island complex are suggestive of crystal accumulation and layering in what is probably an ultramafic magma body, in which case the range of liquid compositions capable of giving rise to layered igneous rocks may be extensive. Certain ultramafic rocks, however, may be drawn from the upper mantle during orogeny; if this is so, then the often-characteristic layering, well shown as olivine–chromite layering in the Appalachian bodies, for example, must be a feature developed within the upper mantle and is less likely, therefore, to have arisen purely by crystal settling through liquid magma. In general, apart from the problematical aspect of upper mantle layering, it would appear that most basalt magmas are in a condition suitable for the formation of layered igneous cumulates, but that occasionally ultramafic or ultrabasic, syenitic, and granitic magmas are in the appropriate condition. The required conditions are necessarily such that the successive magma densities and viscosities, with slow cooling and fractionation, are of an order appropriate to the settling of the crystalline phases successively in equilibrium with those magmas. However, crystal settling is not sufficient in itself to produce the rhythmic pattern of layering, and one or other of the various conditions necessary for such a pattern, discussed below, must also be fulfilled.

Although it has been stressed that basalt magmas are the most likely parent of layered cumulates, a distinction can also be drawn between those broadly of tholeiitic and of alkali basalt affinities. Although it is rare to obtain an undoubted chilled facies representative of the initial parental magma, the assemblage of differentiates provides evidence as to the probable type of parental basalt. By far the greater majority, including those about which most is known (e.g. the Skaergaard, Bushveld, and Stillwater) are of tholeiitic affinities (generally of high-alumina type) in terms of estimated parental magma, late differentiates, and mineralogical trends (especially of the two-pyroxene assemblage). Indeed, the authors are not aware of any major layered intrusion likely to have stemmed from an alkali-basalt parent magma; there are the spectacular layered bodies of felspathoidal syenites, but the bulk composition of these is appropriate more to derivation from a syenitic than from a basaltic magma. It is clear that layered intrusions such as Kiglapait (Labrador) and Insch (N.E. Scotland), where the late differentiates are syenitic and yet the pyroxene fractions of the gabbroic rocks are of the usual tholeiitic type, differ from the majority in relation to the probable parental magma-type involved. They do not give rise to felspathoidal late-differentiates, but in that case the absence of quartz-bearing derivatives is unusual and cannot be explained entirely according to extremely reducing conditions of crystallization. Any consideration of the cumulus mineral assemblage in relation to the likely parental magma, however, needs to be viewed in relation to the likely confining pressures, as well as to the oxidation states of the magmas involved. What does seem clear at this stage is that no described major layered intrusions show the type of differentiation observed in minor intrusions such as the Black Jack and Shiant Isles, picrite–teschenite sills. A requisite for the major layered intrusions is the availability of a huge volume of magma at the time of emplacement. In a closed-system type of intrusion such as the Skaergaard, the encircling sheath of chilled parental magma indicates that only one act of injection was involved, and it has been estimated that the volume of emplaced basaltic magma was about 500 km³ (Ch. II). This is far in excess of the volume involved in individual lava flows (e.g. one of the more voluminous, Laki, was about 15 km³), and would most likely have stemmed from a basalt pool in the upper part of the mantle. Current petrological hypotheses favour the derivation of alkali basalt from deeper levels of the mantle, and although *small* pockets of such magma could reach high levels in the crust, it is suggested that because the slow accumulation of a *large* pool would be more likely to take place near the top of the mantle, the lower pressures may result in alkali basalt liquids precipitating olivine rather than enstatite, and changing towards more siliceous compositions (cf. Yoder & Tilley, 1962, fig. 48). The relationships between tholeiitic and alkali basalts are the subject of intensive current research and speculation, and when the parentage of each or the parental rôle of either is better established, the tendency for the massive emplacement and fractionation at crustal levels only of basalts predominantly of tholeiitic affinities is likely to be a factor of particular significance.

The *intrusive form of layered intrusions* is a topic of importance in relation to the act of injection of a large volume of magma. It is probably not an exaggeration to suggest that most geologists, asked about the form, would answer that large basic intrusions are predominantly lopolithic. By this it is implied that down-sag of a large crustal area resulted in a basin-shaped cavity with a semi-horizontal roof. It is difficult to envisage this required succession of events, if it is to be reconciled with the rapid emplacement and chilling of a huge volume of magma. If the upward movement of magma from the upper mantle initiated crustal down-sag, this would be a slow and steady process of volume-for-volume replacement. Alternatively, a rapid down-sag to provide the necessarily huge cavity would be more unlikely, not only requiring a temporarily stable, empty pocket in the crust, but also, extensive foundering of crustal rocks into a denser medium. Such problems would not arise if the act of injection were a slow and continuous process, but this would apply only to the open-system type of layered intrusion such as the Rhum and, probably, the Bushveld intrusions, where successive magma replenishments can be inferred from the pattern of cryptic and rhythmic layering and where the marginal chilled facies is not, therefore, representative of the composition of the entire bulk of layered rocks. In the Skaergaard case, the single act of injection appears to have been achieved by the upward displacement of crustal rocks along a conical fracture pattern, the resultant intrusion being funnel-shaped. Observations at the margins of other layered intrusions indicate that many dip inwards at angles appreciably steeper than those of the internal layered rocks, so the funnel is probably a form that is far more characteristic than the lopolith. This appears to be true, for example, for the Bushveld and also for the once-standard example of a lopolith, the Duluth intrusion. Appreciable uplift of the crustal rocks is involved in funnel-injection, as in the case of a cone-sheet complex. Apart from some explosive expulsion of material at the surface, the pressure may have been relieved by doming, while crustal down-sag subsequent to magma emplacement would have resulted from the increased load.

It is a fairly straightforward matter to recognize and investigate the pattern of rhythmic and cryptic variation within a layered intrusion, and thence to be fairly confident of the manner in which the layered rocks formed. It is altogether another matter to find evidence for the original form of the magma chamber in which slow cooling and crystallization took place, and to envisage the conditions under which the parental magma was emplaced initially, and cooled against the walls of country rock. The Skaergaard intrusion provides this evidence in the form of a series of marginal and upper border group rocks, but otherwise the major intrusions have a complexity far out of proportion to the relationship between their sizes and those of the relatively uncomplicated sills. For many years the impressive weight of evidence provided by the Skaergaard tended to overshadow and influence ideas on layered intrusions to the extent that if an intrusion, albeit layered, did not show regular cryptic variation and distinctive marginal relationships, then its origin by the bottom-accumulation of crystals was doubted (e.g. Barth, 1962, p. 195). This is particularly the case for the Bushveld, for which origins by separate sheet-injections of contrasted magmas, or by metasomatism of pre-existing rocks, have been proposed by certain investigators over the past few years. Faced with this problem in Rhum, where it was possible to show that the ultrabasic layered features were probably due to crystal accumulation from a basaltic magma, it was necessary to account for the lack of marginal features, the scarcity of evidence for a pore liquid crystallized between the cumulus crystals, and the absence of cryptic layering throughout the thick series of rocks layered chiefly on a macro-scale. None of these features is characteristic of the Skaergaard, but it was proposed that Rhum is an entirely different type of layered intrusion which could usefully be called an ' open system ' rather than a ' closed system '. It was later envisaged that in some other, and perhaps the majority of layered intrusions, the parental magma was not emplaced in a single episode to give a complete series of closely related layered differentiates and a symmetrical border zone of early-cooled rocks. Instead the magma chamber, situated fairly high in the crust where the slow crystallization of basalt could occur, may have lain in the path of basalt moving up from the mantle towards the Earth's surface. Such magma chambers are probably situated below most well-defined volcanic centres (in contrast to the extensive fissure zones), where they would be responsible for the local supply of heat often responsible for melting of the adjacent crustal rocks. Situated at a level where crystallization could occur, they would control the periodicity of the volcanic eruptions in terms of gas-pressure effects, as well as the composition of the extruded lavas, while occasionally they may contribute blocks of layered rocks torn from the crystallization chamber and erupted with the lavas. In the absence of such local, high-level crystallization chambers, there is no reason why lavas erupted directly from the mantle should show appreciable signs of a fractionation trend related to crystal separation at high levels and under low pressures. In regard to the magma chamber itself, the periodic eruptions and the associated replenishments with magma from below, keeping the temperature fairly constant, would be expected to leave their imprint on the character of the crystalline precipitates. Macro-rhythmic layering, the absence of extensive cryptic layering, and the abundance of adcumulates (suggestive of slow cooling and precipitation) are three such records, while in Rhum the abundant slumping and other disruptive features, culminating in upward displacement of part of the consolidated cumulates and destruction of a possible marginal border facies, are further indications of the turbulent history of this type of intrusion. These features have been reiterated here because it is important to appreciate that in many respects the circumstances of igneous activity appear *more likely* to produce this type of layered complex than the Skaergaard type. For the Bushveld, where there is additional evidence of oscillatory cryptic variation and of lateral transgressions of magma at different periods, we have suggested that unstable conditions characterized the earlier part of the Bushveld's crystallization history (the integration stage) but that later (the differentiation stage) the conditions become relatively stable and approximated more to those of the Skaergaard. The same may perhaps be true of the Stillwater, providing an alternative explanation for the macro-rhythmic layering of the Ultramafic Zone, and true of the Bay of Islands and Great Dyke. If so, it would seem logical to propose that the early life of a layered intrusion differed from the later part in that it was less settled, and hence the ultramafic parts may be expected to compare more closely with the layered features of Rhum than with the exposed, gabbroic parts of the Skaergaard. Despite this generalization, it is equally obvious that disruption in the form of magma replenishments could occur at other stages in the history of an intrusion, and the Kap Edvard Holm example shows such evidence for a relatively late event. The Cuillin complex of Skye shows, in particular, how an extremely complicated sequence of faulting and lateral displacement of magma-chamber centres may accompany the formation of a layered intrusion. With these various cases in mind, it is less surprising that most described examples show an incomplete, irregular, or fragmented pattern of structures and rock associations. The Muskox may seem more promising in that the feeder-dyke is observable, but even there it appears that magma emplacement was a continuous, complicated process, while convincing upper border group rocks and late differentiates are apparently not present to complete the picture. If layered intrusions often form under conditions of complex igneous activity, it is all the more important that one should have criteria for establishing the origin of those rocks that show complex inter-relationships. This is why we have stressed those structural and textural features which characterize rocks of an accumulative origin, and believe that in the absence of more extensive evidence one is justified in extrapolating to a certain extent. This extrapolation is difficult in some cases, particularly for certain ultramafic layered bodies from orogenic environments, and it would be unwise at present to postulate that all such cases of mineral segregation are due to the layering process. However, we have been careful to include only those examples which, according to our observations or the published accounts, possess properties analogous to those of other, dominantly ultramafic intrusions where a layered origin can be more convincingly demonstrated; for example, the Great Dyke of Rhodesia.

Layering has been defined in the Preface, and we would wish to repeat, here, that the use of the term should be restricted and not viewed as synonymous either with stratification or with banding. The deposition of crystals from a high-temperature magma, resulting in a conformable series of layers which are inter-related according to the well-established laws of physical chemistry pertaining to crystal–liquid equilibrium and chemical element distribution, is a process sufficiently unique and significant to merit distinction from other geological processes. *Layering is an igneous phenomenon, and moreover it is a planar feature dependent for its origin largely upon the physical effects of gravity and the chemical effects of magmatic fractionation.* These two features could usefully be called ' density layering ' and ' chemical layering ', but the terms ' *rhythmic layering* ' and ' *cryptic layering* ', first used for the Skaergaard, are now in popular use and have been retained. The following discussion will show that in most respects the term rhythmic layering is preferable, while in petrology one can encounter difficulties in speaking of chemical variations, should isochemical phase changes involving mineral polymorphs, for example, be the significant feature.

Practically all the intrusions described in this book consist essentially of rocks whose heterogeneity can be resolved into homogeneous units by superimposing a variable number of imaginary planar boundaries. The boundaries are generally close to the horizontal, and remain almost parallel over vast lateral distances. The compositional differences between adjacent units are often extreme but if more than two or three units are considered together, then the inter-relationship is seen to be cyclic or rhythmic and the extreme differences to be localized. This aspect of layering, clearly visible on rock surfaces in the field, is due to the units varying in the constituent proportions of the minerals of a particular assemblage. Some units we have called *uniform* in that the proportions of each mineral are approximately equal, but it is more common for a particular unit to be rich either in melanocratic or in leucocratic minerals. Frequently a melanocratic unit grades upwards into a leucocratic unit, but it is rare for a leucocratic unit to have other than a sharp boundary against an overlying, melanocratic unit. Gradations are impossible to deal with according to rock nomenclature, and it has been found more convenient to refer to a melanocratic layer grading upwards into a leucocratic layer as a single unit. Similar to graded bedding in sediments, except that density sorting appears to have been more significant than size sorting, this feature is known as *gravity stratification*. The most common repetition of units upwards in a layered sequence, involving (*a*) melanocratic, (*b*) gradational, and (*c*) leucocratic layers, is of the nature *abc . . . abc* and is therefore rhythmic rather than of the cyclic pattern, *abcbabc . . .* Hence the general feature is called *rhythmic layering*, and may be on a coarse scale involving graded units 100 ft or so in thickness (*macro-rhythmic layering*) or on a scale as small as a few millimetres (*fine-scale rhythmic layering*). Graded units may not always be present, however, or may alternate with uniform units, with a group of extreme units or, rarely, with units showing downward grading. Hence the common term, rhythmic layering, must be viewed now as a general, group name for a series of units with contrasted mineralogy and is not to be taken so literally as to exclude the varieties mentioned above. The terms *unit* and *rhythmic unit* of layering were first applied to Rhum where the thick, ultrabasic macro-units dominate the scene, as in the ultrabasic parts of the Bushveld and Stillwater intrusions. In describing the fine-scale layering of the Skaergaard, some difficulties arose through earlier reference to ' gravity-stratified layers ', because the melanocratic and leucocratic parts were also referred to as layers. Hence it has recently been found convenient to apply the term *unit of layering* to any distinct layer, whether extreme or average in mineral proportions, and also to any set of layers which are more conveniently grouped together in considering a repetitive pattern.

The *repetitive layering* of major intrusions is the main feature that distinguishes them from sills which are, in the simplest terms, equivalent to a single macro-unit of a major intrusion. The concept of crystal settling under gravity is inadequate, alone, to explain the presence of a melanocratic layer overlying a leucocratic layer, or the fact that the upper rocks (containing iron-rich silicates) of a thick layered sequence are, on average, denser than the lower rocks (containing magnesium-rich silicates). While the latter is a function of cryptic layering, the former needs to be explained in terms of rhythmic layering. It has often been said, correctly, that igneous layering ' resembles the bedding of sediments, when viewed from a distance '. There are fortunately few geologists who favour distant viewing to the extent of advocating that the rocks are metasomatized sediments or lava flows, and none who, having considered the textures and cryptic variations, would be likely to accept this is an hypothesis based on observable facts. A similar analogy could be drawn with multiple-sill injections, although there are no recorded cases of multiple injections, free from screens or transgressive contacts, occupying a total thickness such as the tens of thousands of feet found in layered sequences. Nevertheless, the hypothesis that layering is due to successive injections of *contrasted* magmas has several advocates, even today, having been the popular hypothesis at the beginning of the present century. It is particularly convenient, where mapping alone is involved, to record each different igneous rock unit as an intrusion, particularly where there are local, complex inter-relationships. The map is more quickly completed, and each rock type is shown according to a particular symbol. If, however, one is concerned also with the genetic history of a complex, certain restrictions mitigate against mapping of this sort. For example, liquids appropriate to

2 M

the composition of the pure olivine, plagioclase, and pyroxene layers would need to be at temperatures in excess of about 1400 °C and in some cases as high as about 1800 °C. Even the combination of these minerals in the uniform layers would require liquidus temperatures well above those of basaltic magmas (*c.* 1200 °C) and it is unlikely that water under pressure would be sufficiently abundant in such magmas to produce the required reduction in melting temperatures. The hypothesis stretches credulity even further in requiring that in the magma chamber being tapped, the available liquids would need to be, alternately, dunitic, pyroxenitic, anorthositic, etc. Whereas successive crystal accumulates may be of that composition, the main bulk of any natural magma can hardly be expected to oscillate in such a way, and is certainly not likely to split into immiscible liquids of that type. It could be argued that regular, cryptically varying series of crystal mushes were successively injected, rather than liquids, which is equivalent to agreeing that layered intrusions existed, but with the unusual reservation that they always lay below the levels of present-day exposures. The regularity in the cryptic variation of many layered intrusions is particularly difficult to explain according to the idea of continuous tapping of a deeper magma reservoir, for it implies that the complete crystallization record of a deep-seated intrusion was transferred, stage by stage, to a higher level intrusion. Finally, it must be emphasized that the most distinctive feature of layering is that extremely thin layers can generally be traced laterally over distances which are enormous relative to the thicknesses involved, and that conformable layering, involving distinctive, repetitive patterns, is the rule rather than the exception. Local irregularities do exist, and we have discussed several of these in the earlier chapters. The same is true for sedimentary rocks, however, in which unconformities, slumped beds and cross-bedded structures, to name a few, are common enough and yet do not, we trust, invalidate against their sedimentary origin. In fact, many irregularities of igneous layering are of particular significance in interpreting the various sedimentation processes involved in the formation of the rocks.

We are convinced that the layered intrusions described in this book are due to bodies of homogeneous magma cooling and crystallizing slowly in large chambers, the crystals accumulating on the floors of these chambers to form distinctive layers. The observed intrusions are the solidified magma bodies, either preserved *in situ* or disrupted after solidification. The repetitive pattern involving hundreds or thousands of layered units has been discussed at length in the preceding chapters. Basically, there are two main hypotheses to explain the phenomenon: either the various minerals were sorted by a mechanical agency, rhythmic in its action, or they were deposited in a definite order according to rhythmically repeated changes in the physico-chemical state of the magma. The mechanical sorting of a variety of minerals could perhaps be effected by settling over great distances, according to the different sinking rates of the crystals, but is more likely due to the action of convection currents.

We have considered mechanical sorting in some detail (Ch. VIII) and concluded that convection currents were probably operative in most large magma chambers, and that slow and continuous, or fast and sporadic currents, could best account for the uniform or graded units, respectively. In fact, the action of convection currents appears to be the likely mechanism to account for most types of fine-scale layering. In contrast, several of the intrusions described here are characterized in part by macro-rhythmic layering, on a scale unlikely to be produced by the sorting action of magma currents. In the formation of a macro-unit containing 100 ft of olivine cumulates, for example, it seems most likely that olivine was the only mineral crystallizing for a considerable period, and although there was probably some current activity the effect would not be observable except for possible size-sorting, igneous lamination or, as in Rhum, the development of olivine crescumulates during stagnant periods. The sequence of minerals in the macro-units is such that they can best be explained according to crystallization in an order, with cooling, predictable from certain synthetic phase-equilibria studies, assuming that the phases appear over shorter temperature intervals in multi-component systems (see Ch. XII & XIII). The repetition of macro-units, requiring slight increases in magma temperatures, would be effected by magma replenishments and it is therefore significant that where macro-units are found (Ch. XII, XIII, XIV & XV) cryptic layering is absent, weak, or oscillatory. In the formation of a macro-unit, a stage is reached when two or three mineral phases apparently were crystallizing together. The fact that such horizons are often layered on a fine-scale suggests that current activity was operative, and possibly throughout the formation of the whole unit.

It has been estimated that in the Skaergaard the faster convection currents probably travelled at about 3 km per day, so that at the stage when the exposed layered series ($2\frac{1}{2}$ km thick) was forming, crystals suspended in such a current could complete a circuit of the contemporary magma chamber in about three days. The rate of deposition would be variable, according to the current velocity, but if for present purposes the calculated average rate of about 20 cm per year is accepted, then about 2 mm would be deposited in the three days. Hence it is possible to explain layering on a very fine-scale according to separate convective overturns. However, it seems unlikely that fast current activity would be so prolonged as to account for the occurrence of dozens of layers, each a few millimetres thick. In many respects this type of layering (a variant found in the Stillwater being called 'inch-scale layering') is similar to that found in certain sills, for which convective overturns cannot be evoked. A convecting cell is unlikely to be much greater in breadth than in depth, in which case the thin layered sequence found in a few sills (e.g. the Shiant) would be extremely limited in lateral extent for any one cell, and the layering should be lenticular and discontinuous. Apart from this, convection requires an appreciable temperature gradient between the top and bottom of the magma body and in sills this is unlikely, heat probably being lost easily through the base of cool country rocks. A comparison with sills reminds one that here the cooling rate would be relatively rapid and that under conditions of incomplete equilibrium, crystallization from slightly supercooled magma would occur, ceasing temporarily after each crystallization event had provided latent heat to the magma. Although the cooling rate of major intrusions would be very slow, the occasional absence of convective stirring of the magma body may result in supercooling, especially if the stagnant magma were locally free from suspended crystals. Nucleation from a supercooled magma would most probably be in a sequence related to the structural complexity of the crystalline phases, and the commonest phases under consideration are cubic (spinels), orthorhombic (olivines and orthopyroxenes), monoclinic (augites), and triclinic (plagio-clases). It so happens that this sequence is similar to that which would be expected according to both of the other hypotheses (density sorting or crystallization from a basaltic magma under equilibrium conditions), so at the moment the supercooling hypothesis can only be offered as a possible alternative for the special case of layering on an extremely fine scale.

Cryptic layering is the term used to describe the regular variations in chemical composition throughout a layered sequence of igneous rocks. It is cryptic in the sense that it cannot be observed directly in the same way as rhythmic layering although this is not always true, especially where a distinctive mineral phase makes its first appearance. It is called layering because it is directly related to the disposition of the layered structures and, although generally impracticable, it would be possible theoretically to subdivide a well-fractionated intrusion according to an almost infinite number of compositional boundaries parallel to the plane of rhythmic layering. The cryptic layering is of two types: firstly, some mineral phases begin and cease to crystallize at specific fractionation stages; and secondly, those minerals belonging to solid-solution series change com-position continuously with fractionation. These types are broadly equivalent to N. L. Bowen's 'Discontinuous' and 'Continuous' Reaction Series, and could perhaps be termed 'discontinuous cryptic layering' and 'continuous cryptic layering' were it not that the first term implies discontinuity in what is, in fact, a regular and continuous chain of events. H. H. Hess's term 'phase-change layering' for the discontinuous type is probably the best available at present, providing it is applied only to a type of cryptic layering and not to the phase changes characteristic of rhythmic layering. We have preferred to use cryptic layering as a term applicable to both types of mineralogical variation, especially as the two are inter-connected in some cases (e.g. the pyroxene phase-changes, partly related to iron-enrichment in the solid-solution series).

There are certain patterns of cryptic variation which characterize most series of rocks developed by the crystallization differentiation of basic magmas. Chief of these are silicon, sodium, and iron enrichment with cooling. In the simplest terms the first is related to a discontinuous series (the system Fo–SiO_2) and the others to continuous series (the systems An–Ab; Fo–Fa; En–Fs, etc.). The complexity of natural magma systems requires analogy with more complex synthetic systems, but those which are pertinent such as CaO–MgO–FeO–SiO_2, Fo–Di–An–Ab, or Fo–Di–Qz–Ne, although of the greatest significance in interpreting the natural relationships, necessarily fall short in regard to certain requirements. It would be impossible, here, to attempt a summary of the extensive, relevant literature on this subject, except to observe that the work of N. L. Bowen, J. F. Schairer, H. S. Yoder, and numerous other experimental petrologists has made it possible to appreciate the significance of most of the major cryptic variation found in layered intrusions.

Cryptic layering in terms of the mineralogical changes has been discussed at length throughout the book, and we have emphasized this critical feature by giving a standardized type of vertical column for the intrusions described in any detail. In view of the lack of relevant experimental data applicable to the often complex mineral assemblages in the alkaline and granitic intrusions (Ch. XVI), discussions of possible cryptic layering are there limited to a generalized account. Excluding these and the type of intrusion showing little or no cryptic layering, the rest, as mentioned earlier in this chapter, are basic and of tholeiitic affinities. Apart from the phase changes of the discontinuous type, and the trace-element behaviour, most of the cryptic layering can be discussed in terms of regular compositional changes in the plagioclase felspars, olivines, and pyroxenes. Some data on the opaque oxides and sulphides are presented (e.g. Ch. III, XIII, XIV) but there is not sufficient information from a variety of intrusions to allow specific comparisons to be drawn.

The earliest *plagioclases* to separate from a basaltic magma, by analogy with phenocryst compositions and synthetic data, are likely to be calcic bytownites close to An_{85} in composition. This is a broad generalization, and a notable exception is the Skaergaard where An_{77} is the earliest detectable plagioclase phase. Also, plagioclases as calcic as An_{90-95} have been recorded from layered plutonic blocks erupted with lavas at St Vincent, the South Sandwich Islands, and Hakone (Japan), so it would seem that subtle differences in magma compositions or in the physical conditions of crystallization can produce marked effects on the earliest plagioclase to separate from broadly basaltic magmas. Extreme fractionation results in a compositional range, from base to top of a layered series, from calcic bytownite to a sodic andesine of about An_{30-35} in composition. Although extreme late-differentiates are rarely preserved, it is probably significant that in both the Bushveld and Skaergaard the plagioclase is of about the composition quoted here, subsequent differentiates generally being granophyres containing an alkali felspar. A more extensive trend for the felspars is shown by the Kiglapait and Aberdeenshire intrusions (Ch. XV), where the late differentiates are syenitic rather than granitic. The *olivines* are similar to the plagioclases in that although one can generalize about the earliest phase being close to Fo_{85}, exceptions are not uncommon (Fo_{81} in the Skaergaard, and up to Fo_{95} in the Great Dyke). Extreme fractionation, as in the Skaergaard, Bushveld, and Kiglapait, leads to pure fayalite in the latest differentiates, as it may in alkali-basalt sills such as the Shiant. However, in intrusions of tholeiitic affinities there is always a stage reached in the fractionation when olivine ceases to crystallize owing to silica enrichment in the magma, and a later stage when it recommences crystallization owing to the continued iron enrichment in the magma (cf. the system $MgO–FeO–SiO_2$). In terms of coexisting plagioclase compositions, these stages differ from one intrusion to another and the compositional ranges of the two series of olivines differ, also.

Apart from this interesting behaviour of olivine, the olivines and plagioclases do not have the same significance in layered intrusions as do the *pyroxenes*.   They are two-component systems in terms of the molecular end-members and thus no latitude is allowed in the pattern of the trends, with cooling and fractionation.   The pyroxenes, however, belong to a system with three molecular end-members, as well as exhibiting significant variations according to aluminium, titanium, and sodium contents.   In terms of the system $CaSiO_3$–$MgSiO_3$–$FeSiO_3$ alone, distinctive pyroxene fractionation trends permit a petrographic means of distinguishing between the products of tholeiitic and alkali-basalt magmas, and of their derivative magmas; of recognizing the general fractionation stages of the host rocks without recourse to compositional measurements; and of distinguishing between equilibrium and non-equilibrium (supercooled) products in certain cases.   The pyroxenes of alkali-basalt derivatives, as mentioned earlier, are not displayed in major layered intrusions but from the evidence provided by sills (Ch. XVIII) it is now well known that a titanaugite or calcic augite→aegirine-augite→ aegirine trend is particularly characteristic.   The tholeiitic trends are more variable, and one can recognize fractionation stages according to the series: magnesian augite+ bronzite→augite+inverted pigeonite→augite+uninverted pigeonite→ferroaugite→ ferrohedenbergite.   Apart from this series, evidence from the exsolution and inversion textures of the Ca-poor pyroxenes permits further sub-division (Ch. III).   The detailed study of the Skaergaard pyroxenes resulted in this type of information, following the classic studies of H. H. Hess on pyroxenes in general, and it was possible to show that available pyroxene analyses from other layered intrusions (e.g. Bushveld, Stillwater, Rhum, Duluth) plotted close to these trends.   More significantly, a recent study of the complete Bushveld pyroxene assemblage (Ch. XIV) has shown a remarkable similarity to the Skaergaard trend.   The irrelevance of ages or environments in considering the petrogenetic processes of crystallization differentiation in basic magmas is well appreciated, but these two intrusions form such a dramatic comparison that is probably worth drawing special attention to them.   Here we have an enormous Precambrian intrusion and a small Tertiary intrusion that can be drawn closely together by a comparison of their fractionation behaviour, in terms of the compositional changes in the olivines and plagioclase felspars.   More dramatic than this, however, are the distinctive pyroxene trends (Ch. III & XIV), either of which might have been viewed as unique if considered separately, but together demonstrating the logicality of the hypothesis that both are due to the extreme fractionation and layered origin of the respective intrusions.   On the other hand the trends show very slight differences (e.g. the extent of ferropigeonite and of iron-wollastonite stability) which appear to be related to differences in load-pressure at the levels of crystallization.

The *opaque minerals* are more problematical than the silicates because of the greater difficulties in distinguishing, optically and from the textures, between the cumulus parts of the crystals and the products of intercumulus crystallization. The available analyses of chromites, for example, may represent cumulus cores together with variable proportions of lower-temperature zones, and this may in part account for oscillations in recorded bulk compositions within the layered sequences. However, the cumulus crystals, indicative of the cryptic variation, may also fluctuate in composition owing to magma replenishments, particularly as the chromites occur in ultramafic layers which generally appear to have formed during an integration stage (e.g. Bushveld, Rhum, Great Dyke and, possibly, Stillwater), while they may vary in composition laterally, owing to oxidation effects (Ch. XIII). Despite these problems, there is in general an indication that cryptic variation, involving an upward decrease in Cr:Fe ratios, is characteristic of the chromite series (e.g. Ch. XIV). A gap in the spinel series appears to exist between the cessation of chromite and the advent of Fe–Ti oxide crystallization, shown particularly in the Bushveld layered sequence. Two phases then coexist, the rhombohedral phase usually being ilmenite-rich and the cubic phase a magnetite–ulvöspinel solid solution. While the total titanium content appears to increase upwards in the layered sequences, there is a progressive decrease in titanium and increase in vanadium in the titaniferous magnetites (Ch. III & XIV). Aspects concerning the effect of oxidation on the exsolved ulvöspinel, giving ilmenite, have been studied in detail by E. A. Vincent and others for the Skaergaard rocks, and it is clear that the Fe–Ti oxides could be of a significance comparable with that of the pyroxenes in interpreting the crystallization history of layered intrusions. The *sulphides* and platinoid minerals do not participate in the cryptic layering, in that they seem to have crystallized from separate, immiscible liquids and therefore do not coexist with the minerals separating from the silicate–oxide liquids. Although the Skaergaard shows a sequence from copper-rich sulphides to iron-rich sulphides upwards in the layered series (Ch. III), the various layered intrusions differ a great deal in the fractionation stages at which separate sulphide liquids have developed (i.e. Bushveld, Stillwater, Duluth, Sudbury) and in the character of the elements concentrated in these liquids.

The *discontinuous phase changes* in cryptically layered sequences are of an importance equal to that of the broadly continuous changes outlined above. A comparison between most of the described intrusions indicates that there are subtle differences, such as in the nature of the minerals that coexist and constitute the lowest exposed layers. In general terms, however, a magnesian olivine, followed closely by magnesian orthopyroxene and chrome-spinel, were the earliest phases to separate from the majority of parental magmas (i.e. the basic tholeiitic type). Later, plagioclase appears and then augite (e.g. Bushveld, Stillwater, and probably Great Dyke, Bay of Islands, and Skaergaard), these five minerals together constituting the ultrabasic and ultramafic parts of the layered intrusions. The disappearance of olivine and chrome-spinel, and the appearance of pigeonite in place of bronzite, are the main changes that succeed the ultrabasic sequence. New phases appearing later in the fractionation sequence are the Fe–Ti oxides, an iron-rich olivine, and apatite, attributed respectively to increases in the magma of Fe^{+++}/Fe^{++} ratios and Ti; Fe^{++}; and P. The effects of saturation in a particular component are also applicable to the development of immiscible sulphide liquids, probably attributable to sulphur contents but perhaps, also, to saturation in particular chalcophile elements. Ultimately, the late tholeiitic differentiates show the appearance of small amounts of quartz, alkali felspar, amphibole, biotite, and zircon.

It is obvious that marked differences in the type of parental magma, in contrast to subtle differences, would result in entirely different mineral assemblages from those listed above, such as the alkali pyroxenes, alkali amphiboles, felspathoids, and eudialytes of the alkaline intrusions discussed in Ch. XVI. However, the products even of a basaltic magma are dependent to a great extent upon the state of oxidation of the initial magma and of the successive residual magmas. As shown by the important studies of E. F. Osborn, the products outlined previously, from tholeiitic basaltic magmas, are chiefly the result of low and decreasing partial pressures of oxygen, this reduced state being reflected in the trend towards silicates rich in ferrous iron, such as the fayalites and ferrohedenbergites. Higher oxidation states would lead to the earlier precipitation of abundant iron oxides, an attendant increase in free silica in the late differentiates, and the abundant development of phases such as amphibole and biotite. Few of the layered intrusions are of this type, except for Guadalupe and, to a less obvious extent, Kaerven and Kap Edvard Holm. The source of most of the world's granites cannot therefore be found in fractionation of the type observed in the majority of the known layered intrusions, these bodies only indicating the feasibility of the process, and the fact that the required oxidation state is rarely attained. Both the Skaergaard and the Bushveld contain a very small proportion of fayalite–hedenbergite granophyre, which appears to have been derived as the end-product of a regular sequence of cryptic variation. In the case of the Skaergaard, isotopic studies have shown that the Sr^{87}/Sr^{86} and O^{18}/O^{16} ratios are in accordance with this hypothesis (Ch. IX), but that the transgressive veins and sheets of leucocratic granophyre are probably derived from the melting of gneiss inclusions. It will be of great interest to see if the same distinction can be drawn between the melanogranophyres and the leucocratic granophyres and felsites of the Bushveld.

The behaviour of *trace elements* can also be viewed in terms of cryptic layering, except that while the element is present in trace amounts it enters the major crystalline phases. The concentration of elements such as Cr and Ni in the early stages, and of Zr and Ba in the later stages of fractionation, for example, are well established features of layered igneous rocks and their constituent minerals. Few intrusions have been studied in sufficient detail, however, for generalizations and comparisons to be made in regard to the distribution of the great number of chemical elements likely to be concentrated in particular mineral phases and in the trapped, contemporary pore liquid, as opposed to their contents in the whole rocks. The Skaergaard has been studied in this way, however, and the information for a total of about 35 minor elements, in varying degrees of detail, is presented in Ch. VII. In regard to the available information, it is apparent that the forthcoming translation of the book by Vlasov *et al.* on the Lovozero alkaline layered intrusion (Ch. XVII) will provide abundant data of the required type, although the differences between Lovozero and Skaergaard will be extreme.

All considerations of rhythmic and cryptic layering are dependent upon a careful, systematic study and interpretation of the *rock textures*. This fundamental aspect of petrology is too often overlooked when geochemical studies are made, and the petrologist is left with expensively accumulated data that are of little use to him because of the uncritical choice of material analyzed. It has been stressed throughout this book, and shown graphically in Ch. VII, that whole-rock analyses need to be considered in relation to the effect of crystal sorting, to the relationship between a mineral assemblage and the likely contemporary magma, and to the ratio of settled minerals to crystallized contemporary magma in a particular rock. For these reasons alone, apart from the need to interpret the textures and so build up a picture of the mechanisms responsible for bringing together the crystals that constitute each rock unit of a layered intrusion, we have stressed the textural relationships throughout this book. Beyond that, we have proposed a new system of textural nomenclature designed specifically for layered igneous rocks (in collaboration with Dr W. J. Wadsworth). If this system is adopted and, where necessary, improved as studies progress, rather than discarded as just another addition to the load of petrographic nomenclature, we are convinced that many of the studies of layered intrusions will be based on a much firmer foundation. The readers of Ch. IV, VIII, XI, XIII, and XIV will have become conversant with the various aspects of this nomenclature and its application to a description of layered intrusions in the subsequent chapters, and only a brief résumé will be given here.

All the layered rocks, known as *igneous cumulates*, consist essentially of minerals which crystallized from the contemporary magma as *primocrysts* and accumulated on the temporary floor of the magma chamber as *cumulus crystals*. These are concentrated in varying proportions to form particular layers, and it is clearly undesirable to attempt to give special, unrelated names to the cumulus assemblages which, rather than having isolated origins, form part of a closely interwoven sequence. This was attempted in the days when the inter-relationships were not fully appreciated, and names such as dunite, peridotite, pyroxenite, and anorthosite were used which still have value, particularly outside layered intrusions. The use of names devised for the numerous variants (e.g. felspathic peridotite, spotted norite, diallagite, granular harzburgite, mafic anorthosite, leucogabbro, etc.), however, requires feats of memory uncompensated by the advantage of then knowing the specific meanings and limitations of the names in terms of mineral assemblages and textural relationships. In metamorphic rock nomenclature, a broadly applicable term of genetic significance (granulite, hornfels, etc.) is prefixed by a list of the particular mineral assemblage, and we have used a similar method for igneous cumulates. A graded unit can then be referred to, say, as an olivine cumulate overlain successively by an olivine–plagioclase, plagioclase–olivine, and plagioclase cumulate, rather than as a composite of dunite, allivalite, and anorthosite.

All the cumulus minerals can be listed in decreasing order of abundance, whereas when more than one is present (and there may be at least three or four) the older 'system' falls short of names to indicate the relative proportions. Also, if required, the cumulus minerals could be listed more specifically, e.g. a chrysolite–bronzite–bytownite cumulate.

The cumulus crystals pack together to form a layer, and the degree of compaction will vary according to the shapes and sizes of the crystals and the effects of contemporary current activity, slumping, and earth tremors. Simple packing experiments suggest that even with close packing, an unconsolidated layer at any one time would consist only of about 65% cumulus crystals, together with 35% of contemporary magma filling the pore spaces and called the *intercumulus liquid*. If this liquid were sealed off from the overlying, main body of magma by the rapid deposition of a thick pile of cumulus crystals, it would crystallize completely to give a *pore material* equal in composition to that reached by the magma at the time it was trapped. The rock, which would not be solid until the trapped liquid had crystallized over an appreciable temperature interval, would have a composition dependent upon that of both the contemporary magma and the particular, local cumulus assemblage, and we have called it an *ortho-cumulate*. Such rocks, formed according to the once *orthodox* concept of 'primary precipitate' and 'interprecipitate' material, are not all that common in layered intrusions. More commonly, it appears that slow diffusion of components between the intercumulus and overlying liquid has resulted in growth of the cumulus crystals by having material added at the temperature operative when they were first deposited, the resultant rocks being called *adcumulates*. Extreme examples are the monomineralic rocks, but adcumulus growth could equally have taken place where as many as four cumulus minerals were present in a layer, the absence of zoning and of extra, late-stage minerals then indicating the adcumulus origin. The diffusion mechanism could operate so as to extend the size of crystals that are not cumulus in origin, but which nucleated locally from the intercumulus liquid, such *heteradcumulutes* being characterized by the presence of large, unzoned, poikilitic minerals with compositions similar to those of the same minerals present as cumulus crystals in adjacent layers. Thus, for example, two 'harzburgites' of the same chemical composition may differ in that one (an adcumulate) contains two cumulus phases (olivine and subsidiary bronzite) and the other (a heterad-cumulate) contains only olivine as the cumulus phase (with poikilitic bronzite). For rocks in which some adcumulus growth was followed by trapping of a small proportion of pore liquid, we have used the name *mesocumulate*. The only layered rocks that are not strictly accumulative in origin are the rare cases where crystal growth has taken place from the floor of the magma chamber, probably during the temporary periods of magma stagnation when convective currents were inoperative. These conditions generally appear to have resulted in upward growth of elongated, branching crystals, and as these have usually grown from a carpet of cumulus crystals we have called the rocks *crescumulates*.

Our nomenclature for igneous cumulates could be cumbersome, leading to names such as hortonolite–andesine–ferroaugite–ferropigeonite–apatite mesocumulate. The alternative, however, would be to call the same rock a ferrodiorite, which is perhaps reasonable if the user were conversant with the coexisting mineral assemblage likely to be present, but which could cover a wide variety of cumulus minerals and their proportions, and would not point to the presence of mesocumulate pore material (such as quartz and alkali felspar in this type of rock). For example, a ferrodiorite could, alternatively, consist only of unzoned andesine and ferroaugite, and should then be distinguished as an andesine–ferroaugite adcumulate. This illustration serves to show that whether the nomenclature is used or not, it is important in any petrological and geochemical study to understand what the analysis will mean in terms of the cumulus mineral assemblage and the pore material. A rock known as an igneous cumulate is a complex unit in itself, and is studied to the best advantage by establishing the compositions of the cumulus minerals present, in order to discover the pattern of cryptic layering throughout the layered series; by considering the extent of adcumulus growth, in order to trace the conditions of deposition; and by examining the pore material, where present, as a guide to the nature of the trapped, contemporary magma.

The intrusions and rocks described in this book have been grouped together under one heading because of their layered characteristics or their affinities with some other feature, discussed in this chapter, involved in the development of layered igneous rocks. Generalizations and comparisons within the group are thus facilitated, as are comparisons that may need to be made with rocks of a non-layered origin. We have remarked that high-temperature magmatic sedimentation is a process of sufficient significance to warrant a place of its own in geological terminology, and have aimed at defining the limitations and characteristics that should identify this placing. As more layered intrusions are discovered, as further work is done on the known examples, and as progress continues in the relevant fields of experimental petrology, geochemistry, and geophysics, it is to be expected that the layered intrusions will continue to provide evidence on the processes of magmatic crystallization and differentiation. Radical ideas of one decade are liable to constitute a reactionary philosophy in another, and the hypothesis that layered igneous rocks originated by crystal sedimentation was once an affront to some geologists. Over recent years the same hypothesis may, through its relative simplicity, have ceased to attract the critical attention of those with new ideas ready to apply to old problems. It is to be hoped that our book will encourage wider interest in layered igneous rocks, particularly amongst the number of petrologists nowadays concerned with the origins of ultramafic, basaltic, and granitic rocks. Advanced studies of the Earth's mantle and crust where, respectively, the magmas of layered intrusions began and ended their existence, require plenty of information on the processes of slow crystallization and fractionation of cooling magmas. In such a context the study of layered intrusions, still posing questions to be answered, will doubtless provide answers to questions not yet formulated.

REFERENCES

ABBOTT, D., 1962. The gabbro cumulates of the Kap Edvard Holm complex, east Greenland.
   Unpublished Ph.D. thesis, Univ. of Manchester.

ADAMS, D. B., 1961. The distribution of silver and thallium in rocks and minerals of the Skaergaard gabbroic intrusion, as determined by neutron activation analysis.
   Unpublished B.A. thesis (Chemistry, Part II), Univ. of Oxford.

AGTERBERG, F. P., 1964. Statistical analysis of X-ray data for olivine.
   Miner. Mag. **33**, 742–8.

AKIMOTO, S., & KUSHIRO, I., 1960. Natural occurrence of titanomaghemite and its relevance to the unstable magnetization of rocks.
   J. Geomag. geolect. **11**, No. 3, 94–110.

ANDERSON, G. E., 1956. Copper–nickel in the Duluth gabbro near Ely, Minnesota.
   Ann. Meeting G.S.A. Guidebook, Field Trip No. 1, 91–5.

ATAMANOV, A. V., LUGOV, S. F., & FEYGIN, YA. M., 1962. New results on the geology of the Lovozero massif (translated from Russian paper, 1961).
   Internat. Geol. Review, **4**, No. 5, 570–7.

ATKINS, D. H. F., & SMALES, A. A., 1960. The determination of tantalum and tungsten in rocks and meteorites by neutron activation analysis.
   Anal. Chim. Acta. **22**, 462–78.

ATKINS, F. B., 1965. The pyroxenes of the Bushveld Igneous Complex, Transvaal.
   Unpublished D. Phil. thesis, Univ. of Oxford.

BAILEY, E. B., 1945. Tertiary igneous tectonics of Rhum (Inner Hebrides).
   Quart. J. Geol. Soc. Lond. **100**, 165–91.

——, CLOUGH, C. T., WRIGHT, W. B., RICHEY, J. E., & WILSON, G. V., 1924. The Tertiary and Post-Tertiary geology of Mull, Loch Aline and Oban.
   Mem. Geol. Surv. Scotland.

BAKER, C. O., & BOTT, M. H. P., 1962. A gravity survey over the Freetown Basic Complex of Sierra Leone.
   Overs. Geol. min. Res. **8**, No. 3, 260–78.

BARTH, T. F. W., 1962. *Theoretical Petrology* (2nd Edn.). John Wiley & Sons: New York and London.

BEATH, C. B., WESTWOOD, R. J., & COUSINS, C. A., 1961. Platinum mining at Rustenburg. The development of operating methods.
   Platin. Metals Rev. **5**, No. 3, 102–8.

BECKER, G. F., 1897a. Some queries on rock differentiation.
   Amer. J. Sci. 4th Ser. **3**, 21–40.

——, 1897b. Fractional crystallization of rocks.
   Amer. J. Sci. 4th Ser. **4**, 257–61.

BERG, J. J. VAN DEN, 1946. Petrofabric analysis of the Bushveld gabbro from Bon Accord.
   Trans. Geol. Soc. S. Afr. **49**, 156–203.

BERNING, J., 1941. The Upper Zone of the Bushveld Complex west of Potgietersrust.
   Unpublished M.Sc. thesis, Univ. of Pretoria.

BEST, M. G., 1963. Petrology of the Guadalupe igneous complex, Southwestern Sierra Nevada foothills, California.
   J. Petrol. **4**, 223–59.

BHATTACHARJI, S., & SMITH, C. H., 1964. Flowage differentiation.
   Science, **145**, 150–3.

BICKFORD, MARION E., 1962. Petrology and structure of layered gabbro, Pleasant Bay, Maine.
   Program Ann. Meeting G.S.A. (Abstr.), p. 17A.

BILJON, S. VAN, 1949.   The transformation of the Pretoria Series in the Bushveld Complex.
      Trans. Geol. Soc. S. Afr. **52**, 1–198.

BOSHOFF, J. C., 1942.   The Upper Zone of the Bushveld Complex at Tauteshoogte.
      Unpublished D.Sc. thesis, Univ. of Pretoria.

BOWEN, N. L., 1913.   The melting phenomena of the plagioclase felspars.
      Amer. J. Sci. 4th Ser. **35**, 577–99.

——, 1915*a*.   Crystallization differentiation in silicate liquids.
      Amer. J. Sci. 4th Ser. **39**, 175–91.

——, 1915*b*.   The later stages of the evolution of the igneous rocks.
      J. Geol. **23**, 1–91.

——, 1921.   Diffusion in silicate melts.
      J. Geol. **29**, 295–317.

——, 1922.   The reaction principle in petrogenesis.
      J. Geol. **30**, 177–98.

——, 1928 (reprinted in 1956).   *The Evolution of the Igneous Rocks.*   Princeton Univ. Press,
      332 pp.

——, & SCHAIRER, J. F., 1935.   The system $MgO–FeO–SiO_2$.
      Amer. J. Sci. **29**, 151–217.

——, ——, & POSNJAK, E., 1933.   The system $CaO–FeO–SiO_2$.
      Amer. J. Sci. **26**, 193–284.

——, & TUTTLE, O. F., 1950.   The system $NaAlSi_3O_8–KAlSi_3O_8–H_2O$.
      J. Geol. **58**, 489–511.

BOWN, M. G., & GAY, P., 1960.   An X-ray study of exsolution phenomena in the Skaergaard
      pyroxenes.
      Miner. Mag. **32**, 379–88.

BRADSHAW, N., 1961.   The mineralogy and petrology of the Eucrites of Centre 3, Ardnamurchan.
      Unpublished Ph.D. thesis, Univ. of Manchester.

BROTHERS, R. N., 1964.   Petrofabric analyses of Rhum and Skaergaard layered rocks.
      J. Petrol. **5**, 255–74.

BROWN, G. M., 1956.   The layered ultrabasic rocks of Rhum, Inner Hebrides.
      Phil. Trans. Roy. Soc. Lond. Ser. B, **240**, 1–53.

——, 1957.   Pyroxenes from the early and middle stages of fractionation of the Skaergaard
      intrusion, East Greenland.
      Miner. Mag. **31**, 511–43.

——, 1960.   The effect of ion substitution on the unit cell dimensions of the common clino-
      pyroxenes.
      Amer. Min. **45**, 15–38.

——, 1963.   Melting relations of Tertiary granitic rocks in Skye and Rhum.
      Miner. Mag. **33**, 533–62.

——, 1967.   Mineralogy, in *Basaltic Rocks* (Ed. H. H. Hess).   Wiley & Sons: New York.

——, & VINCENT, E. A., 1963.   Pyroxenes from late stages of fractionation of the Skaergaard
      intrusion, East Greenland.
      J. Petrol. **4**, 175–97.

BRUYNZEEL, D., 1957.   A petrographic study of the Waterfall Gorge profile at Insizwa.
      Annals Univ. Stellenbosch, **33** (Shand Mem. Vol.), Sect. A, 481–538.

BUCHER, W. H., 1963.   Are cryptovolcanic structures due to meteorite impact?
      Nature, **197**, No. 4874, 1241–5.

BUCKLEY, H. E., 1951.   *Crystal growth.*   John Wiley & Sons: New York.

BUDDINGTON, A. F., 1936.   Gravity stratification as a criterion in the interpretation of the
      structure of certain intrusives of the Northwestern Adironacks.
      Rept. 16th Internat. Geol. Congr. Washington.

——, A. F., FAHEY, J., & VLISIDIS, A.   1955.   Thermometric and petrogenetic significance of
      titaniferous magnetite.
      Amer. J. Sci. **253**, 497–532.

BUDDINGTON, & HESS, H. H., 1937. Layered peridotite laccoliths in the Trout River area, Newfoundland (a Discussion).
　　Amer. J. Sci. **33**, 380–8.

BURRI, C., & NIGGLI, P., 1945. *Die jungen Eruptivgesteine des mediterranen Orogens*, No. 3. Vulkaninstitut Immanuel Friedländer: Zürich. 654 pp.

BUTLER, J. R., & SKIBA, W., 1962. Strontium in plagioclase feldspars from four layered basic masses in Somalia.
　　Miner. Mag. **33**, 213–25.

CABELL, M. J., & SMALES, A. A., 1957. The determination of rubidium and caesium in rocks, minerals and meteorites by neutron activation analysis.
　　The Analyst, **82**, No. 975, 390–406.

CAMERON, E. N., 1963. Structure and rock sequences of the Critical Zone of the Eastern Bushveld Complex.
　　Min. Soc. Amer. Special Paper 1, 93–107.

——, & EMERSON, M. E., 1959. The origin of certain chromite deposits in the eastern part of the Bushveld Complex.
　　Econ. Geol. **54**, 1151–1213.

CARMICHAEL, I. S. E., 1963. The crystallization of feldspar in volcanic acid liquids.
　　Quart. J. Geol. Soc. Lond. **119**, 95–131.

CARR, J. M., 1952. An investigation of the Sgurr na Stri–Druim Hain sector of the basic igneous complex of the Cuillin Hills, Isle of Skye.
　　Unpublished D.Phil. thesis, Univ. of Oxford.

——, 1954a. Contemporaneous slumping and sliding in the banded gabbros of the Isle of Skye, Scotland.
　　Bull. Geol. Soc. Amer. **65**, Abstr. p. 1238.

——, 1954b. Zoned plagioclases in layered gabbros of the Skaergaard intrusion, East Greenland.
　　Miner. Mag. **30**, 367–75.

CHALLIS, G. A., 1965. The origin of New Zealand ultramafic intrusions.
　　J. Petrol. **6**, 322–64.

CHEVALLIER, R., & MARTIN, R., 1959. Le moment magnétique de l'ion ferreux dans une série de pyroxènes monocliniques.
　　Bull. Soc. Chim. France, No. 9, 9–10.

——, & MATHIEU, S., 1958. Susceptibilité magnétique spécifique de pyroxènes monocliniques.
　　Bull. Soc. Chim. France, No. 5, 726–9.

——, ——, & VINCENT, E. A., 1954. Iron–titanium oxide minerals in layered gabbros of the Skaergaard intrusion, East Greenland. Part 2, Magnetic properties.
　　Geochim. et Cosmochim. Acta, **6**, 27–34.

CLARKE, P. D., 1965. The petrology of the Insch basic mass, Aberdeenshire.
　　Unpublished Ph.D. thesis, Univ. of Edinburgh.

COCKBURN, A. M., 1935. The geology of St. Kilda.
　　Trans. Roy. Soc. Edinb. **58**, pt. 2, No. 21.

COERTZE, F. J., 1958. Intrusive relationships and ore deposits in the western part of the Bushveld Igneous Complex.
　　Trans. Geol. Soc. S. Afr. **61**, 387–92.

COLEMAN, A. P., 1905. The Sudbury nickel field.
　　Ontario Bur. Mines, Ann. Rept. **14**, 188 pp.

COOK, A. H., & MURPHY, T., 1952. Gravity survey of Ireland, north of the line Sligo–Dundalk.
　　Geophys. Mem. Dublin, **2**, 1–36.

COOPER, J. R., 1936. Geology of the Southern Half of the Bay of Islands igneous complex.
　　Nfld. Dept. Nat. Res. Geol. Sec. Bull 4, 1–62.

COUSINS, C. A., 1959. The Bushveld igneous complex. The geology of South Africa's platinum resources.
　　Platin. Metals Rev. **3**, No. 3, 94–9.

CURREN, W. D., 1958. Radiochemical methods for the determination of small amounts of phosphorus and sulphur in igneous rocks.
  Unpublished B.Sc. thesis, Univ. of Oxford.

DALY, R. A., 1914. *Igneous Rocks and their Origin.* McGraw-Hill: New York and London, 563 pp.

——, 1926. *Our Mobile Earth.* Scribners: New York. 342 pp.

——, 1928. Bushveld Igneous Complex of the Transvaal.
  Bull. Geol. Soc. Amer. **39**, 703–68.

——, 1933. *Igneous Rocks and the Depths of the Earth.* McGraw-Hill: New York and London. 598 pp.

DANIELS, J. L., 1958. A preliminary investigation of some basic intrusions in the Hargeisa and Borama districts of Somaliland Protectorate.
  Trans. Geol. Soc. S. Afr. **61**, 125–36.

——, SKIBA, J. W., & SUTTON, J., 1965. The deformation of some banded gabbros in the northern Somalia fold-belt.
  Quart. J. Geol. Soc. Lond. **121**, 111–42.

DARWIN, C., 1844. Geological observations on the Volcanic islands, visited during the voyage of H.M.S. *Beagle,* together with some brief notices on the geology of Australia and the Cape of Good Hope. Being the second part of the geology of the voyage of the *Beagle,* under the command of Capt. Fitzroy, R.N., during the years 1832 to 1836. Smith, Elder: London. 175 pp.

DAVIS, B. T. C., & ENGLAND, J. L., 1963. Melting of forsterite, Mg_2SiO_4, at pressures up to 47 kilobars.
  Carnegie Inst. Wash. Yearb. **62**, 119–21.

DEER, W. A., & ABBOTT, D., 1965. Clinopyroxenes of the gabbro cumulates of the Kap Edvard Holm complex, east Greenland.
  Miner. Mag. **34**, 177–93.

——, & WAGER, L. R., 1939. Olivines from the Skaergaard Intrusion, Kangerdlugssuaq, East Greenland.
  Amer. Min. **24**, 18–25.

DODSON, M. H., & LONG, L. E., 1962. Age of Lundy granite, Bristol Channel.
  Nature, **195**, No. 4845, 975–6.

DOUGLAS, G. V., 1953. Notes on localities visited on the Labrador coast in 1946 and 1947.
  Geol. Surv. Canada, Paper 53–1.

DOUGLAS, J. A. V., 1961. A further petrological and chemical investigation of the upper part of the Skaergaard intrusion, East Greenland. Part I, The Basistoppen Sheet. Part II, The Upper Border Group.
  Unpublished D.Phil. thesis, Univ. of Oxford.

——, 1964. Geological investigations in East Greenland: VII. The Basistoppen Sheet—a differentiated basic intrusion into the upper part of the Skaergaard complex, East Greenland.
  Medd. om Grønland, **164**, No. 5, 1–66.

DREVER, H. I., 1953. A note on the field relations of the Shiant Isles' picrite.
  Geol. Mag. **90**, 159–60.

——, 1957. A note on the occurrence of rhythmic layering in the Eilean Mhuire sill, Shiant Isles.
  Geol. Mag. **94**, 277–80.

——, & JOHNSTON, R., 1959. The lower margin of the Shiant Isles' sill.
  Quart. J. geol. Soc. Lond. **114**, 343–65.

DUNHAM, A. C., 1962. The petrology and structure of the northern edge of the Tertiary igneous complex of Rhum.
  Unpublished D.Phil. thesis, Univ. of Oxford.

DUPARC, L., & TIKONOWITCH, M., 1920. *Le platine et les gîtes platinifères de l'Oural et du Monde.* Sonor S.A.: Geneva. 542 pp.

DU TOIT, A. L., 1926. *The Geology of South Africa.* Oliver and Boyd: Edinburgh.

EDWARDS, A. B., 1942. Differentiation of the dolerites of Tasmania.
  J. Geol. **50**, 451–80, 579–610.

EMELEUS, C. H., 1963. Structural and petrographic observations on layered granites from Southern Greenland.
    Min. Soc. Amer. Special Paper 1, 22–9.

——, 1964. The Grønnedal–Ika alkaline complex, South Greenland.
    Medd. om Grønland, **172**, No. 3, 4–73.

ERNST, W. G., 1960. Diabase–granophyre relations in the Endion sill, Duluth, Minnesota.
    J. Petrol. **1**, 286–302.

ESSON, J., 1962. The distribution and analytical chemistry of arsenic, antimony and bismuth in geological materials.
    Unpublished dissertation for Diploma in Geochemistry, Univ. of Oxford.

——, STEVENS, R. H., & VINCENT, E. A., 1965. Aspects of the geochemistry of arsenic and antimony, exemplified by the Skaergaard intrusion.
    Miner. Mag. **35**, 88–107.

FAURE, G., & HURLEY, P. M., 1963. The isotopic compositions of strontium in oceanic and continental basalts: application to the origin of igneous rocks.
    J. Petrol. **4**, 31–50.

FERGUSON, J., 1964. Geology of the Ilimaussaq alkaline intrusion, South Greenland.
    Medd. om Grønland, **172**, No. 4, 5–82.

——, & BOTHA, E., 1963. Some aspects of igneous layering in the basic zones of the Bushveld Complex
    Trans. Geol. Soc. S. Afr. **66** 259–78.

——, & PULVERTAFT, T. C. R., 1963. Contrasted styles of igneous layering in the Gardar Province of South Greenland.
    Min. Soc. Amer. Special paper 1, 10–21.

FERINGA, G., 1959. The geological succession in a portion of the Northwestern Bushveld (Union Section) and its interpretation.
    Trans. Geol. Soc. S. Afr. **62**, 219–33.

FERMOR, L. L., 1925. On the basaltic lavas penetrated by the deep boring for coal at Bhusawal, Bombay Presidency.
    Recd. Geol. Surv. India, **58**, 93–240.

FERSMAN, A. E., 1937. Mineralogiya i geokhimiya Khibinskikh i Lovozerskikh tundr. (Mineralogy and Geochemistry of the Khibina and Lovozero tundras.)
    Trans. 17th Internat. Geol. Congr. Moscow.

FINDLAY, D. C., & SMITH, C. H., 1965. The Muskox drilling project.
    Geol. Surv. Canada, Paper 64–44, 170 pp.

FLEISCHER, M., & STEVENS, R. E., 1962. Summary of new data on rock samples *G*-1 and *W*-1.
    Geochim. et Cosmochim. Acta, **26**, 525–43.

FRANKEL, F. F., & GRAINGER, G. W., 1941. Notes on Bushveld titaniferous iron-ore.
    S. Afr. J. Sci. **37**, 101–10.

FRIEDMAN, G. M., 1957. Structure and petrology of the Caribou Lake intrusive body, Ontario, Canada.
    Bull. Geol. Soc. Amer. **68**, 1531–64.

GASS, I. G., 1958. Ultrabasic pillow lavas from Cyprus.
    Geol. Mag. **95**, 241–51.

——, & MASSON-SMITH, D., 1963. The geology and gravity anomalies of the Troodos massif, Cyprus.
    Phil. Trans. Roy. Soc. Lond. Ser. A, **255**, 417–67.

GAY, P., & MUIR, I. D., 1962. Investigation of the felspars of the Skaergaard intrusion, Eastern Greenland.
    J. Geol. **70**, 565–81.

GEIKIE, A., 1894. On the relations of the basic and acid rocks of the Tertiary Volcanic series of the Inner Hebrides.
    Quart. J. Geol. Soc. Lond. **50**, 212–29.

——, 1897. *Ancient Volcanoes of Great Britain*, **2**. Macmillan & Co.: London.

GEIKIE, A., & TEALL, J. J. H., 1894. On the banded structure of some Tertiary gabbros in the Isle of Skye.
  Quart. J. Geol. Soc. Lond. **1**, 645–59.

GOLDICH, S. S., TAYLOR, R. B., & LUCIA, F. J., 1956. Geology of the Enger Tower area, Duluth, Minnesota.
  Ann. Meeting G.S.A. Guidebook, Field Trip No. 1, 67–90.

GOLDSBROUGH, K., 1962. Scandium in rocks of the Skaergaard intrusion, Greenland.
  Unpublished B.A. thesis (Chemistry, Pt. II), Univ. of Oxford.

GOLDSCHMIDT, V. M., 1954. *Geochemistry*. Oxford Univ. Press, 730 pp.

GOUGH, D. I., & VAN NIEKERK, C. B., 1959. A study of the palaeomagnetism of the Bushveld Gabbro.
  Phil. Mag. **4**, No. 37, 126–36.

GREEN, D. H., 1964. The petrogenesis of the high-temperature peridotite intrusion in the Lizard area, Cornwall.
  J. Petrol. **5**, 134–88.

GROUT, F. F., 1918*a*. The lopolith; an igneous form exemplified by the Duluth gabbro.
  Amer. J. Sci. 4th ser. **46**, 516–22.

——, 1918*b*. A type of igneous differentiation.
  J. Geol. **26**, 626–58.

——, 1918*c*. Two phase convection in igneous magmas.
  J. Geol. **26**, 481–99.

——, 1918*d*. Internal structures of igneous rocks: their significance and origin, with special reference to the Duluth gabbro.
  J. Geol. **26**, 439–58.

GUNN, B. G., 1962. Differentiation in Ferrar dolerites, Antarctica.
  N.Z. J. Geol. Geophys. **5**, 820–63.

——, 1963. Layered intrusions in the Ferrar dolerites, Antarctica.
  Min. Soc. Amer. Special Paper 1, 124–33.

HALL, A. L., 1926. The Bushveld Igneous Complex.
  Johann. J. Chem. Met. Min. Soc. **26**, 160–74.

——, 1932. The Bushveld Igneous Complex of the Central Transvaal.
  Mem. Geol. Surv. S. Afr. No. 28, 554 pp.

——, & DU TOIT, A. L., 1924. On the section across the floor of the Bushveld Complex at the Hartbeestpoort Dam, west of Pretoria.
  Trans. Geol. Soc. S. Afr. **26**, 69–97.

HAMILTON, E. I., 1959. The uranium content of the differentiated Skaergaard intrusion, together with the distribution of the alpha-particle radioactivity in the various rocks and minerals as recorded by nuclear emulsion studies.
  Medd. om. Grønland, **162**, No. 7, 1–35.

——, 1963. The isotopic composition of strontium in the Skaergaard intrusion, East Greenland.
  J. Petrol. **4**, 383–91.

——, 1964. The geochemistry of the northern part of the Ilimaussaq intrusion, S.W. Greenland.
  Medd. om Grønland, **162**, No. 10, 4–104.

HARDING, R. R., 1962. Studies on the Tertiary igneous rocks of St. Kilda.
  Unpublished D.Phil. thesis, Univ. of Oxford.

HARKER, A., 1904. The Tertiary Igneous Rocks of Skye.
  Mem. Geol. Surv. Scotland.

——, 1908. The geology of the Small Isles of Inverness-shire (Sheet 60).
  Mem. Geol. Surv. Scotland.

HARRY, W. T., & EMELEUS, C. H., 1960. Mineral layering in some granite intrusions of S.W. Greenland.
  Rept. 21st Int. Geol. Congr. (Norden), pt. 14, 172–81.

——, & PULVERTAFT, T. C. R., 1963. The Nunarssuit Intrusive Complex, South Greenland.
  Medd. om Grønland, **169**, No. 1, 4–136.

HAWKES, L., 1940. Petrology of East Greenland (a review).
        Nature, **145**, 197–8.
HAWLEY, J. E., 1962. The Sudbury ores: their mineralogy and origin.
        Can. Mineral. **7**, 1–207.
HERZ, N., 1951. Petrology of the Baltimore gabbro, Maryland.
        Bull. Geol. Soc. Amer. **62**, 979–1016.
HESS, H. H., 1939. Extreme fractional crystallization of a basaltic magma; the Stillwater igneous
        complex.
        Trans. Amer. Geophys. Un. Pt. 3, 430–2.
——, 1941. Pyroxenes of common mafic magmas.
        Amer. Min. **26**, 515–35, 573–94.
——, 1949. Chemical composition and optical properties of common clinopyroxenes.
        Amer. Min. **34**, 621–66.
——, 1950. Vertical mineral variation in the Great Dyke of Southern Rhodesia.
        Trans. Geol. Soc. S. Afr. **53**, 159–66.
——, 1952. Orthopyroxenes of the Bushveld type, ion substitutions and changes in unit cell
        dimensions.
        Amer. J. Sci. **Bowen vol**. 173–87.
——, 1956. The magnetic properties and differentiation of dolerite sills: discussion.
        Amer. J. Sci. **254**, 446–51.
——, 1960. Stillwater igneous complex, Montana: a quantitative mineralogical study.
        Mem. Geol. Soc. Amer. **80**, 230 pp.
——, & PHILLIPS, A. H., 1940. Optical properties and chemical composition of magnesian
        orthopyroxenes.
        Amer. Min. **25**, 271–85.
HIEMSTRA, S. A., & VAN BILJON, W. J., 1959. The geology of the upper Magaliesburg stage and
        the lower Bushveld complex in the vicinity of Steelpoort.
        Trans. Geol. Soc. S. Afr. **62**, 239–54.
HOLMES, A., 1931. The problems of the association of acid and basic rocks in central complexes.
        Geol. Mag. **68**, 241–55.
HOTZ, P. E., 1953. Petrology of granophyre in diabase near Dillsburg, Pennsylvania.
        Bull. Geol. Soc. Amer. **64**, 675–704.
HUGHES, C. J., 1956. Geological investigations in East Greenland, VI. A differentiated basic
        sill enclosed in the Skaergaard intrusion, East Greenland, and related sills injecting the
        lavas.
        Medd. om Grønland, **137**, No. 2, 1–27.
——, 1960. The southern mountains igneous complex, Isle of Rhum.
        Quart. J. Geol. Soc. Lond. **116**, 111–38.
HUTCHISON, R., 1964. The Tertiary basic igneous rocks of the Western Cuillin, Isle of Skye.
        Unpublished Ph.D. thesis, Univ. of Glasgow.
INGERSON, E., 1935. Layered peridotitic laccoliths of the Trout River area, Newfoundland.
        Amer. J. Sci. **29**, 422–40.
——, 1937. Layered peridotitic laccoliths of the Trout River area, Newfoundland—A Reply.
        Amer. J. Sci. **33**, 389–92.
IRVINE, T. N., 1963. Origin of the ultramafic complex at Duke Island, Southeastern Alaska.
        Min. Soc. Amer. Special Paper 1, 36–45.
JACKSON, E. D., 1960. X-ray determinative curve for natural olivine of composition Fo_{80-90}.
        U.S. Geol. Surv. Prof. Paper 400-B, 432–4.
——, 1961. Primary textures and mineral associations in the Ultramafic Zone of the Stillwater
        Complex, Montana.
        U.S. Geol. Surv. Prof. Paper 358, 1–106.
——, 1963. Stratigraphic and lateral variation of chromite composition in the Stillwater
        complex.
        Min. Soc. Amer. Special Paper 1, 46–54.

JAEGER, J. C., 1959. The use of complete temperature–time curves for determination of thermal conductivity, with particular reference to rocks.
  Austr. J. Phys. **12**, No. 3, 203–17.

——, & JOPLIN, G., 1956. The magnetic properties and differentiation of dolerite sills : discussion.
  Amer. J. Sci. **254**, 443–6.

JAMBOR, J. L., & SMITH, C. H., 1964. Olivine composition determination with small diameter X-ray powder camera.
  Miner. Mag. **33**, 730–48.

JEFFREYS, H., 1959. *The Earth: its origin, history and physical constitution* (4th edn.). Cambridge Univ. Press. 420 pp.

JOHNSTON, R., 1953. Olivines of the Garbh Eilean sill.
  Geol. Mag. **90**, 161–71.

JONES, W. R., PEOPLES, J. W., & HOWLAND, A. L., 1960. Igneous and tectonic structures of the Stillwater complex, Montana.
  U.S. Geol. Surv. Bull. 1071-H, 281–335.

JUDD, J. W., 1874. The secondary rocks of Scotland. Second paper.
  Quart. J. Geol. Soc. Lond. **30**, 220–301.

——, 1885. On the Tertiary and older Peridotites of Scotland.
  Quart. J. Geol. Soc. Lond. **41**, 354–418.

KEMP, D. M., & SMALES, A. A., 1960*a*. The determination of vanadium in rocks and meteorites by neutron-activation analysis.
  Anal. Chim. Acta, **23**, 397–410.

——, & ——, 1960*b*. The determination of scandium in rocks and meteorites by neutron-activation analysis.
  Anal. Chim. Acta, **23**, 410–18.

KING, L., 1962. *Morphology of the Earth*. Oliver and Boyd: Edinburgh and London. 699 pp.

KORITNIG, S., 1965. Geochemistry of phosphorus—I. The replacement of Si^{4+} by P^{5+} in rock-forming silicate minerals.
  Geochim. et Cosmochim. Acta, **29**, 361–72.

KORN, H., & MARTIN, H., 1954. The Messum igneous complex in Southwest Africa.
  Trans. Geol. Soc. S. Afr. **57**, 83–122.

KROKSTRÖM, T., 1930. The Breven dolerite dyke.
  Bull. Uppsala Univ. Mineralogisk-Geologiska Inst. **13**, 244–330.

——, 1936. The Halleförs dolerite dyke and some problems of basaltic rocks.
  Bull. Uppsala Univ. Mineralogisk-Geologiska Inst. **26**, 113–263.

KUNO, H., 1960. High-alumina basalt.
  J. Petrol. **1**, 121–45.

KUSHIRO, I., 1960. $\gamma \rightarrow \alpha$ Transition in Fe_2O_3 with pressure.
  J. Geomag. geolect. **11**, No. 4, 148–51.

LAUDER, W. R., 1965. The petrology of Dun Mountain.
  N.Z. Jour. Geol. Geophys. **8**, 3–34.

LEAKE, B. E., 1964. New light on the Dawros peridotite, Connemara, Ireland.
  Geol. Mag. **101**, 63–75.

LE BAS, M. J., 1960. The petrology of the layered basic rocks of the Carlingford Complex, Co. Louth.
  Trans. Roy. Soc. Edinb. **64**, No. 8, 169–200.

LEIGHTON, M. W., 1954. Petrogenesis of a gabbro–granophyre complex in northern Wisconsin.
  Bull. Geol. Soc. Amer. **65**, 401–42.

LEWIS, J. V., 1907. Petrography of the Newark igneous rocks of New Jersey.
  New Jersey Geol. Surv. 99–153.

LIEBENBERG, C. J., 1960. The trace elements of the rocks of the Bushveld Igneous Complex.
  Publik. van die Universiteit van Pretoria, Nuwe Reeks, Nr. 12, 1–69; Nr. 13, 1–65.

LIEBENBERG, W. R., 1942. The basal rocks of the Bushveld Igneous Complex in the Marico District, South of Dwarsberg.
>   *Trans. Geol. Soc. S. Afr.* **45**, 81–108.

LIGHTFOOT, B., 1927. Traverses along the Great Dyke of Southern Rhodesia.
>   *Geol. Surv. S. Rhodesia*, Short Report No. 21.

LOBJOIT, W. M., 1957. The petrology of the Ben Buie intrusion, Island of Mull, Argyllshire.
>   Unpublished Ph.D. thesis, Univ. of Manchester.

——, 1959. On the form and mode of emplacement of the Ben Buie intrusion, Island of Mull, Argyllshire.
>   *Geol. Mag.* **96**, 393–402.

LOMBAARD, A. F., 1950. Die Geologie van die Bosveldkompleks langs Bloedrivier.
>   *Trans. Geol. Soc. S. Afr.* **52**, 343–76.

LOMBAARD, B. V., 1934. On the differentiation and relationships of the rocks of the Bushveld Igneous Complex.
>   *Trans. Geol. Soc. S. Afr.* **37**, 5–52.

——, 1956. Chromite and dunite of the Bushveld Complex.
>   *Trans. Geol. Soc. S. Afr.* **59**, 59–74.

LOVERIDGE, B. A., WEBSTER, R. K., MORGAN, J. W., THOMAS, A. M., & SMALES, A. A., 1959. The determination of strontium in rocks and biological materials.
>   *A.E.R.E. Paper*, No. 40, 21.

LWIN, M. T., 1960. Some mineralogical studies of the syenites of the Kap Edvard Holm Complex, east Greenland.
>   Unpublished M.Sc. thesis, Univ. of Manchester.

MACCULLOCH, J., 1819. *A description of the Western Islands of Scotland.* 3 vols. Constable: Edinburgh; Hurst, Robinson: London.

McDOUGALL, I., 1961. Optical and chemical studies of pyroxenes in a differentiated Tasmanian dolerite.
>   *Amer. Min.* **46**, 661–87.

MACGREGOR, A. M., 1951. Some milestones in the Precambrian of Southern Rhodesia.
>   *Trans. Geol. Soc. S. Afr.* **54**, 27–71.

MACKENZIE, D. B., 1960. High temperature, alpine-type peridotite from Venezuela.
>   *Bull. Geol. Soc. Amer.* **71**, 303–18.

McQUILLIN, R., & TUSON, J., 1963. Gravity measurements over the Rhum Tertiary plutonic complex.
>   *Nature*, **199**, 1276–7.

McREATH, I., 1963. The determination of strontium and barium in the rocks of the Skaergaard intrusion, by X-ray fluorescence analysis.
>   Unpublished B.A. thesis (Chemistry, Part II), Univ. of Oxford.

MARTIN, H., MATHIAS, M., & SIMPSON, E. S. W., 1960. The Damaraland sub-volcanic ring complexes in Southwest Africa.
>   *Rept. 21st Internat. Geol. Congr. (Norden)*, Pt. 13, 156–74.

MATHIAS, M., 1956. The petrology of the Messum igneous complex, South-West Africa.
>   *Trans. Geol. Soc. S. Afr.* **59**, 23–58.

MATTHEWS, W. H., 1957. Petrology of Quaternary volcanics of the Mount Garibaldi map-area, southwestern British Columbia.
>   *Amer. J. Sci.*, **255**, 400–15.

MIERS, H. A., & ISAAC, F., 1907. The spontaneous crystallization of binary mixtures—experiments on salol and bethol.
>   *Proc. Roy. Soc. Lond.* Ser. A, **79**, 322–51.

MOLENGRAAFF, G. A. F., 1901. Géologie de la République Sud Africaine du Transvaal.
>   *Bull. Soc. Géolog. de France*, 4 Sér. Tome 1, 13–92.

——, 1905. *Inter-Colonial Irrigation Commission.* Pretoria. 90 pp.

MOORBATH, S., & BELL, J. D., 1965. Strontium isotope abundance studies and rubidium–strontium age determinations on Tertiary igneous rocks from the Isle of Skye, North-west Scotland.
  J. Petrol. **6**, 37–66.

MORSE, S. A., 1961. The geology of the Kiglapait layered intrusion, coast of Labrador, Canada. Unpublished Ph.D. thesis, Univ. of McGill.

MUIR, I. D., 1951. The clinopyroxenes of the Skaergaard intrusion, East Greenland.
  Miner. Mag. **29**, 690–714.

——, 1954. Crystallization of pyroxenes in an iron-rich diabase from Minnesota.
  Miner. Mag. **30**, 376–88.

NESBITT, R. W., & KLEEMAN, A. W., 1964. Layered intrusions of the Giles Complex, central Australia.
  Nature, **203**, 391–3.

NICOLAYSEN, L. O., DE VILLIERS, J. W. L., BURGER, A. J., & STRELOW, F. W. E., 1958. New measurements relating to the absolute age of the Transvaal System and of the Bushveld Igneous Complex.
  Trans. Geol. Soc. S. Afr. **61**, 137–63.

NIGHTINGALE, G., 1962. Gallium in rocks of the Skaergaard intrusion, Greenland. Unpublished B.A. thesis (Chemistry, Pt. II), Univ. of Oxford.

OJHA, D. N., 1960. Petrological studies on the Kaerven layered basic intrusion, East Greenland. Unpublished D.Phil. thesis, Univ. of Oxford.

OOSTEROM, M. G., 1963. The ultramafites and layered gabbro sequences in the granulite facies rocks on Stjernöy (Finnmark, Norway).
  Leid. Geol. Meded. **28**, 177–296.

OSBORN, E. F., 1959. Role of oxygen pressure in the crystallization and differentiation of basaltic magma.
  Amer. J. Sci. **257**, 609–47.

——, & TAIT, D. B., 1952. The system diopside–forsterite–anorthite.
  Amer. J. Sci. **Bowen vol.** 413–33.

OSBORNE, F. F., & ROBERTS, E. J., 1931. Differentiation in the Shonkin Sag Laccolith, Montana.
  Amer. J. Sci. **22**, 331–53.

PEOPLES, J. W., 1936. Gravity stratification as a criterion in the interpretation of the structure of the Stillwater Complex, Montana.
  Rept. 16th Internat. Geol. Congr. Washington, 353–60.

PHEMISTER, T. C., 1937. A review of the problems of the Sudbury irruptive.
  J. Geol. **65**, 91–116.

PIRSSON, L. V., 1905. Petrology and geology of the igneous rocks of the Highwood Mountains.
  U.S. Geol. Surv. Bull. **237**, 1–208.

POLDERVAART, A., 1944. The petrology of the Elephant's Head dike and the New Amalfi sheet (Matatiele).
  Trans. Roy. Soc. S. Afr. **30**, 85–119.

——, 1946. The petrology of the Mount Arthur complex (East Griqualand).
  Trans. Roy. Soc. S. Afr. **31**, 83–110.

——, & ELSTON, W. E., 1954. The calc-alkaline series and the trend of fractional crystallization of basaltic magma: A new approach at graphic representation.
  J. Geol. **62**, 150–62.

——, & HESS, H. H., 1951. Pyroxenes in the crystallization of basaltic magma.
  J. Geol. **59**, 472–89.

PULVERTAFT, T. C. R., 1965. The Eqaloqarfia layered dyke, Nunarssuit, South Greenland.
  Medd. om Grønland, **169**, 1–39.

RAMDOHR, P., 1953. Ulvöspinel and its significance in iron ores.
  Econ. Geol. **48**, 677–88.

——, 1956. Die Beziehungen von Fe–Ti-erzen aus magmatischen gesteinen.
  Bull. comm. géol. Finlande, **29**, No. 173, 1–18.

RAVICH, M. G., & CHAIKA, L. A., 1956. Differentsirovanaya intruziya trappovoy formatsii Taymrskoy skladchatoy oblasti (A differentiated intrusion of trap formation in the Taimyr folded region).
Izv. Akad. Nauk. SSSR, Ser. Geol. No. 1, 50–64.

READ, H. H., 1919. The two magmas of Strathbogie and Lower Banffshire.
Geol. Mag. **56**, 364–71.

——, 1923. The geology of the country around Banff, Huntly and Turriff.
Mem. Geol. Surv. Scotland.

——, & HAQ, B. T., 1963. The distribution of trace-elements in the dunite–syenite differentiated series of the Insch complex, Aberdeenshire.
Proc. Geol. Assoc. **74**, 203–12.

——, SADASHIVAIAH, M. S., & HAQ, B. T., 1961. Differentiation in the Olivine-Gabbro of the Insch Mass, Aberdeenshire.
Proc. Geol. Assoc. **72**, 391–413.

——, ——, & ——, 1965. The Hypersthene-gabbro of the Insch complex, Aberdeenshire.
Proc. Geol. Assoc. **76**, 1–11.

REUNING, E., 1927. Verbandsverhältnisse und Chemismus der Gesteine des Bushveld Igneous Complex, Transvaal, und das Problem seiner Enstehung.
Nevus Jb. Min. Geol. Pälaont. Abt. A, 631–64.

RICHEY, J. E., 1932. Tertiary ring structures in Britain.
Trans. Geol. Soc. Glasg. **19**, 42–140.

——, 1948. British Regional Geology: The Tertiary Volcanic Districts, Scotland.
Geol. Surv. Gt. Britain.

——, THOMAS, H. H., *et al.*, 1930. The geology of Ardnamurchan, Northwest Mull and Coll.
Mem. Geol. Surv. Scotland, 393 pp.

ROBSON, G. R., & SPECTOR, J., 1962. Crystal fractionation of the Skaergaard type in modern Icelandic magmas.
Nature, **193**, 1277–8.

ROTHSTEIN, A. T. V., 1957. The Dawros peridotite, Connemara, Eire.
Quart. J. Geol. Soc. Lond. **113**, 1–25.

——, 1958. Pyroxenes from the Dawros peridotite.
Geol. Mag. **95**, 456–62.

——, 1961. Phase relations in the peridotites of Dawros (Eire) and Belhelvie (Scotland). (In Russian.)
Izvest. Akad. Nauk SSSR. Ser. Geol. **3**, 69–86.

——, 1964. New light on the Dawros peridotite.
Geol. Mag. **101**, 283–5.

RUCKMICK, J. C., & NOBLE, J. A., 1959. Origin of the ultramafic complex at Union Bay, southeastern Alaska.
Bull. Geol. Soc. Amer. **70**, 981–1018.

SADASHIVAIAH, M. S., 1954. The form of the eastern end of the Insch igneous mass, Aberdeenshire.
Geol. Mag. **91**, 137–43.

SAMPSON, E., 1932. Magmatic chromite deposits in Southern Africa.
Econ. Geol. **27**, 113–44.

SANDBERG, C. G. S., 1926. On the probable origin of the members of the Bushveld Igneous Complex (Transvaal).
Geol. Mag. **43**, 210–19.

SCHMIDT, E. R., 1952. The structure and composition of the Merensky Reef and associated rocks on the Rustenburg Platinum Mine.
Trans. Geol. Soc. S. Afr. **55**, 233–80.

SCHOLTZ, D. L., 1936. The magmatic nickeliferous ore deposits of East Griqualand and Pondoland.
Trans. Geol. Soc. S. Afr. **39**, 81–210.

SCHREINER, G. D. L., 1958. Comparison of the ^{87}Rb–^{87}Sr ages of the Red Granite of the Bushveld Complex from measurements of the total rock and separated mineral fractions.
　　Proc. Roy. Soc. Lond. Ser. A, **245**, 112–7.

SCHWANDER, H., 1953. Bestimmung des relativen Sauerstoffisotopen-Verhältnisses in Silikatgesteinen und Mineralien.
　　Geochim. et Cosmochim. Acta, **4**, 261–91.

SCHWARTZ, G. M., 1956. Summary of the Precambrian geology of Northeastern Minnesota.
　　Ann. Meeting G.S.A. Guidebook, Field Trip No. 1, 1–9.

SCHWELLNUS, C. M., & WILLEMSE, J., 1944. Titanium and vanadium in the magnetic iron ores of the Bushveld Complex.
　　Trans. Geol. Soc. S. Afr. **46**, 23–38.

SCHWELLNUS, J. S. I., 1956. The Basal portion of the Bushveld Igneous Complex and the adjoining metamorphosed sediments in the northeastern Transvaal.
　　Unpublished D.Sc. thesis, Univ. of Pretoria.

SHACKLETON, R. M., 1948. Overturned rhythmic banding in the Huntly gabbro of Aberdeenshire.
　　Geol. Mag. **85**, 358–60.

SHAW, H. R., 1965. Comments on viscosity, crystal settling, and convection in granitic magmas.
　　Amer. J. Sci. **263**, 120–52.

SHIMAZU, Y., 1959a. A thermodynamical aspect of the earth's interior—physical interpretation of magmatic differentiation process.
　　J. Earth Sci. Nagoya Univ. **7**, 1–34.

——, 1959b. A physical interpretation of crystallization differentiation of the Skaergaard intrusion.
　　J. Earth Sci. Nagoya Univ. **7**, 35–48.

SIMPSON, E. S. W., 1954a. On the graphical representation of differentiation trends in igneous rocks.
　　Geol. Mag. **91**, 238–44.

——, 1954b. The Okonjeje igneous complex, South-West Africa.
　　Trans. Geol. Soc. S. Afr. **57**, 125–72.

——, & OTTO, J. D. T., 1960. On the Pre-Cambrian anorthosite mass of Southern Angola.
　　Rept. 21st Internat. Geol. Congr. (Norden), Pt. 13, 216–27.

SKIBA, W., & BUTLER, J. R., 1963. The use of Sr–An relationships in plagioclases to distinguish between Somalian metagabbros and country-rock amphibolites.
　　J. Petrol. **4**, 352–66.

SMALES, A. A., SMIT, J. VAN R., & IRVING, H., 1957. Determination of indium in rocks and minerals by radioactivation.
　　The Analyst, **82**, No. 977, 539–49.

SMITH, C. H., 1958. Bay of Islands igneous complex, Western Newfoundland.
　　Mem. Geol. Surv. Canada, No. 290, 1–132.

——, 1962. Notes on the Muskox Intrusion, Coppermine River area, District of Mackenzie.
　　Geol. Surv. Can. Paper 61–25, 16 pp.

——, & KAPP, H. E., 1963. The Muskox Intrusion, a recently discovered layered intrusion in the Coppermine River area, Northwest Territories, Canada.
　　Min. Soc. Amer. Special Paper 1, 30–5.

——, & MACGREGOR, I. D., 1960. Ultrabasic intrusive conditions illustrated by the Mt. Albert Ultrabasic Pluton, Gaspe, Quebec.
　　Bull. Geol. Soc. Amer. (Abstr.), **71**, 1978.

SØRENSEN, H., 1958. The Ilimaussaq batholith: a review and discussion.
　　Medd. om Grønland, **162**, No. 3, 4–48.

SPEERS, E. C., 1957. The age relation and origin of common Sudbury breccia.
　　J. Geol. **65**, 497–514.

STEVENS, R. H., 1961. Aspects of the geochemistry of antimony.
　　Unpublished B.A. thesis (Chemistry, Part II), Univ. of Oxford.

STEVENSON, J. S., 1963.   The upper contact phase of the Sudbury micropegmatite.
        Can. Mineral. **7**, 413–9.
STEWART, F. H., 1946.   The gabbroic complex of Belhelvie in Aberdeenshire.
        Quart. J. Geol. Soc. Lond. **102**, 465–98.
——, & JOHNSON, M. R. W., 1960.   The structural problem of the Younger Gabbros of North-
        east Scotland.
        Trans. Edinb. Geol. Soc. **18**, 104–12.
——, & WAGER, L. R., 1947.   Gravity stratification in the Cuillin gabbro of Skye.
        Geol. Mag. **84**, 374.
STONE, P., & BROWN, G. M., 1958.   The Quihita–Cunene layered gabbroic intrusion of South-
        west Angola.
        Geol. Mag. **95**, 195–206.
STUMPFL, E. F., 1961.   Some new platinoid-rich minerals, identified with the electron micro-
        analyser.
        Miner. Mag. **32**, 833–47.
TAUBENECK, W. H., & POLDERVAART, A., 1960.   Geology of the Elkhorn mountains, Northeastern
        Oregon: Part 2.   Willow Lake intrusion.
        Bull. Geol. Soc. Amer. **71**, 1295–1322.
TAYLOR, H. P., JR., & EPSTEIN, S., 1962.   The relationship between O^{18}/O^{16} ratios in coexisting
        minerals of igneous and metamorphic rocks.   Part 1, Principles and experimental
        results.   Part 2, Application to petrologic problems.
        Bull. Geol. Soc. Amer. **73**, 461–80, 675–94.
——, & ——, 1963. O^{18}/O^{16} ratios in rocks and co-existing minerals of the Skaergaard Intrusion,
        east Greenland.
        J. Petrol. **4**, 51–74.
——, & NOBLE, J. A., 1960.   Origin of the ultramafic complexes in southeastern Alaska.
        Rept. 21st Int. Geol. Congr. (*Norden*), Pt. 13, 175–87.
TAYLOR, R. B., 1955.   Petrology and petrography of the Duluth gabbro complex near Duluth
        Minnesota.
        Unpublished Ph.D. thesis, Univ. of Minnesota.
——, 1956.   The Duluth gabbro complex, Duluth, Minnesota.
        Ann. Meeting G.S.A. Guidebook, Field Trip No. 1, 42–66.
THAYER, T. P., 1960.   Some critical differences between Alpine-type and stratiform peridotite–
        gabbro complexes.
        Rept. 21st Internat. Geol. Congr. (*Norden*), Pt. 13, 247–59.
THOMSON, J. E., 1957.   Geology of the Sudbury Basin.
        Ontario Dept. Mines Ann. Rept. **65**, 1–56.
THORNTON, C. P., & TUTTLE, O. F., 1960.   Chemistry of igneous rocks.   I. Differentiation
        index.
        Amer. J. Sci. **258**, 664–84.
TILLEY, C. E., 1950.   Some aspects of magmatic evolution.
        Quart. J. Geol. Soc. Lond. **106**, 37–61.
——, YODER, H. S., JR., & SCHAIRER, J. F., 1963.   Melting relations of basalts.
        Carnegie Inst. Wash. Yearb. **62**, 77–84.
TRUTER, F. C., 1955.   Modern concepts of the Bushveld Igneous Complex.
        C.C.T.A. South reg. Com. Geol. **1**, 77–92.
TUSON, J., 1959.   A geophysical investigation of the Tertiary volcanic districts of Western
        Scotland.
        Unpublished Ph.D. thesis, Univ. of Durham.
TUTTLE, O. F., & BOWEN, N. L., 1958.   Origin of granite in the light of experimental studies in
        the system $NaAlSi_3O_8–KAlSi_3O_8–SiO_2–H_2O$.
        Mem. Geol. Soc. Amer. **74**, 1–153.
TYNDALE-BISCOE, R., 1949.   The geology of the country around Gwelo.
        Bull. Geol. Surv. S. Rhodesia, No. 39, 1–145.

UPTON, B. G. J., 1960. The alkaline igneous complex of Kûngnât Fjeld, South Greenland.
    Medd. om Grønland, **123**, No. 4, 5–145.

——, 1961. Textural features of some contrasted igneous cumulates from South Greenland.
    Medd. om Grønland, **123**, No. 6, 6–29.

——, 1962. Geology of Tugtutôq and neighbouring islands, South Greenland, Part I.
    Medd. om Grønland, **169**, 1–60.

——, 1964. Geology of Tugtutôq and neighbouring islands, South Greenland, Part III.
    Medd. om Grønland, **169**, 1–47.

USSING, N. V., 1912. Geology of the country around Julianehaab, Greenland.
    Medd. om Grønland, **38**, 1–426.

VINCENT, E. A., 1960. Ulvöspinel in the Skaergaard intrusion, East Greenland.
    N. Jb. Miner. Abh. Bd. **94** (Festband Ramdohr), 993–1016.

——, & BILEFIELD, L. I., 1960. Cadmium in rocks and minerals from the Skaergaard intrusion, East Greenland.
    Geochim. et Cosmochim. Acta, **19**, 63–9.

——, & CROCKET, J. H., 1960. Studies in the geochemistry of gold—I. The distribution of gold in rocks and minerals of the Skaergaard intrusion, East Greenland.
    Geochim. et Cosmochim. Acta, **18**, 130–42.

——, & PHILLIPS, R., 1954. Iron–titanium oxide minerals in layered gabbros of the Skaergaard intrusion, East Greenland, Part I. Chemistry and ore-microscopy.
    Geochim. et Cosmochim. Acta, **6**, 1–26.

——, & SMALES, A. A., 1956. The determination of palladium and gold in igneous rocks by radioactivation analysis.
    Geochim. et Cosmochim. Acta, **9**, 154–60.

——, DOUGLAS, J. A. V., & BOWN, M. G., 1964. Note on the revised composition of two olivines from the Skaergaard intrusion, East Greenland.
    Amer. Min. **49**, 805–6.

——, WRIGHT, J. B., CHEVALLIER, R., & MATHIEU, S., 1957. Heating experiments on some natural titaniferous magnetites.
    Miner. Mag. **31**, 624–55.

VISHER, G. S., 1961. Petrology of the Moxie pluton, west-central Maine.
    Program Ann. Meeting G.S.A. (Abstr.), p. 165A.

VLASOV, K. A., KUZ'MENKO, M. K., & ES'KOVA, E. M., 1959. *Lovozerskiy shchelochnoy massiv.*
    Izd. Akademii Nauk SSSR. (*The Lovozero alkali massif*: English translation, Oliver and Boyd, Edinburgh, 1966.)

VOGT, J. H. L., 1916–8. Die Sulfid–Silikatschmelzlosungen.
    Norsk Geol. Tidsskr. **4**, p. 151.

VOROBYEVA, O. A., 1940. *O pervichnoy poloschatnosti Lovozerskogo shchelochnogo massiva: Sborn. Proizvoditel'nye sily Kol'skogo poluostrova.* (Primary banding in the Lovozero alkali massif: collection of papers on the resources of the Kola Peninsula.)
    Izd. Akademii Nauk SSSR.

WADSWORTH, W. J., 1961. The ultrabasic rocks of southwest Rhum.
    Phil. Trans. Roy. Soc. Lond. Ser. B, **244**, 21–64.

——, 1963. The Kapalagulu layered intrusion of Western Tanganyika.
    Min. Soc. Amer. Special Paper 1, 108–15.

——, STEWART, F. H., & ROTHSTEIN, A. T. V., 1966. Cryptic layering in the Belhelvie intrusion, Aberdeenshire.
    Scot. J. Geol. **2**, 54–66.

WAGER, L. R., 1934. Geological investigations in East Greenland, Pt. I. General geology from Angmagssalik to Kap Dalton.
    Medd. om Grønland, **105**, 1–46.

——, 1947. Geological investigations in east Greenland, Pt. IV. The stratigraphy and tectonics of Knud Rasmussens Land and the Kangerdlugssuaq region.
    Medd. om Grønland, **134**, No. 5, 1–62.

WAGER, L. R., 1953. Layered Intrusions (Notes on three lectures given to the Danish Geological Society in Nov. 1952).
Medd. om Grønland, **12**, 335–49.

——, 1956. A chemical definition of fractionation stages as a basis for comparison of Hawaiian, Hebridean, and other basic lavas.
Geochim. et Cosmochim. Acta, **9**, 217–48.

——, 1959. Differing powers of crystal nucleation as a factor producing diversity in layered igneous intrusions.
Geol. Mag. **96**, 75–80.

——, 1960. The major element variation of the layered series of the Skaergaard intrusion and a re-estimation of the average composition of the hidden layered series and of the successive residual magmas.
J. Petrol. **1**, 364–98.

——, 1961. A note on the origin of ophitic texture in the chilled olivine gabbro of the Skaergaard intrusion.
Geol. Mag. **98**, 353–66.

——, 1963. The mechanism of adcumulus growth in the layered series of the Skaergaard intrusion.
Min. Soc. Amer. Special Paper 1, 1–9.

——, 1965. The form and internal structure of the alkaline Kangerdlugssuaq intrusion, East Greenland.
Miner. Mag. **34**, 487–97.

——, 1967. Layering in plutonic rocks, in *Basaltic Rocks* (Ed. H. H. Hess). Wiley & Sons: New York.

——, & BROWN, G. M., 1951. A note on rhythmic layering in the ultrabasic rocks of Rhum.
Geol. Mag. **88**, 166–8.

——, & ——, 1957. Funnel-shaped layered intrusions.
Bull. Geol. Soc. Amer. **68**, 1071–4.

——, ——, & WADSWORTH, W. J., 1960. Types of igneous cumulates.
J. Petrol. **1**, 73–85.

——, & DEER, W. A., 1939 (re-issued in 1962). Geological Investigations in East Greenland, Pt. III. The Petrology of the Skaergaard Intrusion, Kangerdlugssuaq, East Greenland.
Medd. om Grønland, **105**, No. 4, 1–352.

——, & HAMILTON, E. I., 1964. Some radiometric rock ages and the problem of the southward continuation of the East Greenland Caledonian Orogeny.
Nature, **204**, No. 4963, 1079–80.

——, & MITCHELL, R. L., 1951. The distribution of trace elements during strong fractionation of basic magma—a further study of the Skaergaard intrusion, East Greenland.
Geochim. et Cosmochim. Acta, **1**, 129–208.

——, SMIT, J. VAN R., & IRVING, H., 1958. Indium content of rocks and minerals from the Skaergaard intrusion, East Greenland.
Geochim. et Cosmochim. Acta, **13**, 81–6.

——, & VINCENT, E. A., 1962. Ferrodiorite from the Isle of Skye.
Miner. Mag. **33**, 26–36.

——, ——, BROWN, G. M., & BELL, J. D., 1965. Marscoite and related rocks of the Western Red Hills complex, Isle of Skye.
Phil. Trans. Roy. Soc. Lond. Ser. A, **257**, 273–307.

——, ——, & SMALES, A. A., 1957. Sulphides in the Skaergaard intrusion, East Greenland.
Econ. Geol. **52**, 855–903.

WAGNER, P. A., 1924. On the magmatic nickel deposits of the Bushveld Complex in the Rustenburg District, Transvaal.
Mem. Geol. Surv. S. Afr. No. 21, 181 pp.

——, 1929. *Platinum deposits and mines of South Africa.* Oliver and Boyd: Edinburgh. 326 pp.

WALKER, F., 1930. The geology of the Shiant Isles (Hebrides).
    Quart. J. Geol. Soc. Lond. 86, 355–98.
——, 1940. Differentiation of the Palisades Diabase, New Jersey.
    Bull. Geol. Soc. Amer. 51, 1059–1106.
——, 1953. The pegmatitic differentiates of basic sheets.
    Amer. J. Sci. 251, 41–60.
——, 1956. The magnetic properties and differentiation of dolerite sills—a critical discussion.
    Amer. J. Sci. 254, 433–43.
——, 1957. The causes of variation in dolerite intrusions.
    Symposium on dolerites, Univ. of Tasmania, 1–25.
——, & POLDERVAART, A., 1949. Karroo dolerites of the Union of South Africa.
    Bull. Geol. Soc. Amer. 60, 591–706.
WALT, C. F. J. VAN DER, 1942. Chrome ores of the Western Bushveld Complex.
    Trans. Geol. Soc. S. Afr. 44, 79–112.
WEEDON, D. S., 1956. Studies concerning the Tertiary igneous rocks of the Southern Cuillins, Isle of Skye.
    Unpublished D.Phil. thesis, Univ. of Oxford.
——, 1960. The Gars-Bheinn sill, Isle of Skye.
    Quart. J. Geol. Soc. Lond. 116, 37–54.
——, 1961. Basic igneous rocks of the southern Cuillin, Isle of Skye.
    Trans. Geol. Soc. Glasg. 24, 190–212.
——, 1965. The layered ultrabasic rocks of Sgurr Dubh, Isle of Skye.
    Scot. J. Geol. 1, 42–68.
WEISS, O., 1940. Gravimetric and Earth magnetic measurements of the Great Dyke of Southern Rhodesia.
    Trans. Geol. Soc. S. Afr. 43, 143–51.
WELLS, M. K., 1954. The structure and petrology of the hypersthene-gabbro intrusion, Ardnamurchan, Argyllshire.
    Quart. J. Geol. Soc. Lond. 109, 367–95.
——, 1962. Structure and petrology of the Freetown Layered Basic Complex of Sierra Leone.
    Overs. Geol. min. Res. (London), Bull. Suppl. 4, 1–115.
——, & BAKER, C. O., 1956. The anorthosites in the Colony Complex near Freetown, Sierra Leone.
    Col. Geol. min. Res. 6, No. 2, 137–58.
WESTHUIZEN, J. M. VAN DER, 1945. Die geologie van Bothasberg, Bosveldkompleks.
    Unpublished M.Sc. thesis, Univ. of Pretoria.
WET, J. F. DE, 1952. Chromite investigations—Part III, Variations in the composition of the pure chromite mineral from the eastern chrome belt, Lydenburg district.
    J. Chem. Metall. Min. Soc. S. Afr. 52, 143–53.
WHEELER, E. P., 2nd, 1942. Anorthosite and associated rocks about Nain, Labrador.
    J. Geol. 50, 611–42.
WILKINSON, J. F. G., 1956. The olivines of a differentiated teschenite sill near Gunnedah, New South Wales.
    Geol. Mag. 93, 441–55.
——, 1958. The petrology of a differentiated teschenite sill near Gunnedah, New South Wales.
    Amer. J. Sci. 256, 1–39.
WILLEMSE, J., 1959. The floor of the Bushveld Igneous Complex and its relationships, with special reference to the Eastern Transvaal.
    Proc. Geol. Soc. S. Afr. 62, 21–80.
——, 1964. A brief outline of the geology of the Bushveld Igneous Complex.
    From The Geology of some ore deposits of Southern Africa, 2, Geol. Soc. S. Afr. 91–128.
WILLIAMS, H., 1957. Glowing avalanche deposits of the Sudbury basin.
    Ontario Dept. Mines, Ann. Rept. 65, 57–89.

WILSON, H. D. B., 1956. Structure of lopoliths.
     Bull. Geol. Soc. Amer. **67**, 289–300.

WINCHELL, A. N., 1951. *Elements of optical mineralogy*, Pt. II. John Wiley & Sons: New York.

WOODE, R. D. A., 1961. The determination of traces of zinc in igneous rocks and minerals.
     Unpublished B.Sc. thesis, Univ. of Oxford.

WORST, B. G., 1956. The differentiation and structure of the Great Dyke of Southern Rhodesia.
     Unpublished D.Sc. thesis, Univ. of Pretoria.

——, 1958. The differentiation and structure of the Great Dyke of Southern Rhodesia.
     Trans. Geol. Soc. S. Afr. **61**, 283–354.

WRIGHT, J. B., 1961. Solid solution relationships in some iron oxide ores of basic igneous rocks.
     Miner. Mag. **32**, 778–89.

YELISEYEV, N. A., 1941. O proiskhozhdenii pervichnoy poloschatnosti v Lovozerskom plutone.
     (Origin of the primary banding in the Lovozero massif.)
     Zap. Vses. Mineral. Obshch. Pt. 70, No. 1.

——, & FEDOROV, E. E., 1953. *Lovozerskiy pluton i yego mestorozhdeniya.* (The Lovozero massif and its surroundings.)
     Izd. Akademii Nauk SSSR.

YODER, H. S., JR., 1952. Change of melting point of diopside with pressure.
     J. Geol. **60**, 364–74.

——, & SAHAMA, TH. G., 1957. Olivine X-ray determinative curve.
     Amer. Min. **42**, 475–91.

——, STEWART, D. B., & SMITH, J. R., 1957. Ternary feldspars.
     Carnegie Inst. Wash. Yearb. **56**, 206–14.

——, & TILLEY, C. E., 1962. Origin of basalt magmas: an experimental study of natural and synthetic rock systems.
     J. Petrol. **3**, 342–532.

——, ——, & SCHAIRER, J. F., 1963. Pyroxenes and associated minerals in the crust and mantle.
     Carnegie Inst. Wash. Yearb. **62**, 84–95.

ZINOVIEFF, P., 1958. The basic layered intrusion and the associated igneous rocks of the Central and Eastern Cuillin Hills, Isle of Skye.
     Unpublished D.Phil. thesis, Univ. of Oxford.

ZYL, C. VAN, 1959. An outline of the geology of the Kapalagulu complex, Kungwe Bay, Tanganyika Territory, and aspects of the evolution of layering in basic intrusives.
     Trans. Geol. Soc. S. Afr. **62**, 1–31.

SUBJECT INDEX

Subjects are listed under the names printed in **bold** face,
of each intrusion except for general references

2 0

AUTHOR INDEX